AF379856

Modern Electroencephalographic Assessment Techniques

Theory and Applications

Edited by

Vangelis Sakkalis

Institute of Computer Science (ICS)
Foundation for Research and Technology — Hellas (FORTH)
Heraklion, Crete, Greece

 Humana Press

Editor
Vangelis Sakkalis
Institute of Computer Science (ICS)
Foundation for Research and Technology – Hellas (FORTH)
Heraklion, Crete, Greece

ISSN 0893-2336   ISSN 1940-6045 (electronic)
ISBN 978-1-4939-1297-1   ISBN 978-1-4939-1298-8 (eBook)
DOI 10.1007/978-1-4939-1298-8
Springer New York Heidelberg Dordrecht London

Library of Congress Control Number: 2014954624

Humana Press is a brand of Springer
Springer is part of Springer Science+Business Media (www.springer.com)

Preface to the Series

Experimental life sciences have two basic foundations: concepts and tools. The *Neuromethods* series focuses on the tools and techniques unique to the investigation of the nervous system and excitable cells. It will not, however, shortchange the concept side of things as care has been taken to integrate these tools within the context of the concepts and questions under investigation. In this way, the series is unique in that it not only collects protocols but also includes theoretical background information and critiques which led to the methods and their development. Thus it gives the reader a better understanding of the origin of the techniques and their potential future development. The *Neuromethods* publishing program strikes a balance between recent and exciting developments like those concerning new animal models of disease, imaging, in vivo methods, and more established techniques, including, for example, immunocytochemistry and electrophysiological technologies. New trainees in neurosciences still need a sound footing in these older methods in order to apply a critical approach to their results.

Under the guidance of its founders, Alan Boulton and Glen Baker, the *Neuromethods* series has been a success since its first volume published through Humana Press in 1985. The series continues to flourish through many changes over the years. It is now published under the umbrella of Springer Protocols. While methods involving brain research have changed a lot since the series started, the publishing environment and technology have changed even more radically. Neuromethods has the distinct layout and style of the Springer Protocols program, designed specifically for readability and ease of reference in a laboratory setting.

The careful application of methods is potentially the most important step in the process of scientific inquiry. In the past, new methodologies led the way in developing new disciplines in the biological and medical sciences. For example, Physiology emerged out of Anatomy in the nineteenth century by harnessing new methods based on the newly discovered phenomenon of electricity. Nowadays, the relationships between disciplines and methods are more complex. Methods are now widely shared between disciplines and research areas. New developments in electronic publishing make it possible for scientists that encounter new methods to quickly find sources of information electronically. The design of individual volumes and chapters in this series takes this new access technology into account. Springer Protocols makes it possible to download single protocols separately. In addition, Springer makes its print-on-demand technology available globally. A print copy can therefore be acquired quickly and for a competitive price anywhere in the world.

Saskatoon, Canada   *Wolfgang Walz*

Preface

Understanding the human brain has been one of the most challenging topics in neuroscience and science in general. Different and heterogeneous paths to extracting valuable information and knowledge have been proposed over the past years in aggregate trying to solve one of the biggest mysteries: How does the brain work? But is our brain capable of understanding itself? Fortunately, in this book, we do not attempt to complete such an obscure quest, but we can certainly contribute in this endeavor by reviewing modern quantitative analysis methodologies.

Data acquisition and analysis techniques have developed in parallel, enabling the measurement of connectivity between brain regions. Anatomic and functional neuroimaging methods for detecting brain activity using both electroencephalography (EEG) and magnetoencephalography (MEG) have developed dramatically during the past two decades. Although neuroimaging studies based on positron emission tomography (PET) and functional magnetic resonance imaging (fMRI) are especially popular mostly due to the prospective possibility of being able to measure the neuronal activity on a high temporal as well as spatial scale, there is no physical way (several hundreds of milliseconds needed to reflect blood oxygen level dependent—BOLD—effects) of reaching the ultrahigh temporal precision of the EEG and MEG techniques that are already widely used. In this sense, EEG and MEG remain of paramount importance in the neuroimaging community and may be used as single methods or in combination with simultaneous fMRI scans as discussed in the final two chapters of this book.

A number of connectivity analysis methodologies addressing a wide variety of clinical applications including epilepsy, schizophrenia, Alzheimer's disease, and alcoholism, as well as cognitive studies are presented in this book, each one having its specific strengths and weaknesses. Our intention is to provide a comprehensive overview of the most modern and widely established approaches mainly applied in, but not limited to, decomposing high-resolution multichannel EEG and MEG signals into functional interconnected brain regions, as the functional segregation and integration concept suggests. We wish to aid interested researchers by including manuscripts describing the theoretical basis of each presented method along with prosperous application domains, in the form of a balanced mixture of theoretical tutorials, comprehensive reviews, and original research. Emphasis is given on the underlying assumptions, on technical matters that greatly affect the outcome of each proposed method, on the ambitions, and on the domain of application of each method. Furthermore, links to graph theory and visualization of connectivity motifs are also presented in an attempt to better describe the functional characteristics of brain networks.

Chapter Operational Architectonics Methodology for EEG Analysis: Theory and Results presents the strengths and drawbacks of the EEG signals. Only a deep understanding of brain spatiotemporal dynamics guaranties genuine long-term progress in psychophysiological, cognitive, and medical science. Andrew Fingelkurts and Alexander Fingelkurts introduce various aspects of *operational architectonics*, a methodology addressing the peculiarities of spatial and temporal EEG nonstationarity. Such a novel technique is sensitive to the underlying quasi-stationary nature of EEG signal and plausible for a better understanding of the functional organization of the neocortex and its relation to consciousness. Thus, there is a wide application domain, including *epilepsy, schizophrenia, major depression, sleep disorders,* and *chronic opioid abuse,* in the clinical field reflecting mostly neuropsychiatric disorders. Design and administration of psychotropic drugs are also discussed, such as the methadone effects as a maintenance treatment for heroin-dependent patients and Lorazepam administration. Cognitive neuroscience applications include *Working Memory* experiments and *ontological and personality development studies,* while "operational architectonics" touches also upon *neurophilosophy* and *artificial intelligence.*

Chapters Clinical Electroencephalography in the Diagnosis and Management of Epilepsy, Effective Brain Connectivity from Intracranial EEG Recordings: Identification of Epileptogenic Zone in Human Focal Epilepsies, On the Effect of Volume Conduction on Graph Theoretic Measures of Brain Networks in Epilepsy, and Methods for Seizure Detection and Prediction: An Overview focus on different methodologies applied in epilepsy. More specifically, Chapter Clinical Electroencephalography in the Diagnosis and Management of Epilepsy provides an introductory *clinical perspective of EEG* in the diagnosis and management of *epilepsy,* overviewing the present status and emerging approaches.

Chapter Effective Brain Connectivity from Intracranial EEG Recordings: Identification of Epileptogenic Zone in Human Focal Epilepsies provides an overview of the different intracranial EEG signal processing methods used to identify the *epileptogenic zone* that may be resected surgically to suppress seizures. Particular attention is being given to the methods aimed at characterizing effective brain connectivity using intracranial EEG recordings. G. Varotto and colleagues present connectivity pattern analysis associated with a particular form of focal epilepsy (type II focal cortical dysplasia), based on multivariate *autoregressive parametric models* and measures derived from *graph theory.*

Chapter On the Effect of Volume Conduction on Graph Theoretic Measures of Brain Networks in Epilepsy evaluates and compares two standard and most commonly used linear connectivity measures—*cross-correlation* in the time domain and *coherence* in the frequency domain—with measures that account for volume conduction, namely *corrected cross-correlation, imaginary coherence, phase lag index,* and *weighted phase lag index.* M. Christodoulakis et al. focus mostly on the way connectivity measures are affected by both volume conduction and the choice of recording reference (montage) in the time and frequency domain. Graph-theoretic indices are again used to assess network topology changes in epileptic subjects.

Chapter Methods for Seizure Detection and Prediction: An Overview discusses another application of *linear/nonlinear analysis, chaos,* and *information-based analysis,* in detecting and even predicting *epilepsy.* G. Giannakakis et al. review feature selection procedures from the aforementioned methods that are able to detect and classify epileptic states. Each method's accuracy is evaluated through performance measures indicating the strengths of each proposed technique.

Functional connectivity measures and, more specifically, *coherence, phase synchronization,* and *nonlinear state-space generalized synchronization* assessment methods are further discussed in Chapter Graph-theoretic Indices of Evaluating Brain Network Synchronization: Application in an Alcoholism Paradigm. Synchronization matrices define graphs whose topological structure and properties are characterized using measures for graphs and weighted networks. Graph-theoretic measures are also used as the tools to visualize and characterize the topology of a brain network as in a working memory task-related *alcoholism* paradigm.

Chapter Time-Varying Effective Connectivity for Investigating the Neurophysiological Basis of Cognitive Processes describes the methodological advancements developed during the last 20 years in the field of *effective connectivity* based on Granger causality and linear autoregressive modeling. Apart from introducing the most widely accepted methodologies, a graph-theoretic approach demonstrates their potential application in a *motor imagery* process.

Chapter Assessment of Sensory Gating Deficit in Schizophrenia Using a Wavelet Transform Methodology on Auditory Paired-Click Evoked Potentials discusses the use of EEG in the search for biomarkers of *schizophrenia* and, more specifically, the assessment of the *sensory gating process* in patients with schizophrenia via time-frequency decomposition of the EEG signals.

G. Zouridakis and colleagues in Chapter Schizophrenia Assessment Using Single-Trial Analysis of Brain Activity discuss the clinical application of *Independent Component Analysis* (*ICA*) with a focus on schizophrenia. The methodology presented is based on *single-trial analysis using an iterative independent component analysis procedure.* This method is capable of identifying and measuring the amplitude, latency, and overall morphology of individual components in single trials, and as such, permits the study of phase characteristics among single trials, while preserving known features of the average evoked potentials.

Phase synchronization and two recent adaptations of the *Common Spatial Patterns method* are further investigated in Chapter Phase Variants of the Common Spatial Patterns Method, provided by T. Camilleri et al. There is no one best methodology for estimating *phase synchronization.* The selection of the techniques presented complements each other and is evaluated in a motor imagery task.

Chapter Estimation of Regional Activation Maps and Interdependencies from Minimum Norm Estimates of Magnetoencephalography (MEG) Data describes the development of a minimally supervised pipeline for the analysis of event-related MEG recordings that preserves the temporal resolution of the data and enables estimation of patterns of regional interdependencies between activated regions in an object naming task. The proposed approach is based on a *spatiotemporal source-clustering* algorithm initially applied to identify extended regions of significant activation, and subsequently regional interdependencies are estimated through *cross-lag correlation analysis* between time series representing the time course of activity within each cluster.

Chapter Blind Signal Separation Methods in Computational Neuroscience presents a survey of *Blind Source Separation* methods based on *Independent* and *Sparse Component Analysis.* The theoretical basis and mathematical formulation of the most widely adopted methods are described. M. N. Syed et al. evaluate the strengths and weaknesses of different formulations in an experimental setup capturing simultaneous electrical activity over the scalp (EEG) and over the exposed surface of the cortex (ECoG). Since the data from this experiment is simultaneously collected from above the scalp and under the scalp, it opens the door to understand the mixing mechanism across the brain. Further applications in fMRI, MRI, and finger prints are also mentioned.

Chapter Current Trends in ERP Analysis Using EEG and EEG/fMRI Synergistic Methods introduces methods and measures used for the analysis of *Event Potentials* but combines the excellent temporal resolution of the EEG with the spatial details provided by the fMRI. Such an approach allows moving beyond isolation and connection of specific EEG features to specific cognitive processes. Fusion between EEG and fMRI looks very promising since the strong feature of the one is the weakness of the other. Reality though proved more complex, and *EEG and fMRI fusion* is still an open area for research.

Finally, Chapter Computer-Based Assessment of Alzheimer's Disease Employing fMRI and/or EEG: A Comprehensive Review focuses on assessing *Alzheimer's disease using both fMRI and EEG* approaches. In addition, fusion of both approaches is also presented.

Over the years, EEG and more recently MEG have evolved and are widely accepted in today's clinical practice. However, only traditional analysis techniques are mostly used. We believe that in the coming years most of the emerging and promising techniques presented in this book could become more established and reach the clinical research community.

Hopefully this book, touching upon both the biomedical and computational aspects of this exciting and rapidly evolving field, will be an interesting and enjoyable read and will allow for a more in-depth understanding of the brain's underlying mechanisms.

Heraklion, Crete, Greece   *Vangelis Sakkalis*

Acknowledgments

I would like to thank the authors for their valuable contributions to this volume. I would also like to thank the referees for their help in achieving high clarity despite the complexity of the presented topics.

Finally, I gratefully acknowledge Wolfgang Walz and the Springer team for their patience, support, and guidance during the different preparation phases of this book.

About the Editor

Vangelis Sakkalis, Ph.D. currently holds a Principal Researcher position in the Institute of Computer Science—Foundation for Research and Technology (ICS—FORTH). He received his Ph.D. in Electronic and Computer Engineering after completing his master's degree at Imperial College of Science, Technology and Medicine, UK. His background falls in Biomedical Engineering, Atomic-Molecular Physics, Optoelectronics, and Laser. His research interests include biosignal and image analysis, visualization, classification algorithms and biostatistics applied in computational medicine, cognitive neuroscience, and biomedical informatics. He currently coordinates four European and national projects related to brain disorders and cancer research. He has published more than 100 papers in scientific archival journals, proceedings of international conferences and workshops, and scientific newsletters related to his fields of expertise. He has given numerous invited lectures worldwide, and his research has been funded by numerous funding agencies and companies.

About the Editor

Contents

Contributors

ANDREAS V. ALEXOPOULOS • *Cleveland Clinic Epilepsy Center, Cleveland, OH, USA*

MARIA ANASTASIADOU • *Department of Electrical and Computer Engineering, University of Cyprus, Nicosia, Cyprus*

LAURA ASTOLFI • *Department of Computer, Control, and Management Engineering "Antonio Ruberti", "Sapienza" University, Rome, Italy; Neuroelectrical Imaging and BCI Lab, IRCCS Fondazione Santa Lucia, Rome, Italy*

FABIO BABILONI • *Department of Physiology and Pharmacology, "Sapienza" University, Rome, Italy*

NIKOLAOS BOURBAKIS • *Wright State University, Dayton, OH, USA*

NASH N. BOUTROS • *Department of Psychiatry and Neurosciences, University of Missouri, Kansas City, MO, USA*

KENNETH P. CAMILLERI • *Department of Systems and Control Engineering, Faculty of Engineering and Centre for Biomedical Cybernetics, University of Malta, Msida, Malta*

TRACEY CAMILLERI • *Faculty of Engineering, Department of Systems and Control Engineering, University of Malta, Msida, Malta*

MANOLIS CHRISTODOULAKIS • *Department of Electrical and Computer Engineering, University of Cyprus, Nicosia, Cyprus*

JAVIER DIAZ • *Department of Electronics and Communications, CEIT, Donostia-San Sebastián, Spain*

NICHOLAS-TIBERIO ECONOMOU • *Cleveland Clinic Epilepsy Center, Cleveland, OH, USA*

SIMON G. FABRI • *Faculty of Engineering, Department of Systems and Control Engineering, University of Malta, Msida, Malta*

OWEN FALZON • *Centre for Biomedical Cybernetics, University of Malta, Msida, Malta*

CRISTINA FARMAKI • *Computational Medicine Laboratory (CML), Institute of Computer Science (ICS), Foundation of Research and Technology Heraklion (FORTH), Heraklion, Crete, Greece*

ALEXANDER A. FINGELKURTS • *BM-Science—Brain and Mind Technologies Research Centre, Espoo, Finland*

ANDREW A. FINGELKURTS • *BM-Science—Brain and Mind Technologies Research Centre, Espoo, Finland*

DIMITRIOS I. FOTIADIS • *Unit of Medical Technology and Intelligent Information Systems, Department of Materials Science and Engineering, University of Ioannina, Ioannina, Greece*

SILVANA FRANCESCHETTI • *Neurophysiology and Diagnostic Epileptology Operative Unit, Epilepsy Centre, C. Besta Neurological Institute, IRCCS Foundation, Milan, Italy*

PANDO G. GEORGIEV • *Center for Applied Optimization, University of Florida, Gainesville, FL, USA*

GIORGOS GIANNAKAKIS • *Institute of Computer Science (ICS), Foundation of Research and Technology Heraklion (FORTH), Heraklion, Crete, Greece*

KLEVEST GJINI • *Department of Psychiatry and Neurosciences, University of Missouri, Kansas City, MO, USA*

AVGIS HADJIPAPAS • *St George's University of London Medical School, University of Nicosia, Nicosia, Cyprus*

DARSHAN IYER • *Respiratory and Monitoring Solutions, Covidien, Boulder, CO, USA*

DONATELLA MATTIA • *Neuroelectrical Imaging and BCI Lab, IRCCS Fondazione Santa Lucia, Rome, Italy*

KOSTAS MICHALOPOULOS • *Wright State University, Dayton, OH, USA*

GEORGIOS D. MITSIS • *Department of Electrical and Computer Engineering, University of Cyprus, Nicosia, Cyprus*

ABDOU MOUSAS • *Department of Basic Sciences, School of Medicine, University of Crete, Heraklion, Crete, Greece*

THEOFANIS OIKONOMOU • *Informatics and Telematics Institute (ITI), Centre for Research and Technology Hellas (CERTH), Thessaloniki, Greece*

FERRUCCIO PANZICA • *Neurophysiology and Diagnostic Epileptology Operative Unit, "C. Besta" Neurological Institute IRCCS Foundation, Milan, Italy*

SAVVAS S. PAPACOSTAS • *The Cyprus Institute for Neurology and Genetics, Nicosia, Cyprus*

ANDREW C. PAPANICOLAOU • *Division of Clinical Neurosciences, Department of Pediatrics, College of Medicine & Neuroscience Institute—Le Bonheur Children's Hospital, University of Tennessee Health Science Center, Memphis, TN, USA*

ELEFTHERIOS S. PAPATHANASIOU • *The Cyprus Institute for Neurology and Genetics, Nicosia, Cyprus*

PANOS M. PARDALOS • *Center for Applied Optimization, University of Florida, Gainesville, FL, USA*

MATTHEW PEDIADITIS • *Institute of Computer Science (ICS), Foundation of Research and Technology Heraklion (FORTH), Heraklion, Crete, Greece*

MANUELA PETTI • *Department of Computer, Control, and Management Engineering "Antonio Ruberti", "Sapienza" University, Rome, Italy; Neuroelectrical Imaging and BCI Lab, IRCCS Fondazione Santa Lucia, Rome, Italy*

ROOZBEH REZAIE • *Division of Clinical Neurosciences, Department of Pediatrics, College of Medicine & Neuroscience Institute—Le Bonheur Children's Hospital, University of Tennessee Health Science Center, Memphis, TN, USA*

FABIO ROTONDI • *Neurophysiology and Diagnostic Epileptology Operative Unit, "C. Besta" Neurological Institute IRCCS Foundation, Milan, Italy*

VANGELIS SAKKALIS • *Computational Medicine Laboratory (CML), Institute of Computer Science (ICS), Foundation of Research and Technology Heraklion (FORTH), Heraklion, Crete, Greece*

PANAGIOTIS G. SIMOS • *Department of Psychiatry, School of Medicine, University of Crete, Heraklion, Crete, Greece*

ROBERTO SPREAFICO • *Clinical Epileptology and Experimental Neurophysiology Operative Unit, C. Besta Neurological Institute, IRCCS Foundation, Milan, Italy*

MUJAHID N. SYED • *Center for Applied Optimization, University of Florida, Gainesville, FL, USA*

LAURA TASSI • *"C. Munari" Centre of Epilepsy Surgery, Niguarda Hospital, Milan, Italy*

IOANNIS TOLLIS • *Institute of Computer Science (ICS), Foundation of Research and Technology Heraklion (FORTH), Heraklion, Crete, Greece*

JLENIA TOPPI • *Department of Computer, Control, and Management Engineering "Antonio Ruberti", "Sapienza" University, Rome, Italy; Neuroelectrical Imaging and BCI Lab, IRCCS Fondazione Santa Lucia, Rome, Italy*

EVANTHIA E. TRIPOLITI • *Unit of Medical Technology and Intelligent Information Systems, Department of Materials Science and Engineering, University of Ioannina, Ioannina, Greece*

VASSILIS TSIARAS • *Institute of Computer Science (ICS), Foundation of Research and Technology Heraklion (FORTH), Heraklion, Crete, Greece*

MANOLIS TSIKNAKIS • *Institute of Computer Science (ICS), Foundation of Research and Technology Heraklion (FORTH), Heraklion, Crete, Greece*

GIULIA VAROTTO • *Unit of Biomedical Engineering, Neurophysiology and Diagnostic Epileptology Operative Unit, Epilepsy Centre, C. Besta Neurological Institute, IRCCS Foundation, Milan, Italy*

MICHALIS ZERVAKIS • *Department of Electronic and Computer Engineering, Technical University of Crete, Chania, Crete, Greece*

GEORGE ZOURIDAKIS • *Department of Engineering Technology, University of Houston, Houston, TX, USA; Department of Computer Science, University of Houston, Houston, TX, USA; Department of Electrical and Computer Engineering, University of Houston, Houston, TX, USA; Basque Center on Cognition, Brain, and Language (BCBL), Donostia-San Sebastián, Spain*

Neuromethods (2015) 91: 1–59
DOI 10.1007/7657_2013_60
© Springer Science+Business Media New York 2013
Published online: 18 December 2013

Operational Architectonics Methodology for EEG Analysis: Theory and Results

Andrew A. Fingelkurts and Alexander A. Fingelkurts

Abstract

This chapter discusses various aspects of operational architectonics methodology for EEG analysis that have been developed over the course of last 17 years in relation to nonstationarity of brain functioning. At first we detail the peculiarities and evidence for a spatial and temporal nonstationarity in the EEG signal, then we review a theoretical framework that could integrate the existing data with a focus on theoretical advantages provided by an operational architectonics framework, and finally we describe the experimental results related to methodology. In the last part of the chapter we outline the application of OA methodology to clinical, pharmacological, cognitive, and neurophilosophical studies.

Key words Electroencephalography (EEG), Nonstationarity, Spontaneous brain activity, Neuronal assemblies, Brain operations, Operational module (OM), Rapid transitional process (RTP), Operational synchrony (OS), Functional synchrony, Operational architectonics (OA)

We will have to face the fact that our models will have to be much more complex if we take into account the topographical differentiation and the temporal variability, i.e. the dynamics of the EEG. And as long as we are not able to do this we cannot claim to be talking about a model that can even remotely describe EEG activity Kunkel [1]

---

1 Introduction

Ongoing spontaneous brain activity at the cortical level (electroencephalogram—EEG) is the result of dendritic and post-synaptic currents of many cortical neurons (see Note 1) firing in nonrandom partial synchrony [8, 9]. It was demonstrated that neural activity patterns measurable at the macro-level by EEG are correlated with underlying neural computations [10–15] and accompany specific behavior and cognition [16, 17]. These studies suggest that EEG provides a direct measure of cortical activity with millisecond temporal resolution.

EEG studies in their totality revealed the following important features of the EEG signal:

- EEG characteristics were found to be independent from cultural and ethnic factors [18] which may reflect a common genetic heritage of the human brain development.

- EEG is characterized by high inherent stability. Indeed, it was reported that repeated 20-s samples of EEG were about 82 % reliable, 40-s samples were about 90 % reliable, and 60-s samples were approximately 92 % reliable [19].

- EEG is highly reproducible [20–26]. For example, Burgess and Gruzelier [22] reported average reliabilities of 0.81 and 0.86 for theta and alpha bands (see Note 2) in resting, eyes open EEG with a test–retest interval of about 1 h. Test–retest correlation coefficients for EEG power, after a 12–16-week interval between measurements, are also high ~0.8 for both absolute and relative power [19, 21, 27]. For longer intervals (with an average 10-month interval), the test–retest reliability stays ~0.7 [20]. Even over a time period of 5 years the EEG parameters demonstrated very high stability [28].

- EEG characteristics are mostly explained by heritability [29, 30] (for a review and meta-analysis, see [31]). The authors conclude that EEG is one of the most heritable characteristics in humans.

- The existence of EEG phenotypes has been demonstrated [32]. EEG phenotypes are clusters of commonly occurring EEG patterns found in the general population that are believed to be the result of underlying genetics. Several studies have begun to identify genes associated with certain EEG phenotypes (for the review see [33]).

- EEG oscillations are phylogenetically preserved [34] and provide basic links to brain functions, especially for communication and associative functions [35]. EEG oscillations define short temporal windows for flexible communication between widely distributed neuronal ensembles, which are associated with different types of sensory and cognitive processes [11, 36–39].

- EEG oscillations are causally implicated in cognitive functions: Transcranial-magnetic stimulation at physiologically meaningful rhythms has domain-specific effects on cognitive activities [40–42].

- As a field, EEG may act (top down) on networks (neuronal assemblies) [17]: An electric field may induce electrophoretic redistribution of charged ions both intracellularly and extracellularly and thereby directly modulate neuronal physiology; additionally, various structures in the brain are sensitive to electromagnetic fields [43]. Indeed, it was demonstrated that the firing rate of a spontaneously active single neuron depends strongly on the instantaneous spatial pattern of ongoing population activity in a large cortical area [44] (for the review see [45]).

Here EEG characteristics may be considered as order parameters which modulate the behavior of neurons or neuronal assemblies [46].

Thanks to these important features, EEG is extensively used in brain research and for clinical purposes. It appears that *as a neurophysiological phenomenon EEG has its own structure, regularities, and rules of organization* [16, 47–50] (for the reviews see [51–57]). Only when one knows the structural peculiarities of an EEG signal, it is possible to make proper use of EEG as a tool and provide adequate data interpretation. In fact, it is impossible to design a cognitive EEG experiment that is not biased by assumptions (explicit or implicit) regarding brain dynamics and the statistical characteristics of EEG, particularly with respect to its temporal and spatial dynamics. Therefore, a much deeper understanding of brain spatiotemporal dynamics (which is reflected in EEG) is essential for genuine long-term progress in psychophysiological, cognitive, and medical sciences.

A great number of EEG studies have been accumulated since the first EEG report published by Berger in 1929 [58] with a remarkably consistent fact emerging: *the spatiotemporal dynamics of EEG is hidden in its nonstationary structure.*

2 EEG Nonstationarity

The available evidence suggests that neurons self-organize into transient networks (neuronal assemblies) that synchronize in *time* and *space* to produce a mixture of short bursts of oscillations that are observable in the EEG signal [11, 38, 39, 59]. The dynamics of EEG spatiotemporal variability is characterized by abrupt alteration of relatively stable periods, the duration and size of which are significantly different from the respective characteristics of a random process [51, 56, 57, 60–67].

In this context EEG nonstationarity is expressed in both temporal (*temporal nonstationarity*) and spatial (*spatial heterogeneity*) dimensions. In the following two subsections we consider these two features separately.

2.1 Temporal Nonstationarity of EEG

At present it is well established that an EEG is a highly nonstationary signal [48, 51, 53, 56, 64, 65, 68–77]. This means that EEG signal has different characteristics in various points in time. It was demonstrated that in the phenomenon of EEG temporal variability, not only the stochastic (noise) fluctuations of the EEG parameters but also the *temporal structure* of the signal itself are reflected [48, 78, 79] (for a review see [52, 53, 56]). It is assumed that EEG variability or nonstationarity is the reflection of structural or piecewise stationary organization of the signal. Piecewise stationary structure of EEG is considered to be the result of "gluing" of short-term stationary causal processes with different probability characteristics [61, 64, 65, 80–82] (for the reviews see [52–54, 56, 67]).

Considering that each local EEG signal (registered from a given cortical area) is characterized by three major components (amplitude, frequency, and phase), one may assume that each of them can exhibit nonstationary behavior in relation to time. Indeed, it has been demonstrated that all three EEG characteristics change *abruptly* (not necessarily simultaneously) with the progression of time (for EEG amplitude see [53, 56, 83–86]; for EEG frequency see [49, 50]; for EEG phase see [62, 87–90]). In other words, the values of EEG amplitude, frequency, and phase persist for some time around some stable average and then abruptly "jump" up or down to a new stable average which after some time is replaced by another average level. These "jumps" in separate local EEG characteristics mark the discontinuities of relatively stable functioning of local neuronal networks. It has been proposed that during these stationary periods a particular brain system executes separate operations [51, 81, 82] (for a review see [91]). This suggests that ongoing brain activity occurs in discrete steps [62, 64, 85, 87, 89, 92] and confirms the view that the cerebral cortex is always active, even during rest [44, 57, 93–98].

The functional significance of EEG temporal structure was confirmed by numerous studies where it was demonstrated that different EEG temporal structures were associated with (a) different degrees of psychophysiological and social adaptivity of individuals [99–101], (b) different cognitive loading [26, 49–52], and (c) different psychopathologies (for schizophrenia [102]; for epilepsy [103]; for major depression disorder [104]; for opioid dependence [105]; for a review see [106]).

In this context each local EEG signal can be reduced to a temporally organized sequence of nearly stationary segments of various types. It can be suggested that the EEG activity within each type (or class) of segment is generated by the same or similar dynamics and driving force [51, 81, 82, 107]. However, EEG activity from different classes of segments has, in effect, different driving forces and is therefore generated by different dynamics. In their turn, consecutive EEG segments comprise a new sequence in a particular time scale. Such functional EEG structure comprises hierarchical multivariability which reflects the poly-operational structure of brain activity [55, 108].

2.2 Spatial Heterogeneity of EEG

Besides the temporal nonstationarity of EEG signal there is another striking feature of EEG which is the spatial difference in electrical activity from electrode to electrode [109] indicating that the brain generates a highly structured in space extracellular electric field [85, 87].

Experimental findings suggest the existence of statistical *heterogeneity* (anisotropy) of the electromagnetic field in relation to local field potential (LFP) processes [110] and local EEGs [53, 56, 57, 64, 111, 112]. It was demonstrated that such electromagnetic spatial heterogeneity relates to large-scale morpho-functional organization of the cortex:

- Dynamic baseline of intrinsic (not stimulus- or task-evoked) brain activity during resting wakefulness is topographically organized in discrete brain networks—resting state networks [57, 113, 114]. Each of these functional networks is characterized by a specific electrophysiological signature that involves a combination of different EEG rhythms. Thus, topographic EEG structure is frequency specific [115].

- Each local EEG recording or small group of local EEG recordings has its own activity (indexed by a set of spectral patterns of different types) [49].

- The same type of EEG activity (indexed by spectral pattern of a given type) occurs in approximately 70 % of all observations in no more than two functionally homologous EEG locations: for example O1–O2 [106], thus suggesting functionally heterogeneous topology of EEG.

- Topographic variability of different types of EEG short-term spectral patterns revealed that spatial EEG map is expressed as a mosaic of cortical structures that are functionally integrated into clusters of different size [57, 63, 66].

- A high number of mutual intercorrelations between various piecewise descriptors of EEG when all local EEG recordings are accounted for and, at the same time, the absence of such intercorrelations when each local EEG recording was taken in isolation suggest that functional dynamics of neuronal assemblies (indexed by various piecewise EEG descriptors) take place within a rigid and narrow morpho-functional range, which constrains topological (among locations) relations between these piecewise descriptors of EEG [53].

- Spatial heterogeneity of neuronal synchrony has been repeatedly demonstrated at the level of LFP [110] as well as at large-scale level of EEG [55–57, 116–122].

- Cortical areas separated by distances exceeding the diameters of "wave packets" (see Note 3) have differing wave forms and therefore different spectral pattern types [85, 87].

- Covariance between neighboring electrodes across cortex functional boundaries (e.g., parietal to temporal areas) is much higher than covariance within functional regions (e.g., left parietal to midline parietal area), indicating that multiple functionally distinct areas are reliably assessed by EEG topographic heterogeneity [124, 125]. This morpho-functional heterogeneity of EEG was also confirmed in an independent study in which the spatial heterogeneity of scalp-recorded EEG synchronicity was measured along longitudinal (anterior-to-posterior and posterior-to-anterior directions) and transversal (right-to-left anterior and right-to-left posterior directions) electrode arrays with scalp electrodes equally spaced in all these arrays [53, 73]. Data from actual EEG was compared with the so-called surrogate EEG in which a mixing of actual local EEG recordings was done so that the natural time relations between all local EEG recordings were completely destroyed, but the number, duration, and sequence of segments within each local recording remained the same as in the natural EEG. For longitudinal electrode arrays (despite all testing pairs of EEG electrodes having the exact same interelectrode distance) the synchronicity index exhibited a notable topological landscape: significantly decreasing ($p < 0.05$) in locations of EEG electrode pairs on the head that crossed functional cortex boundaries [53, 73]. This data clearly indicates that temporal consistency of segmental architectonics of the electrical field becomes weak at the boundaries of well-outlined functional cortical areas.

Additionally (a) the relationship between synchronicity index and interelectrode distance was not monotonous for both longitudinal electrode arrays: stepwise dependency was observed, and (b) forward (posterior-to-anterior) and backward (anterior-to- posterior) dependences of synchronicity index from the interelectrode distance varied significantly from each other [56]. These results suggest that in such conditions the contribution of volume conduction is insignificant.

For transversal electrode arrays it was demonstrated that (a) anterior and posterior cortex areas had opposite tendencies in the dynamics of synchronicity index (notice that anterior and posterior cortex areas have different morpho-functional organization) and (b) maximal synchronicity index values in the posterior cortical areas were obtained for homological lateral EEG locations (which have similar morpho-functional organization) despite the largest interelectrode distance in the electrode array [56].

Summarizing, these findings suggest notable topological peculiarities of the EEG signal along the cortex, thus reflecting a morphologically and functionally heterogeneous organization of the cortex.

Taking together the aforementioned findings, one may suggest the existence of statistical heterogeneity (anisotropy) of the electromagnetic brain field in regard to neurodynamics within and between regional EEGs, which is best described by the so-called clustered functional networks [57, 126]. Simulations of clustered networks suggested that such spatial heterogeneity of EEG may have a number of advantages when compared to a random or a homogeneous structure. It was demonstrated that clustered networks are more easily activated than random networks of the same size [126]. This is due to the higher density of connections within the clusters which facilitated local activation. At the same time, the sparser connectivity between clusters prevents the spreading of activity across the whole network. Thus, in contrast to random networks, clustered networks possess an expanded critical functional range for which initial activations resulted in persistent but nonglobal network activity [126–128]. Such topology is consistent with the so-called *small-world* networks [129–134], which are characterized by predominantly local coupling with a small number of long-range connections [130, 135].

Functional significance of spatial heterogeneity of EEG was also confirmed by numerous studies where it was demonstrated that different EEG spatial structures were associated with (a) encoding of various sensory information and the meaning of this information [73, 136], (b) different levels of complexity in conscious thinking [61, 137, 138], (c) different cognitive loading [61, 73, 86, 139–141], (d) reconfiguration between rest and task conditions [142], (e) changes across age [143], and (f) different psychopathologies [86, 144] (for Alzheimer's disease [145]; for epilepsy [66, 146]; for major depression disorder [147, 148]; for drug intake [83, 84, 149–154]; for schizophrenia [155–157]).

Summarizing this subsection we can conclude that as opposed to being a continuous process, the EEG (within each local signal/channel and as a global whole-cortex field) represents a series of consecutive short-term segments, each of which is produced from synchronized activity of a different group of cells—neuronal assemblies (for a review see [57, 158]). Surprisingly, even though such knowledge about the structure of EEG signal has been available for decades [159], it has been largely neglected in most methods used for modern-day EEG analysis [51, 63]. Usually, manifestations of nonstationarity in the EEG signal are either carefully eliminated or are considered as an unavoidable "noise," which should be diminished [53]. To minimize this so-called noise, various mathematical procedures of smoothing and averaging are applied to the raw data. The key assumption underlying such statistical analyses is the "stationarity" of the EEG signal. At the same time, in view of the overwhelming evidence for the disjointed nature of EEG activity presented above, it would be unwise to treat the EEG as a stationary statistical process [53, 63]. Since we are essentially dealing with a sequence of discrete processes (short-term segments), methods presupposing a continuous process are likely to have a rather limited application in EEG analysis [4, 48, 53, 56, 60, 68, 69].

Even though approaches based on the assumption of EEG stationarity have revealed some important signal characteristics (for example, the functional significance of different EEG frequency bands [35, 160, 161]), the initially rapid temporal resolution of the EEG signal is usually lost under such conditions. In the meantime, it is obvious that regardless of how powerful or statistically significant the different estimations of averaged EEG characteristics may be, there might be substantial difficulties in arriving at a meaningful neurophysiological interpretation of these if they are not matched to their inherent piecewise stationary structure [4, 50, 53, 63, 162]. For example, invariants such as the mean power/amplitude spectrum, average ERP and ERD/ERS, coherency, fractal dimensions, Lyapunov exponents, and many other indexes have an interpretation only for stationary dynamics [163]. Furthermore, the nonstationarity of the EEG signal usually does not allow researchers to construct a global dynamical model for the whole observable phenomenon [164]. Therefore, there is a need for a *novel methodology of EEG analysis* that is sensitive to the underlining quasi-stationary nature of EEG signal (in both temporal and spatial dimensions) and plausible for a better understanding of the functional organization of the neocortex and its relation to a cognition and eventually consciousness.

One of such novel methodologies for EEG analysis is an *operational architectonics* (*OA*) framework (see Note 4) which (a) considers the nonstationarity of EEG (is sensitive to spatial–temporal structure of EEG), (b) does not contain temporal averaging procedures, (c) is model independent, (d) has special tests for nonrandom and non-occasional nature of the results, (e) produces results which are easy to interpret in terms of their neurophysiological correlates, and (f) permits to measure postulated entities in practice [55, 56].

---

3   Brief Introduction to the Operational Architectonics Methodology

In this subsection we overview the main concepts of OA theory and the related methodological approaches to EEG analysis. The basic operational units of brain activity within OA theory are presented by the local field activity of neuronal assemblies "*hidden*" in the complex nonstationary structure of brain EEG field [53, 56, 63, 91]. Such local fields are organized spatially and temporally within a *nested hierarchy* (see Note 5) of increasing in complexity discrete metastable states that serve as the basis for functioning of such a multivariable and complex system like the brain [108].

3.1   Neuronal Assemblies and EEG Signal

A neuronal assembly is generally defined as a set of neurons that are able to synchronize their sub-threshold oscillations (excitatory/inhibitory postsynaptic potentials—EPSPs/IPSPs), leading to a coherent activity of the whole assembly [16, 34] in order to perform a specific physiologic or cognitive operation [173–175]. Historically, this notion goes back to Hebb [176], though the classical (Hebbian) neuronal assemblies are too slow and rigid and therefore would not be suitable for fast cognitive operations (for a discussion, see [55]).

A modern understanding of neuronal assemblies stresses their functional and highly transient nature, which is at scales both coarser (for *spatial* dimension) and finer (for *temporal* dimension) compared to classical assemblies [177]. The idea is that large masses of individual neurons can quickly become functionally self-organized (associated or disassociated), thus giving rise to *transient* assemblies [178, 179]. The overall pattern of emergent neuronal assembly's correlated activity—local field or wave packet [65]—persists over some temporal interval, during which it is thought to execute the basic physiologic or cognitive operation [180, 181] which may also be subjectively experienced as a presence of simple phenomenal feature/qualia (see Note 6) [54, 91]. This pattern is very sensitive to fluctuations and it can be swiftly rearranged during rapid transitional period [91, 186, 187]. As has been demonstrated in vitro, such intervals of correlated activity are manifested in oscillatory waves [34, 188] accessible through local EEG measurements [59, 85, 87, 189]. Thus, the behavior of neuronal assemblies is highly dynamic and nonlinear [53, 65, 158, 169] and arises as a consequence of propagating local synchrony in the form of avalanches of neuronal activity [190, 191] (for a brief explanation of avalanches see below).

EEG waves recorded from the scalp are integrated EPSPs/IPSPs of neuronal membranes [85, 87, 189]. Since they reflect extracellular currents caused by synchronized neural activity within a given brain volume [192], the local EEG signal within quasi-stationary (nearly stable) segments represents an envelope of the probability of nonrandom coherence (the so-called common mode or wave packet [65]) in the neuronal assemblies near to the recording electrode [158, 169]. Even though the neuronal cells that comprise an assembly under the electrode may be spatially intermixed with neuronal cells from other neuronal assemblies performing different computational tasks, they are separated by different temporal scales reflected in a set of particular narrow EEG frequencies (see Note 7) [39, 193]. Therefore, it is possible to consider each local EEG quasi-stationary segment of particular type as the single event in the EEG phenomenology, which reflects the particular operation of a related neuronal assembly [54, 56, 106].

Of course, there is no simple (one-to-one) relation between an EEG quasi-stationary segment of particular type and the actual state of the neurons in the underlying network: many different configurations of firing neurons can give rise to the same type of short-term EEG activity (many-to-one relation). At the same time, the same configuration of firing neurons cannot give rise to two (or more) different types of short-term EEG activity. Thus, two different types of short-term EEG activities are likely to originate from two different configurations of firing neurons [107]. Consequently, short-term EEG activity of a particular type reflects a specific class of neuronal activity, where each of the activities has something common with the others within the same class (one class-to-one relation). Moreover, two classes of neuronal activities do not overlap (otherwise the same configuration of firing neurons could give rise to two or more different short-term EEG activities). Thus, a given type of short-term EEG activity may be considered as a single event (which reflects a particular class of neuronal activity) in EEG phenomenology. Additionally, considering the aforementioned spatial heterogeneity of EEG (see Section 2.2 above) one may suggest that local EEG short-term activity is mainly determined by underlying neurodynamic (functional state) and that this type of activity mainly reflects the large-scale morpho-functional organization of the cortex rather than the effect of volume conduction, at least for the 10/20 system (used in the majority of EEG studies) which measures the main cortex lobes.

Within the duration of one quasi-stationary EEG segment, the neuronal assembly that generates the oscillations is supposed to be in a steady quasi-stationary state [51, 73, 81]. The *rapid transition processes* (RTPs) occurring in the continuous EEG activity mark the boundaries between quasi-stationary segments for this activity. The transition from one segment to another reflects the changes of the neuronal assembly microstate or, if multiple frequencies are considered, changes in the activity of several neuronal assemblies [53–56, 65]. This is the first (*low*) level of operational architectonics framework of brain functioning (for a complete description see [55, 56, 91]).

3.1.1   EEG Signal Segmentation Approach: RTP Estimation

Because the major contributor to temporal modulation of the variance and power of the EEG signal is the sharp change in its amplitude [194], the identification of RTPs can be reduced to detecting the moments of rapid statistically significant decrease or increase of EEG amplitude [56]. In this sense, RTP is defined as an abrupt change (see Note 8) in the analytical amplitude of the local EEG signal above a particular threshold derived from a statistical procedure which was established experimentally for each local EEG in previous modelling and empirical studies [56].

There are several mathematical approaches to segment a multichannel EEG signal. Unfortunately, the vast majority of them are based on the assumption that an EEG signal can be sufficiently described by using a mathematical model characterized by a finite collection of parameters [81]. In mathematical statistics such methods are known as parametric [196]. One example of such methods used for EEG segmentation is autoregressive moving average modelling [69, 197, 198]. Linear regression modelling is another example of used EEG segmentation method [199]. Even though parametric methods are quite effective if the phenomenological model of the analyzed process is known [196], in case of EEG signal no such generally accepted phenomenological model has been ever suggested [81]. In fact, very different mathematical models can be fitted to the same EEG signal, thus resulting in vastly different estimates of segments. This makes the usage of parametric methods for EEG segmentation unjustified.

Another drawback of nearly all EEG segmentation methods is the fact that they are designed for stationary signals. Such algorithms can be applied correctly only to stationary intervals of the signal, but the presence and location of these stationary intervals are unknown before the procedures are actually applied—the so-called vicious circle [81]. Therefore, nonparametric approaches that do not require any model of the signal and a priori information about its distribution are needed.

Fingelkurts and Fingelkurts [200] were among the first who applied such nonparametric segmentation procedure which has been proposed by Brodsky and Darkhovsky to EEG signal. Since then, the initial procedure has been substantially modified and improved reaching its current form [56]. The general statistical principles of it have been described extensively elsewhere [54–56]. Therefore, here we provide only a brief overview of this approach. This adaptive segmentation procedure is used for the parallel automatic segmentation of local EEG signals within a multichannel EEG record. The method is based on an automated algorithm that moves a double-window screening along each separate EEG signal/channel. The following steps are taken to estimate RTPs:

(1) Comparisons are made between ongoing EEG amplitude absolute values averaged in two windows—"test" and "level" (duration of test window $\ll$ duration of level window), both starting from the first data point. The durations of test and level windows are identical across different subjects and EEG channels for each particular frequency band (but different between the bands) to guarantee the best conditions for RTP evaluation.

(2) If the absolute maximum of the averaged amplitude values in the test window is less or equal to the averaged amplitude values in the level window, then the hypothesis of EEG homogeneity is accepted.

(3) If the absolute maximum of the averaged amplitude values in the test window exceeds the averaged amplitude values in the level window, according to the threshold of "false alerts" (first condition—the Student criteria), its time instant becomes the preliminary estimate of the RTP.

(4) A second condition must be fulfilled to eliminate "false alerts" associated with possible anomalous peaks in the amplitude: several points of the digitized EEG recording following this preliminary RTP must have a statistically significant difference between averaged amplitude values in the test and level windows (Student's t-test).

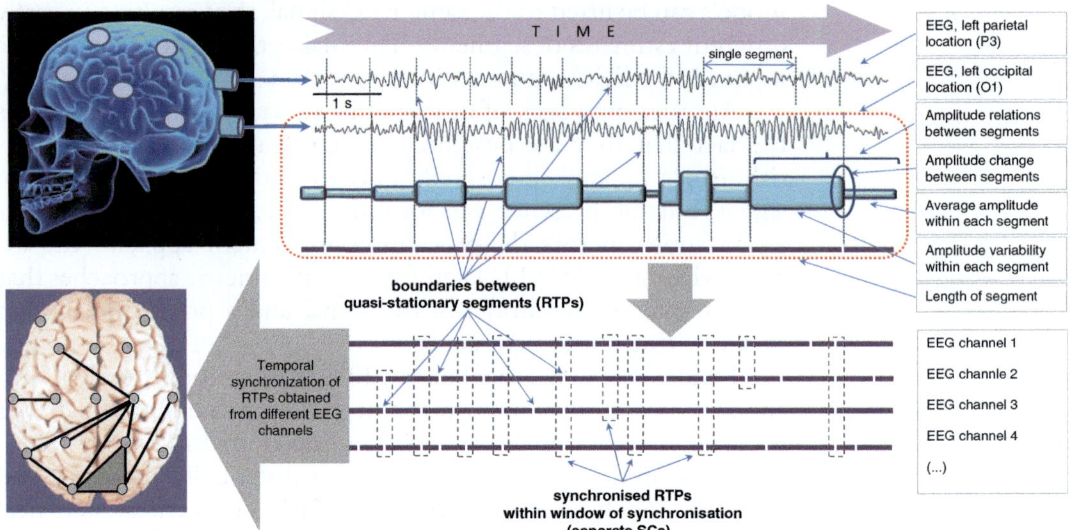

Fig. 1 Schematic illustration of the operational architectonics methodology including steps for EEG segmentation, extraction of segments' features, and synchronization of segments. Further explanations are provided in the text. RTP—rapid transitional processes (boundaries between quasi-stationary EEG segments); SC—momentary synchro-complexes (synchronization of RTPs between several but always the same local EEGs at the particular time instants)

(5) If these two criteria are met, the preliminary RTP is considered real.

(6) Thereafter, both windows are shifted from this RTP to the next time point, and the procedure is repeated to detect the next RTP.

With this technique, the sequence of RTPs with statistically proven ($p < 0.05$, Student's t-test) time coordinates is determined individually for each EEG channel and for each epoch of analysis (Fig. 1, upper panel).

The choice of window durations and of the "false alerts" rule was made during a few preliminary modelling studies (for a review see [56]) to guarantee the following two goals: (a) to detect as many real RTPs as possible and (b) to keep the level of "contamination" by "false" RTPs in the results reasonably low.

The identified quasi-stationary segments of EEG activity could be characterized by several attributes that reflect different characteristics of transient neuronal assemblies [83, 86]. These attributes are (Fig. 1, middle panel) as follows:

(1) Average amplitude within each segment (microvolts)—as generally discussed above, mainly indicates volume or size of the neuronal assembly; the more the neurons recruited into an assembly through local synchronization of their activities, the

higher the amplitude corresponding to this assembly's oscillation in the EEG [16, 201].

(2) Average length of segments (milliseconds)—illustrates the functional life-span of a neuronal assembly or the duration of operation produced by this assembly; because a transient neuronal assembly functions during a particular time interval, this period is reflected in EEG as a stabilized interval of quasi-stationary activity [180].

(3) Coefficient of amplitude variability within segments (%)—shows the stability of local neuronal synchronization within a neuronal assembly [194].

(4) Average amplitude relation among adjacent segments (%)—indicates neuronal assembly growth (recruitment of new neurons) or disassembling (functional elimination of neurons) [180].

(5) Average steepness among adjacent segments, estimated in the close area of RTP (%)—shows the speed of neuronal assemblies growing or disassembling [180].

3.1.2 Methodological Results Related to EEG Segmentation

Experimental results revealed a relatively large number of EEG segments per minute for nearly all narrow-frequency bands [73, 83, 139, 180, 200, 202]. For example, there are about 60 quasi-stationary segments for delta activity (~1,000 ms average duration), 120 quasi-stationary segments for theta activity (~500 ms), 200 quasi-stationary segments for alpha activity (~300 ms), 250 quasi-stationary segments for beta-1 activity (~250 ms), 300 quasi-stationary segments for beta-2 activity (~200 ms), and 350 quasi-stationary segments for gamma activity (~170 ms) per each 1-min local EEG during restful wakefulness with closed eyes. Similar relations between durations of quasi-stable states within different frequency bands have been shown in independent research [57]. This range of temporal durations fits very well with the biophysical processes underlying the assemblies' formation, often iteratively on time scales corresponding to 100–900 ms range [34]. As it has been pointed out by David et al. [203] a neuronal assembly's operation should be longer than the time for spike transmission between neurons either directly or through a small number of synapses (from few milliseconds to several tens of milliseconds), and it must be shorter than the time it takes for a cognitive act to be completed (which is from several hundreds of milliseconds to a few seconds).

Within such segments, the degree of synchronization (alignment of EPSP/IPSP times in multiple neurons making up the neuronal assembly) can vary in time, so it is important to consider how much variability can be tolerated and still be able to consider assembly member neurons to be "acting together" [204]. Our data have shown that the coefficient of within-segment amplitude variability was relatively low for all studied frequency bands (thus indicating the formation of neuronal assemblies [91]; rest with closed or open eyes), and it was substantially higher (up to 30 %) for the randomly altered EEG (see Note 9) (see also [139]). Such a considerable increase in the values of within-segment amplitude variability in the "random" EEG indicates the absence of stabilized epochs and thus lack of neuronal assembly formation [56]. More precisely it represents an estimation of the maximum possible rate of relative alterations in the amplitude variability for a given EEG [53].

Additionally, it has been shown that average amplitude relations among adjacent segments were substantially lower for the randomly altered EEG when compared to real EEG (rest with closed or open eyes; for an example, see [139]), thus indicating that RTPs are true authentic boundaries of EEG quasi-stationary segments which reflect the episodes of relative stabilization of neuronal activity within separate neuronal assemblies separated by such RTPs [53, 56, 139].

OA theory proposes that the simple operations of neuronal assemblies that we experience subjectively as simple patterns/features of complex thought/image/scene and/or observe objectively as elemental operations of behavior are created and executed in emergent neural fields of self-organized activity of neuronal assemblies [91]. If this is so, then one should observe an RTP in local EEGs (reflecting a reset of a new operation) every time a novel task (or experimental input) is induced. In other words, if RTPs are true EEG markers of transitional moments between local neuronal assembly operations (indicating their start and end), then under experimental conditions the number of RTPs should be much higher at the exact moment of systematic change from one cognitive/mental operation to another (induced start of a new task) when compared to other task-free time coordinates within the same condition or a control condition that does not include systematic change of operations according to an experimental protocol. Indeed, experimental studies have confirmed this hypothesis [73, 200] (for illustration see Fig. 4 in [56]). These findings clearly indicate that RTPs are the true markers of beginnings and ends of brain–mind operations.

Generally, it has been shown that segmental description of local EEG signals within the multichannel EEG recording is very sensitive to a both spontaneous (stimulus independent) and induced (stimulus dependent) brain states [73, 139, 180, 200, 205, 206] (for a review see [86]) as well as pharmacological influence [83, 151, 153, 154] and various pathological conditions [73, 102, 152] (for a review see [86]).

Summarizing this subsection we would like to stress that application of OA methodology to an EEG signal allows researchers to monitor local electric fields of transient neuronal assemblies with millisecond time resolution via extracellularly placed (at the scalp) electrodes (see Note 10) and can be used to interpret many facets of neuronal communication and computation [180].

For example, such parameters as size, life-span, stability, and speed of neuronal assembly growth or disassembly can now be studied robustly as a function of subject's states, behavioral acts, pharmacology usage, and neurological or psychiatric pathology (for a review see [86]). The main advantage of such methodology over others, used for the investigation of EEG activity, is that not only are the underlying biophysical mechanisms taken into account but also the neurophysiologic processes related to the EEG and reflected in its spatial and temporal nonstationarity are considered [91].

From the first studies of RTP dynamics [52, 73, 139, 200, 208] it has been noticed that when an RTP in EEG amplitude occurs at any one site (separate EEG channel), it tends to simultaneously occur at several sites of the multichannel EEG recording, marking the onset of a spatiotemporal EEG pattern (see Note 11). Peculiarities of such synchronization are discussed in the following subsection.

3.2 Synchronization of Operations of Neuronal Assemblies and EEG Signals

The second (*high*) level of OA theory describes the temporal synchronization of different brain operations (*operational synchrony*) that are executed simultaneously by different local and transient neuronal assemblies [91]. Indeed, as it has been observed in the previous subsection, individually each neuronal assembly presents only a partial aspect of the whole object/scene/concept [211], while the wholeness of "perceived" or "imagined" is brought into existence by joint (synchronized) operations of many functional and transient neuronal assemblies in the brain (see Note 12) (for a thorough discussion see [91, 217]; see also [218]). The recombination of neuronal assemblies in new configurations makes it possible to present subjectively a nearly infinite number of different qualities, patterns, objects, scenes, and concepts—even those, with which we have never been acquainted before.

Such synchronization of operations gives rise to a completely new level of brain abstractness—*metastable* brain states (see Note 13) [108]. These metastable brain states or functional *operational modules* (OMs), as we name them, accompany the realization of brain complex (and composite) macro-operations (e.g., cognitive percepts, phenomenal objects, and reflective thoughts; for a review see [91, 217]), whereas each of them is instantiated by the particular volumetric spatial– temporal pattern in the electromagnetic field [54–56]. Within the OA theory, the complex operation or operational act has internal structure where each element in its turn also has its own internal structure and so on until the simplest elemental operations are reached (Fig. 2). In other words, there is a more complex operation/operational act that subsumes the simplest ones. It is critical that any complex operation/operational act is not just a conjunction of simplest operations (or operational acts)—but it is also an operation (or operational act) in its own right with emergent properties that are not evident in the subordinate constituents. Such architecture has a clear *nested hierarchy* [224] and thus could serve as the needed

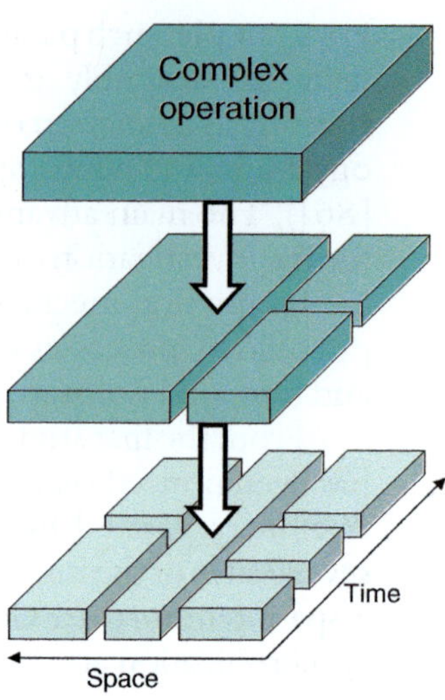

Fig. 2 Schematic presentation of the compositionality of operations. Further explanations are provided in the text

ingredient of brain organization that would allow a conscious mind to be expressed and present the features of consciousness (referred to as neural states, mental unity, qualia, and mental causation), which are discussed in detail elsewhere [54, 91, 217] (see also [170–172]).

3.2.1 Synchronization of EEG Segments: Structural Synchrony Measure

Because the beginning and end of discrete operations of local neuronal assemblies are marked by the sharp changes in local EEG amplitude (RTPs), the simultaneous occurrence of these RTPs found in different local EEG signals within the multichannel EEG recording could provide evidence of neuronal assembly synchronization (located in different brain areas) that participate in the same functional act as a *group* (Fig. 1, bottom panel), e.g., executing a particular complex operation responsible for a subjective presentation of complex objects, scenes, concepts, or thoughts [54, 91, 217]. Fingelkurts and Fingelkurts [200] were the first who came up with this hypothesis as well as suggested the quantitative method for estimation of such unique type of brain synchrony. The method is named "*structural synchrony*" (or index of structural synchrony, ISS) since it refers to a synchrony between structures of local EEG signals, though qualitatively it estimates the synchrony among operations of multiple neuronal assemblies; thus, it corresponds to an "*operational synchrony*" [73]. Later this method has been modified to expand its functionality [56].

The details of this technique are beyond the scope of this chapter; therefore, we concentrate on the essential aspects only. In brief, each RTP in the reference EEG channel (the channel with the minimal number of RTPs from any pair or set of EEG channels) is surrounded by a short "window" (Δt, ms). Any RTP from another (test) EEG channel is considered to coincide if it fell within this window (Fig. 1, bottom panel). It is important to note that such coincidence of RTPs is related to a specific type of signal coupling—the *structural synchronization of discrete events*—which completely ignores the level of signal synchronization in the intervals (segments) between the coinciding RTPs (see Note 14) [122, 200]. Therefore, this approach explicitly uses the definition of functional connectivity agreed upon in neuroimaging community [226]: Functional connectivity is defined as the temporal correlation between spatially remote neurophysiological *events* [227]. As it has been noted by Plenz [191], such measure would in fact represent an unbiased estimate of synchronization, because in the absence of any knowledge about the underlying spatial organization of the involved neuronal assemblies, it uses temporal proximity of RTPs to estimate nonrandom interaction [56]. Importantly, ISS could capture both the near-instantaneous synchronous RTPs, comparable to zero phase-locked synchronization between two sites, as well as successively synchronized RTPs, comparable with delayed phase-locked synchronization [122].

To arrive at a direct estimate at the 5 % level of statistical significance ($p < 0.05$) of the ISS, computer simulation of RTP coupling is undertaken based on random shuffling of time segments marked by RTPs (500 independent trials). As a result of this procedure, the stochastic (random) levels of RTP coupling (ISS_{stoch}), together with the upper and lower thresholds of ISS_{stoch} significance (5 %), are calculated. The ISS tends towards zero if there is no synchronization between EEG segments derived from different EEG channels and has positive or negative values where such synchronization exists. Positive values (higher than upper stochastic level) indicate "active" coupling of EEG segments where synchronization of EEG segments is observed significantly more often than expected by chance (as a result of random shuffling during a computer simulation), whereas negative values (lower than low stochastic level) mark "active" decoupling of segments where synchronization of EEG segments is observed significantly less than expected by chance (as a result of random shuffling during a computer simulation). Dynamics of ISS and its positive and negative levels are shown in Fig. 3. The strength of EEG structural synchrony is proportional to the actual value of ISS: the higher this value, the greater the strength of functional connection.

Since most methods that estimate synchronization look at pairs of recording channels only, a global picture of synchronization in multichannel EEG recording is usually obtained by averaging the pairwise synchronization between every possible pair of channels. The main problem with such an approach is its inability to detect the naturally arising synchronization patterns across space and time [228] (for a review see [91]). In contrast to the majority of other synchronization methods (that have a number of other limitations) (see Note 15) the OA methodology uses algorithmic instruments

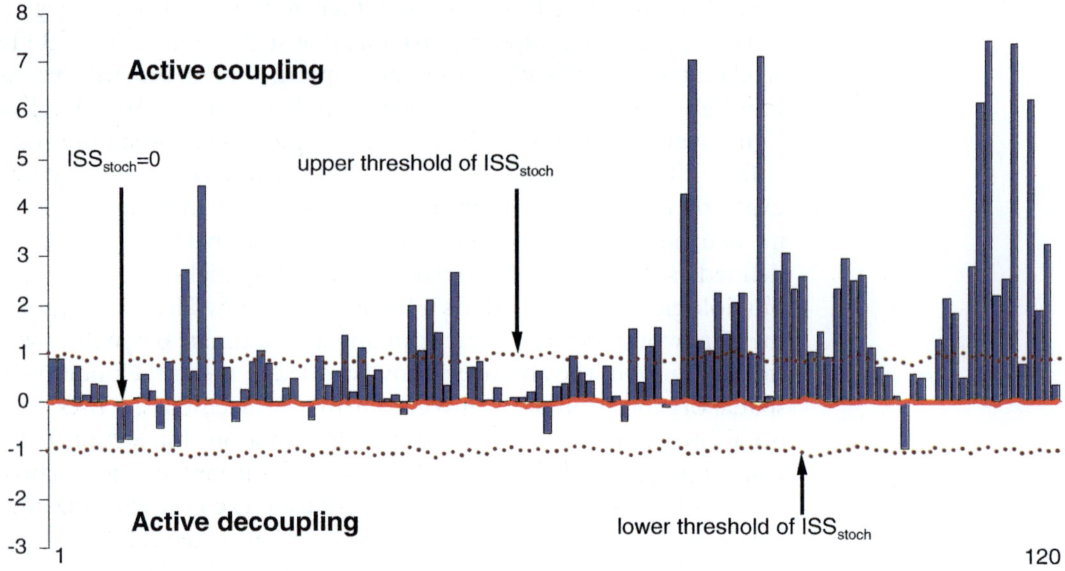

Fig. 3 Schematic illustration of the index of structural synchrony (ISS) and its stochastic levels. As an example, the calculations of ISS are shown for 16 EEG channels. The *Y*-axis displays the ISS values found in the experiment (illustrated as bars). The *X*-axis displays the 120 possible pair combinations of 16 EEG channels. The data presented for a healthy subject in a rest condition with eyes closed

that permit it to estimate synchrony of events in more than two EEG channels.

Structural synchrony can be identified in several (more than two) EEG channels, thus identifying synchrocomplexes (SC). An SC is a set of EEG channels in which each channel forms a paired combination (with valid values of ISS) with all other EEG channels within the same SC, meaning that all pairs of channels in an SC have to have statistically significant ISS values linking them together (Fig. 1, bottom panel). The number of cortical areas (indexed by the synchronized EEG channels) recruited in an SC is described as "the order of areas recruitment" [73, 200, 210]. Therefore, all SCs could be divided into a set of categories based on the number of cortex areas involved: SC_2—SC with second order of areas recruitment, SC_3—SC with third order of areas recruitment, SC_4—SC with forth order of areas recruitment, and so on. Notice that any given SC is considered as a member of its own category (for example, SC_3) only if the correspondent RTPs coincided in time among correspondent number of EEG channels (in this example, 3). However, any three SCs_2 which could comprise the particular SC_3 but which do not coincide in time between each other are not considered as producing SC_3 type and, therefore, are not counted as SC_3 but instead as SC_2. The same logic is applied to any other SC category larger than SC_2.

Notice that synchronized RTPs mark transitions between different segments in the EEG signal, which are usually of different type in the various EEG channels. Therefore, the described RTP-based measure of functional connectivity, in contrast to conventional approaches, is free from similarities of the EEG signals in different channels. In this context, the simultaneous stabilization of RTPs across several cortical areas is supposed to reflect the formation of a steady cooperation of operations among those scattered in different cortical areas neuronal assemblies independent of the particular characteristics of these operations [91]. At the more conceptual level, such synchrony presents the relevant suboperations or phenomenal qualia about the external objects (or scenes) reflected in the wave packets of quasi-stationary EEG segments derived from different cortical areas for integration and formation of a unified macroscopic (complex) phenomenal object, scene, or thought as the culminating act of perception, imagination, or thinking (for a detailed discussion see [91, 217]). In this context SCs have a direct relation to the OM estimation.

3.2.2 Operational Module (OM) Estimation

The criterion for defining an OM is a sequence of the same type of SCs of the same category during given epoch of analysis. In other words, the constancy and continuous existence of each OM persist across a sequence of discrete and concatenated segments of stabilized (coupled) local EEG activities (indexed by the same SCs) that constitute that particular OM (Fig. 4; [55, 56]). Conceptually, *the continuity of any given OM exists as long as the set of neuronal assemblies located in different brain areas maintains synchronicity between their discrete operations* (see Note 16). Analogous to SCs, OMs can be of different order of areas recruitment (indexed by the number of synchronized EEG channels).

The notion of *operational space–time* (OST) applies here [229]. Intuitively, OST is the abstract (*virtual*) space and time which is "self-constructed" in the brain each time a particular OM emerges (see Note 17). Formally, the OST concept holds that for a particular complex operation, the spatial distribution of the locations of a *particular set* of neuronal assemblies together with their synchronous activity at repetitive instants of time (indexed by SCs) comprises the OM. These distributed locations of that set of neuronal assemblies are discrete, and their proximity as well as the activity in the "in-between area(s)," delimited by the known locations of neuronal assemblies, are not considered in the definition (only the exact locations of a particular set of neuronal assemblies are relevant). Also, between the moments in time that particular locations of the neuronal assemblies synchronize, there can be smaller subset (s) of these locations synchronized between themselves or with other neural locations, though these do not relate to the same space–time of the same OM (although they may relate to some other OM). Therefore several OMs, each with its own OST, can coexist at the same time within the same volumetric

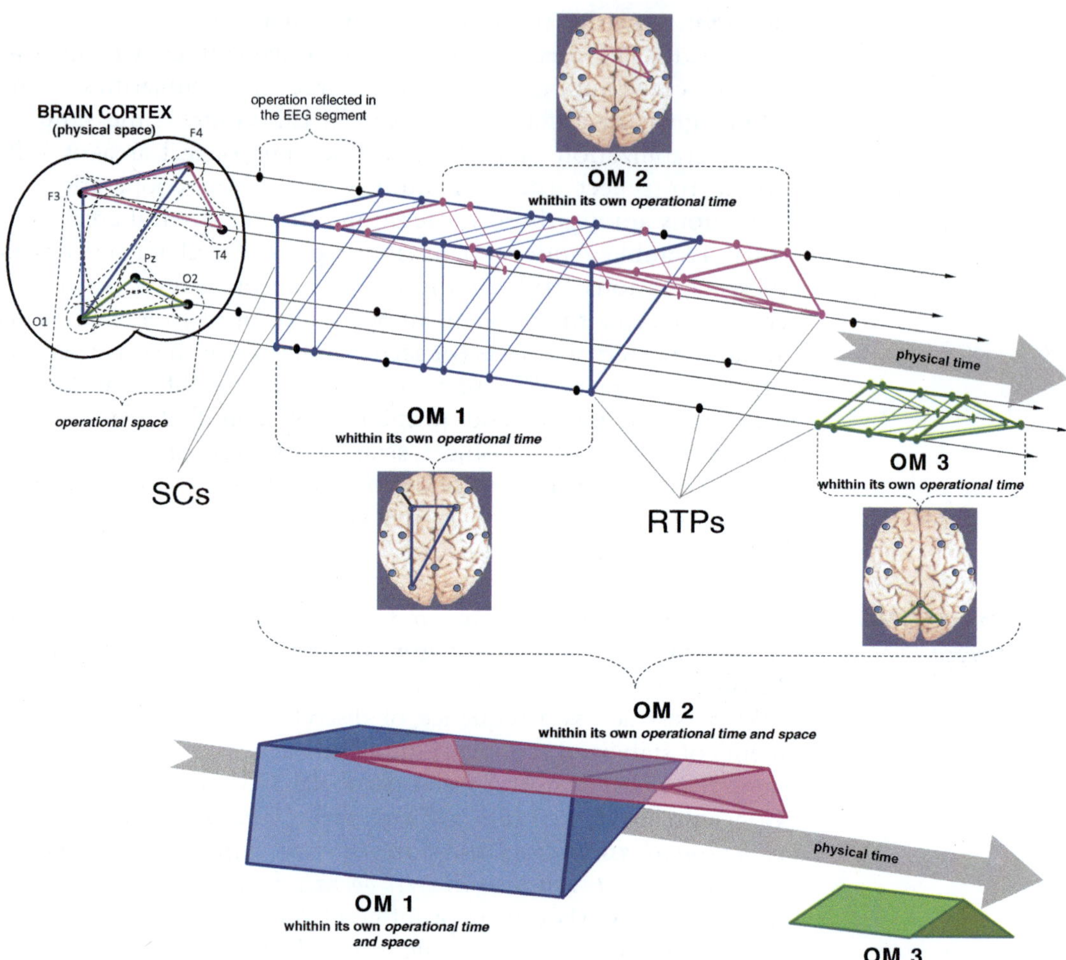

Fig. 4 Schematic illustration of operational modules (OMs) and operational space–time (OST). Each OM exists in its own OST, which is "blind" to other possible time and space scales present simultaneously in the brain "system": all neural assemblies that do not contribute to a particular OM are temporarily and spatially "excluded" from the OST of that particular OM. Further explanations are provided in the text. RTP—rapid transitional processes (boundaries between quasi-stationary EEG segments); SC—momentary synchro-complexes (synchronization of RTPs between several but always the same local EEGs at the particular time instants); F3—the left frontal cortical area; F4—the right frontal cortical area; O1—the left occipital cortical area; O2—the right occipital cortical area; T4—the right temporal cortical area; Pz—the central parietal cortical area. As an example, it is shown that neural assemblies in these areas could synchronize their operations on three different (even though partially intertwined) spatial–temporal scales, thus forming three separate OMs each having its own operational space–time (*bottom panel*, 3D objects)

electromagnetic field [91]. The sketch of this general idea (based on experimental data) is presented in Fig. 4.

According to OA theory, the metastable OMs at an OST level somehow "freeze" and "classify" the ever-changing and multiform stream of our cognition and conscious experiences [91]. In this

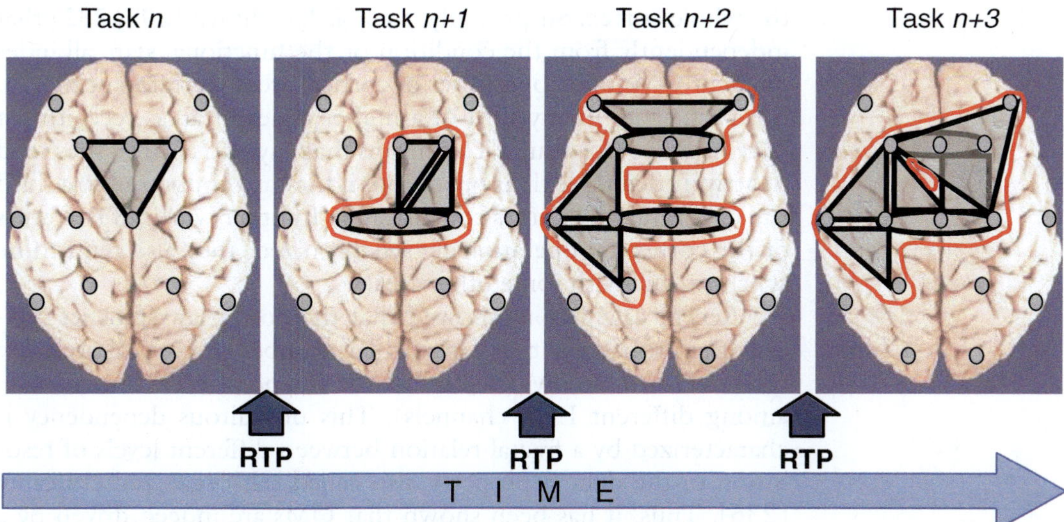

Fig. 5 Schematic diagram depicting dynamics of operational modules (OMs). Relatively stable complex OMs (outlined by the *red line*) undergo abrupt changes simultaneously with changes in cognitive task. Such abrupt changes marked as rapid transitional periods (RTPs). *Grey* shapes illustrate individual OMs. *Red line* illustrates complex OMs. For interpretation of the references to color in this figure legend, the reader is referred to the Web version of this chapter. This scheme is based on data published in ref. 247

sense, the succession of complex cognitive operations, phenomenal images, or thoughts is presented by the succession of discrete and relatively stable OMs, which are separated by RTPs, i.e., abrupt changes of OMs (see Note 18) (see Fig. 5). As it has been shown experimentally [52, 83, 84, 139, 148, 153, 202, 208, 232], at the critical point of transition in a mental state, e.g., during changes from one task/thought to another, the OM undergoes a profound reconfiguration which is expressed through the following process [91]: A set of local bioelectrical fields (which constitute an OM) produced by transient neuronal assemblies located in several brain areas rapidly loses functional couplings with one another and establishes new couplings within another set of local bioelectrical fields, thus demarcating a new OM in the volumetric OST continuum of the brain. Using this mechanism the brain determines who "talks" to whom at any particular moment. In this sense, the brain can "rewire" itself dynamically and functionally on a milliseconds time scale without changing the synaptic hardware (see Note 19) [57, 235].

3.2.3 Methodological Results Related to EEG Structural Synchrony

Already first experimental studies [52, 73, 139, 200, 208, 232] had shown that OMs (specific spatiotemporal configuration of stabilized segments among local EEG fields) indeed exist, thus confirming the discovery of a new and previously unknown type of brain functional connectivity. It was found that such OMs are characterized by different order of recruitment of cortical areas: from any two to the whole cortex. Surprisingly, analysis has shown [200, 232] that independently from the condition or the functional state all independent OMs of second order of cortical areas recruitment that do not participate in any OM of a higher order of areas recruitment have negative ISS values. This means that any two cortical sites tend to actively decouple their operations if they are not bound to a third site or several other areas concurrently. In other words, if two areas of the cortex are operationally synchronized, they also tend to be synchronized with some other area(s).

Recent calculations showed that the power-law statistics governs the probability that a particular number of cortical areas are recruited into an OM (defined as the temporal RTP coincidences among different EEG channels). This ubiquitous dependency is characterized by a fractal relation between different levels of resolution of the data, a property also called self-organized criticality [236]. Thus, it has been shown that OMs are indeed driven by a renewal process with power index $\mu \approx 2$ [237–239] (see Note 20), which is in line with Beggs and Plenz's [244] avalanche research. An avalanche is defined as a spontaneous and abrupt burst of activity observed on variable numbers of electrodes for different periods of time separated by silent or quasi-stable periods [191]. Neuronal avalanche behavior is an indication that the cortex is working in a critical condition. Critical systems exhibit scale-free, power-law dynamics at many levels of description during transitions between order and disorder, and their universal behavior is independent of the specific realization of system components [245].

It has been shown that independent OMs of the second order of areas recruitment which have negative ISS values constitute only 14 % from total number of all possible OMs formed by two cortical areas that are part of triple, quadruple, and higher degree OMs [73, 200]. These higher degree OMs have only positive values of ISS. Thus, one may conclude that the brain "prefers" to synchronize three or more cortical areas in order to execute ongoing complex operations; otherwise, neuronal assemblies tend to work autonomously. Our analysis also revealed that such stabilized spatiotemporal OM configurations have transient dynamics which is expressed as a series of sudden transitions between OMs [73, 139, 200, 208]. What is the life-span of OMs with different order of areas recruitment?

From an information-theory point of view, one may suppose that large (covering most or the whole cortex) long-lasting OMs are not efficient in the healthy brain [178] because context-dependent information transfer is necessarily more transitory and would require dynamic reconfiguration of OMs as well as existence of many OMs. This intuition has been confirmed experimentally [73]: it has been shown that the average life-span of OMs is longest for OMs with second order of areas recruitment (~30 s in average) and shortest for OMs that span most or the whole cortex (~100 ms in average) (see Note 21). Thus, one may conclude that the brain operates as a highly dynamic system where large spatial–temporal patterns of stabilized activity (indexed as OMs) are formed only for a very brief episode and quickly dissolve allowing the brain (as a whole) to have more degrees of freedom to form new OMs needed to execute newly immediately emerged and ever-changing operations. Therefore, the dynamics of cortex functional organization is usually dominated by local interactions between brain regions. Here it is important to note that some types of OMs of particular categories are surprisingly stable and persist across all studied experimental conditions in all studied subjects [73, 232, 247]. Initially we have interpreted such OMs as responsible for long-lasting, complex brain operations and body "housekeeping" tasks [229]. However, it later became evident [98] that, as a whole, these highly stable OMs constituted a set of cortical areas that has been named the "default mode network" (DMN) [248]. Nowadays researchers tend to associate the DMN either with stimulus-independent thought [249, 250] or with the "autobiographical" self [249, 251], or being a "self," or having self-consciousness [170, 252]. Indeed, as we have discussed in other work [98], a subject that experiences phenomenal self-consciousness always feels directly present in the center of an externalized multimodal perceptual reality [253, 254]. This is why the set of particular OMs that constitute a DMN and altogether specify the sense (probably in an implicit form) of "being a self" is always active, even during realization of some cognitive (or other) tasks and conditions (including dreaming), independently of their complexity [98].

Coupling of neuronal assemblies is usually examined within single-frequency bands (meaning on a particular temporal scale). At the same time, as noted by Buzsáki and Draguhn [34], different oscillatory frequencies might carry different dimensions of brain integration, and the coupling of operations of neuronal assemblies between two or more temporal scales (frequency bands) could provide enhanced combinatorial opportunities for storing/processing/presenting complex spatiotemporal patterns [91, 255]. The central assumption behind such a view is that oscillations within different frequency bands reflect different types of cognitive processes [11, 39, 106, 256] and that different oscillations become synchronized if the task (or condition) demands require a co-activation or an integration of the respective cognitive processes [256, 257].

Considering the polyphonic character (mixture of different frequency oscillations) of the EEG field [11, 16, 39] and the hierarchical nature (different time scales) of segmental descriptions of local EEG fields [52–54, 56], OMs could coexist on and between different temporal (frequency oscillations) scales. In addition to topographical relationships, cross-frequency synchronization is capable of describing synchronous activity between different neuronal assemblies that may spatially overlap under the same EEG electrode.

The OA methodology enables researches to study such cross-frequency coupling in a direct manner (see Note 22) [56]. In contrast to many other methods (for a critique see [258]), it (a) requires no a priori assumptions about which frequency bands should be synchronized but rather relies on the natural statistical properties of the data to reveal cross-frequency synchronization; (b) is sensitive to transient changes in cross-frequency coupling over both time and frequency; (c) can simultaneously assess multiple cross-frequency synchronizations (i.e., different synchronizations in different frequency bands), and (d) is well suited for investigating possible cross-frequency coupling during dynamic and/or brief cognitive or perceptual events [210].

From the first studies in this field [73, 200, 255] it became evident that segmental flows among the EEG frequency components are more or less synchronized and depend on the character of information processing of brain activity. Interestingly, such synchrony is independent of narrow frequency bands' closeness within broad EEG spectral pattern [73, 255]. For example, the ISS is not always higher in the alpha1–alpha2 pair of frequency bands (neighboring oscillations) when compared to delta–beta1 pair of frequency bands (non-neighboring oscillations). The principal finding is that ISS between basic EEG rhythms (delta, theta, alpha1, alpha2, beta1, beta2) decreases in each local cortical area as cognitive loading increases (see Note 23) [73]. Another finding concerns the occipital-frontal gradient: ISS values for the cross-frequency synchrony increase during rest condition in the direction from occipital to frontal cortical areas, while they decrease along the same direction during cognitive activity (see Note 24) [73] (for an illustration see [56]).

Together, these findings pointed to the existence of not only spatially nested neuronal assemblies, but also to a dynamically organized nested structure of OMs between different temporal scales [224] (see also [167, 256]). In such functionally nested architecture, the ongoing brain activity during an *awake resting state* is characterized by (1) increased coupling between operations of neuronal assemblies located within the same cortical area, though performing their operations on different temporal scales, and at the same time (2) decreased coupling between operations of neuronal assemblies located in different cortical areas, though performing their operations on the same temporal scale. In response to *cognitive loading* such dependency reverses: local cross-scale synchrony decreases while the topographical same-scale synchrony increases [73, 232].

Our research has shown [73, 200, 255] that OMs (being themselves the result of synchronized operations produced by distributed transitive neuronal assemblies) are capable of further operational synchronization between each other both within the same and across different temporal scales, thus forming a more abstract and complex OM which constitutes an integrated experience [91, 217]. In such operational architecture, each of the complex OMs is not just a sum of simpler OMs but rather a natural union of abstractions about simpler OMs. Therefore, OMs have a rich combinatorial complexity and the ability to reconfigure themselves rapidly, which is crucially important for the presentation of highly dynamic cognitive and phenomenal (subjective) experience [91, 217]. Yet the opposite process is also possible, where any complex OM can be partitioned into a set of sub-modules, and to that effect each sub-module may be further decomposed into sub-sub-modules, stripping these processes all the way down to basic operations. Such decomposition would be responsible for a segmentation of our subjective experience and focused attention. In other words, such operational architecture has the fractal property of hierarchical modularity, multi-scale modularity, or, as Meunier and co-workers called it, "Russian doll" modularity [259].

Taking into account the hierarchy of segmental description of EEG in different temporal and spatial scales, it could be suggested that the discrete structure of brain activity depicted in the EEG piecewise spatiotemporal stationary structure is the operational framework within which a variety of rapid "microscopic" variables can obey the "macroscopic" operational structure of brain activity [91]. Thus, the spatial and temporal nested hierarchy of discrete metastable states of neuronal assemblies can serve as the basis for functioning of such a potentially multivariable system like the brain [108].

What are the functional benefits that make such nested hierarchy of virtual modules of brain activity so attractive for brain functioning? One of the earliest propositions formulated by Simon [260] states that a "nearly decomposable" system that comprises multiple and sparsely interconnected modules allows more rapid adaptation of the system in response to changing environmental conditions (see also [261]). In respect to a central nervous system, there could be several important advantages: (a) Networks of functional and dynamic modules have the property of small-worldness which is advantageous because it favors locally segregated processing (with low wiring cost) of specialized functions (for example visual or auditory detection) due to the high clustering of couplings between nodes in the same module while at the same time support globally integrated operations of more generic functions (for example working memory) due to the short path length [259]. Such double tendencies intervened within the same modules demonstrate the principle of metastability [108]. (b) The topology of functional and dynamic modules is associated with rich nonlinear dynamical behavior: temporal scale separation due to fast intra-modular processes and slow inter-modular processes [262] as well as high dynamical complexity due to the coexistence of both segregated and integrated activity [108, 211, 263]. Meunier et al. [259] observed that the presence of modules allows some neuronal activity to remain locally encapsulated while at the same time maintaining a dynamical balance where dynamical activity is maintained between the extremes of rapidly "dying out" and invading the whole network, as marginally stable modules can be combined or divided while preserving stability [264–266]. (c) Another advantage of virtual modules is their optimality at performing tasks in a changing environment [259], where a certain set of simple (basic) operations are required and where a combination (coupling) of these "building operations" is needed to solve complex tasks [91, 217].

4 Implications of OA Methodology

In the previous section we have briefly observed the main tenets and general methodological results of operational architectonics framework for EEG analysis. The main advantage of such methodology is the fact that it allows researchers to study the inherent local and global processes in the EEG field at a common level of formal analysis. According to this analysis all multivariability and complexity of brain operations are reflected in the nested hierarchy of dynamic local fields of neuronal assemblies [91, 256]. In the following subsections we overview the implications of OA methodology for specific domains.

4.1 Implications for Cognitive Neuroscience

There are several major problems in cognitive neuroscience that attract many resources, with researchers aiming to resolve or clarify them. One such problem is a working memory. Even though it has been understood already for some time that memory (and working memory in particular) reflects a distributed property of large-scale cortical systems [35, 218, 267, 268], the particular contribution of neuronal assemblies located in different cortical areas during episodic encoding and retrieval processes is still uncertain and only partially understood [269].

The application of OA methodology to memorizing conditions revealed rich data about the neurophysiological mechanisms behind different stages of the memory task [73, 139, 200, 208, 232, 247]. The experiments were designed in such a way that it was possible to test resting, waiting, encoding, keeping-in-mind (retaining), and identification (retrieval) short-term periods of the memory task separately and in their natural chronology.

The most straightforward result was that bursts of abrupt OM formations were observed at the exact moments when one stage of a memory task was switching over to another in a controllable manner, for example, transition from encoding to retention [73, 200]. This finding clearly indicates that formation of virtual OMs is guided by the need to execute complex cognitive or mental operations. Since each complex operation is a coupling of several simpler ones [91, 217], one may expect to find many OMs of different

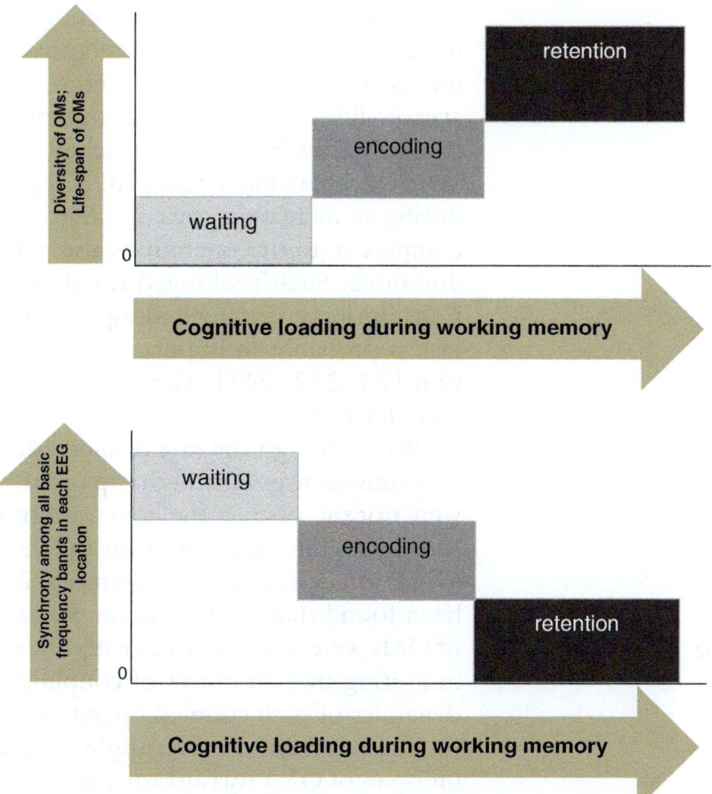

Fig. 6 Schematic presentation of the dynamics of diversity and life-span of OMs (*upper panel*) and average operational synchrony among all basic frequency bands within each EEG channel (*bottom panel*) as a function of cognitive loading during different stages of working memory task. This scheme is based on data published in ref. 247

complexity present within each stage of the memory task. This is indeed the case. The diversity of OMs grows as cognitive loading increases, reaching its maximum at the retention stage (Fig. 6, upper panel). It were OMs with the small and medium order of areas recruitment that contributed the most to such increase in OM diversity [73, 200, 232, 247]. This observation can be easily interpreted as an increased need for independent brain operations that parallel the need to anticipate, perceive, encode, and then keep-in-mind external stimuli during the studied increase of cognitive loading of a memory task. Indeed, in agreement with the general understanding, many neuronal assemblies with distributed parallel processing are active and must synchronize their operations in order to be able to execute the needed cognitive operations as a basis for successful memory performance [247] (see also [16, 257, 270, 271]).

Interestingly, the increase in diversity of sets of OMs along the increase of cognitive loading is also accompanied by the increase of life-span of respected OMs (Fig. 6, upper panel). For example, the average life-span of OMs was minimal during resting stage (before stimuli presentation) and reached its maximum at the retention stage of the memory task [73, 232, 247]. It can be explained that during an increase in the cognitive loading, the growing number of complex cognitive operations also paralleled by an increase of their durations. Such prolonged duration of complex cognitive operations during encoding and especially retention stages of a memory task is probably an important condition for successful memorization [73, 232, 247]. However, detailed analysis revealed a more complex picture.

Even though the rate of successfully memorized stimuli in our experiments was quite high (up to 90 %) some trials for each subject were not successful—the item was not memorized correctly. Therefore, it was interesting to study the diversity and life-span of OMs as a function of high and low memorability in the same subjects. It has been found that the maximum diversity as well as longest life-span of OMs were found in trials with low memorization [73, 232], thus indicating that an excess of couplings of operations and too rigid dynamics of such coupled operations is incompatible with successful memorization. Interestingly, such an excess in the number and life-span of OMs was already present at the "waiting" stage (before the stimulus has been presented), where subjects only prepare to memorize the future coming stimuli [73, 232]. Thus, such an excess is not an event-related phenomenon but reflects a process mode that most likely can be interpreted in terms of a top-down process, for example the mentation (association thoughts) not related to the current task. The fact that such mentation could directly contribute to the process of forgetting has been documented by Klimesch et al. [272].

Thus, the findings presented here suggest that a particular number of specific medium-sized and with medium life-span OMs (that "cover" certain cortical areas) seems necessary to achieve successful memorization [73, 232, 247]. Indeed, although memory encoding, retention, and retrieval often share common regions of the cortex, the operational synchrony of these areas is always unique and presented as a mosaic of nested OMs for each stage of the short-term memory task [73, 232, 247]. When there are too few or too many OMs and their life-span is either too short or too long, then such conditions lead to cessation of efficient cognitive operation. Such dynamic balance of integrated and segregated processes as a necessary and sufficient condition to produce efficient cognitive activity (and eventually consciousness) is becoming increasingly recognized [91, 169, 184, 217, 220–222, 273].

As we have discussed above, operational synchrony can exist across a number of functional domains, with different frequency rhythms associated with each domain, thus indicating cross-frequency synchronization [256–258, 274, 275]. Application of OA methodology to different stages of a working memory task reveals the progressive decrease of average values of operational synchrony among all basic frequency bands (delta, theta, alpha1, alpha2, beta1, and beta2) in a cognitive load-dependent manner (Fig. 6, bottom panel) in each cortical location [73, 200, 232]. These data corroborated with the fact of parallel increase of operational synchrony between different cortical locations—spatial synchrony within the same frequency domain (Fig. 6, upper panel). Together these results clearly indicate that the decrease of cross-frequency synchrony in *each cortical location* releases the needed degrees of freedom of the neuronal assemblies (located in such cortical areas), allowing them to perform synchronously their operations on a particular temporal scale (frequency) *between different cortical locations* as a function of cognitive loading [73, 232].

At the same time, operational synchrony between some particular pairs of EEG frequencies increased as a function of cognitive loading during the various stages of working memory task [73, 200, 232]. These frequencies were mostly alpha1–beta1 and alpha2–beta1 and to a lesser extent delta–theta and theta–beta2 pairs that were positively correlated with memory loading in all frontal, central, temporal, parietal, and occipital regions. On the contrary, operational synchrony in the delta–alpha2 and alpha1–alpha2 pairs decreased as a function of cognitive loading during working memory task. A similar wide-band alpha–beta coupling has been observed using the method of a very-narrow-band EEG analysis to characterize the working memory encoding and retention [276] and also during pharmacological influences [277, 278]. Thus, the current findings support earlier suggestions for complementary roles of alpha and beta rhythms during working memory and cognitive loading [256, 275, 279, 280]. In line with this understanding, alpha and beta band network topologies (neuronal assemblies) have similar characteristics when compared to other frequency bands: they are more clustered and small-world-like [281]. Hence, operational synchrony between them is much more likely.

Coupling between theta and beta frequency bands has been previously observed in animals [282] and humans [283] during states of increased vigilance. These observations are in line with our more recent finding on increased coupling (indexed by operational synchrony) between theta and beta1 frequencies during increased cognitive load of a memory task [73, 200, 232]. The decreased coupling between alpha1 and alpha2 frequency bands during working memory most likely reflects a non-synchronous suppression of both components of alpha activity, which also have different functional meaning ("attention" for slow alpha and "engram consolidation" for the fast alpha; see [272, 284]).

Interestingly, the trials with high memorability were characterized by a very strong coupling between alpha2 and beta1 frequency bands, while trials with low memorability (in the same subjects) had strong synchrony in the alpha1–beta1 band pair [73, 200, 232]. These findings are in line with earlier studies of Klimesch [272] who demonstrated that subjects with faster alpha rhythm performed better on a memory task compared to subjects with slower alpha activity. An important difference though of data presented here is that the subjects were the same, but when the trial happened to be unsuccessful (subjects failed to memorize a given stimuli) it was alpha1 that was operationally synchronized with beta1, while during the successful trials (subjects memorized well) it was alpha2 that was synchronized with beta1 [73, 232].

There was also another characteristic feature of the unsuccessful trials: topographical operational synchrony between segmental structure of alpha activity in any given cortical area and segmental structure of theta activity in all other cortical areas had negative values of ISS, thus indicating *unbinding* of operations performed by neuronal assemblies operating on different temporal scales [73, 232]. For the successful trials such topographical alpha–theta operational synchrony on the contrary increased as a function of cognitive loading throughout the working memory task [73]. The reported results suggest that forgetting is associated with some sort of unbinding of operations performed by particular neuronal assemblies (for a similar conclusion see [285]). Most interestingly, such negative values of ISS (that indicate unbinding) between alpha and theta rhythms were observed over the entirety of the unsuccessful trials: pre-stimulus as well as post-stimulus. Thus, one may conclude that such alpha–theta cross-frequency decoupling is not an event-related phenomenon but rather reflects a process that can be interpreted in terms of top-down controlled processes [257]. Other situations, where operational synchrony exhibits negative ISS values, include (a) the failing of conscious binding of multisensory features into a coherent mental "object" [286] and (b) fading of consciousness altogether as in vegetative and to a smaller degree in minimally conscious patients [287, 288].

To summarize this subsection, findings on OM dynamics clearly point to the fact that the binding of sensory feature representations into phenomenal (subjective) "objects" and active encoding, maintenance, and retrieval of these mental "objects" during working memory are critically dependent on dynamic millisecond-range synchronization of multiple operations performed by local neuronal assemblies that operate on different temporal (oscillations) scales nested within the same operational hierarchy [91, 256].

4.2 Implications for Clinical Practice

After years of research, modern neuroscience conceptualizes the human brain as a complex nested hierarchy of functionally specialized neuronal assemblies that interact with each other in a spatially and temporally coherent fashion [170–172, 256]. By means of such interactions, these neuronal assemblies and their larger synchronized conglomerates (OMs) shape physiological (normal) and pathological (diseased) behaviors [86]. Despite the wide spread of neuropsychiatric disorders [289, 290] and the progress in the basic neuroscience, there is only little advance in understanding the pathophysiology of such disorders and correspondent to it delay in the development of effective therapies [291]. We believe that this is due to a lack of a consistent paradigm of psycho- and neuropathology that would incorporate the novel knowledge from basic neuroscience. At the same time, such a novel paradigm that stresses the dynamic balance between isolated functions of local neuronal assemblies and globally coordinated activity between them is beginning to emerge [291]. It has been suggested that the loss of such a metastable balance in favor of either independent or hyper-ordered processing leads to pathological states that give rise to neuropsychiatric syndromes constituting a particular disorder [86, 151] (see also [291, 292]). This subsection reviews current knowledge about OMs' dynamics in different pathologies derived from application of OA methodology in neurological and psychiatric conditions.

4.2.1 Chronic Opioid Addiction

It has been found that the number of OMs and strength of coupling within such OMs (estimated by an ISS) were significantly lower in chronic opioid abusers than in healthy controls [152]. This dependence was observed in alpha and beta frequency bands. Such disruption in the formation of virtual OMs is consistent with disorganization syndrome of cortex processes (suggested by Bressler [292]). Most likely, the disrupted operational connectivity between local and distributed neuronal assemblies is a candidate mechanism for the well-documented pattern of impairment in addicts, expressed as lack of integration of different cognitive functions for effective problem solving, deficits in abstract concept formation, set maintenance, set shifting, behavioral control, and problems in the regulation of affect and behavior [293–296]. From this perspective then, disorganization in the cortex activity of chronic opioid addicts is in favor of independent processing that was paralleled by increased volume, stability, and life-span of local neuronal assemblies in such addicts [152]. Furthermore, it has been found that the more years the subjects abuse opioids on a daily basis the more disintegration (indexed by ISS) takes place within the posterior section (low central, parietal, low temporal, and occipital areas) of the cortex. Hence, these findings give ground to suppose that longitudinal opioid abuse preferentially impairs the formation of OMs in the posterior cortex section.

Areas located in this part of cortex are suggested to be responsible for a number of important functions [297]: perception, differentiation and somatosensory functions, memory functions and autobiographical records, as well as visual perception. Therefore one may suggest that the absence of appropriate relations between all these functions in opioid abuse patients due to a broken mechanism of OM formation leads such patients to specific maladaptive actions such as those directed at drug-seeking and drug-taking behavior which characterize addiction [152].

4.2.2 Opioid Withdrawal/Abstinence

Based on clinical and cognitive psychology studies it is known that when addicts crave for the drug, anxiety, nervousness, lack of inhibitory control, positive drug-related expectancies, and intrusive thoughts related to drugs are all simultaneously present [298, 299]. Consistent with the OA methodology framework, these complex cognitive functions are critically based on the dynamical interactions of operations performed by many cortical neuronal assemblies (see also [64, 271, 300, 301]). Considering that withdrawal/abstinence initiates a widespread parallel activation of cortical regions responsible for the cognitive operations mentioned above, one can expect to see a significant increase in formation of many OMs. Such increased operational synchrony among many transient neuronal assemblies would then explain the strong motivation of abstinent patients for the excessive drug craving, where the dynamics of local brain operations (functions) would be restrained by the large-scale context (removal of the aversive state) of mutually connected cortical areas. Application of OA methodology to EEG of abstinent patients fully confirms this hypothesis: a significant increase in formation of many OMs of different complexity (with a focus in the anterior part of the cortex) was found in opioid-dependent patients during withdrawal as compared to healthy controls [153]. It has also been documented that the complexity of OMs and strength of coupling within OMs (indexed by ISS) have had a predictive force towards the severity of abstinence: patients with strong withdrawal symptoms had higher number and strength of couplings among neuronal assemblies than patients with mild symptoms [153].

4.2.3 Major Depression

Even though the modern model of major depression stresses the key role of anterior asymmetry—the so-called cognitive anterior model of depression [302], from the point of view of metastable balance hypothesis outlined above, major depression could generally be viewed as a disorder of disturbed neuronal assemblies' plasticity, leading to an inadequate relationship between multiple operations produced by these neuronal assemblies. Application of OA methodology to EEG of medication-free patients with major depression reveals a widespread and significant increase in formation of OMs when compared to healthy subjects, with greater number of small OMs in the left hemisphere and large OMs in the right hemisphere [148]. Such peculiarities of operational synchrony could be interpreted as signs of adaptive (over)compensation of the "depressed" brain in an attempt to achieve more adequate semantic context which is presented differently in left and right hemispheres [303] through establishing a new overall metastable brain state [108]. It has been suggested that in the process of such overcompensation, connections between neuronal representations of negative affect and different semantic concepts become strongly activated [148]. This hypothesis found confirmation in a semantic network model study [304, 305] in which both semantic and affective features are represented as nodes in the network. It was found that people who are depressed suffer from strongly activated connections between negative affective nodes and multiple semantic concepts, creating feedback loops that maintain depressive affect and mentation [306–308]. This is why depressed individuals tend to see even positive information as negative because it becomes associated with personally relevant negative information [309]. The observed type of a new metastable brain state with increased operational synchrony could be a possible mechanism underlying the maintenance of a depressive state [148].

4.2.4 Schizophrenia

Schizophrenia is another mental disorder whose pathophysiological mechanisms could be better understood if it is viewed as a disorder of metastable balance between large-scale integration (formation of OMs) and independent processing (local transient neuronal assemblies) in the cortex, favoring independent operations [156, 157, 291, 292]. Application of OA methodology to patients with schizophrenia [155] reveals a significant decrease of operational synchrony among remote neuronal assemblies (indexed as OM formation) in schizophrenic patients when compared to healthy subjects [86]. Such low level of operational synchrony may signify a well-documented pattern of mental impairment in schizophrenics that expresses a lack of integration of different cognitive functions for effective problem solving, deficits in abstract concept formation, set maintenance, set shifting, behavioral control, and problems in the regulation of affect and behavior [310, 311]. Since in this study negative and positive symptomology were mixed, one could not make a firm conclusion of how different symptomology would be reflected in the operational architectonics of EEG. One may suppose though that high values of positive symptomology in patients with schizophrenia would be paralleled by rather increased level of OM formation; this hypothesis needs to be tested in a separate study.

4.2.5 Epilepsy

The epileptic brain presents a unique pathological condition that, despite the fact that it is the most common disorder of the nervous system [312], is still poorly understood. A recently emerging view that brain disorders and associated psychiatric problems are accompanied by disruption in the spatiotemporal structure of integrative brain activity [313], where this structure is either more irregular (uncorrelated randomness) or more regular (excessive order) than in normal healthy brain, could be of a particular help [151]. Application of OA methodology to an epileptic brain during the interictal (without signs of epileptiform abnormalities) periods in medication-free patients with chronic idiopathic generalized epilepsy reveals that (a) interictal EEG is characterized by OMs which contain more cortex areas than healthy control EEG and (b) brain oscillations of cortex regions involved in such OMs tend to have longer periods of temporal stabilization in interictal EEG when compared with control EEG [66]. Both findings suggest less dynamic performance of cooperative brain operations (dynamic rigidity) in patients with chronic epilepsy than in healthy subjects. Generally these findings pointed to a conclusion that chronic epilepsy should be determined not by a focus of pathological activity, but rather by an epileptic system, that contains a set of coupled distributed oscillatory states (or resonances) involved in a common activity [66].

Summarizing this subsection we could express the hope that the effort to look at many chronic disabling mental or neurological disorders as conditions with dynamical disruption in the spatiotemporal functional structure of integrative brain activity [64, 271, 291, 292, 300, 301, 313] would result in a new generation of evidence-based diagnosis and treatment strategies [314, 315].

4.3 Implications for Somnology Studies

In mammals sleep is associated with specific cortical EEG patterns, generally divided into rapid eye movement (REM) and non-rapid eye movement (NREM) (see Note 25) sleep [317]. During REM sleep EEG is desynchronized and hippocampal theta rhythms are present, while during NREM sleep EEG is characterized by the presence of slow (delta) waves, sleep spindles, and K-complexes [318]. As recently observed by Mignot [318], it is likely that sleep is a distributed process with some neuronal systems being the primary drivers. Interactions between sleep- or wake-specific neuronal assemblies must ensure that the processes occur in synchrony and in exclusion of each other to create stable states of wake, NREM, and REM, with limited time spent in transition states. This view is highly compatible with OA methodology, according to which many transient neuronal assemblies rapidly synchronize their operations in order to present the stability of a particular complex state/operation [91].

So far the OA methodology has been applied only to EEG from NREM sleep. Results show that during NREM sleep the size of neuronal assemblies gets smaller, while the life-span increased compared to wakefulness state with closed eyes [86]. Furthermore, NREM sleep was characterized with increased formation of OMs, though coupling strength within such modules was not strong

[86]. These findings mark a weak communication among neuronal assemblies located in different cortex areas during NREM sleep, allowing some level of flexibility with quick reorganization of synchronized neuronal assemblies into different combinations (OMs) in order to present a large number of possible cognitive operations needed for proper learning and memory consolidation which are posited to take place during NREM sleep [319–321].

Additionally, dividing NREM sleep on the dream-present and dream-absent epochs reveals that NREM dream condition (which is usually presented with simple, static, isolated image(s) or thought (s) and by one modality [322]) is characterized by short-lived small neuronal assemblies, long-lived large neuronal assemblies, and a significant increase in OM formation when compared to dreamless NREM condition [91].

Further studies are needed to detail OA organization of the brain during REM and NREM conditions as well as during sleep pathologies.

4.4 Implications for Neuropharmacology

As we have discussed above, recent research emphasizes that the majority of brain disorders, and psychiatric/mental problems are accompanied by disruption in the temporal and metastable structure of brain activity [313], where this temporal structure could be either more irregular (uncorrelated randomness) or more regular (excessive order) than normal [291, 292, 323, 324]. From this perspective, it has been suggested that the future of psychopharmacology lies in its ability to design psychotropic drugs that can restore the normal temporal structure and metastable structure of brain activity [151]. This approach seems more physiologically adequate to integrative, nonstationary, and self-organized nature of brain processes and fits in with a novel understanding of the dynamical nature of brain diseases, where the so-called lesions in time become more evident especially in the early stages of the disease than lesions in structure [325]. In this context, it is important to study how different psychotropic drugs can modify temporal/metastable structure of brain activity in healthy subjects and patients.

Application of OA methodology to EEGs recorded under the influence of lorazepam ($GABA_A$ agonist) reveals that the number of OMs and the strength of coupling within such OMs significantly increased under the lorazepam administration [84]. It is important to note that it was a randomized, double-blind, cross-over, placebo-controlled study. In the same study, it was found that different-sized neuronal assemblies in alpha and beta frequency bands performed differently under lorazepam when compared with placebo [83]: For the alpha-generated neuronal assemblies, it was observed that large neuronal populations exhibited a total decrease in size, functional life-span, and stability under lorazepam administration when compared to placebo. In contrast, small neuronal assemblies were very stable. The functional life-span of all beta-generated neuronal assemblies was prolonged [83]. These findings are likely to reflect the prolongation of inhibitory neuronal operations that manifest in the form of slowing of cognitive performance under the lorazepam administration [326]. It has also been suggested that, probably, strong enhancement of $GABA_{ergic}$ function by lorazepam "mimics" the conditions of an immature brain [84], where excitatory actions of GABA provide most of the initial activity, a primitive signal with poor information content that propagates to all brain structures [327].

In contrast, heroin imposes a full disruption in the OMs' formation with a parallel decrease in coupling strength within the OMs [152]. As we have discussed above, such disruption of operational synchrony among neuronal assemblies located in different cortical areas is consistent with cortex disorganization syndrome (suggested by Bressler [292]). It is likely that heroin-induced disruption of operational synchrony may constitute the candidate mechanism for the well-documented pattern of cognitive impairment in heroin users: lack of integration of different cognitive functions, deficits in abstract concept formation and behavioral control, and problems in the regulation of affect and behavior [293, 294, 296].

Since methadone is sometimes used as a maintenance treatment for heroin-dependent patients [328], it was interesting to study its effect on the EEG operational architectonics in such patients. Application of the OA methodology to EEGs of patients who were on methadone treatment during many months showed that methadone restored the normal values of number and coupling strength of OMs [154] and consequently normalized the metastable organization of the cortex in all patients (for a discussion see [108]). This conclusion is consistent with the current theoretical view that normal brain function is the product of a large-scale network of coupled neuronal assemblies exhibiting transient and inherently metastable dynamics [122, 168, 219, 329].

Summarizing this subsection we hope that the application of OA approach for EEG analysis will contribute to a current effort in developing a more rational psychopharmacology that takes into consideration the novel views on brain/mental disorders.

4.5 Implications for Ontological Development Research

There are several critical questions in the developmental EEG studies: What is the physiological basis for the synchrony within oscillations in human EEG over the developmental life-span? If oscillations reflect important cognitive processes then do different synchronicity modes within and between oscillations contribute to the development of different aspects of human cognition? In order to get some insight into these questions, we reanalyzed results of OA methodology which was applied to EEGs of healthy 12-year-old teenagers [155] and healthy 30-year-old adults [86]. The results revealed that the formation of OMs and the strength of coupling within the observed OMs were smaller in teenagers when compared to adults ($p < 0.01$, Wilcoxon test). These findings are in agreement with current understanding that with age the repertoire of the states of individual neuronal assemblies located in different cortical regions becomes bigger due to increased specialization and that at the same time there is increased integration between distributed neuronal populations responsible for different cognitive operations in adults [330]. In general, these findings are in line with modern neo-Piagetian models of cognitive development from childhood to adulthood [331–335]. They are also consistent with previous EEG coherence studies [118, 336–338] which show that as the number of connections increases with age then there is accompanied increased coherence (synchronicity), increased stability, and decreased "chaos" [339].

4.6 Implications for Personality Feature Studies

The dynamic balance between synchronization and desynchronization of cognitive operations is considered essential for normal brain function [86, 291, 292], as we have discussed above. If this is so, then such balance paralleled by EEG operational architecture balance should (a) have noticeable stability within individuals and (b) exhibit variability between individuals—thus being the "signature" of personality. This issue has not studied at a proper detail yet; however, some preliminary data is emerging. Applying the OA methodology to EEGs of monozygotic twins and comparing the OA indexes within the monozygotic pairs (genetically related subjects) and between members from different pairs (genetically unrelated subjects) revealed that the formation of OMs, their diversity, and types were nearly identical within members of the monozygotic pairs but differ significantly in participants from different monozygotic twin pairs [73, 200]. Specifically, the average Spearman rank correlation in the intra-pair and inter-pair similarities of OMs for the resting condition was 0.53 for genetically related and -0.13 for genetically unrelated subjects. These findings point to the existence of particular dynamic OM profiles that are individually predetermined and most likely have high heritability. The possibility of high genetic heritability for EEG functional connectivity has been documented in other independent studies [29, 337–340].

It has also been shown that such OM profiles correlate significantly with different cognitive, affective styles of subjects and their personality features [200]: for example, verbal type of thinking, normal level of anxiety and neuroticism, and phlegmatic temperament type (see Note 26) altogether characterize subjects with a high number of large OMs in the left hemisphere with a focus in anterior part of the cortex. Subjects with nonverbal/abstract type of thinking, diminished level of situational anxiety and neuroticism, and sanguine temperament type were characterized by a diverse set of large OMs in the right hemisphere with a focus in the posterior part of the cortex (Fig. 7, upper plane). Interestingly, such dependence

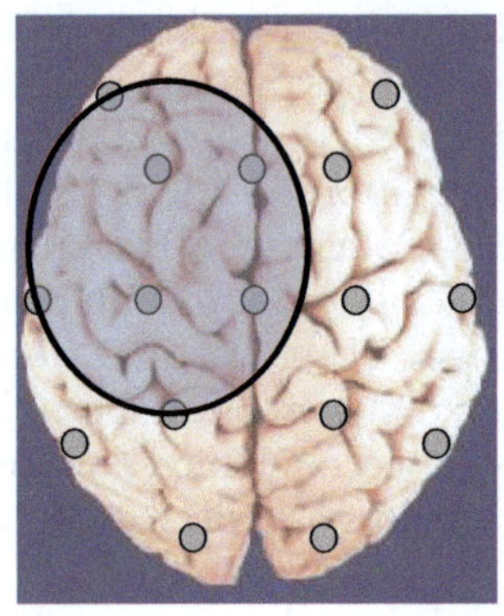
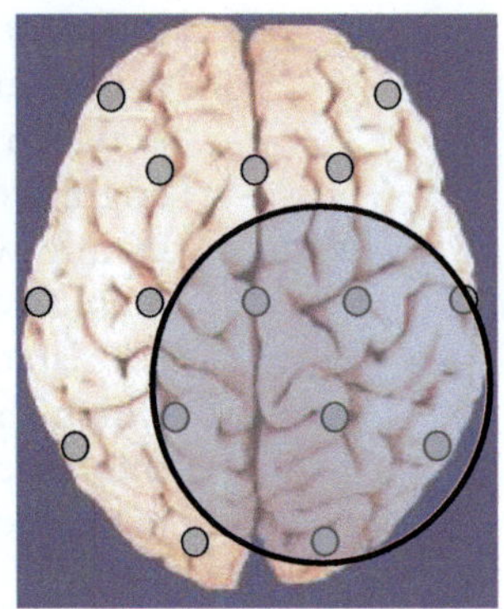

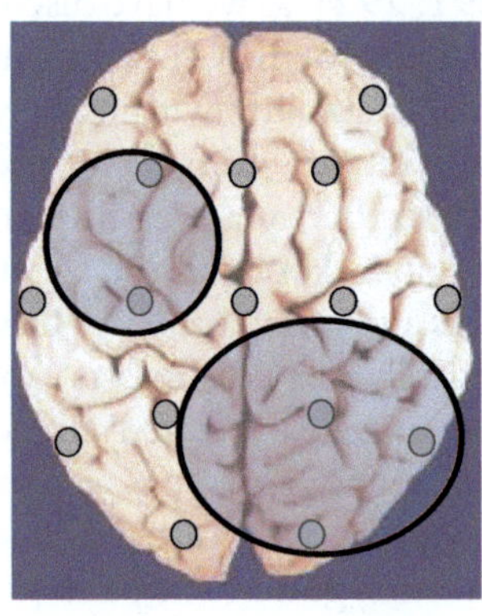
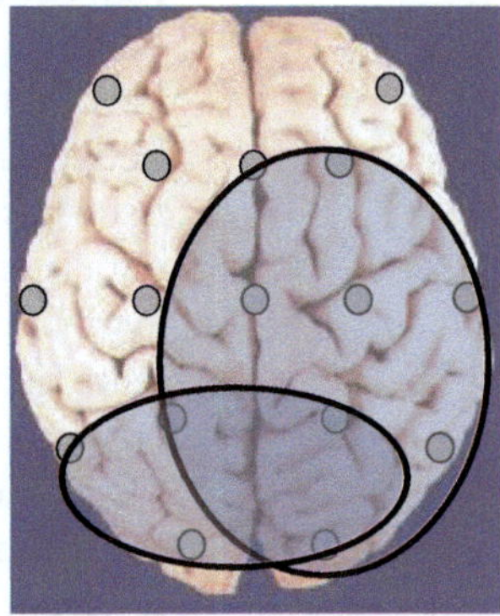

Fig. 7 The position of OMs typical for individuals with different personality features. *Circles* indicate the locations of most frequently found OMs. Further explanations are provided in the text

between OM profiles and personality features was "hardwired" during rest condition only. During a particular cognitive task, the given OM profile tends to transform in order to execute a particular task. For example, if the initial personal OM profile was verbal but the cognitive task was visual (nonverbal), then during such visual task the verbal OM profile characteristics became similar to a non-verbal OM profile (Fig. 7, bottom plane), though the efficiency of subjects with such transformed profile during the visual cognitive task was slightly lower than the efficiency of subjects with initially present nonverbal OM profile [200]. These findings have been confirmed in the test–retest study which took place after 2 weeks.

Generally these data suggest that human individual cognitive abilities relate to a particular operational architectonics of the brain resting-state activity that is expressed as a nested spatial and temporal nonstationarity of EEG field which though possesses some degree of freedom that allows it to reorganize itself in order to execute a particular complex task or operation. Such a view is congruent with the hypothesis of Bekhtereva [343], who proposed that brain activity during the resting state may reflect the brain's potential processing abilities and is correlated with individual differences in the cognitive process (for recent research see [344, 345]).

4.7 Implications for Neurophilosophy

The greatest unresolved theoretical issue that neuroscientists would hope to gain some understanding of is "consciousness" (the entity that none can still easily define but nearly all accept that it exists [346]) and its neural (brain) constitutes. To date, no one has provided a complete explanation as to how the subjective experience (phenomenality) could arise from the multiple actions of neurons in the brain. Nevertheless, currently, it is agreed that further understanding of phenomenal consciousness will obviously rely upon the view according to which phenomenal consciousness is grounded to material carrier processes that take place in the brain [43, 54, 55, 158, 301, 347]. It is even suggested that phenomenal consciousness is a higher level of biological organization in the brain [253]. OA theory is highly compatible with these views.

The main tenets of the OA theory [91] are as follows: The brain generates a highly structured and dynamic extracellular electric field in spatial and temporal domains [43] and over a range of frequencies [35]. This field exists within brain internal physical space–time (IPST) and is best captured by the EEG measurement [16]. Detailed analysis of the structure of EEG's complex hierarchical architecture reveals a specific operational space–time (OST) which literally resides within the IPST and is isomorphic to phenomenal space–time (PST) and, as it has been proposed, may serve as the potential neurophysiological constituent of the phenomenal consciousness' architecture [91, 217]. Therefore, to test whether consciousness is indeed an emergent phenomenon of coherent dynamic binding of operations performed by multiple neuronal assemblies (organized within a nested hierarchical brain architecture), further experimental work was needed to demonstrate that the attributes and operational synchrony of local EEG segments would change in circumstances when awareness expression is either weakened or lost completely. The OA theory predicts that both low and high levels of operational synchrony among neuronal assemblies would result in a dramatic fading of consciousness [91].

Application of OA methodology to EEGs of patients who are in a permanent vegetative state (VS) or in a minimally conscious state (MCS) pointed to one among the two alternatives predicted by the OA theory in relation to a specific case of (un)consciousness expression in minimally communicated and non-communicated patients with severe brain injuries [287, 288]. Taken together, the results of these studies support the view according to which it is an intact coordinated activity (operational synchrony) among relatively large, long-lived, and stable neuronal assemblies that is important for enabling routine representational processes to be integrated within a coherent phenomenal world from the first-person perspective [253, 348, 349]. Additionally, and as predicted by the OA theory, transient operational integrity of neuronal assemblies allows discrete moments of "phenomenal present" to be bundled in larger units (formation of OMs) making it possible to present mentally a practically infinite number of different qualities, patterns, objects, scenes, concepts, and decisions [91]. Impairment in such operational integrity (disruption up to an active decoupling) among neuronal assemblies may underlie the fading of consciousness until its complete absence, if such impairment reaches a critical level as it does in the patients in VS, who have complete unawareness of self and the environment [287, 288].

Another important model, where subjective experience is presented in a contrasted form (which could be easily manipulated), is hypnosis. In a pure hypnotic state the subject experiences an altered background state of consciousness different from the normal baseline state of consciousness [350]. This subjective state is characterized by some sort of "emptiness" or "absorption" brought about by dissociations in the cognitive system, such that separate cognitive modules and subsystems may be temporarily incapable of normal communication with one another [351, 352]. Additionally, it has been shown that the sensation of time passing is stretched during hypnosis, because internal events are subjectively slowed [353, 354]. Adhering to the tenets of OA framework, these subjective experiences should be reflected in the operational architectonics of the EEG brain field. In a pilot study applying the OA methodology to hypnotic EEG data [202] it was indeed shown that the functional life-span of all neuronal assemblies (indexed by the EEG quasi-stationary segments) was significantly longer during hypnosis when compared to normal baseline conscious condition of the same subject. It was further found that the number and strength of synchronized operations among different neuronal assemblies were significantly lower during hypnosis than during the baseline, thus limiting the possibility of any OMs with order of areas recruitment higher than two to emerge. As a result they were absent [202]. Since OMs represent the formation of integrated conscious experiences, their absence may explain such unusual subjective experiences during pure hypnosis as amnesia, timelessness, detachment from the self, a "willingness" to accept distortions of logic or reality, and the lack of initiative or wilful movement [355].

This research program aims to reveal the neural mechanisms of phenomenal consciousness and to evaluate whether such findings are leading towards a neuroscientific explanation of consciousness within the OA framework. Phenomenal consciousness refers to a higher level of organization in the brain [253] and captures all immediate and undeniable (from the first-person perspective) phenomena of subjective experiences (concerning hearing, seeing, touching, feeling, embodiment, moving, and thinking) that present to any person right now (subjective present) and right here (subjective space) [91]. According to this definition someone possesses phenomenal consciousness if there is any type of subjective experiences that are currently present for him/her. The well known in the brain–mind research philosophical "hard problem" is the problem of understanding how the brain (or, more generally, any physical system) could produce any subjective, phenomenal experiences at all [356]. To make progress in solving this "hard problem," the neural counterparts directly constituting phenomenal consciousness must be localized and identified. The suggested research program utilizing OA methodology may contribute to this goal.

4.8 Implications for Artificial Intelligence

The idea of machine consciousness first considered by Angel [357] had recently progressed from being an interesting philosophical diversion to a real possibility [358]. Several researchers, engineers, and computer scientists have already begun to address the subject by designing and implementing models for artificial consciousness referred to as "machine consciousness" or "synthetic consciousness" [359–365]. However, almost all of them take a more or less conventional computational or low-level neurally inspired (anatomical) approach [366]. At the same time, as noted by Indiveri et al. [367], such an approach stands before a large conceptual challenge: to bridge the gap from systems (machines) that merely mimic or simulate cognitive processes usually correlated with consciousness (the so-called weak artificial consciousness [363, 368]) to ones that are genuinely conscious (strong artificial consciousness [363, 368]). An important practical matter related to this challenge is that a true "conscious machine" should be seen as a man-made artificial system (e.g., robot) that enjoys subjective phenomenal experiences and related rational thinking [366].

The OA theory of brain and mind functioning offers a conceptual–theoretical architecture that computing system (robot) could implement (after appropriate mathematical formalization and engineering would be achieved) to simulate the operational level in which consciousness and thinking would self-emerge [369]. In this context the problem of producing man-made "machine" consciousness and "artificial" thought is the problem of duplicating all levels of nested OA hierarchy (with its inherent rules and mechanisms) found in the EEG field, which can constitute the neurophysiological basis of phenomenal level of brain organization [369]. The aim should be to abstract and formalize the principles of the nested hierarchy of operations which constitute phenomenal consciousness and thought, rather than attempting to directly mimic the whole diversity of chemical and physiological mechanisms of brain functioning or the whole diversity of consciousness states, which is a quite unrealistic enterprise [370]. In this case one could expect that by reproducing one architecture (brain operational) we can observe the self-emergence of the other (mind phenomenal) [369].

5 Conclusion

In the 17 years of its existence, OA methodology of EEG analysis has evolved into a valuable technique which can be useful for clinicians, basic researchers, and philosophers of mind. It has a set of tool that provides means to examine different features and dynamics of cortical neuronal assemblies and topological coupling of their operations in normal and abnormal human brains as well as in altered states of consciousness. The specific approaches of EEG analysis within the OA methodology [56] are especially well suited for studies of nonstationary signals and uniquely capable of investigating the dynamic and metastable changes of brain spatial–temporal patterns that are isomorphic with cognitive and phenomenal levels [371]. Essentially these tools take repetitions of spatial–temporal patterns (indexed as OMs) into account at all functional levels, thus capturing both dynamic as well as hierarchical complexities of brain activity which is nested within a multi-scale operational architecture [91].

6 Notes

1. One also needs to consider the possible role of glial cells as sources of extracellular currents, especially since these cells are known to be sensitive to changes in extracellular ions [2, 3]. Therefore some researchers suggest that glial cells might "convert" ongoing spike activity into wave potentials. However, glial cells have an extremely slow time course, requiring several seconds to reach their peak. For this reason, it appears likely that the major glial contribution is to "steady cortical potentials" rather than to the EEG from which potentials under 1 Hz are generally excluded [4]. There is another drawback with such a suggestion: extracellular potassium concentrations determine the membrane potential of astrocytes [5]; however, during spontaneous activity it is very unlikely that changes in extracellular potassium concentration would be sufficient to produce any glial potentials; in fact, glial activity reveals no wave activity at all [6, 7]. These, and pharmacological data [4], indicate that it is the activity of neurons that is essential and critical for EEG activity.

2. EEG bands represent electromagnetic oscillations in different frequency ranges arising from synchronous and coherent electrical activity of neurons in the brain: delta: 1.5–3.5 Hz, theta: 4–7.5 Hz, alpha: 8–13 Hz, beta: 13–30 Hz, and gamma: >30 Hz.

3. The coordinated activity manifests a "wave packet" that requires synchronization of a shared carrier wave of the outputs of a large number of neurons over the area [85, 87, 123].

4. For other major methodologies of EEG analysis which explicitly utilize EEG nonstationarity see the work of Lehman and co-workers [60, 61, 67, 80, 138, 147, 149, 150, 165], Freeman and co-workers [62, 64, 65, 85, 87, 89, 90, 112, 123, 158, 166], and Breakspear and Stam with co-workers [75, 134, 145, 167–169].

5. According to Feinberg [170–172], in a non-nested hierarchy the entities at higher levels of the hierarchy are physically independent from the entities at lower levels and there is strong constraint of higher levels upon the lower levels, whereas in a nested hierarchy the higher levels are physically composed of lower levels, and there is no central control of the system resulting in weak high-to-low-level constraint.

6. Indeed, it has been shown that distinct neuronal assemblies display preferential processing for certain features such as color, shape, motion, and smell [182]. Moreover, it has been documented that the local fields of various neuronal assemblies correlate with different conscious percepts [158, 183, 184] and if cognitive processing does not take place, such transient neuronal assemblies do not form [185].

7. Experimental studies have shown that different frequencies appear to be related to the timing of different neuronal assemblies, which are associated with different types of sensory and cognitive processes (for a detailed overview, analysis, and discussion see [91]).

8. In mathematical statistics this phenomenon is known as the change-point problem [195].

9. The procedure of random mixing of amplitude values within each local EEG signal for a multichannel EEG recording was used [73].

10. When recorded from the scalp it is referred to as an EEG, an electrocorticogram (ECoG) when recorded by electrodes on the cortical surface, and as the LFP when recorded by a small-size electrode in the brain [189]. The magnetic field induced by the same activity is referred to as magnetoencephalogram (MEG) [207].

11. Similar dependence has also been found in the phase [166, 209] and frequency [210] characteristics of EEG signals.

12. The idea that synchrony of EEG potentials reflects functional connectivity at the neuronal assembly level has been proposed by Livanov [212] and later proven by him in a direct neurophysiological experiment [213]. In this experiment the correlation coefficient between EEGs in visual and motor cortical areas of the rabbit was estimated. It appeared that, if the correlation coefficient exceeded a particular level, the visual signal triggered paw movements, and if this coefficient was lower than the established level, no motor reaction occurred. Further important work in this domain was developed by Lazarev with co-workers [214–216].

13. Metastability in the brain [108] refers to competition of complementary tendencies of cooperative integration and autonomous fragmentation among many distributed brain areas [219–221]. The interplay of these two tendencies (autonomy and integration) constitutes the metastable regime of brain functioning [108, 219], where local (autonomous) and global (integrated) processes coexist as a complementary pair, not as antagonists [222, 223]. The OMs are metastable because of intrinsic differences in the activity between neuronal assemblies, which constitute OMs, each doing its own job while at the same time still retaining a tendency to be coordinated together within the same OM in order to execute the macro-operation [91].

14. It should be noted here that the original rather narrow neurophysiological definition of synchronization as two or many subsystems sharing specific common *continuous* frequencies is replaced here by the broader notion of a process, whereby two or many subsystems adjust some of their time-varying properties to a common temporal scale due to *coupling of events* or common external forcing [225].

15. Such methods (1) are predominantly designed for EEG analysis only in pairs of derivations; (2) do not take into consideration the nonstationary nature of the signal; (3) indicate only the linear statistical link between time-series curves in a frequency band; (4) require long time epochs of analysis; (5) can only be applied to a homogeneous medium, which is an unrealistic assumption for the brain; (6) borrow complex methodologies and conceptual frameworks from physics, mathematics, and engineering but use them loosely when applying to the analysis of physiological signal; (7) being averaged indices lose a substantial part of their diagnostic value for studying discrete functional states of the brain; (8) use local EEGs that participate in the formation of the resultant dipole vector far from

equally, and this is unjustified from the viewpoint of indubitable neurobiological equivalence of cortical areas; (9) are often difficult to interpret in terms of their physiological correlates; and (10) do not directly estimate metastability in the brain (see also [56, 122, 210]).

16. The presented concept of the OM suggests that both parallel and serial processing may be just different sides of the same one underlying mechanism—synchrony of operations of neuronal assemblies. Parallel processing is performed by individual neuronal assemblies, while serial processing emerges as a result of formation of OMs and their changes along with shifts in the process of actualization of objects in the physical or the mental world [122].

17. In this way, OMs lie, in some sense, between classical, connectionist, and process architectures. They resemble connectionist networks [230] in many respects: they may serve as associative, content-addressable memories, and they are *distributed* across many neural assemblies. Yet, the specific spatial–temporal patterns (OMs) per se are *unitary*, like symbols of classical logics [231]. And yet, each OM is a *process*, since it lasts as long as several operations that have some continuity in time (and which are produced by different neuronal assemblies) are synchronized among each other during a given temporal interval [91].

18. First substantial experimental support for this formulation was obtained within the framework of the brain microstates concept: it has been shown that momentary cortical electric field distributions are abruptly upgraded and constantly replaced [61, 80, 138, 165].

19. This is why an OM is relatively independent from the underlying small-scale neurophysiological processes in the brain [91]. More specifically it is independent from intrinsic brain anatomical topology that determines which single neuron of a given anatomical circuit produces a particular spike pattern of a given temporal signature (for similar argumentation see [43, 233, 234]).

20. In physics the value $\mu = 2$ indicates a transition between two kinds of ergodicity breakdown, stationary and nonstationary, respectively [240, 241]. Moreover, it has been shown that having complex networks at $\mu \approx 2$ provides ideal conditions for transmitting and receiving information [242]. Recently it has been confirmed that neuronal avalanches provide optimal condition for the mutual information transfer between stimulus and response [243] and allow the realization of a large diversity of activity patterns [191].

21. These life-span accounts of OMs, including variations in duration, are consistent with known estimates of cognitive

processes and of highly dynamic "moments of experience" or "thoughts," which may vary between ~100 ms and several seconds depending on circumstances [246]. These values are also consistent with independent EEG studies applying different methods of data analysis [57, 67, 80].

22. It is worth noting that within the OA framework there are no methodological restrictions for relations between frequency bands, that is because the ISS measure is not associated with phase relations of the EEG signal as it is the case for more conventional methods of estimating EEG synchronization [56, 255].

23. The reader is reminded that the topographical (between different cortex locations) operational synchrony within the single-frequency band, on the contrary, increases in response to cognitive loading [73, 210, 232].

24. Importantly, this phenomenon does not depend on the EEG alpha activity expression, since it has been demonstrated that it is identically present in subjects with both very high and very low ("flat") alpha activity [73].

25. The only known exception so far includes the primitive egg-laying mammals echidnas, in which a REM/NREM mixed state has been observed [316].

26. For personality features and temperament types see Eysenck and Eysenck [341] and Diamond [342].

Acknowledgments

This work was supported by BM-Science Centre, Finland. The authors would like to thank anonymous reviewers who provided thoughtful comments and constructive criticism. Special thanks for English editing to Dmitry Skarin.

References

1. Kunkel H (1980) Elektroenzephalographie und psychiatrie. In: Kisker KP, Meyer JE, Müller C, Strömgren E (eds) Psychiatrie der Gegenwart Bd. Springer, Berlin, pp 115–196

2. Kuffler SW, Potter DD (1964) Glia in the leech central nervous system: physiological properties and neuron-glia relationship. J Neurophysiol 27:290–320

3. Orkand RK, Nicholls JG, Kuffler SW (1966) Effect of nerve impulses on the membrane potential of glial cells in the central nervous system of amphibia. J Neurophysiol 29 (4):788–806

4. Elul R (1972) The genesis of the EEG. Int Rev Neurobiol 15:227–272

5. Kuffler SW (1967) Neuroglial cells: physiological properties and a potassium mediated effect of neuronal activity on the glial membrane potential. Proc R Soc Lond B Biol Sci 168:1–21

6. Elul R (1967) Statistical mechanisms in generation of the EEG. Program Biomed Eng 1:131–150

7. Elul R (1968) Brain waves: intracellular recording and statistical analysis help clarify their physiological significance. In: Enslein K (ed) Data acquisition and processing in

biology and medicine. Pergamon Press, Oxford, pp 93–115

8. Lopes da Silva F (1991) Neural mechanisms underlying brain waves: from neural membranes to networks. Electroencephalogr Clin Neurophysiol 79:81–93

9. Nunez PL (1995) Neocortical dynamics and human EEG rhythms. Oxford University Press, New York

10. Freeman WJ (1975) Mass action in the nervous system. Academic, New York

11. Başar E (1998) Brain function and oscillations I Brain oscillations: principles and approaches. Springer, Berlin

12. Andras P, Wennekers T (2007) Cortical activity pattern computation. Biosystems 87:179–185

13. Moran RJ, Stephan KE, Kiebel SJ et al (2008) Bayesian estimation of synaptic physiology from the spectral responses of neural-masses. NeuroImage 42:272–284

14. Hadjipapas A, Casagrande E, Nevado A, Barnes GR, Green GG, Holliday IE (2009) Can we observe collective neuronal activity from macroscopic aggregate signals? NeuroImage 44:1290–1303

15. van Albada SJ, Kerr CC, Chiang AKI, Rennie CJ, Robinson PA (2010) Neurophysiological changes with age probed by inverse modelling of EEG spectra. Clin Neurophysiol 121 (1):21–38

16. Nunez PL (2000) Toward a quantitative description of large-scale neocortical dynamic function and EEG. Behav Brain Sci 23 (3):371–437

17. Nunez PL, Srinivasan R (2006) A theoretical basis for standing and traveling brain waves measured with human EEG with implications for an integrated consciousness. Clin Neurophysiol 117(11):2424–2435

18. Hughes JR, John ER (1999) Conventional and quantitative electroencephalography in psychiatry. J Neuropsychiatry Clin Neurosci 11:190–208

19. Salinsky MC, Oken BS, Morehead L (1991) Test-retest reliability in EEG frequency analysis. Electroencephalogr Clin Neurophysiol 79(5):382–392

20. Gasser T, Bacher P, Steinberg H (1985) Test–retest reliability of spectral parameters of the EEG. Electroencephalogr Clin Neurophysiol 60:312–319

21. Pollock VE, Schneider LS, Lyness SA (1991) Reliability of topographic quantitative EEG amplitude in healthy late-middle-aged and elderly subjects. Electroencephalogr Clin Neurophysiol 79:20–26

22. Burgess A, Gruzelier J (1993) Individual reliability of amplitude distribution in topographical mapping of EEG. Electroencephalogr Clin Neurophysiol 86:219–223

23. Harmony T, Fernandez T, Rodriguez M, Reyes A, Marosi E, Bernal J (1993) Test–retest reliability of EEG spectral parameters during cognitive tasks: II. Coherence. Int J Neurosci 68:263–271

24. Lund TR, Sponheim SR, Iacono WG, Clementz BA (1995) Internal consistency reliability of resting EEG power spectra in schizophrenic and normal subjects. Psychophysiology 32:66–71

25. Corsi-Cabrera M, Solis-Ortiz S, Guevara MA (1997) Stability of EEG inter- and intrahemispheric correlation in women. Electroencephalogr Clin Neurophysiol 102:248–255

26. Fingelkurts AA, Fingelkurts AA, Ermolaev VA, Kaplan AY (2006) Stability, reliability and consistency of the compositions of brain oscillations. Int J Psychophysiol 59:116–126

27. Stassen HH, Bomben G, Propping P (1987) Genetic aspects of the EEG: an investigation into the within-pair similarity of monozygotic and dizygotic twins with a new method of analysis. Electroencephalogr Clin Neurophysiol 66:489–501

28. Stassen HH, Bomben G, Hell D (1998) Familial brain wave patterns: study of a 12-sib family. Psychiatr Genet 8:141–153

29. van Beijsterveldt CEM, Molenaar PC, de Geus EJ, Boomsma DI (1996) Heritability of human brain functioning as assessed by electroencephalography. Am J Hum Genet 58:562–573

30. Smit DJA, Posthuma D, Boomsma DI, de Geus EJC (2005) Heritability of background EEG across the power spectrum. Psychophysiology 42:691–697

31. van Beijsterveldt CEM, van Baal GCM (2002) Twin and family studies of the human electroencephalogram: a review and a meta-analysis. Biol Psychol 61:111–138

32. Johnstone J, Gunkelman J, Lunt J (2005) Clinical database development: characterization of EEG phenotypes. Clin EEG Neurosci 36 (2):99–107

33. Ehlers CL, Phillips E, Gizer IR, Gilder DA, Wilhelmsen KC (2010) EEG spectral phenotypes: heritability and association with marijuana and alcohol dependence in an American Indian community study. Drug Alcohol Depend 106:101–110

34. Buzsáki G, Draguhn A (2004) Neuronal oscillations in cortical networks. Science 304:1926–1929

35. Başar E, Başar-Eroglu C, Karakas S, Schurmann M (2001) Gamma, alpha, delta, and theta oscillations govern cognitive processes. Int J Psychophysiol 39:241–248

36. Engel AK, Fries P, Singer W (2001) Dynamic predictions: oscillations and synchrony in top-down processing. Nat Rev Neurosci 2:704–716

37. Fries P (2005) A mechanism for cognitive dynamics: neuronal communication through neuronal coherence. Trends Cogn Sci 9:474–480

38. Başar E (1992) Brain natural frequencies are causal factors for resonances and induced rhythms. (Epilogue). In: Başar E, Bullock TH (eds) Induced rhythms in the brain. Birkhauser, Boston, MA, pp 425–467

39. Başar E (2008) Oscillations in "brain–body–mind"—a holistic view including the autonomous system. Brain Res 1235:2–11

40. Pascual-Leone A, Walsh V, Rothwell J (2000) Transcranial magnetic stimulation in cognitive neuroscience—virtual lesion, chronometry, and functional connectivity. Curr Opin Neurobiol 10:232–237

41. Ilmoniemi RJ (2006) Transcranial magnetic stimulation. Wiley Encyclopedia of Biomedical Engineering, New York

42. Thut G, Miniussi C (2009) New insights into rhythmic brain activity from TMS–EEG studies. Trends Cogn Sci 13(4):182–189

43. McFadden J (2002) Synchronous firing and its influence on the brain's electromagnetic field: evidence for an electromagnetic field theory of consciousness. J Conscious Stud 9(4):23–50

44. Tsodyks M, Kenet T, Grinvald A, Arieli A (1999) Linking spontaneous activity of single cortical neurons and the underlying functional architecture. Science 286:1943–1946

45. Weiss SA, Faber DS (2010) Field effects in the CNS play functional roles. Front Neural Circuits 4:15

46. Haken H (2006) Synergetics of brain function. Int J Psychophysiol 60:110–124

47. Bodunov MV (1988) The EEG "alphabet": the typology of human EEG stationary segments. In: Rusalov VM (ed) Individual and psychological differences and bioelectrical activity of human brain. Nauka, Moscow, pp 56–70 (in Russian)

48. Jansen BH, Cheng W-K (1988) Structural EEG analysis: an explorative study. Int J Biomed Comput 23:221–237

49. Fingelkurts AA, Fingelkurts AA, Kaplan AY (2003) The regularities of the discrete nature of multi-variability of EEG spectral patterns. Int J Psychophysiol 47(1):23–41

50. Fingelkurts AA, Fingelkurts AA, Krause CM, Kaplan AY (2003) Systematic rules underlying spectral pattern variability: experimental results and a review of the evidences. Int J Neurosci 113:1447–1473

51. Kaplan AY (1998) Nonstationary EEG: methodological and experimental analysis. Usp Fiziol Nauk 29(3):35–55 (in Russian)

52. Kaplan AY, Shishkin SL (2000) Application of the change-point analysis to the investigation of the brain's electrical activity. In: Brodsky BE, Darhovsky BS (eds) Non-parametric statistical diagnosis. Problems and methods. Kluwer Academic Publishers, Dordrecht, pp 333–388

53. Kaplan AY, Fingelkurts AA, Fingelkurts AA, Borisov SV, Darkhovsky BS (2005) Nonstationary nature of the brain activity as revealed by EEG/MEG: methodological, practical and conceptual challenges. Signal Process 85:2190–2212

54. Fingelkurts AA, Fingelkurts AA (2001) Operational architectonics of the human brain biopotential field: towards solving the mind-brain problem. Brain Mind 2(3):261–296, http://www.bm-science.com/team/art18.pdf

55. Fingelkurts AA, Fingelkurts AA (2005) Mapping of the brain operational architectonics. In: Chen FJ (ed) Focus on brain mapping research. Nova Science Publishers Inc, New York, pp 59–98, http://www.bm-science.com/team/chapt3.pdf

56. Fingelkurts AA, Fingelkurts AA (2008) Brain-mind operational architectonics imaging: technical and methodological aspects. Open Neuroimag J 2:73–93

57. Betzel RF, Erickson MA, Abell M, O'Donnell BF, Hetrick WP, Sporns O (2012) Synchronization dynamics and evidence for a repertoire of network states in resting EEG. Front Comput Neurosci 6:74

58. Berger H (1929) Über das Elektroenkephalogramm des Menschen. Arch Psychiatr 87:527–570, Translated and reprinted in Pierre Gloor, Hans Berger on the electroencephalogram of man. Electroencephalogr Clin Neurophysiol 1969; Supp. 28. Elsevier, Amsterdam

59. Miwakeichi F, Martinez-Montes E, Valdes-Sosa PA, Nishiyama N, Mizuhara H, Yamaguchia Y (2004) Decomposing EEG data into space–time–frequency components using Parallel Factor Analysis. NeuroImage 22:1035–1045

60. Lehmann D (1987) Principles of spatial analysis. In: Gevins AS, Remont A (eds) Methods

of analysis of brain electrical and magnetic signals. Elsevier, Amsterdam, pp 309–354

61. Lehmann D, Strik WK, Henggeler B, Koenig T, Koukkou M (1998) Brain electric microstates and momentary conscious mind states as building blocks of spontaneous thinking: I. Visual imagery and abstract thoughts. Int J Psychophysiol 29:1–11

62. Freeman WJ (1990) On the problem of anomalous dispersion in chaoto-chaotic phase transitions of neural masses, and its significance for the management of perceptual information in brains. In: Haken H, Stadler M (eds) Synergetics of cognition, vol 45. Springer, Berlin, pp 126–143

63. Kaplan AY, Fingelkurts AA, Fingelkurts AA, Ermolaev VA (1999) Topographic variability of the EEG spectral patterns. Fiziol Cheloveka 25(2):21–29 (in Russian)

64. Freeman WJ, Holmes MD (2005) Metastability, instability, and state transition in neocortex. Neural Netw 18:497–504

65. Freeman WJ, Vitiello G (2005) Nonlinear brain dynamics and many-body field dynamics. Electromagn Biol Med 24:233–241

66. Fingelkurts AA, Fingelkurts AA, Kaplan AY (2006) Interictal EEG as a physiological adaptation. Part II. Topographic variability of composition of brain oscillations in interictal EEG. Clin Neurophysiol 117(4):789–802

67. Van de Ville D, Britz J, Michel CM (2010) EEG microstate sequences in healthy humans at rest reveal scale-free dynamics. Proc Natl Acad Sci U S A 107:18179–18184

68. Bodenstein G, Praetorius HM (1977) Feature extraction from the electroencephalogram by adaptive segmentation. Proc IEEE 65:642–652

69. Barlow JS (1985) Methods of analysis of nonstationary EEGs, with emphasis on segmentation techniques: a comparative review. J Clin Neurophysiol 2:267–304

70. Gersch W (1987) Non-stationary multichannel time series analysis. In: Gevins A (ed) EEG Handbook, Revised Series, vol 1. Academic, New York

71. Oken BS, Chiappa KH (1988) Short-term variability in EEG frequency analysis. Electroencephalogr Clin Neurophysiol 69(3):191–198

72. Shishkin SL, Brodsky BE, Darkhovsky BS, Kaplan AY (1997) EEG as a nonstationary signal: an approach to analysis based on nonparametric statistics. Fiziol Cheloveka 23(4):124–126 (in Russian)

73. Fingelkurts AA (1998) Time-spatial organization of the human EEG segmental structure.

Ph.D. Dissertation. MSU, Moscow, Russian Federation, 401 p, (in Russian)

74. Klonowski W (2009) Everything you wanted to ask about EEG but were afraid to get the right answer. Nonlinear Biomed Phys 3:2. doi:10.1186/1753-4631-3-2

75. Freyer F, Aquino K, Robinson PA, Ritter P, Breakspear M (2009) Bistability and non-gaussian fluctuations in spontaneous cortical activity. J Neurosci 29:8512–8524

76. Latchoumane CFV, Jeong J (2010) Quantification of brain macrostates using dynamical non-stationarity of physiological time series. IEEE Trans Biomed Eng 58:1084–1093

77. Chu CJ, Kramer MA, Pathmanathan J et al (2012) Emergence of stable functional networks in long-term human electroencephalography. J Neurosci 32:2703–2713

78. Rusinov VS (1973) The dominant focus: electrophysiological investigations. Consultants Bureau, New York, p 220 (Translated from Russian)

79. Burov IV, Kaplan AY (1993) The effect of amiridin on the spectral characteristics of the human EEG. Eksp Klin Farmakol 56(5):5–8 (in Russian)

80. Lehmann D, Ozaki H, Pal I (1987) EEG alpha map series: brain micro-states by space oriented adaptive segmentation. Electroencephalogr Clin Neurophysiol 67:271–288

81. Brodsky BE, Darkhovsky BS, Kaplan AY, Shishkin SL (1999) A nonparametric method for the segmentation of the EEG. Comput Methods Programs Biomed 60:93–106

82. Fell J, Kaplan A, Darkhovsky B, Röschke J (2000) EEG analysis with nonlinear deterministic and stochastic methods: a combined strategy. Acta Neurobiol Exp 60:87–108

83. Fingelkurts AA, Fingelkurts AA, Kivisaari R, Pekkonen E, Ilmoniemi RJ, Kähkönen SA (2004) Local and remote functional connectivity of neocortex under the inhibition influence. NeuroImage 22(3):1390–1406

84. Fingelkurts AA, Fingelkurts AA, Kivisaari R, Pekkonen E, Ilmoniemi RJ, Kähkönen SA (2004) Enhancement of GABA-related signalling is associated with increase of functional connectivity in human cortex. Hum Brain Mapp 22(1):27–39

85. Freeman WJ (2004) Origin, structure, and role of background EEG activity. Part 1. Analytic amplitude. Clin Neurophysiol 115:2077–2088

86. Fingelkurts AA, Fingelkurts AA (2010) Alpha rhythm operational architectonics in the continuum of normal and pathological brain states: current state of research. Int J Psychophysiol 76:93–106

87. Freeman WJ (2004) Origin, structure, and role of background EEG activity. Part 2. Analytic phase. Clin Neurophysiol 115:2089–2107

88. Wallenstein GV, Kelso JSA, Bressler SL (1995) Phase transitions in spatiotemporal patterns of brain activity and behaviour. Physica D 84(3–4):626–634

89. Kozma R, Freeman WJ (2002) Classification of EEG patterns using nonlinear dynamics and identifying chaotic phase transitions. Neurocomputing 44:1107–1112

90. Puljic M, Kozma R (2003) Phase transitions in a probabilistic cellular neural network model having local and remote connections. International joint conference on neural networks IJCNN'2003, Portland, OR, 14–19 July 2003, p 831–835

91. Fingelkurts AA, Fingelkurts AA, Neves CFH (2010) Natural world physical, brain operational, and mind phenomenal space-time. Phys Life Rev 7(2):195–249

92. Kozma R, Puljic M, Balister P, Bollobas B, Freeman WJ (2005) Phase transitions in the neuropercolation model of neural populations with mixed local and nonlocal interactions. Biol Cybern 92:367–379

93. Thatcher RW, John ER (1977) Functional neuroscience. Vol. 1: foundations of cognitive processes. Lawrence Erlbaum, New York

94. Herscovitch P (1994) Radiotracer techniques for functional neuroimaging with positron emission tomography. In: Thatcher RW, Halletr M, Zeffro T, John ER, Huerta M (eds) Functional neuroimaging: technical foundations. Academic, San Diego

95. Arieli A, Sterkin A, Grinvald A, Aertsen A (1996) Dynamics of ongoing activity: explanation of the large variability in evoked cortical responses. Science 273:1868–1871

96. Raichle ME, MacLeod AM, Snyder AZ, Powers WJ, Gusnard DA, Shulman GL (2001) A default mode of brain function. Proc Natl Acad Sci U S A 98:676–682

97. Raichle ME, Snyder AZ (2007) A default mode of brain function: a brief history of an evolving idea. NeuroImage 37(4):1083–1090

98. Fingelkurts AA, Fingelkurts AA (2011) Persistent operational synchrony within brain default-mode network and self-processing operations in healthy subjects. Brain Cogn 75(2):79–90

99. Soroko SI, Suvorov NB, Bekshaev SS (1977) Voluntary control of the level of brain bioelectrical l activity as a method to study autoregulatory properties of CNA. In: Vasilevskii NN (ed), Adaptive self-regulation of functions. Moscow: Meditsina, pp. 206–248

100. Bekshaev SS, Vasilevskii NN, Suvorov NB, Kutuev VB, Soroko SI (1978) Combined approach for analysis of temporal statistical structure of EEG rhythms. In: Adaptive reactions of the brain and their prognosis, p 117–123 (in Russian)

101. Soroko SI, Bekshaev SS (1981) EEG rhythms' statistical structure and individual properties of brain self-regulation mechanisms. Fiziologich J 67:1765–1773 (in Russian)

102. Borisov SV, Kaplan AY, Gorbachevskaia NL, Kozlova IA (2005) Segmental structure of the EEG alpha activity in adolescents with disorders of schizophrenic spectrum. Zh Vyssh Nerv Deiat Im I P Pavlova 55(3):329–335 (in Russian)

103. Fingelkurts AA, Fingelkurts AA, Kaplan AY (2006) Interictal EEG as a physiological adaptation. Part I. Composition of brain oscillations in interictal EEG. Clin Neurophysiol 117:208–222

104. Fingelkurts AA, Fingelkurts AA, Rytsala H, Suominen K, Isometsa E, Kahkonen S (2006) Composition of brain oscillations in ongoing EEG during major depression disorder. Neurosci Res 56:133–144

105. Fingelkurts AA, Fingelkurts AA, Kivisaari R et al (2006) Reorganization of the composition of brain oscillations and their temporal characteristics in opioid dependent patients. Prog Neuropsychopharmacol Biol Psychiatry 30:1453–1465

106. Fingelkurts AA, Fingelkurts AA (2010) Short-term EEG spectral pattern as a single event in EEG phenomenology. Open Neuroimag J 4:130–156

107. Manuca R, Savit R (1996) Stationarity and nonstationarity in time series analysis. J Phys D 99:134–161

108. Fingelkurts AA, Fingelkurts AA (2004) Making complexity simpler: multivariability and metastability in the brain. Int J Neurosci 114:843–862

109. Speckmann EJ, Elger CE (1998) Introduction to the neurophysiological basis of the EEG and DC potentials. In: Niedermeyer E, Lopes da Silva F (eds) Electroencephalography. Williams and Wilkins, Baltimore

110. Bullock TH (1997) Signals and signs in the nervous system: the dynamic anatomy of electrical activity. Proc Natl Acad Sci U S A 94:1–6

111. Towle VL, Carder RK, Khorasani L, Lindber D (1999) Electro-corticographic coherence patterns. J Clin Neurophysiol 16:528–547

112. Freeman WJ, Holmes MD, West GA, Vanhatalo S (2006) Fine spatiotemporal structure of phase in human intracranial EEG. Clin Neurophysiol 117:1228–1243

113. Sviderskaya NE, Shlitner LM (1990) Coherent cortical electric activity structures in the

human brain. Fiziol Cheloveka 16(3):12–19 (in Russian)

114. Mantini D, Perrucci MG, Del Gratta C, Romani GL, Corbetta M (2007) Electrophysiological signatures of resting state networks in the human brain. Proc Natl Acad Sci U S A 104(32):13170–13175

115. Chorlian DB, Rangaswamy M, Porjesz B (2009) EEG coherence: topography and frequency structure. Exp Brain Res 198:59–83

116. Hori H, Hayasaka K, Sato K, Harada O, Iwata H (1969) A study on phase relationship in human alpha activity. Correlation of different regions. Electroencephalogr Clin Neurophysiol 26:19–24

117. Ozaki H, Suzuki H (1986) Transverse relationships of the alpha rhythm on the scalp. Electroencephalogr Clin Neurophysiol 66:191–195

118. Thatcher RW, Krause P, Hrybyk M (1986) Corticocortical associations and EEG coherence: a two compartmental model. Electroencephalogr Clin Neurophysiol 64:123–143

119. Bullock TH, Achimowicz JZ (1994) A comparative survey of oscillatory brain activity, especially gamma-band rhythms. In: Pantev C, Elbert T, Lukenhoner B (eds) Oscillatory event related brain dynamics. Plenum Publishing Corp, New York, pp 11–26

120. Bullock TH, McClune MC, Achimowicz JZ, Iragui-Madoz VJ, Duckrow RB, Spencer SS (1995) EEG coherence has structure in the millimeter domain: subdural and hippocampal recordings from epileptic patients. Electroencephalogr Clin Neurophysiol 95:161–177

121. Shen B, Nadkarni M, Zappulla RA (1999) Spectral modulation of cortical connections measured by EEG coherence in humans. Clin Neurophysiol 110(1):115–125

122. Fingelkurts AA, Fingelkurts AA, Kähkönen SA (2005) Functional connectivity in the brain—is it an elusive concept? Neurosci Biobehav Rev 28(8):827–836

123. Freeman WJ (2003) The wave packet: an action potential for the 21st Century. J Integr Neurosci 2:3–30

124. Kooi KA (1971) Fundamentals of electroencephalography. Harper & Row Publishers, New York

125. Bullock TH, McClune MC (1989) Lateral coherence of the electroencephalogram: a new measure of brain synchrony. Electroencephalogr Clin Neurophysiol 73:479–498

126. Kaiser M, Görner M, Hilgetag CC (2007) Criticality of spreading dynamics in hierarchical cluster networks without inhibition. New J Phys 9:110

127. Braitenberg V, Schüz A (1998) Cortex: statistics and geometry of neuronal connectivity, 2nd edn. Springer, Berlin

128. Hellwig B (2000) A quantitative analysis of the local connectivity between pyramidal neurons in layers 2/3 of the rat visual cortex. Biol Cybern 82:111–121

129. Watts DJ, Strogatz SH (1998) Collective dynamics of 'small-world' networks. Nature 393:440–442

130. Barabasi AL, Albert R (1999) Emergence of scaling in random networks. Science 286:509–512

131. Sporns O, Tononi G, Edelman GM (2002) Theoretical neuroanatomy and the connectivity of the cerebral cortex. Behav Brain Res 135:69–74

132. Sporns O, Chialvo D, Kaiser M, Hilgetag CC (2004) Organization, development and function of complex rain networks. Trends Cogn Sci 8:418–425

133. Bassett DS, Bullmore E (2006) Small-world brain networks. Neuroscientist 12:512–523

134. Stam C, Reijneveld J (2007) Graph theoretical analysis of complex networks in the brain. Nonlinear Biomed Phys 1(1):3

135. Boccaletti S, Latora V, Moreno Y, Chavez M, Hwang DU (2006) Complex networks: structure and dynamics. Phys Rep 424:175–308

136. Freeman WJ (1991) The physiology of perception. Scientific Am, p 78–85

137. Molle M, Marshall L, Lutzenberger W, Pietrowsky R, Fehm HL, Born J (1996) Enhanced dynamic complexity in the human EEG during creative thinking. Neurosci Lett 208:61–64

138. Lehmann D, Koenig T (1997) Spatio-temporal dynamics of alpha brain electric fields, and cognitive modes. Int J Psychophysiol 26:99–112

139. Borisov SV (2002) Studying of a phasic structure of the alpha activity of human EEG. Ph. D. dissertation, MSU, Moscow, Russian Federation, 213 p, (in Russian)

140. Muller TJ, Koenig T, Wackermann J et al (2005) Subsecond changes of global brain state in illusory multistable motion perception. J Neural Transm 112:565–576

141. Bassett DS, Bullmore ET, Meyer-Lindenberg A, Apud JA, Weinberger DR, Coppola R (2009) Cognitive fitness of cost-efficient brain functional networks. Proc Natl Acad Sci U S A 106:11747–11752

142. Bassett DS, Meyer-Lindenberg A, Achard S, Duke T, Bullmore ET (2006) Adaptive reconfiguration of fractal small-world human brain functional networks. Proc Natl Acad Sci U S A 103:19518–19523

143. Micheloyannis S, Vourkas M, Tsirka V, Karakonstantaki E, Kanatsouli K, Stam CJ (2009) The influence of ageing on complex brain networks: a graph theoretical analysis. Hum Brain Mapp 30:200–208

144. Boldyreva GN, Zhavoronkova LA, Sharova EV, Dobronravova IS (2007) Electroencephalographic intercentral interaction as a reflection of normal and pathological human brain activity. Span J Psychol 10(1):167–177

145. Stam CJ, Jones BF, Nolte G, Breakspear M, Scheltens P (2007) Small-world networks and functional connectivity in Alzheimer's disease. Cereb Cortex 17:92–99

146. Lantz G, Michel CM, Seeck M et al (2001) Space-oriented segmentation and 3-dimensional source reconstruction of ictal EEG patterns. Clin Neurophysiol 112:688–697

147. Strik WK, Dierks T, Becker T, Lehmann D (1995) Larger topographical variance and decreased duration of brain electric microstates in depression. J Neural Transm 99:213–222

148. Fingelkurts AA, Fingelkurts AA, Rytsala H, Suominen K, Isometsä E, Kähkönen S (2007) Impaired functional connectivity at EEG alpha and theta frequency bands in major depression. Hum Brain Mapp 28(3):247–261

149. Lehmann D, Wackermann J, Michel CM, Koenig T (1993) Space-oriented EEG segmentation reveals changes in brain electric field maps under the influence of a nootropic drug. Psychiatry Res 50:275–282

150. Kinoshita T, Strik WK, Michel CM, Yagyu T, Saito M, Lehmann D (1995) Microstate segmentation of spontaneous multichannel EEG map series under diazepam and sulpiride. Pharmacopsychiatry 28:51–55

151. Fingelkurts AA, Fingelkurts AA, Kähkönen S (2005) New perspectives in pharmaco-electroencephalography. Prog Neuropsychopharmacol Biol Psychiatry 29:193–199

152. Fingelkurts AA, Fingelkurts AA, Kivisaari R et al (2006) Increased local and decreased remote functional connectivity at EEG alpha and beta frequency bands in opioid-dependent patients. Psychopharmacology 188(1):42–52

153. Fingelkurts AA, Fingelkurts AA, Kivisaari R et al (2007) Opioid withdrawal results in an increased local and remote functional connectivity at EEG alpha and beta frequency bands. Neurosci Res 58(1):40–49

154. Fingelkurts AA, Fingelkurts AA, Kivisaari R et al (2009) Methadone may restore local and remote EEG functional connectivity in opioid-dependent patients. Int J Neurosci 119(9):1469–1493

155. Borisov SV, Kaplan AY, Gorbachevskaya NL, Kozlova IA (2005) Analysis of EEG structural synchrony in adolescents with schizophrenic disorders. Fiziol Cheloveka 31:16–23 (in Russian)

156. Micheloyannis S, Pachou E, Stam CJ et al (2006) Small-world networks and disturbed functional connectivity in schizophrenia. Schizophr Res 87:60–66

157. Rubinov M, Knock SA, Stam CJ et al (2009) Small-world properties of nonlinear brain activity in schizophrenia. Hum Brain Mapp 30:403–416

158. Freeman WJ (2007) Indirect biological measures of consciousness from field studies of brains as dynamical systems. Neural Netw 20:1021–1031

159. Elul R (1969) Gaussian behavior of the electroencephalogram: changes during performance of mental task. Science 164(3877):328–331

160. Klimesch W (1999) Event-related band power changes and memory performance. In: Pfurtscheller G, da Silva FH L (eds) Event-Related desynchronization. Handbook of electroencephalography and clinical neurophysiology. Elsevier, Amsterdam, pp 161–178

161. Başar E, Özgören M, Karakas S, Başar–Eroglu C (2004) Super-synergy in the brain: the grandmother percept is manifested by multiple oscillations. Int J Bifurcat Chaos 14:453–491

162. Fingelkurts AA, Fingelkurts AA, Krause CM, Sams M (2002) Probability interrelations between pre-/post-stimulus intervals and ERD/ERS during a memory task. Clin Neurophysiol 113:826–843

163. Landa P, Gribkov D, Kaplan A (2000) Oscillatory processes in biological systems. In: Malik SK, Chandrashekaran MK, Pradhan N (eds) Nonlinear phenomena in biological and physical sciences. Indian National Science Academy, New Delhi, pp 123–152

164. Skinner JE, Molnar M (2000) "Response cooperativity": a sign of a nonlinear neocortical mechanism for stimulus-binding during classical conditioning in the act. In: Malik SK, Chandrashekaran MK, Pradhan N (eds) Nonlinear phenomena in biological and physical sciences. Indian National Science Academy, New Deli, pp 223–248

165. Lehmann D (1971) Multichannel topography of human alpha EEG fields. Electroencephalogr Clin Neurophysiol 31:439–449

166. Freeman W, Burke B, Holmes M (2003) Aperiodic phase re-setting in scalp EEG of beta-

gamma oscillations by state transitions at alpha-theta rates. Hum Brain Mapp 19 (4):248–272

167. Breakspear M, Stam CJ (2005) Dynamics of a neural system with a multiscale architecture. Philos Trans R Soc Lond B Biol Sci 360:1051–1074

168. Breakspear M, Terry JR (2003) Topographic organization of nonlinear interdependence in multichannel human EEG. NeuroImage 16:822–835

169. Stam CJ (2006) Nonlinear brain dynamics. Nova Science Publishers Inc., New York

170. Feinberg TE (2000) The nested hierarchy of consciousness: a neurobiological solution to the problem of mental unity. Neurocase 6 (2):75–81

171. Feinberg TE (2009) From axons to identity: neurological explorations of the nature of the self. WW Norton & Company, New York

172. Feinberg TE (2012) Neuroontology, neurobiological naturalism, and consciousness: a challenge to scientific reduction and a solution. Phys Life Rev 9(1):13–34

173. Palm G (1990) Cell assemblies as a guideline for brain research. Concepts Neurosci 1:133–147

174. Eichenbaum H (1993) Thinking about brain cell assemblies. Science 261:993–994

175. Buzsáki G (2006) Rhythms of the brain. Oxford University Press, Oxford

176. Hebb DO (1949) The organization of behavior. Wiley, New York

177. von der Malsburg C (1999) The what and why of binding: the modeler's perspective. Neuron 24:95–104

178. Friston K (2000) The labile brain. I. Neuronal transients and nonlinear coupling. Philos Trans R Soc Lond B Biol Sci 355:215–236

179. Triesch J, von der Malsburg C (2001) Democratic integration: self-organized integration of adaptive cues. Neural Comput 13 (9):2049–2074

180. Kaplan AY, Borisov SV (2003) Dynamic properties of segmental characteristics of EEG alpha activity in rest conditions and during cognitive load. Zh Vyssh Nerv Deiat Im I P Pavlova 53:22–32 (in Russian)

181. Averbeck BB, Lee D (2004) Coding and transmission of information by neural ensembles. Trends Neurosci 27:225–230

182. Zeki S (2004) Insights into visual consciousness. In: Frackowiak RSJ, Friston KJ, Frith CD et al (eds) Human brain function. Academic, San Diego

183. Singer W (2001) Consciousness and the binding problem. Ann N Y Acad Sci 929:123–146

184. van Leeuwen C (2007) What needs to emerge to make you conscious? J Conscious Stud 14:115–136

185. Pulvermueller F, Preissl H, Eulitz C, et al (1994) Brain rhythms, cell assemblies and cognition: evidence from the processing of words and pseudowords. Psycoloquy 5(48): brain-rhythms.1.pulvermueller

186. Kirillov AB, Makarenko VI (1991) Metastability and phase transition in neural networks: statistical approach. In: Holden AV, Kryukov VI (eds) Neurocomputers and attention, vol 2. Manchester University Press, Manchester, pp 825–922

187. Fujisawa S, Matsuki N, Ikegaya Y (2006) Single neurons can induce phase transitions of cortical recurrent networks with multiple internal states. Cereb Cortex 16:639–654

188. Leznik E, Makarenko V, Llinas R (2002) Electrotonically mediated oscillatory patterns in neuronal ensembles: an in vitro voltage-dependent dye-imaging study in the inferior olive. J Neurosci 22:2804–2815

189. Buzsáki G, Anastassiou CA, Koch C (2012) The origin of extracellular fields and currents—EEG, ECoG, LFP and spikes. Nat Rev Neurosci 13:407–420

190. Plenz D, Thiagarajan TC (2007) The organizing principles of neuronal avalanches: cell assemblies in the cortex? Trends Neurosci 30:101–110

191. Plenz D (2012) Neuronal avalanches and coherence potentials. Eur Phys J Spec Top 205:259–301

192. John ER (2002) The neurophysics of consciousness. Brain Res Brain Res Rev 39:1–28

193. Başar E (2005) Memory as the "whole brain work". A large-scale model based on "oscillations in super-synergy". Int J Psychophysiol 58:199–226

194. Truccolo WA, Ding M, Knuth KH, Nakamura R, Bressler S (2002) Trial-to-trial variability of cortical evoked responses: implications for analysis of functional connectivity. Clin Neurophysiol 113:206–226

195. Brodsky BE, Darkhovsky BS (1993) Nonparametric methods in change-point problems. Kluwer, Dordrecht

196. Geisser S, Johnson WM (2006) Modes of parametric statistical inference. Wiley, Hoboken, NJ

197. Lopes da Silva FH, Mars NJI (1987) Parametric methods in EEG analysis. In: Gevins AS, Remond A (eds) EEG handbook (revised series): methods of analysis of brain electrical

and magnetic signals, vol 1. Elsevier Science, Amsterdam, pp 243–260

198. Pardey J, Roberts S, Tarassenko L (1996) A review of parametric modelling techniques for EEG analysis. Med Eng Phys 18:2–11

199. Deistler M, Prohaska O, Reschenhofer E, Vollrner R (1986) Procedure for identification of different stages of EEG background activity and its application to the detection of drug effects. Electroencephalogr Clin Neurophysiol 64:294–300

200. Fingelkurts AA, Fingelkurts AA (1995) Microstructural analysis of active brain EEG: general characteristics and synchronization peculiarities of change-point process. Diploma Project. MSU, Moscow, Russian Federation, 207 p, (in Russian)

201. Klimesch W, Schack B, Sauseng P (2005) The functional significance of theta and upper alpha oscillations. Exp Psychol 52:99–108

202. Fingelkurts AA, Fingelkurts AA, Kallio S, Revonsuo A (2007) Cortex functional connectivity as a neurophysiological correlate of hypnosis: an EEG case study. Neuropsychologia 45:1452–1462

203. David O, Cosmelli D, Lachaux J-P, Baillet S, Garnero L, Martinerie J (2003) A Theoretical and experimental introduction to the noninvasive study of large-scale neural phase synchronization in human beings. Int J Comput Cogn 1(4):53–77

204. Ainsworth M, Lee S, Cunningham MO et al (2012) Rates and rhythms: a synergistic view of frequency and temporal coding in neuronal networks. Neuron 75:572–583

205. Verevkin E, Putilov D, Donskaya O, Putilov A (2007) A new SWPAQ's scale predicts the effects of sleep deprivation on the segmental structure of alpha waves. Biol Rhythm Res 39(1):21–37

206. Putilov DA, Verevkin EG, Donskaya OG, Putilov AA (2007) Segmental structure of alpha waves in sleep-deprived subjects. Somnologie 11(3):202–210

207. Hämäläinen M, Hari R, Ilmoniemi RJ, Knuutila J, Lounasmaa OV (1993) Magnetoencephalography—theory, instrumentation, and applications to noninvasive studies of the working human brain. Rev Mod Phys 65:413–497

208. Shishkin SL (1997) A study of synchronization of instants of abrupt changes in human EEG alpha activity. PhD dissertation. MSU, Moscow, Russian Federation, (in Russian)

209. Freeman W, Rogers L (2003) A neurobiological theory of meaning in perception. Part 5. Multicortical patterns of phase modulation in gamma EEG. Int J Bifurc Chaos 13:2867–2887

210. Fingelkurts AA, Fingelkurts AA (2010) Topographic mapping of rapid transitions in EEG multiple frequencies: EEG frequency domain of operational synchrony. Neurosci Res 68:207–224

211. Singer W, Engel AK, Kreiter AK, Munk MHJ, Neuenschwander S, Roelfsema PR (1997) Neural assemblies: necessity, signature and detectability. Trends Cogn Sci 1:252–261

212. Livanov MN, Gavrilova NA, Aslanov AS (1964) Intercorrelations between different cortical regions of human brain during mental activity. Neuropsychologia 2:281–289

213. Livanov MN (1977) Spatial organization of cerebral processes. Wiley, New York

214. Lazarev VV, Sviderskaya NE, Khomskaya ED (1977) Changes in spatial synchronization of biopotentials during various types of intellectual activity. Hum Physiol 3:187–194, a translation from Fiziol Chelov

215. Lazarev VV (1978) Changes of functional state of the brain during motor and intellectual activity. In: Psychological aspects of human activity. II. Industrial psychology and psychology of labour. Institute of Psychology, USSR Academy of Sciences, Moscow, p 103–114

216. Lazarev VV (1998) On the intercorrelation of some frequency and amplitude parameters of the human EEG and its functional significance. Com. I. Multidimensional neurodynamic organization of functional states of the brain during intellectual, perceptive and motor activity in normal subjects. Int J Psychophysiol 28:77–98

217. Fingelkurts AA, Fingelkurts AA, Neves CFH (2009) Phenomenological architecture of a mind and operational architectonics of the brain: the unified metastable continuum. New Math Nat Comput 5:221–244

218. Bressler SL, McIntosh AR (2007) The role of neural context in large-scale neurocognitive network operations. In: Jirsa VK, McIntosh AR (eds) Handbook of brain connectivity. Springer, Berlin, pp 403–419

219. Kelso JAS (1995) Dynamics patterns: the self-organization of brain and behaviour. MIT Press, Cambridge, MA

220. Bressler SL, Kelso JAS (2001) Cortical coordination dynamics and cognition. Trends Cogn Sci 5(1):26–36

221. Kelso JAS, Tognoli E (2007) Toward a complementary neuroscience: Metastable coordination dynamics of the brain. In: Kozma R, Perlovsky L (eds) Neurodynamics of higher-level cognition and consciousness. Springer, Heidelberg

222. Kelso JAS, Engstrøm D (2006) The complementary nature. MIT Press, Cambridge

223. Kelso JAS (2009) Coordination dynamics. In: Meyers RA (ed) Encyclopedia of complexity and systems sciences. Springer, Berlin, pp 1537–1564

224. Fingelkurts AA, Fingelkurts AA (2012) Mind as a nested operational architectonics of the brain. Comment on "Neuroontology, neurobiological naturalism, and consciousness: a challenge to scientific reduction and a solution" by Todd E. Feinberg. Phys Life Rev 9:49–50

225. Brown R, Kocarev L (2000) A unifying definition of synchronization for dynamical systems. Chaos 10:344–349

226. Horwitz B (2003) The elusive concept of brain connectivity. NeuroImage 19:466–470

227. Friston KJ, Frith CD, Liddle PF, Frackowiak RSJ (1993) Functional connectivity: the principal component analysis of large (PET) data sets. J Cereb Blood Flow Metab 13:5–14

228. Le Van Quyen M, Bragin A (2007) Analysis of dynamic brain oscillations: methodological advances. Trends Neurosci 30(7):365–373

229. Fingelkurts AA, Fingelkurts AA (2006) Timing in cognition and EEG brain dynamics: discreteness versus continuity. Cogn Process 7:135–162

230. Churchland PS, Sejnowski T (1992) The computational brain. MIT, Cambridge

231. Fodor JA, Pylyshyn ZW (1988) Connectionism and cognitive architecture: a critical analysis. Cognition 28:3–71

232. Fingelkurts AA, Fingelkurts AA, Ivashko RM, Kaplan AY (1998) EEG analysis of operational synchrony between human brain cortical areas during memory task performance. Vestn Moskovsk Univ, Series 16. Biol 1:3–11 (in Russian)

233. Köhler W, Held R (1947) The cortical correlate of pattern vision. Science 110:414–419

234. Dresp-Langley B, Durup J (2009) A plastic temporal brain code for conscious state generation. Neural Plast 2009:482696

235. Izhikevich EM, Desai NS, Walcott EC, Hoppensteadt FC (2003) Bursts as a unit of neural information: selective communication via resonance. Trends Neurosci 26:161–167

236. Bak P, Tang C, Wiesenfeld K (1987) Self-organized criticality: an explanation of 1/f noise. Phys Rev Lett 59:364–374

237. Allegrini P, Menicucci D, Bedini R et al (2009) Spontaneous brain activity as a source of ideal 1/f noise. Phys Rev E Stat Nonlin Soft Matter Phys 80:061914

238. Allegrini P, Menicucci D, Bedini R, Gemignani A, Paradisi P (2010) Complex intermittency blurred by noise: theory and application to neural dynamics. Phys Rev E Stat Nonlin Soft Matter Phys 82:015103

239. Allegrini P, Paradisi P, Menicucci D, Gemignani A (2010) Fractal complexity in spontaneous EEG metastable state transitions: new vistas on integrated neural dynamics. Front Physiol 1:128

240. Lee MH (2007) Birkhoff's theorem, many-body response functions, and the ergodic condition. Phys Rev Lett 98:110403

241. Silvestri L, Fronzoni L, Grigolini P, Allegrini P (2009) Event-driven power-law relaxation in weak turbulence. Phys Rev Lett 102:014502

242. West BJ, Geneston EL, Grigolini P (2008) Maximizing information exchange between complex networks. Phys Rep 468:1–99

243. Shew WL, Yang H, Yu S, Roy R, Plenz D (2011) Information capacity and transmission are maximized in balanced cortical networks with neuronal avalanches. J Neurosci 31(1):55–63

244. Beggs JM, Plenz D (2003) Neuronal avalanches in neocortical circuits. J Neurosci 23:11167–11177

245. Stanley HE (1987) Introduction to phase transitions and critical phenomena. Oxford University Press, Oxford, UK

246. Pöppel E (1988) Mindworks: time and conscious experience. Harcourt Brace Jovanovich, Boston

247. Fingelkurts AA, Fingelkurts AA, Krause CM, Kaplan AY, Borisov SV, Sams M (2003) Structural (operational) synchrony of EEG alpha activity during an auditory memory task. NeuroImage 20(1):529–542

248. Gusnard DA, Raichle ME (2001) Searching for a baseline: functional imaging and the resting human brain. Nat Rev Neurosci 2:685–694

249. Gusnard DA, Akbudak E, Shulman GL, Raichle ME (2001) Medial prefrontal cortex and self-referential mental activity: relation to a default mode of brain function. Proc Natl Acad Sci U S A 98:4259–4264

250. Mason MF, Norton MI, Van Horn JD, Wegner DM, Grafton ST, Macrae CN (2007) Wandering minds: the default network and stimulus independent thought. Science 315:393–395

251. Buckner RL, Carroll DC (2007) Self-projection and the brain. Trends Cogn Sci 11:49–57

252. Northoff G, Heinzel A, de Greck M et al (2006) Self-referential processing in our brain—a meta-analysis of imaging studies on the self. NeuroImage 31:440–457

253. Revonsuo A (2006) Inner presence: consciousness as a biological phenomenon. MIT Press, Cambridge

254. Trehub A (2007) Space, self, and the theater of consciousness. Conscious Cogn 16:310–330

255. Kaplan AY, Fingelkurts AA, Fingelkurts AA, Ivashko RM (1998) The temporal consistency of phasic conversions in the basic frequency components of the EEG. Zh Vyssh Nerv Deiat Im I P Pavlova 48(5):816–826 (in Russian)

256. Monto S (2012) Nested synchrony—a novel cross-scale interaction among neuronal oscillations. Front Physiol 3:384

257. Klimesch W, Freunberger R, Sauseng P (2010) Oscillatory mechanisms of process binding in memory. Neurosci Biobehav Rev 34:1002–1014

258. Cohen MX (2008) Assessing transient cross-frequency coupling in EEG data. J Neurosci Methods 168:494–499

259. Meunier D, Lambiotte R, Bullmore ET (2010) Modular and hierarchically modular organization of brain networks. Front Neurosci 4:200. doi:10.3389/fnins.2010.00200

260. Simon HA (1962) The architecture of complexity. Proc Am Philos Soc 106:467–482

261. Simon HA (1995) Near-decomposability and complexity: how a mind resides in a brain. In: Morowitz H, Singer J (eds) The mind, the brain, and complex adaptive systems. Addison-Wesley, Reading, MA, pp 25–43

262. Pan RK, Sinha S (2009) Modularity produces small-world networks with dynamical time-scale separation. Europhys Lett 85:68006

263. Shanahan M (2010) Metastable chimera states in community-structured oscillator networks. Chaos 20:013108

264. Müller-Linow M, Hilgetag CC, Hütt MT (2008) Organization of excitable dynamics in hierarchical biological networks. PLoS Comput Biol 4:e100019. doi:10.1371/journal.pcbi.1000190

265. Robinson PA, Henderson JA, Matar E, Riley P, Gray RT (2009) Dynamical reconnection and stability constraints on cortical network architecture. Phys Rev Lett 103:108104

266. Kaiser M, Hilgetag CC (2010) Optimal hierarchical modular topologies for producing limited sustained activation of neural networks. Front Neuroinform 4:8. doi:10.3389/fninf.2010.00008

267. Fuster JM (1997) Network memory. Trends Neurosci 20:451–459

268. McIntosh AR (1999) Mapping cognition to the brain through neural interactions. Memory 7:523–548

269. Vincent C, Thou N, Ferguson E et al (2001) Scene specific memory in humans: neural activity associated with the detection of novelty prior to memory formation. NeuroImage 13:S758

270. Mesulam MM (1990) Large-scale neurocognitive networks and distributed processing for attention, language, and memory. Ann Neurol 28:597–613

271. Varela F, Lachaux JP, Rodriguez E, Martinerie J (2001) The brainweb: phase synchronization and large-scale integration. Nat Rev Neurosci 2:229–239

272. Klimesch W (1997) EEG-alpha rhythms and memory processes. Int J Psychophysiol 26:319–340

273. Bressler SL, Tognoli E (2006) Operational principles of neurocognitive networks. Int J Psychophysiol 60:139–148

274. Palva JM, Palva S, Kaila K (2005) Phase synchrony among neuronal oscillations in the human cortex. J Neurosci 25:3962–3972

275. Palva S, Palva M (2012) Discovering oscillatory interaction networks with M/EEG: challenges and breakthroughs. Trends Cogn Sci 16(4):219–230

276. Kaplan AY, Fingelkurts AA, Fingelkurts AA, Ivashko RM (1998) Probability patterns of the human EEG narrow-band differential spectra during memory processes. Fisiol Chelov (Hum Physiol) 24(4):453–461 (in Russian)

277. Herrmann WM (1982) Development and critical evaluation of an objective procedure for the electroencephalographic classification of psychotropic drugs. In: Herrmann WM (ed) EEG in drug research. Gustav Fisher, Stuttgart, NY, pp 249–351

278. Kaplan AY, Kochetova AG, Nezavibathko VN, Rjasina TV, Ashmarin IP (1996) Synthetic ACTH analogue SEMAX displays nootropic-like activity in humans. Neurosci Res Commun 19(2):115–123

279. Knyazev GG, Slobodskaya HR (2003) Personality trait of behavioural inhibition is associated with oscillatory systems reciprocal relationships. Int J Psychophysiol 48:247–261

280. Klimesch W, Sauseng P, Hanslmayr S (2007) EEG alpha oscillations: the inhibition-timing hypothesis. Brain Res Rev 53:63–88

281. Palva S, Monto S, Palva JM (2010) Graph properties of synchronized cortical networks during visual working memory maintenance. NeuroImage 49:3257–3268

282. Kuperstein M, Eichenbaum H, VanDeMark T (1986) Neural group properties in the rat hippocampus during the theta rhythm. Exp Brain Res 61:438–442

283. Sem-Jacobsen CW, Petersen MC, Dodge HW, Lazarte JA, Holman CB (1956) Electroencephalographic rhythms from the depths of parietal, occipital and temporal lobes in man. Electroencephalogr Clin Neurophysiol 8:263–278

284. Klimesch W, Doppelmayr M, Schimke H, Ripper B (1997) Theta synchronization and alpha desynchronization in a memory task. Psychophysiology 34:169–176

285. Bäuml K-H, Hanslmayr S, Pastötter B, Klimesch W (2008) Oscillatory correlates of intentional updating in episodic memory. NeuroImage 41:596–604

286. Fingelkurts AA, Fingelkurts AA, Krause CM, Möttönen R, Sams M (2003) Cortical operational synchrony during audio–visual speech integration. Brain Lang 85:297–312

287. Fingelkurts AA, Fingelkurts AA, Bagnato S, Boccagni C, Galardi G (2012) Toward operational architectonics of consciousness: basic evidence from patients with severe cerebral injuries. Cogn Process 13:111–131

288. Fingelkurts AA, Fingelkurts AA, Bagnato S, Boccagni C, Galardi G (2012) DMN operational synchrony relates to self-consciousness: evidence from patients in vegetative and minimally conscious states. Open Neuroimag J 6:55–68

289. Collins PY, Patel V, Joestl SS et al (2011) Scientific advisory board and the executive committee of the grand challenges on global mental health. Nature 475:27–30

290. Wittchen HU, Jacobi F, Rehm J et al (2011) The size and burden of mental disorders and other disorders of the brain in Europe 2010. Eur Neuropsychopharmacol 21:655–679

291. Uhlhaas PJ, Singer W (2012) Neuronal dynamics and neuropsychiatric disorders: toward a translational paradigm for dysfunctional large-scale networks. Neuron 75:963–980

292. Bressler SL (2003) Cortical coordination dynamics and the disorganization syndrome in schizophrenia. Neuropsychopharmacology 28:S35–S39

293. Miller L (1990) Neuropsychodynamics of alcoholism and addiction: personality, psychopathology, and cognitive style. J Subst Abuse Treat 7:31–49

294. Ornstein TJ, Iddon JL, Baldacchino AM et al (2000) Profiles of cognitive dysfunction in chronic amphetamine and heroin abusers. Neuropsychopharmacology 23:113–126

295. Robinson TE, Berridge KC (2000) The psychology and neurobiology of addiction: an incentive-sensitization view. Addiction 95:S91–S117

296. Davis PE, Liddiard H, McMillan TM (2002) Neuropsychological deficits and opiate abuse. Drug Alcohol Depend 67:105–108

297. Damasio AR (2000) The feeling of what happens. Body, emotion and the making of consciousness. Vintage, London

298. De Vries TJ, Shippenberg TS (2002) Neural systems underlying opiate addiction. J Neurosci 22:3321–3325

299. Franken IHA (2003) Drug craving and addiction: integrating psychological and neuropsychopharmacological approaches. Prog Neuropsychopharmacol Biol Psychiatry 27:563–579

300. Bressler SL (2002) Understanding cognition through large-scale cortical networks. Curr Dir Psychol Sci 11:58–61

301. Edelman GM, Tononi G (2000) A universe of consciousness: how matter becomes imagination. Basic Books, New York

302. Davidson RJ (1998) Anterior electrophysiological asymmetries, emotion, and depression: conceptual and methodological conundrums. Psychophysiology 35:607–614

303. Rotenberg VS (2004) The peculiarity of the right-hemisphere function in depression: solving the paradoxes. Prog Neuropsychopharmacol Biol Psychiatry 28:1–13

304. Collins A, Loftus E (1975) A spreading-activation theory of semantic processing. Psychol Rev 82:407–428

305. LeDoux JE (2003) The self: clues from the brain. Ann N Y Acad Sci 1001:295–304

306. Ingram R (1984) Towards an information processing analysis of depression. Cogn Ther Res 8:443–478

307. Fossati P, LeB G, Ergis AM, Allilaire JF (2003) Qualitative analysis of verbal fluency in depression. Psychiatry Res 117:17–24

308. Moore BJ, Singh KD, Kinderman P, Bentall RP, Morriss RK, Roberts N (2001) Neuroanatomical basis of semantic processing in relation to personality descriptors of self: an fMRI study in healthy subjects. Poster HBM 2001, Brighton

309. Siegle GJ (1999) A neural network model of attention biases in depression. In: Reggia J, Ruppin E (eds) Disorders of brain, behavior, and cognition: the neurocomputational perspective. Elsevier, New York, pp 415–441

310. Tononi G, Edelman GM (2000) Schizophrenia and the mechanisms of conscious integration. Brain Res Rev 31:391–400

311. Rotarska-Jagiela A, van de Ven V, Oertel-Knöchel V, Uhlhaas PJ, Vogeley K, Linden DEJ (2010) Resting-state functional

network correlates of psychotic symptoms in schizophrenia. Schizophr Res 117:21–30

312. Joensen P (1986) Prevalence, incidence and classification of epilepsy in the Faroes. Acta Neurol Scand 76:150–155

313. Dawson KA (2004) Temporal organization of the brain: neurocognitive mechanisms and clinical implications. Brain Cogn 54:75–94

314. Insel TR (2010) Rethinking schizophrenia. Nature 468:187–193

315. Tost H, Bilek E, Meyer-Lindenberg A (2012) Brain connectivity in psychiatric imaging genetics. NeuroImage 62:2250–2260

316. Siegel JM, Manger PR, Nienhuis R, Fahringer HM, Pettigrew JD (1996) The echidna *Tachyglossus aculeatus* combines REM and non-REM aspects in a single sleep state: Implications for the evolution of sleep. J Neurosci 16:3500–3506

317. Zeplin H, Siegel J, Tobler I (2005) Mammalian sleep. In: Kryger MH, Roth T, Dement WC (eds) Principles and practice of sleep medicine. Elsevier Saunders, Philadelphia

318. Mignot E (2008) Why we sleep: the temporal organization of recovery. PLoS Biol 6(4): e106. doi:10.1371/journal.pbio.0060106

319. Karni A, Tanne D, Rubenstein BS, Askenasy JJ, Sagi D (1994) Dependence on REM sleep of overnight improvement of a perceptual skill. Science 265:679–682

320. Stickgold R (2005) Sleep-dependent memory consolidation. Nature 437:1272–1278

321. Tononi G, Cirelli C (2006) Sleep function and synaptic homeostasis. Sleep Med Rev 10:49–62

322. Noreika V, Valli K, Lahtela H, Revonsuo A (2009) Early-night serial awakenings as a new paradigm for studies on NREM dreaming. Int J Psychophysiol 74:14–18

323. Glass L (2001) Synchronization and rhythmic processes in physiology. Nature 410:277–284

324. Buchman TG (2002) The community of the self. Nature 420:246–251

325. Tirsch WS, Stude P, Scherb H, Keidel M (2004) Temporal order of nonlinear dynamics in human brain. Brain Res Brain Res Rev 45:79–95

326. Volkow ND, Wang GJ, Hitzemann R et al (1995) Depression of thalamic metabolism by lorazepam is associated with sleepiness. Neuropsychopharmacology 12:123–132

327. Ben-Ari Y (2002) Excitatory actions of GABA during development: the nature of the nurture. Nat Rev Neurosci 3:728–739

328. Maremmani I, Reisinger M (1995) Methadone treatment in Europe. European Methadone Association Forum, Oct 13, Phoenix, AZ, USA

329. Friston KJ (1997) Transients, metastability, and neuronal dynamics. NeuroImage 5:164–171

330. Vakorin VA, Lippe S, McIntosh AR (2011) Variability of brain signals processed locally transforms into higher connectivity with brain development. J Neurosci 31 (17):6405–6413

331. Fischer KW (1980) A theory of cognitive development: the control and construction of hierarchies of skills. Psychol Rev 87:477–531

332. Case R (1985) Intellectual development: birth to adulthood. Academic, New York

333. Case R (1987) The structure and process of intellectual development. Int J Psychol 22:571–607

334. Pascual-Leone J (1976) A view of cognition from a formalist's perspective. In: Riegel KF, Meacham J (eds) The developing individual in a changing world. Mouton, The Hague

335. van Geert P (1991) A dynamic systems model of cognitive and language growth. Psychol Rev 98:3–53

336. Thatcher RW, North DM, Biver CJ (2008) Development of cortical connections as measured by EEG coherence and phase delays. Hum Brain Mapp 29(12):1400–1415

337. van Beijsterveldt CE, Molenaar PC, de Geus EJ, Boomsma DI (1998) Genetic and environmental influences on EEG coherence. Behav Genet 28(6):443–453

338. van Baal GC, Boomsma DI, de Geus EJ (2001) Longitudinal genetic analysis of EEG coherence in young twins. Behav Genet 31 (6):637–651

339. Thatcher RW, North DM, Biver CJ (2009) Self-organized criticality and the development of EEG phase reset. Hum Brain Mapp 30 (2):553–574

340. van Baal C (1997) A genetic perspective on the developing brain. Ph.D. Dissertation, VRIJE University, The Netherlands Organization for Scientific Research

341. Eysenck HJ, Eysenck SBG (1976) Psychoticism as a dimension of personality. Hodder & Stoughton, London

342. Diamond S (1957) Personality and temperament. Harper, New York

343. Bekhtereva NP (1978) The neurophysiological aspects of human mental activity, 2nd edn. Oxford University Press, New York

344. Ramos-Loyo J, Gonzalez-Garrido AA, Amezcua C, Guevara MA (2004) Relationship

between resting alpha activity and the ERPs obtained during a highly demanding selective attention task. Int J Psychophysiol 54:251–262

345. Kounios J, Fleck JI, Green DL et al (2008) The origins of insight in resting-state brain activity. Neuropsychologia 46:281–291

346. Zeki S (2005) The Ferrier Lecture 1995 behind the seen: the functional specialization of the brain in space and time. Philos Trans R Soc Lond B Biol Sci 360:1145–1183

347. Pockett S (2000) The nature of consciousness: a hypothesis. Writers Club Press, Lincoln, NE

348. Metzinger T (2003) Being no one. The self-model theory of subjectivity. MIT Press, Cambridge

349. Metzinger T (2007) The self-model theory of subjectivity (SMT). Scholarpedia 2:4174

350. Kallio S, Revonsuo A (2003) Hypnotic phenomena and altered states of consciousness: a multilevel framework of description and explanation. Contemp Hypn 20:111–164

351. Hilgard ER (1986) Divided consciousness: multiple controls of human thought and action, revisedth edn. Wiley, New York

352. Gruzelier JH (2000) Redefining hypnosis: theory, methods and integration. Contemp Hypn 17:51–70

353. Von Kirchenheim C, Persinger M (1991) Time distortion: a comparison of hypnotic induction and progressive relaxation procedures. Int J Clin Exp Hypn 39:63–66

354. Naish P (2001) Hypnotic time perception: busy beaver or tardy timekeeper. Contemp Hypn 18:87–99

355. Dietrich A (2003) Functional neuroanatomy of altered states of consciousness: the transient hypofrontality hypothesis. Conscious Cogn 12:231–256

356. Chalmers D (1995) Facing up to the problem of consciousness. J Conscious Stud 2:200–219

357. Angel L (1989) How to build a conscious machine. Westview Press, Boulder, CO

358. Holland O (2003) Editorial introduction. J Conscious Stud 10:1–6

359. Minsky M (1991) Conscious machines. Machinery of consciousness. National Research Council of Canada, Montreal

360. Minsky M (2006) The emotion machine: commonsense thinking, artificial intelligence, and the future of the human mind. Simon and Schuster, New York

361. McCarthy J (1995) Making robot conscious of their mental states. In: Muggleton S (ed) Machine intelligence. Oxford University Press, Oxford

362. Aleksander I (2001) The self 'out there'. Nature 413:23

363. Holland O (2003) Machine consciousness. Imprint Academic, Exeter, UK

364. Adami C (2006) What do robots dreams of? Science 314:1093–1094

365. Chella A, Manzotti R (2007) Artificial consciousness. Imprint Academic, Exeter, UK

366. Fingelkurts AA, Fingelkurts AA, Neves CFH (2009) Brain and mind operational architectonics and man-made "machine" consciousness. Cogn Process 10:105–111

367. Indiveri G, Chicca E, Douglas R (2006) A VLSI array of low-power spiking neurons and bistable synapses with spike-timing dependent plasticity. IEEE Trans Neural Network 17:211–221

368. Seth AK (2009) The strength of weak artificial consciousness. Int J Mach Conscious 1:71–82

369. Fingelkurts AA, Fingelkurts AA, Neves CFH (2012) "Machine" consciousness and "artificial" thought: an operational architectonics model guided approach. Brain Res 1428:80–92

370. Koch C, Tononi G (2008) Can machines be conscious? IEEE Spectr 6:47–51

371. Fingelkurts AA, Fingelkurts AA, Neves CFH (2013) Consciousness as a phenomenon in the operational architectonics of brain organization: Criticality and self-organization considerations. Chaos Solitons Fractals 55:13–31

Neuromethods (2015) 91: 61–86
DOI 10.1007/7657_2014_73
© Springer Science+Business Media New York 2014
Published online: 6 September 2014

Clinical Electroencephalography in the Diagnosis and Management of Epilepsy

Nicholas-Tiberio Economou and Andreas V. Alexopoulos

Abstract

This chapter deals with the clinical electroencephalography (EEG) and its application on epilepsy. EEG, its origin, its technical aspects comprising also the electrode placement, and the related montages are addressed. Then, the clinical applications of EEG and mainly its use in epilepsy are reviewed, while cardinal EEG findings in interictal and ictal recordings, such as the epileptiform discharges and the ictal patterns in seizures, respectively, are mentioned. The rest of the chapter deals with the epilepsies, their classification as focal or generalized with their sub-divisions, and the related clinical manifestation. Then, EEG and epilepsy in specific populations, such as neonates and infants, are addressed, and finally, main pitfalls in the interpretation of EEG are also reviewed.

Key words Electroencephalography (EEG), Technical aspects, Clinical application, EEG in epilepsy, The epilepsies

1 Introduction

Electroencephalography (EEG) constitutes the single most important and frequently performed neurodiagnostic study in patients with seizures and related paroxysmal events. It is essential to state at the outset that epilepsy remains a clinical diagnosis, and therefore the results of any EEG study need to be carefully examined individually and under the prism of each patient's clinical presentation. The descriptions of the seizure by witnesses (observer-report) when available, and by the patient (self-report) when possible, are key elements of the diagnostic evaluation. Preictal signs and behaviors, presence of aura, and ictal and postictal signs and symptoms are cardinal functional features. These features can be further elucidated with additional laboratory investigations, the majority of which rely on the use of EEG, e.g., outpatient or ambulatory EEG recordings and inpatient continuous bedside EEG or video-EEG monitoring. Despite tremendous progress in structural and functional neuroimaging techniques and improved spatial and temporal resolution of functional neurodiagnostic studies, EEG remains the paramount tool to assess ictal and interictal electrical brain signals.

2   History

Hans Berger (1873–1941), a professor of psychiatry in Jena, was the first to record the human EEG in 1924. His discovery that the brain develops a low-level subaudio-frequency electrical activity gave birth to a new neurophysiological specialty [1]. Berger coined the term "electroencephalography" and described for the first time the alpha wave (occipital dominant) background rhythm and its relationship to eye closure. Prior to this discovery quantification or description of the ongoing brain activity had not been possible. Later in the 1930s, Gibbs, Lennox, and Jasper defined the patterns of generalized spike-and-wave activity and focal spikes or sharp waves as markers of epilepsy [2].

As time went by the techniques of EEG recording became more sophisticated with the addition of more channels and emergence of the first ambulatory EEG systems in the 1970s [3]. The development of video technology allowed for the patient's peri-ictal behavior to be captured and time-locked with the EEG and facilitated the creation and steady increase of long-term video-EEG monitoring units. Multichannel intracranial long-term EEG recordings appeared together with the advancement of the epilepsy surgery.

3   Technical Aspects: Electrode Placement

Although there is no single best method for recording EEGs under all circumstances, the various guidelines and consensus statements that have been published and revised over the years [4, 5] have largely been adopted by most EEG laboratories. The current guideline of the American Clinical Neurophysiology Society [5] on the minimal technical requirements for performing clinical scalp EEG recordings suggests that 16 electrodes (rather than only 8 electrodes as had been stated in the previously published guidelines [6]) are required at a minimum to adequately analyze the areas producing most of the normal and abnormal EEG patterns [5]. Additional electrodes (including closely spaced "10–10" electrodes) are often needed for monitoring other physiologic activities and resolving subtle epileptiform patterns.

Appropriate alternating current (AC) wiring and adequate grounding of all EEG equipment to a common point are mandatory. For standard recordings, the low-cut frequency filter should be no higher than 1 Hz (corresponding to a time constant of at least 0.16 s), while the high-cut filter must be at least at 70 Hz. It is important to emphasize that a setting lower than 70 Hz for the high-cut frequency filters can distort or attenuate spikes and other pathologic discharges into unrecognizable forms and can cause muscle artifact to resemble spikes.

Occasionally a 60 Hz filter (also called notch filter) is used when other measures to eliminate 60 Hz interference have failed, but this can also distort or attenuate important EEG features such as spikes. For many years EEGs were obtained with analog recordings that were then charted on paper. The available amplifiers and the frequency response of the EEG pens hampered our ability to record higher frequency activity [7]. With the advent of digital recordings we are now able to display and explore much higher frequencies, when the record has been digitized at a high enough rates (at least twice as high as the frequency of interest).

Electrode impedances should not exceed 5 kΩ and should be checked as a routine prerecording procedure and rechecked during the recording, whenever there is concern that an artifactual pattern has appeared as the study progresses. Appropriate calibrations should be made at the beginning and end of every EEG recording. The calibration procedure is an integral part of every EEG recording, provides a scaling factor for the interpreter, and tests the EEG machine for sensitivity, high- and low-frequency response, and noise level among other parameters. The sensitivity of the EEG for routine recordings must be within 5–10 µV/mm of pen deflection, while for young children sensitivity is usually decreased to 10–15 µV/mm due to the higher voltage of EEG activity in this age [8]. Sensitivity is defined as the ratio of input voltage to pen deflection. It is expressed in microvolts per millimeter (µV/mm). A commonly used sensitivity is 7 µV/mm, which, for a calibration signal of 50 µV, results in a deflection of 7.1 mm.

For routine recordings, a paper speed of 3 cm/s or digital display of 10 s per page is usually utilized in the US. A lower paper speed of 1.5 cm/s or more condensed digital display of 20 s/page is sometimes used for EEG recordings in newborns or in other special situations. Silver–silver chloride or gold disk electrodes held on by collodion are mostly recommended, but other electrode materials and pastes may be used effectively especially with contemporary amplifiers which allow for high input impedances.

Routine EEG consists of at least 20 min of technically satisfactory recording. In the usual 20-min recording specific tasks should be included. Thus, it must be ensured that the recording contains periods of eye opening/eye closure. Moreover, activation procedures, which might elicit the appearance of pathologic electroclinical features (such as spikes or ictal patterns), should be routinely performed save for medical or other contraindications. These maneuvers include hyperventilation, which is performed for a minimum of 3 min, and intermittent photic stimulation. Sleep deprivation prior to the recording is another activation technique that increases the probability of recording the states of drowsiness and/or sleep, which are critical segments of time especially in patients with suspected or known seizure disorders [6].

It is the responsibility of the EEG technologist to specify on the EEG recording the patient's level of consciousness (awake, drowsy, sleeping, lethargic, stuporous, or comatose) and any related changes. In patients with depressed level of consciousness or those showing invariant EEG patterns of any kind the technologist should systematically apply visual, auditory, and somatosensory stimuli and document the patient's responses [5]. Careful observation of the patient with frequent notations is essential particularly when unusual waveforms are observed in the tracing or paroxysmal behavioral changes occur during the period of the recording. Every effort should be made to obtain a technically satisfactory record. Production of a clinical EEG record with lost or inaccurate information due to technical oversights is poor medical practice.

The 10-20 system or International 10-20 system is an internationally recognized method that describes the application and location of EEG electrodes on the scalp. This largely accepted method of electrode placements was proposed by Jasper in 1958 to ensure standardized reproducibility within and between subjects [9]. The designation of each electrode takes into account the underlying area of cerebral cortex that is covered by the particular electrode. Odd or even numbers (i.e., F3, F4) refer to left and right hemisphere, respectively. The measurement technique is based on standard landmarks on the skull (namely, the nasion, inion and the left and right preauricular points) and allows for measurements to be proportional to the size and shape of the skull. The numbers "10" and "20" refer to the fact that the actual distances between adjacent electrodes are either 10 or 20 % of the total front-to-back (nasion to inion) or right-to-left distances of the skull [10].

The first measurement is in the anterior-posterior plane through the vertex (taken from the nasion to the inion). This measurement divides the skull from anterior to posterior into five separate regions, expressed as Fp, F, C, P, and O with the letters representing the expected location of the underlying frontopolar, frontal, central, parietal, and occipital areas. The first mark (Fp) is placed above the nasion at 10 % of the nasion–inion distance. The second (F), third (C), and fourth (P) marks—corresponding to the position of the three midline electrodes Fz, Cz, and Pz, respectively—are placed at 20 % intervals of the total nasion–inion distance. Accordingly, the midline O mark would be located above the inion at 10 % of the total distance.

Regarding the lateral measurements across the central coronal plan, 10 % of the entire distance is taken for the temporal (T) point (positions of T7 and T8 electrodes) up from the preauricular point on either side. The corresponding left and right central points (C3 and C4 electrodes) are then marked 20 % of the distance above the temporal points. As described above, the center of central-coronal line corresponds to the C vertex (Cz) location. The anterior-posterior line of electrodes over the temporal lobe, frontal to occipital, is assessed by measuring the distance between the Fp midline point through the T position of the central-coronal line and back to the mid-occipital spot. The left and right Fp electrode position (Fp1 and Fp2) is marked 10 % of the distance from the midline in front, while the occipital electrode position (O1 and O2, respectively) is marked 10 % of the distance from the midline (O position) in the back. The inferior frontal (F7 and F8) and posterior temporal (P7 and P8) positions then fall 20 % of the distance from the left and right Fp and occipital electrodes, respectively. Finally, the midfrontal and midparietal electrodes (F3, F4 and P3, P4, respectively) are placed along the frontal and parietal coronal lines, respectively, equidistant between the temporal and the midline. The 10-20 system is the only one recommended by the International Federation of Clinical Neurophysiology (IFCN). All 21 electrodes and placements of the 10-20 system as recommended by IFCN should be used [9, 11].

EEG montages may vary among laboratories. It is very desirable for all EEG laboratories to use at least some comparable montages and include the classical montages (longitudinal bipolar, transverse bipolar, and a referential montage, for example an ipsilateral ear reference montage) to facilitate communication and comparisons. The ACNS recommends that EEG montages should be designed in conformity with Guideline 6: A Proposal for Standard Montages to Be Used in Clinical Electroencephalography [12]. Both bipolar and referential montages should be used. In general the longitudinal bipolar (LB or "double banana") and the referential (R) montages consist of a series of leads grouped in anatomical proximity and extending sequentially across the head from the left to the right side. In this fashion hemispheric differences can be readily appreciated and blocks of homologous derivations involving homologous brain regions can be compared.

4 Clinical Applications

The main indication for performing an EEG is clinical suspicion of epilepsy, although a single routine EEG study may be normal in up to 50 % of individuals with epilepsy [4, 13]. Repeat or prolonged EEG studies with the use of appropriate activation maneuvers (such as hyperventilation, sleep deprivation, and photic stimulation) may yield epileptiform abnormalities in up to 80–90 % of individuals with epilepsy [7, 14]. Therefore, a normal EEG study by itself can never be used in isolation to rule out the diagnosis of epilepsy.

The purpose of EEG recordings is to detect interictal epileptiform abnormalities, which include spikes and sharp waves. When such abnormalities are present, the EEG can help regionalize the brain areas that are involved to produce the observed interictal and/or ictal patterns on scalp electrodes. The ability of the scalp

EEG to detect interictal epileptic discharges depends on the extent of the irritative zone, the location and proximity of the generator in relation to the scalp, and the orientation of the dipole [15]. In the presence of epileptiform activity, the EEG recording can also help determine the type of seizure or epilepsy syndrome. The two main EEG hallmarks of epilepsy are the focal interictal spike or sharp wave in focal epilepsies and the bifrontocentral spike–wave discharges in the generalized epilepsies.

EEGs may also be used in the assessment of acute or chronic encephalopathies (e.g., secondary to metabolic derangements, infectious conditions, or degenerative disorders) and focal brain lesions (such as cerebral infarction, hemorrhage, neoplasms). In pediatrics the EEG is useful in assessing the level of maturation of the brain. Lastly, the only clear indication for use of the EEG as a tool to monitor the effect of treatment with antiepileptic drugs is in patients with absence epilepsy [4].

Quantitative electroencephalography (qEEG) allows for a compressed view of several hours of EEG and has been used by some investigators to enhance the clinical diagnosis and treatment planning provided to individuals with mild traumatic brain injury and postconcussive symptoms [16]. Finally continuous bedside EEG monitoring and qEEG in the intensive care unit (ICU) setting have been shown to significantly increase the detection of nonconvulsive seizures (NCS) and nonconvulsive status epilepticus (NCSE) in critically ill patients, who typically have subtle if any clinical manifestations [17].

Fast sampling rates (up to 2,000 Hz or higher) and wideband intracranial EEG recordings have allowed investigators to explore high-frequency oscillations (HFOs) in the 100–600 Hz frequency range, which are more clearly defined on intracranial recordings but have also been detected with scalp electrodes. Some investigators have reported that HFOs represent a promising neurophysiological biomarker of epileptogenic foci [18], but their clinical significance is not well established at this time, and their discussion is beyond the scope of this chapter.

5   The EEG in Epilepsy

5.1   Interictal Recording: Epileptiform Discharges

Interictal epileptiform discharges (IEDs)—which include spikes and sharp waves (SWs)—are extremely useful in the diagnosis of epilepsy and in the classification of the epileptic condition, because they occur much more frequently than clinical seizures during routine EEG recordings. In the appropriate clinical setting and depending on the patient's age and type of epilepsy syndrome, up to 98 % of patients whose EEG demonstrates clear epileptiform discharges may have a history of epilepsy [19]. Moreover, by defining the "irritative zone" (the area of cortical hyperexcitability responsible for the generation of localized epileptiform activity during the interictal state) these epileptiform abnormalities can be extremely helpful in regionalizing the epileptogenic focus, which is a key step in patients with medically intractable focal epilepsy who are potential candidates for epilepsy surgery [20, 21]. Thus, by combining the patient's clinical and semeiological history with the observed IEDs the clinician is often able to correctly identify and classify the electroclinical syndrome [21]. Electroclinical correlations are essential. It is important, for example, to differentiate IEDs in the setting of benign rolandic epilepsy from IEDs that involve a similar brain region but occur in a different subset of patients who may manifest difficult-to-control focal motor seizures. This differentiation has clinical implications, which may include considerations for or against resective epilepsy surgery.

The presence or the absence of IEDs is influenced by the duration of the EEG recording and the need to sample EEG activities at different periods of time. Although a single routine EEG study in a person with epilepsy may only capture IEDs in 50 % of cases, the yield increases with repeat EEG recordings performed at different times and plateaus around 92 % by the fourth routine EEG study [14]. Similarly, recordings performed in epilepsy monitoring units (EMU) increase the probability of capturing IEDs over a period of several days although approximately 10–19 % of patients with epilepsy do not demonstrate IEDs on scalp EEG regardless of the number and duration of recordings [13]. The frequency of IEDs can vary for many reasons. Their frequency may increase near the seizure onset [22, 23], and/or after a seizure [24, 25], during NREM sleep [26], or following withdrawal of antiepileptic medications [25, 27] and activation maneuvers (i.e., hyperventilation). Lastly, IEDs may (rarely) be observed in people with no seizure history. Even though the presence of IEDs may be an incidental finding an increased epileptogenic tendency cannot be excluded, and avoidance of known seizure precipitants may be advised in some cases (such as individuals with abnormal photoparoxysmal responses during intermittent photic stimulation).

Typical spikes and sharp waves should be easily distinguishable from the background EEG activity [19, 28]. However, the experienced electroencephalographer should also be familiar with numerous "sharp transients" such as vertex or lambda waves which are features of normal EEG studies or wicket spikes and other variants of benign nature that should not be confused with epileptiform patterns [19]. Spikes are epileptiform discharges of 40–80 ms of duration, while sharp waves have a duration of 80–200 ms (both are measured at the midline between the baseline and peak of the spike or the sharp wave). As a rule the main component of epileptiform discharges has a negative polarity of scalp EEG. There is no difference between spikes and sharp waves with regard to the diagnostic value of these abnormalities [19, 28]. Both are sharply contoured transients that stand out from the background (their amplitude is usually at least 30 % higher than the amplitude of the background EEG) and are usually followed by a slow wave.

Spike and SW discharges are more commonly focal in terms of distribution but can also be generalized and appear either as independent waveforms or as components of complexes that encompass various different waves [20]. The background often shows slowing in the same regions that are implicated by the epileptiform discharges or may have periods of slowing with a more diffuse distribution. Focal and/or diffuse slowing is seen mostly in patients with focal or symptomatic generalized epilepsies, while epileptiform discharges in idiopathic generalized epilepsy are usually accompanied by an otherwise normal background activity [29]. The typical epileptiform discharges of IGE are bisynchronous, symmetrical, generalized, and fairly regular spike–wave discharges, polyspikes, and polyspike–wave complexes [30]. Other abnormalities such as periodic lateralized epileptiform discharges have important clinical implications and may be seen in the setting of acute or subacute structural lesions, such as stroke, necrotizing focal encephalitis, and rapidly growing neoplasms, or may represent a transient postictal finding following a seizure in some patients with focal epilepsy [28].

5.2 Ictal Recording: Ictal Patterns

Focal and generalized seizures usually manifest with distinctly different ictal patterns. In contrast to the paroxysmal epileptiform discharges occurring during the interictal state, ictal patterns are characterized by an EEG onset, a distinct evolution, and an end (EEG offset). The EEG onset in focal epilepsies may be nonspecific manifesting with either focal or generalized desynchronization, low-voltage fast activity, or irregular focal or bilateral delta activity. The evolution of the seizure is often the clearest part of the ictal EEG, which may demonstrate progressive changes from lower amplitude, faster activity to activities of higher amplitude, slower frequencies, and wider distribution. Postictal slowing may be seen at the end of the electrographic seizure marking the end of the ictal event and the transition to the postictal period. Postictal slowing, when present, is definitely less rhythmic than the ictal pattern itself.

Some simple partial seizures, such as auras or focal motor seizures, may be accompanied by subtle EEG changes without clear evolution of amplitude or frequency. Importantly, up to 70 % of simple partial seizures may not be associated with a definite EEG correlate. Therefore, more often than not, focal seizures that only activate a small or a deep area of cortex have no EEG signature on scalp recordings [7]. It is critical to understand that the absence of scalp EEG changes during such paroxysmal events cannot rule out an epileptic etiology for these events. On the other hand, a clear ictal pattern is always present in secondarily generalized tonic-clonic (GTC) seizures. In cases characterized by rapid spread of the ictal discharge, a focal onset may not be readily apparent. However, significant postictal slowing may be seen in some of these cases following secondary generalization. The EEG pattern itself during the period of a secondarily GTC seizure is similar to that seen with GTC seizures in primary generalized epilepsy. The usual pattern is rapid low-amplitude spiking evolving to a slower spike–slow wave discharges. These patterns correlate with the observed tonic and clonic activity, respectively [29, 31].

Other distinct ictal EEG patterns may be observed in the generalized epilepsies. Generalized myoclonic seizures are associated with a high-amplitude, generalized, polyspike activity with frontocentral accentuation occurring in isolation or repeating itself at a frequency of 10–16 Hz. Ictal discharges are sometimes preceded by irregular 2–5 Hz generalized spike–wave complexes and are followed by irregular 1–3 Hz slow waves. In myoclonic-astatic seizures, simultaneous myoclonic jerks of the flexor and extensor muscles of the body are associated with fast (>2.5–3 Hz), generalized, polyspike–wave discharges or spike–wave complexes. Typical absence seizures exhibit a characteristic monomorphic ictal rhythm of abrupt onset and offset, which consists of bilateral, symmetrical, and synchronous 3 Hz spike–wave discharges on a normal background. In epilepsy with myoclonic absences the ictal EEG shows regular, rhythmic, bisynchronous, 3 Hz spike–wave discharges, which are also commonly associated with an abrupt onset and end. Admixed polyspikes may also be seen occasionally during the ictal discharge in these particular seizures [30, 32].

6 The Epilepsies

6.1 Focal Epilepsies

The definition of focal seizures according to International League Against Epilepsy (ILAE) is "a seizure whose initial semeiology indicates, or is consistent with, initial activation of only part of one cerebral hemisphere" [33]. The ictal symptomatology and the EEG features vary among the different locations of focal seizure onset.

6.1.1 Mesial Temporal Epilepsy

The temporal lobe is the most epileptogenic region of the human brain. Within the temporal lobe the three-layered archicortex of the hippocampal formation is the central component of the highly epileptogenic electrophysiologic network that includes the amygdalohippocampal complex as well as the subiculum and entorhinal cortex [34]. The most common cause of mesial temporal epilepsy (mTLE) is hippocampal sclerosis (HS) which is oftentimes associated with an initial precipitating injury, such as complex febrile seizures, perinatal insults, head trauma, and infection of the central nervous system. Not infrequently HS is a bilateral disease [35], with bilateral hippocampal neuropathological abnormalities present in autopsy tissue from patients with a clinical diagnosis of temporal lobe seizures [36].

Clinically, the most frequent seizures in mTLE are the so-called complex partial seizures with or without automatisms. 80 % of patients report auras or a history of auras, which are mainly associated with experiential (i.e., deja vu, fear, anxiety) and viscerosensory symptoms (i.e., nausea, or indescribable rising sensation from the epigastrium) [37]. Prolonged seizures and status epilepticus are uncommon in most patients with mTLE.

Typical complex partial seizures are characterized by a fixed motionless stare and limited if any motor manifestations with the exception of oroalimentary or distal manual automatisms. These seizures are commonly accompanied by autonomic manifestations such as tachycardia, hyperventilation, and piloerection. Lateralizing ictal signs include contralateral hand dystonia, which is usually accompanied by ipsilateral automatisms. Ictal speech, ictal emesis, and automatisms with preserved awareness suggest nondominant lateralization, whereas postictal aphasia implicates the dominant temporal lobe. Prominent motor phenomena such as tonic manifestations or contralateral clonic jerking and contralateral postictal paresis (Todd's phenomenon) suggest propagation outside the mesial temporal lobe to involve the ipsilateral motor neocortex [37, 38].

More than 90 % of patients with mTLE exhibit anterior temporal spikes with maximal amplitude on scalp EEG over the F7, F8 electrodes (in the standard 10-20 system described earlier in this chapter) or over additional anterior temporal surface electrodes [such as the commonly added FT9, FT10 electrodes, which are placed in the FT position (between the F and T positions) and inferior—by 10 % of the measured distance—to the standard infrasylvian chain of temporal electrodes, i.e., below the FT7 and FT8 marks] and over the so-called basal temporal electrodes, which may be utilized during video-EEG monitoring in some centers [such as sphenoidal (Sp1 and Sp2) or less commonly oropharyngeal electrodes] [31].

The sphenoidal electrode consists of a flexible wire with an uninsulated tip and is placed near the sphenoidal wing over a cannula that is first inserted between the zygoma and mandibular notch traversing the temporal and masseter muscles. The guiding cannula is removed after advancing the wire to a depth of 4–5 cm. This is a simple bedside procedure, although some authors favor precise placement under fluoroscopic guidance. Sphenoidal electrodes are mainly of value for distinguishing between mesial and lateral temporal epileptogenic sources. Whether sphenoidal electrodes should be used in the presurgical evaluation of selected patients with medically intractable focal epilepsy remains a matter of debate [39, 40].

Anterior temporal sharp waves may be seen bilaterally in approximately one-third of patients with unilateral mTLE particularly during sleep. Strongly lateralized (>90 %) or exclusively unilateral anterior temporal IEDs are predictive of the side of seizure onset in patients with unilateral mTLE. IEDs that are maximum over the mid-temporal (T7 and T8) electrodes may be seen in some patients with documented mTLE, but such finding may also signify the presence of an extramesial temporal generator or a more extensive irritative zone. Subtle or no definite EEG changes may be seen during auras or at the time of the initial behavioral change. Therefore, the clinical onset of the seizure precedes the EEG onset in these patients. Ictal (as well as interictal) activities arising from the mesial temporal lobe will not produce visible changes on scalp EEG, if they remain confined to the mesial temporal structures, unless or until the epileptiform activities have also recruited the inferolateral temporal neocortex. As the seizure evolves, a regional anterior temporal or lateralized ictal pattern, which consists of rhythmical theta or alpha activity (typically 5–9 Hz), is typically seen within 10–40 s after clinical onset in up to 80 % of patients with mTLE [41]. If present this activity, which is considered to be a hallmark of mTLE, correctly lateralizes the seizure focus in up to 95 % of patients [42].

The postictal record should be examined for the presence of lateralized focal postictal slowing that involves the temporal region. Postictal slowing may be seen in approximately 70 % of seizures and has strong lateralizing value offering correct lateralization in up to 90 % of cases [31, 42]. The probability of a favorable surgical outcome following anterior temporal lobectomy is highest in patients, who exhibit exclusively unilateral anterior temporal IEDs combined with scalp ictal EEGs that remain regionalized without contralateral propagation.

In patients with suspected pharmacoresistant mTLE, in whom the scalp EEG provides discordant information and/or non-localizable ictal patterns, further evaluation with intracranial recordings (preferably using depth electrodes) may be necessary. As expected, intracranial electrodes detect a higher number of localized discharges and focal hypersynchronous activities as compared to scalp EEG [43]. In fact, most patients with mTLE (>60 %) respond to antiepileptic medications and enjoy long periods of sustained seizure freedom. However, a significant number of patients, and in particular the majority of patients (up to 60–90 % in some studies) with the substrate of HS, continue to experience recurrent seizures despite multiple trials of antiepileptic medications at high doses.

Mesial TLE is the prototype surgically remediable epilepsy syndrome. Therefore, it is imperative for these patients to be carefully identified and referred to a comprehensive epilepsy center early on in the course of the intractability. Standard anterior temporal lobectomy (ATL) is the most common surgical procedure performed in approximately two-thirds of patients with medically intractable focal epilepsies [42, 43]. Indisputable class I evidence supports the effectiveness and superiority of ATL, as compared to continued medical therapy, in patients with intractable mTLE [44].

6.1.2 Neocortical Temporal Lobe Epilepsy

According to the current epilepsy classification the entity of neocortical temporal lobe epilepsy (nTLE), together with that of mTLE, is classified under the diagnosis of TLE. nTLE compared to mTLE is a rather heterogeneous entity as far as the etiologies and the electroclinical features are concerned. nTLE is clearly less common and usually presents later in life. As compared to mTLE, individuals with nTLE are usually older by an average of 5–10 years when they first start having seizures. The causes of nTLE are variable and include the so-called long-term epilepsy-associated tumors such as gangliogliomas and dysembryoplastic neuroepithelial tumors along with vascular malformations (mainly caveromas), focal cortical dysplasia, and trauma.

Motionless staring and unresponsiveness may represent the first objective clinical manifestations in nTLE and are often followed by contralateral clonic jerks. Patients with nTLE are more likely to report auditory, vertiginous, or experiential auras. Other clinical symptoms suggestive of nTLE include cephalic/indescribable sensations, ictal speech and vocalizations, whole-body movements, and rapid onset of version. nTLE seizures are generally shorter in duration. Moreover, evolution to secondarily GTC seizures is seen more frequently in nTLE as compared to mTLE patients [45–47].

Regarding scalp EEG features, the presence of interictal rhythmic slowing over one temporal region may raise the question of nTLE. Ipsilateral interictal epileptiform abnormalities are seen in up to 80–90 % of nTLE patients [48, 49]. However, there is little evidence to support the use of interictal EEG alone in differentiating nTLE from mTLE. The ictal pattern in NTLE usually consists of an irregular, polymorphic delta slowing with lateralized rather than regionalized (temporal) distribution at seizure onset [46]. This can be preceded by repetitive spiking or followed by a more rhythmic theta or alpha activity [50].

Invasive recordings with stereotactically placed depth electrodes allow for the exploration of larger epileptogenic networks within the temporal lobe and surrounding perisylvian and insular cortex [51]. Approximately 20 % of TLE surgeries involve neocortical resections sparing the mesial temporal structures. Overall, the postoperative seizure outcome does not differ significantly in nTLE as compared to mTLE with the exception of patients with incomplete resection. The prognosis of surgical resection in patients with medically intractable nTLE depends on the underlying cause: developmental tumors and cavernomas have a better outcome than focal cortical dysplasia, trauma, or nonlesional TLE [49].

6.1.3 Frontal Lobe
Epilepsy

The frontal lobe occupies almost one-third of the human brain. Thus, it is not surprising that the frontal lobe is the second most common origin of focal seizures. Approximately 20 % of patients with pharmacoresistant focal epilepsy referred to tertiary epilepsy centers have frontal lobe epilepsy (FLE) [52]. The prevalence in nonsurgical cohorts is probably higher, given that samples of patients with uncontrolled epilepsy evaluated for epilepsy surgery is highly selected. Although seizure semeiology can be nonspecific or misleading, typical FLE seizures are associated with one or more of the following signs: unilateral clonic or focal tonic phenomena, early head version or asymmetric tonic posturing, motor agitation, and hyperkinetic behavior—oftentimes consciousness is preserved [53]. Careful analysis of ictal semeiology is of particular importance in FLE given the limitations of ictal EEG, especially when seizures are brief and/or arise from a restricted region in the frontal lobe (in which case scalp EEG may not show any appreciable changes) or characterized by prominent hyperkinetic manifestations (resulting in EEG recordings that are obscured by copious movement and electrode artifacts). Interictal EEG in FLE may be less informative as compared to TLE given the anatomical peculiarities of the frontal lobe.

EEG and clinical features obviously depend on the sub-compartment(s) of the frontal lobe involved in seizure generation and subsequent propagation. The following subregions are considered briefly in this review.

Firstly, FLE arising from the dorsolateral frontal lobe (dorsolateral FLE) can be further subdivided into central, premotor, and prefrontal lobe epilepsy [54]. The central (perirolandic) area comprises the primary motor and primary sensory cortex. Semeiology consists of focal somatosensory auras (unilateral paraesthesias with discrete somatotopic representation restricted to the face/mouth or hand) contralateral to the epileptic perirolandic cortex. Motor phenomena are mostly clonic or myoclonic, rather than tonic, and usually involve distal segments of the contralateral extremities or hemiface. The premotor cortex functionally includes the secondary motor area and the frontal eye field as well as the Broca's area within the language-dominant hemisphere. Ictal semeiology includes early versive seizures (forced and unnatural lateral version of the eyes, head, and trunk) often followed by other motor manifestations. Aphasic seizures may occur if Broca's language area is involved. Finally, seizures of the prefrontal cortex, which is implicated in behavioral, memory, and learning processes, are commonly associated with hypermotor manifestations, i.e., complex movements involving the trunk and proximal limb segments.

Secondly, mesial frontal epilepsy arises from the interhemispheric surface of each frontal lobe, which comprises the primary sensory and motor cortex of the lower extremity, the supplementary sensorimotor area (SSMA a.k.a. supplementary motor area,

SMA), the anterior cingulate cortex, and the mesial prefrontal cortex. Seizures arising from (or involving) the SSMA are characterized by bilateral, usually asymmetric, proximal tonic posturing. Some seizures may exhibit "fencing posture" and/or "sign of four" with tonic extension of one elbow (usually but not always contralateral to the epileptic SSMA) combined with tonic flexion of the other elbow. Indeed, electrical cortical stimulation of the SSMA elicits predominantly tonic contractions involving the proximal limbs or combined proximal and distal limb portions. These responses mainly involve the lower and/or upper extremities bilaterally but can be restricted to the contralateral or even the ipsilateral extremities. Moreover, SSMA seizures may be associated with bilateral or mostly contralateral sensory phenomena, such as numbness, tingling, or sensation of pressure that is not well localized (non-discrete somatosensory auras) [52]. Other seizure types are observed in mesial FLE depending on the sub-lobar region activated by the epileptic discharge, such as focal clonic or myoclonic seizures involving the contralateral lower extremity (posterior mesial frontal lobe) or hypermotor seizures (anteromesial frontal lobe). Dialeptic seizures may also be seen in patients with mesial frontal and/or frontopolar epilepsies. Dialeptic seizures are defined as episodes of loss of consciousness characterized by paucity of (or minimal) motor manifestations, behavioral arrest, and absence of reactivity (or only limited reactivity) to external stimuli [55]. Dialeptic seizures are more common in TLE but do occur infrequently in patients with anterior FLE ("frontal absences").

Finally, basal frontal epilepsy arises from the mesial or the lateral orbitofrontal region. Seizure semeiology is variable and includes olfactory auras (with involvement of the gyrus rectus), autonomic manifestations, hyperkinetic phenomena, or seizures with typical TLE semeiology suggestive of ictal propagation to the intimately interconnected ipsilateral temporal lobe [15].

Interictal EEGs are often normal; in fact IEDs are found only in 65–70 % of patients, despite repeated recordings. Despite their focal onset IEDs in FLE may appear as bilateral or midline discharges. Bilateral spikes or spike-and-wave discharges may exhibit consistent amplitude asymmetry and/or brief (oftentimes imperceptible) lag between the frontal lobe of origin and the contralateral hemisphere. These abnormalities are thought to represent secondary bilateral synchrony, as a result of rapid interhemispheric propagation, rather than truly generalized discharges that have a synchronous bihemispheric onset.

As stated above, the prominent physical activity associated with many frontal lobe seizures results in excessive myogenic and movement artifact to further complicate EEG interpretation. Subtle differences between the two hemispheres may be difficult to appreciate under these circumstances rendering the ictal EEG non-localizable, "pseudo-generalized," falsely lateralized, or obscured.

Due to these limitations, scalp EEG localization is often problematic in patients with FLE. False localization may occur especially to the ipsilateral temporal lobe. Ictal onset and evolution may appear generalized as a result of secondary bilateral synchrony, further limiting our confidence in lateralizing the side of seizure onset based on scalp EEG alone. Moreover seizures and postictal periods tend to be shorter when compared to TLE [31].

6.1.4 Parietal Epilepsy (PLE)

Relatively few series of patients with well-documented parietal lobe epilepsy (PLE) have been reported in the literature as compared to patients with TLE or FLE. In a large cohort of patients undergoing epilepsy surgery at the Montreal Neurological Institute between 1929 and 1988 only 6 % were thought to have PLE [56, 57]. The parietal lobe comprises large areas of association cortex and plays a central role in multisensory integration. As a result this lobe is elaborately intertwined with numerous extralobar brain regions. Oftentimes, the clinical semeiology of seizures originating from the parietal lobe may be misleading (or falsely localizing and/or lateralizing) as it only becomes apparent when the seizure propagates to these extralobar brain regions. For similar reasons, interictal scalp EEG rarely demonstrates well-localized IEDs. Rather, as compared to TLE and FLE, IEDs in PLE show a much more variable scatter and may localize to several extraparietal regions within the ipsilateral or the contralateral hemisphere or have the appearance of bilateral discharges presumably due to interhemispheric propagation. That is why PLE has been described as "the great imitator among focal epilepsies" [58].

The suspicion of PLE is highest in patients who report discrete somatosensory auras such as tingling or numbness and less commonly pain and thermal sensations involving a particular segment of the contralateral hemibody. Less frequent auras localizing to the parietal lobe include episodic disturbances of body image (sensation of movement of one extremity or feeling that an extremity has been displaced in space) and vertiginous sensations. Depending on the pathways of seizure spread complex visual or auditory hallucinations, aphasia, and conscious confusion may accompany the symptoms reported by patients with PLE. Spread patterns that involve the sensorimotor cortex, the SSMA, or the premotor frontal eye field give rise to semeiological manifestations of FLE, rather than PLE, such as contralateral clonic and tonic activity, bilateral asymmetric tonic posturing, or head deviation [56]. Generally, it has been suggested that seizures from the superior parietal lobe spread toward the frontal eye field (dorsal fronto-parietal network) and the supplementary sensorimotor area. Seizures arising from temporoparietal junction may spread to the ventral prefrontal cortex or to the ipsilateral temporo-limbic network [58, 59].

As stated above, IEDs in PLE are often non-localizing or falsely localizing. Similarly ictal patterns are scarcely localizable (as low as

10 % of patients with PLE) and may falsely localize to the ipsilateral temporal or frontal lobe. The presence of interictal and ictal EEG patterns localized to the parietal lobe or the history of typical parietal lobe auras facilitates correct identification of patients with PLE. When such features are absent correct clinical and neurophysiologic localization is extremely challenging for patients with PLE and even more so in patients without a focal lesion on MRI [31, 57, 58].

6.1.5 Occipital Lobe Epilepsy

Visual and/or oculomotor phenomena are the main manifestations of seizures arising from the occipital lobe [60]. Epileptic activation of the primary visual cortex gives rise to elementary visual hallucinations characterized by colored, usually multicolored, and circular features. Other features include shapes (usually triangular, rectangular, squares, or star-like) and/or flashing or flickering achromatic lights. These visual hallucinations are often unilateral, at least at the onset, involving only one hemifield (or quadrant). Their components tend to multiply and become larger (increase in number and size) as the seizure continues to develop. Movement toward the center of vision or toward the other side is less common and in rare occasions may have a spinning, circling, or rotatory character. Other visual symptoms include blindness (ictal amaurosis), visual illusions and palinopsia. Oculomotor symptoms include tonic deviation or nystagmoid movements of the eyes (ictal nystagmus). Repetitive eyelid closures or eyelid fluttering may also occur.

Regarding seizure propagation, infracalcarine occipital foci propagate to the ipsilateral (medial) temporal lobe and manifest with complex partial seizures (which by themselves may be indistinguishable from TLE seizures), while seizures from supracalcarine foci tend to propagate to the ipsilateral parietal and frontal lobes producing various sensory and motor manifestations. Seizure duration is relatively short (1–3 min on average), but sustained seizures restricted to the occipital region do occur in some cases [61].

Again, scalp EEG in OLE is less helpful for seizure localization as compared to TLE and FLE. IEDs restricted to the occipital lobe are seen in a minority of cases (<20 % of OLE patients in some series). A rhythmical, well-localized, ictal pattern involving the occipital region at onset is present in 15–20 % of patients [31, 61].

6.2 Generalized Epilepsies

We already discussed one of the two distinct types of epilepsy, namely, the focal epilepsies (i.e., epilepsies presenting with focal seizures), as defined by the 1989 ILAE classification of epilepsies and epileptic syndromes. On the other hand, epilepsies manifesting with seizures that do not have a localized origin, and involve both hemispheres in a bisynchronous manner at onset, are classified as generalized. This basic dichotomy cannot be clearly established in a minority of patients, in whom an overlap spanning both generalized and focal features can be observed—as is the case in some of the "symptomatic generalized epilepsies" [33]. It has also been recognized that these two distinct types (i.e., generalized and focal epilepsies) may coexist in a small number of patients [62].

Moreover, the 1989 ILAE classification broadly divides epilepsies with respect to their etiology to idiopathic (i.e., genetically determined and unrelated to any structural brain pathology, which are usually associated with normal neurological and neuropsychological status), symptomatic (i.e., resulting from a known cerebral pathology or disorder of the central nervous system), and cryptogenic (i.e., presumed to be symptomatic but have a yet unidentified or "hidden" cause) [33]. Even though the ILAE has since revised the terminology and concepts for organization of seizures [63], the major divisions of focal versus generalized and idiopathic versus symptomatic epilepsies remain widely used in clinical practice.

6.2.1   Idiopathic Generalized Epilepsy

Idiopathic generalized epilepsy (IGE), also known as primary generalized epilepsy (PGE), is usually present after the age of 4 years (age related) and manifests with three major seizure types: absence, myoclonic, and GTC—including tonic-clonic or clonic-tonic-clonic—seizures.

Absence seizures in patients with childhood absence epilepsy are characterized electroencephalographically by the classic interictal pattern of repetitive, bisynchronous, and symmetric 3 Hz spike-and-wave complexes (SWCs), which usually have a superior frontal maximum [19]. During sleep, these discharges tend to become more fragmented with the appearance of bursts of polyspike-and-wave activities [30]. As a general rule, the background EEG activity is expected to be normal in IGE patients. Exceptions to this rule do exist in the appropriate clinical setting; for example some children with absence seizures exhibit an unusual paroxysmal pattern of occipital intermittent rhythmic delta activity (OIRDA) in their interictal EEG recordings.

Activation maneuvers (such as hyperventilation) or hypoglycemia may provoke the emergence of SWCs or typical absence seizures in these patients [64]. In general, seizures become clinically apparent when the duration of SWCs exceeds a period of 3 s or longer and are characterized by abrupt arrest of movement and vacant staring. These symptoms are usually brief in duration (<10 s) and resolve abruptly upon termination of the ictal discharge, at which time there is also immediate recovery of the normal background EEG activities. Occasionally, mild or subtle motor activities (myoclonus involving the eyelids or other facial muscles, increase or decrease of muscle tone, and elementary automatisms) or autonomic phenomena may be observed during absence seizures (especially with seizures of relatively longer duration). Beyond the common syndrome of childhood absence epilepsy, absence seizures occurring in different settings may be associated with other types of EEG discharges, such as generalized irregular spike-and-wave bursts or slow (<2.5 Hz) spike-and-wave activity or even generalized paroxysmal fast activity [30, 32, 65].

Generalized myoclonic seizures are typically characterized by high-amplitude spikes, which are usually (but not always) followed by slow waves. They may present as single or repetitive irregular myoclonic jerks—the latter sometimes precede evolution to GTC seizures (myoclonic crescendo)—involving the extremities (predominantly the arms) in a bilateral and relatively symmetric fashion. Repetitive, forceful jerks may cause some patients to fall suddenly. When occurring in isolation these seizures are not associated with alteration of consciousness.

The prototypical syndrome of juvenile myoclonic epilepsy (JME) is characterized by bursts of frontocentral dominant generalized spike-and-wave and polyspike-and-wave discharges. The frequency of these bursts tends to be more irregular than the typical 3 Hz spike-and-wave bursts and varies between 3 and 5 Hz. Generalized spikes as well as "fragments" of generalized spikes are commonly present and may or may not be followed by irregular 1–3 Hz slow waves. As with other types of IGE the background EEG activities are usually normal.

Sleep recordings, and especially periods of arousal from sleep, are of major importance in JME. During sleep, the bursts of spike-and-wave and polyspike-and-wave activity may decrease, particularly during stages of deep slow-wave sleep, and may even disappear during rapid eye movement (REM) sleep. Arousal from sleep is an extremely potent activator of spike-and-wave and polyspike-and-wave discharges. In some patients, this may be the only time epileptic activity takes place. Likewise, clinical seizures often occur on awakening. Hyperventilation and intermittent photic stimulation may increase or induce epileptic discharges [30, 32, 65].

The interictal EEG in patients with GTC seizures is variable. Specifically, in patients with IGE manifesting with GTC seizures, interictal recording may reveal epileptiform abnormalities in the form of brief bursts of generalized 2.5–3.5 Hz spike-and-wave and/or polyspike-and-wave activity. The duration of these bursts (which occur in approximately half of the patients with IGE associated with GTC seizures) is usually <2.5 s. Notably, such bursts may rarely be observed in individuals without a documented history of seizures.

Again, the background EEG activity is usually normal, and the bursts tend to appear mostly during slow-wave sleep (primarily at the beginning of this stage) or exclusively during transition from wakefulness to sleep. That is why sleep deprivation is an important activation technique for this type of seizures that should always be considered, especially in cases of diagnostic uncertainty [30, 32, 65].

The ictal EEG during a GTC seizure in patients with IGE is usually obscured by high-voltage myogenic artifact. Recurrent high-voltage polyspike-and-wave bursts, which are clinically associated with repetitive myoclonic jerks, may be seen at the time of seizure onset. Subsequently, the EEG recording may exhibit generalized voltage attenuation with superimposed low-voltage fast activity (20–40 Hz). This period of diffuse attenuation is short-lived (only few seconds) and is replaced by an evolving rhythmic generalized activity (10–12 Hz), which progressively builds up in amplitude in the next 8–10 s during the tonic phase of the seizure. At this point, the ictal pattern transitions to slower activities of moderate-to-high amplitude and gradual decrease in frequency from 7 to 8 Hz down to 1–2 Hz. As the frequency of these waveforms decreases below 4–5 Hz, the predominant slow-wave activity becomes intermixed with polyspikes and polyspike-and-wave complexes, which give rise to generalized myoclonic jerks and dominate the recording during the clonic phase of the seizures. As the seizure progresses, polyspike–wave complexes are seen intermittently in bursts interrupted by periods of diffuse suppression and are gradually replaced by irregular, low-voltage, generalized delta slowing of increasing amplitude and frequency [30, 32, 65].

6.2.2 Symptomatic Generalized Epilepsies

The typical EEG feature of symptomatic generalized epilepsies (SGE) is the so-called slow spike–wave discharge whose frequency does not exceed 2.5 Hz (ranging usually ranges from 1.5 to 2.5 Hz). This activity demonstrates irregularities in frequency, amplitude, morphology, as well as distribution within and between paroxysms, as compared to the classic and fairly stereotyped spike–wave discharge of IGE. In contrast to IGE, slowing of background EEG activities is observed in SGE patients. Other interictal EEG features include generalized paroxysmal fast activity as well as coexistent focal/multifocal and generalized discharges [30].

Various syndromes fall within the SGE spectrum including the more commonly observed West (infantile spasms) and Lennox–Gastaut syndromes (LGS), along with the less common epilepsies manifesting with myoclonic absences (EMA) or myoclonic-astatic seizures (Doose syndrome). LGS is an epileptic condition characterized by mental retardation in the setting of multiple generalized seizure types that are often difficult to control with medications. Slow spike–wave discharges are the hallmark of EEG studies in this population of patients with SGE. There is a variety in the type of seizures in this syndrome, which include tonic seizures as well as atypical absence, atonic, and myoclonic seizures. Other seizure types may also be present including akinetic, clonic, and GTC seizures [30, 32, 65].

7   Electroencephalography and Epilepsy in Neonates and Infants

The neonatal electroencephalogram (EEG) remains one of the oldest, yet most valuable, diagnostic and prognostic tests. It has several peculiarities and characteristics given that the normal electroencephalographic activities continue to evolve and are a reflection of the ongoing brain maturation at this stage of life.

Decreases in the total power and component power of the background delta activity during burst in quiet (non-REM) and active (REM) sleep are cardinal signs of neonatal EEG maturation, as is the gradual emergence of discrete sleep states.

A pediatric EEG can only be determined to be normal by assessing whether the EEG patterns are appropriate for maturational age. A hallmark of cerebral function of prematurity is the discontinuity of electrical activity. Thus, periods of short bursts are interspaced with periods of hypoactivity that can last for up to 60 s. As brain matures, the periods of sustained activity become longer and more concordant to active sleep-and-wake states. Discontinuous tracing with alternating periods of hypoactivity and higher amplitude activity becomes concordant to quiet sleep states. As the infant matures, the hypoactive periods shorten leading to the appearance of a characteristic pattern termed tracé alternant [66–68].

Sleep in infants at term exhibits the REM, and partly the non-REM, features. The dominant posterior background rhythm of relaxed wakefulness, which is typically observed occipitally but can also be seen more anteriorly (involving the central leads), increases in frequency with age: 3.5–4.5 Hz in 75 % of normal infants by 3–4 months post-term; 5–6 Hz in infants 5–6 months post-term; and 6 Hz in 70 % of normal children by 2 months of age. Sleep spindles appear second to third months post-term. Approximately 50 % of sleep spindles remain asynchronous before the age of 6 months of age; the percentage of asynchronous spindles in infants decreases to 30 % or less at 1 year. Sleep spindles along with the physiological high-amplitude graphoelements known as K-complexes are defining characteristics of stage 2 sleep. K-complexes first appear 5 months post-term and are mostly reflected in the frontal leads (Fz, F3, F4). On the other hand, well-defined vertex sharp waves appear at 16 months post-term and are best detected over the central electrodes (Cz, C3, C4). The slow-wave activity of slow-wave sleep is first seen as early as 2–3 months post-term and is uniformly present in normal infants at 4–4.5 months post-term [66, 68, 69].

When investigating the neonatal EEG, several parameters are taken into account (i.e., abnormalities of amplitude, continuity, frequency, symmetry, synchrony, sleep states, and maturation). Certain abnormalities, such as cerebral electrical inactivity, persistent low-voltage amplitude activities, or burst suppression, are usually associated with unfavorable outcome.

Febrile seizures are the most common seizures in childhood affecting 2–5 % of all children between the ages of 6 months and 5 years. These seizures are accompanied by fever (temperature ≥ 100.4 °F or 38 °C by any method) occurring in the absence of a central nervous system infection [70]. Febrile seizures, by themselves, do not represent a form of epilepsy and are considered among the so-called acute symptomatic seizures. This is a matter of debate in certain cases given the limited understanding of the underlying pathophysiology and its relationship to inherent biological and genetic susceptibility [71]. Simple, as opposed to complex, febrile seizures mostly present with generalized convulsions of shorter duration (<15 min) and generally do not recur within the same febrile illness. The majority of children with simple febrile seizures have a benign prognosis; their risk of developing epilepsy is estimated at around 1–1.5 %, only slightly higher than an estimated incidence of 0.5 % in the general population [72].

Complex febrile seizures, on the other hand, are defined as focal, prolonged (>15 min), and/or recurrent seizures within a period of 24 h. The risk of developing epilepsy in these patients is significantly higher than the normal population, estimated between 4 and 15 % or higher, especially in children with a positive family history of epilepsy and/or abnormal neurodevelopmental status before the first episode [71–73].

West syndrome is classified among the SGE and is characterized by the triad of axial spasms occurring in clusters (referred to as infantile spasms) along with psychomotor regression and hypsarrhythmia on EEG recordings. The onset is in up to 90 % of cases in the first year of life, typically between 3 and 6 months of age (ranging from 1 day to 5 years). The typical interictal pattern is that of hypsarrhythmia, which consists of diffuse giant slow waves (of high voltage >400 µV) with a chaotic background of random, irregular, multifocal spikes and sharp waves and little synchrony between the cerebral hemispheres. Modified hypsarrhythmia is characterized by synchronous bursts of generalized slow spike–wave discharges along with some preservation of background rhythms, in the presence of a significant asymmetry between the two hemispheres or a suppression–burst pattern [74]. Occasionally, some spikes and sharp waves present with a generalized pattern but are arrhythmic and do not occur in repetitive sequences unlike the slow spike–slow wave discharges seen in LGS.

Infantile spasms are associated with marked suppression of the background that lasts for the duration of the spasm. This characteristic response is also known as a diffuse electrodecremental response. Numerous other ictal patterns may be observed in children with West syndrome. It is important to note that the appearance of asymmetric ictal patterns should raise suspicion for an underlying lateralized structural abnormality (or abnormalities) [32].

8 Pitfalls in the Interpretation of Clinical EEG Recordings

A critical responsibility of individuals interpreting clinical EEG data is the proper identification of diagnostic IEDs and ictal activities on scalp EEG recordings. Indeed, there are many EEG transients that morphologically resemble epileptiform discharges and need to be distinguished from true epileptiform abnormalities to avoid overdiagnosis or misdiagnosis [29]. Such activities include artifacts (for example "muscle spicules" reflecting electromyographic activity, eyelid movements, electrocardiogram), physiological activity (such as vertex sharp waves, posterior occipital sharp transients of sleep), or benign variants associated with sharply contoured waveforms (e.g., wicket spikes, benign epileptiform transients of sleep, and others).

A major contributor to the misdiagnosis of epilepsy is the overinterpretation of normal EEG patterns as epileptiform. "Overinterpretation" of EEG findings is more common, and perhaps more harmful, than "underreading" the EEG record. Erroneous diagnosis of epilepsy is a serious problem given the multiple consequences and, often unnecessary, interventions associated with this diagnosis. In one study of 37 adult patients, who were misdiagnosed as having epilepsy on the basis of falsely classified EEG features, the most common patterns overread as epileptiform were nonspecific fluctuations of background in the temporal regions, which were misconstrued for temporal sharp waves [75]. Strict criteria should be applied systematically to determine the epileptiform significance of any questionable sharply contoured transient and avoid confusion with nonspecific patterns or activities that fall within the range of normal variation.

Next, it is important to re-emphasize that the scalp localization of IEDs and ictal activities may be misleading (falsely localizing or lateralizing) with respect to the true origin of the epileptic discharge. For example, scalp EEG recordings may exhibit epileptiform patterns involving the temporal electrodes in patients with certain types of extratemporal epilepsy [76, 77]. These "pseudotemporal" EEG activities can commonly be observed in patients, whose seizures arise from the orbitofrontal, temporo-occipital, cingulate, insular, or perisylvian cortex.

As discussed earlier, IEDs may be absent in up to 20 % of patients with an affirmative diagnosis of focal epilepsy [77]. Several factors account for scalp EEG-negative studies in patients with epilepsy. For example, this phenomenon is not uncommon in patients with deeply seated epileptogenic foci (such as the cingulate gyrus or insular cortex) or in patients with a restricted epileptogenic generator, whose extent does not exceed the minimum amount of gyral cortex required to produce epileptiform activities that can be resolved at the level of the scalp.

It has been estimated that at least 6 cm^3 of gyral cortex is required to produce spikes or sharp waves that are visible on scalp EEG [78]. Importantly, the extent of the minimum cerebral source area may be significantly larger, exceeding 10 cm^3 of gyral cortex in some cases, depending on the localization, orientation, and synchrony of the epileptic generator. Intracranial electrodes (subdural or depth electrodes) implanted in patients with intractable focal epilepsies commonly reveal several populations of IEDs that were not captured by scalp EEG electrodes. Therefore, more often than not, focal seizures that only activate a small or a deep area of cortex have no EEG signature on scalp recordings [7]. It is critical to understand that the absence of any scalp EEG changes during certain paroxysmal events cannot rule out an epileptic etiology.

References

1. Walter WG (1953) The living brain. Duckworth, London
2. Shipton HW (1975) EGG analysis: a history and a prospectus. Annu Rev Biophys Bioeng 4:1–13
3. Ives JR, Schomer DL (2005) A brief history of ambulatory EEG. In: Chang BS, Schachter SC, Schomer DL (eds) Atlas of ambulatory EEG. Elsevier, Burlington, MA, pp 3–10
4. Flink R, Pedersen B, Guekht AB, Malmgren K, Michelucci R, Neville B, Pinto F, Stephani U, Ozkara C, Commission of European Affairs of the International League Against Epilepsy: Subcommission on European Guidelines (2002) Guidelines for the use of EEG methodology in the diagnosis of epilepsy. International League Against Epilepsy: commission report. Commission on European Affairs: Subcommission on European Guidelines. Acta Neurol Scand 106:1–7
5. American Clinical Neurophysiology Society (2006) Guideline 1: minimum technical requirements for performing clinical electroencephalography. J Clin Neurophysiol 23:86–91
6. (1994) Guideline one: minimum technical requirements for performing clinical electroencephalography. American Electroencephalographic Society. J Clin Neurophysiol 11:2–5
7. Maganti RK, Rutecki P (2013) EEG and epilepsy monitoring. Continuum (Minneap Minn) 19(3 Epilepsy):598–622
8. American Clinical Neurophysiology Society (2006) Guideline 2: minimum technical standards for pediatric electroencephalography. J Clin Neurophysiol 23:92–96
9. Jasper HH (1958) The ten-twenty electrode system of the International Federation. Electroencephalogr Clin Neurophysiol 10:371–375
10. Klem GH, Luders HO, Jasper HH, Elger C (1999) The ten-twenty electrode system of the International Federation. The International Federation of Clinical Neurophysiology. Electroencephalogr Clin Neurophysiol Suppl 52:3–6
11. Jasper HH (1983) The ten-twenty electrode system of the International Federation. In: International Federation for Societies of Electroencephalography and Clinical Neurophysiology (ed) Recommendations for the practice of clinical electroencephalography. Elsevier, Amsterdam, pp 3–10
12. American Clinical Neurophysiology Society (2006) Guideline 6: a proposal for standard montages to be used in clinical EEG. J Clin Neurophysiol 23:111–117
13. Mendiratta A (2003) Clinical neurophysiology of epilepsy. Curr Neurol Neurosci Rep 3:332–340
14. Salinsky M, Kanter R, Dasheiff RM (1987) Effectiveness of multiple EEGs in supporting the diagnosis of epilepsy: an operational curve. Epilepsia 28:331–334
15. Alexopoulos AV, Tandon N (2008) Basal frontal lobe epilepsy. In: Lüders HO (ed) Textbook

of epilepsy surgery. Informa, London, pp 285–313

16. Arciniegas DB (2011) Clinical electrophysiologic assessments and mild traumatic brain injury: state-of-the-science and implications for clinical practice. Int J Psychophysiol 82:41–52

17. Kennedy JD, Gerard EE (2012) Continuous EEG monitoring in the intensive care unit. Curr Neurol Neurosci Rep 12:419–428

18. Zijlmans M, Jiruska P, Zelmann R, Leijten FS, Jefferys JG, Gotman J (2012) High-frequency oscillations as a new biomarker in epilepsy. Ann Neurol 71:169–178

19. Lüders HO, Noachtar S (2000) Atlas and classification of electroencephalography. Saunders, Philadelphia, PA

20. So EL (2010) Interictal epileptiform discharges in persons without a history of seizures: what do they mean? J Clin Neurophysiol 27:229–238

21. Dworetzky BA, Reinsberger C (2011) The role of the interictal EEG in selecting candidates for resective epilepsy surgery. Epilepsy Behav 20:167–171

22. Palmini A, Gambardella A, Andermann F, Dubeau F, da Costa JC, Olivier A, Tampieri D, Gloor P, Quesney F, Andermann E et al (1995) Intrinsic epileptogenicity of human dysplastic cortex as suggested by corticography and surgical results. Ann Neurol 37:476–487

23. Asano E, Muzik O, Shah A, Juhasz C, Chugani DC, Sood S, Janisse J, Ergun EL, Ahn-Ewing J, Shen C, Gotman J, Chugani HT (2003) Quantitative interictal subdural EEG analyses in children with neocortical epilepsy. Epilepsia 44:425–434

24. Gotman J (1991) Relationships between interictal spiking and seizures: human and experimental evidence. Can J Neurol Sci 18:573–576

25. Spencer SS, Goncharova II, Duckrow RB, Novotny EJ, Zaveri HP (2008) Interictal spikes on intracranial recording: behavior, physiology, and implications. Epilepsia 49:1881–1892

26. Sammaritano M, Gigli GL, Gotman J (1991) Interictal spiking during wakefulness and sleep and the localization of foci in temporal lobe epilepsy. Neurology 41:290–297

27. Marciani MG, Gotman J, Andermann F, Olivier A (1985) Patterns of seizure activation after withdrawal of antiepileptic medication. Neurology 35:1537–1543

28. Novotny EJ Jr (1998) The role of clinical neurophysiology in the management of epilepsy. J Clin Neurophysiol 15:96–108

29. Markand ON (2003) Pearls, perils, and pitfalls in the use of the electroencephalogram. Semin Neurol 23:7–46

30. Seneviratne U, Cook M, D'Souza W (2012) The electroencephalogram of idiopathic generalized epilepsy. Epilepsia 53:234–248

31. Verma A, Radtke R (2006) EEG of partial seizures. J Clin Neurophysiol 23:333–339

32. Hrachovy RA, Frost JD Jr (2006) The EEG in selected generalized seizures. J Clin Neurophysiol 23:312–332

33. (1989) Proposal for revised classification of epilepsies and epileptic syndromes. Commission on Classification and Terminology of the International League Against Epilepsy. Epilepsia 30:389–399

34. Schwartzkroin PA (1986) Hippocampal slices in experimental and human epilepsy. Adv Neurol 44:991–1010

35. Wieser HG, ILAE Commission on Neurosurgery of Epilepsy (2004) ILAE Commission Report. Mesial temporal lobe epilepsy with hippocampal sclerosis. Epilepsia 45:695–714

36. Margerison JH, Corsellis JA (1966) Epilepsy and the temporal lobes. A clinical, electroencephalographic and neuropathological study of the brain in epilepsy, with particular reference to the temporal lobes. Brain 89:499–530

37. French JA, Williamson PD, Thadani VM, Darcey TM, Mattson RH, Spencer SS, Spencer DD (1993) Characteristics of medial temporal lobe epilepsy: I. Results of history and physical examination. Ann Neurol 34:774–780

38. So EL (2006) Value and limitations of seizure semiology in localizing seizure onset. J Clin Neurophysiol 23:353–357

39. Blume WT (2003) The necessity for sphenoidal electrodes in the presurgical evaluation of temporal lobe epilepsy: con position. J Clin Neurophysiol 20:305–310

40. Sperling MR, Guina L (2003) The necessity for sphenoidal electrodes in the presurgical evaluation of temporal lobe epilepsy: pro position. J Clin Neurophysiol 20:299–304

41. Risinger MW, Engel J Jr, Van Ness PC, Henry TR, Crandall PH (1989) Ictal localization of temporal lobe seizures with scalp/sphenoidal recordings. Neurology 39:1288–1293

42. Tatum WO 4th (2012) Mesial temporal lobe epilepsy. J Clin Neurophysiol 29:356–365

43. Engel J Jr (1996) Surgery for seizures. N Engl J Med 334:647–652

44. Wiebe S, Blume WT, Girvin JP, Eliasziw M, Effectiveness, Efficiency of Surgery for Temporal Lobe Epilepsy Study Group (2001) A randomized, controlled trial of surgery for

temporal-lobe epilepsy. N Engl J Med 345:311–318

45. Pacia SV, Devinsky O, Perrine K, Ravdin L, Luciano D, Vazquez B, Doyle WK (1996) Clinical features of neocortical temporal lobe epilepsy. Ann Neurol 40:724–730

46. Foldvary N, Lee N, Thwaites G, Mascha E, Hammel J, Kim H, Friedman AH, Radtke RA (1997) Clinical and electrographic manifestations of lesional neocortical temporal lobe epilepsy. Neurology 49:757–763

47. Lee SY, Lee SK, Yun CH, Kim KK, Chung CK (2006) Clinico-electrical characteristics of lateral temporal lobe epilepsy; anterior and posterior lateral temporal lobe epilepsy. J Clin Neurol 2:118–125

48. Bercovici E, Kumar BS, Mirsattari SM (2012) Neocortical temporal lobe epilepsy. Epilepsy Res Treat 2012:103160

49. Kennedy JD, Schuele SU (2012) Neocortical temporal lobe epilepsy. J Clin Neurophysiol 29:366–370

50. Ebersole JS, Pacia SV (1996) Localization of temporal lobe foci by ictal EEG patterns. Epilepsia 37:386–399

51. Bartolomei F, Cosandier-Rimele D, McGonigal A, Aubert S, Regis J, Gavaret M, Wendling F, Chauvel P (2010) From mesial temporal lobe to temporoperisylvian seizures: a quantified study of temporal lobe seizure networks. Epilepsia 51:2147–2158

52. Unnwongse K, Wehner T, Foldvary-Schaefer N (2012) Mesial frontal lobe epilepsy. J Clin Neurophysiol 29:371–378

53. Manford M, Fish DR, Shorvon SD (1996) An analysis of clinical seizure patterns and their localizing value in frontal and temporal lobe epilepsies. Brain 119(Pt 1):17–40

54. Lee RW, Worrell GA (2012) Dorsolateral frontal lobe epilepsy. J Clin Neurophysiol 29:379–384

55. Lüders H, Acharya J, Baumgartner C, Benbadis S, Bleasel A, Burgess R, Dinner DS, Ebner A, Foldvary N, Geller E, Hamer H, Holthausen H, Kotagal P, Morris H, Meencke HJ, Noachtar S, Rosenow F, Sakamoto A, Steinhoff BJ, Tuxhorn I, Wyllie E (1998) Semiological seizure classification. Epilepsia 39:1006–1013

56. Salanova V, Andermann F, Rasmussen T, Olivier A, Quesney LF (1995) Parietal lobe epilepsy. Clinical manifestations and outcome in 82 patients treated surgically between 1929 and 1988. Brain 118(Pt 3):607–627

57. Salanova V (2012) Parietal lobe epilepsy. J Clin Neurophysiol 29:392–396

58. Ristić AJ, Alexopoulos AV, So N, Wong C, Najm IM (2012) Parietal lobe epilepsy: the great imitator among focal epilepsies. Epileptic Disord 14:22–31

59. Pandya DN, Yeterian EH (1985) Architecture and connections of cortical association areas. In: Peters A, Jones EG (eds) Cerebral cortex, vol 4. Plenum, New York, pp 3–61

60. Tandon N, Alexopoulos AV, Warbel A, Najm IM, Bingaman WE (2009) Occipital epilepsy: spatial categorization and surgical management. J Neurosurg 110:306–318

61. Panayiotopoulos CP (2010) Symptomatic and cryptogenic (probably symptomatic) focal epilepsies. In: Panayiotopoulos CP (ed) A clinical guide to epileptic syndromes and their treatment. Springer, London, pp 435–496

62. Nicolson A, Chadwick DW, Smith DF (2004) The coexistence of idiopathic generalized epilepsy and partial epilepsy. Epilepsia 45:682–685

63. Berg AT, Berkovic SF, Brodie MJ, Buchhalter J, Cross JH, van Emde Boas W, Engel J, French J, Glauser TA, Mathern GW, Moshe SL, Nordli D, Plouin P, Scheffer IE (2010) Revised terminology and concepts for organization of seizures and epilepsies: report of the ILAE Commission on Classification and Terminology, 2005-2009. Epilepsia 51:676–685

64. Tsiptsios DI, Howard RS, Koutroumanidis MA (2010) Electroencephalographic assessment of patients with epileptic seizures. Expert Rev Neurother 10:1869–1886

65. Drury I, Henry TR (1993) Ictal patterns in generalized epilepsy. J Clin Neurophysiol 10:268–280

66. Nunes ML, da Costa JC (2010) Sleep and epilepsy in neonates. Sleep Med 11:665–673

67. Shany E, Berger I (2011) Neonatal electroencephalography: review of a practical approach. J Child Neurol 26:341–355

68. Tsuchida TN, Wusthoff CJ, Shellhaas RA, Abend NS, Hahn CD, Sullivan JE, Nguyen S, Weinstein S, Scher MS, Riviello JJ, Clancy RR, American Clinical Neurophysiology Society Critical Care Monitoring Committee (2013) American clinical neurophysiology society standardized EEG terminology and categorization for the description of continuous EEG monitoring in neonates: report of the American Clinical Neurophysiology Society critical care monitoring committee. J Clin Neurophysiol 30:161–173

69. Grigg-Damberger M, Gozal D, Marcus CL, Quan SF, Rosen CL, Chervin RD, Wise M, Picchietti DL, Sheldon SH, Iber C (2007) The visual scoring of sleep and arousal in infants and children. J Clin Sleep Med 3:201–240

70. Subcommittee on Febrile Seizures, American Academy of Pediatrics (2011) Neurodiagnostic evaluation of the child with a simple febrile seizure. Pediatrics 127:389–394

71. Alam S, Lux AL (2012) Epilepsies in infancy. Arch Dis Child 97:985–992

72. Sapir D, Leitner Y, Harel S, Kramer U (2000) Unprovoked seizures after complex febrile convulsions. Brain Dev 22:484–486

73. Capovilla G, Mastrangelo M, Romeo A, Vigevano F (2009) Recommendations for the management of "febrile seizures": Ad Hoc Task Force of LICE Guidelines Commission. Epilepsia 50(Suppl 1):2–6

74. Covanis A (2012) Epileptic encephalopathies (including severe epilepsy syndromes). Epilepsia 53(Suppl 4):114–126

75. Benbadis SR, Lin K (2008) Errors in EEG interpretation and misdiagnosis of epilepsy. Which EEG patterns are overread? Eur Neurol 59:267–271

76. Quesney LF (1992) Extratemporal epilepsy: clinical presentation, pre-operative EEG localization and surgical outcome. Acta Neurol Scand Suppl 140:81–94

77. Walczak TS, Jayakar P, Mizrahi EM (2008) Interictal electroencephalography. In: Engel J Jr, Pedley TA (eds) Epilepsy: a comprehensive textbook, vol 1. Lippincott Williams & Wilkins, Philadelphia, PA, pp 809–823

78. Tao JX, Ray A, Hawes-Ebersole S, Ebersole JS (2005) Intracranial EEG substrates of scalp EEG interictal spikes. Epilepsia 46:669–676

Neuromethods (2015) 91: 87–101
DOI 10.1007/7657_2013_61
© Springer Science+Business Media New York 2013
Published online: 12 February 2014

Effective Brain Connectivity from Intracranial EEG Recordings: Identification of Epileptogenic Zone in Human Focal Epilepsies

Giulia Varotto, Laura Tassi, Fabio Rotondi, Roberto Spreafico, Silvana Franceschetti, and Ferruccio Panzica

Abstract

In the context of focal and drug-resistant epilepsy, surgical resection of the epileptogenic zone (EZ) may be the only therapeutic option for reducing or suppressing seizures. The aim of epilepsy surgery is the exeresis of the EZ, which is assumed to be the cortical region responsible for the onset, early organization, and propagation of seizures. EZ represents the minimum amount of cortex that must be resected in order to achieve seizure freedom; therefore, the correct identification of its extent and organization is a crucial objective. Nevertheless, the rather high rate of failure in epilepsy surgery in extra-temporal epilepsies highlights that the precise identification of the EZ is still an unsolved problem and that more sophisticated methods of investigation are required.

In many patients, intracranial stereo-EEG recordings still represent the gold standard for the epilepsy surgery work-up, and, over the last 10 years, considerable efforts have been made to develop advanced signal analysis techniques able to improve the identification of the EZ. Since it is widely assumed that epileptic phenomena are associated with abnormal changes in brain synchronization mechanisms, particular attention has been paid to those methods aimed at quantifying and characterizing the interactions and causal relationships of neuronal populations, and initial evidence has shown that this can be a suitable approach to localizing the EZ.

The aim of this review is to provide an overview of the different intracranial EEG signal processing methods used to identify the EZ, with particular attention being given to the methods aimed at characterizing effective brain connectivity using intracranial EEG recordings. Then, we briefly present our studies of the connectivity pattern associated with a particular form of focal epilepsy (type II focal cortical dysplasia), based on multivariate autoregressive parametric models and measures derived from graph theory.

Key words Focal epilepsy, Surgery, Intracranial EEG, Effective connectivity, Causality, EZ localization

---

1 Effective Brain Connectivity Assessed Using Intracranial EEG Recordings: Identifying the Epileptogenic Zone in Human Focal Epilepsies

Focal epilepsies, in which the seizures originate from a region limited to a part of one cerebral hemisphere, are common and account for more than 50 % of all epilepsies [1]. However, despite the great improvement in pharmacological research, approximately

30 % of patients with focal epilepsies experience seizures that are resistant to antiepileptic drugs (AEDs) [2]. A subset of these patients can be considered candidates for epilepsy surgery in order to suppress the seizures or at least considerably reduce their frequency and thus significantly improve the patients' quality of life. The aim of epilepsy surgery is the exeresis of the epileptogenic zone (EZ), which is assumed to be the cortical region responsible for the onset, early organization, and propagation of seizures [3]. The EZ represents the minimum amount of cortex that must be resected (inactivated or completely disconnected) in order to achieve seizure freedom [4].

Successful surgical treatment requires the precise identification of the EZ. It can sometimes be adequately localized by means of noninvasive investigations including a clinical neurological examination, a detailed description of ictal signs and symptoms, imaging findings by means of magnetic resonance imaging (MRI) and positron emission tomography (PET), and interictal and ictal EEG scalp recordings associated with a video recording of ictal events. However, when the region of seizure onset cannot be precisely identified noninvasively, intracranial EEG recordings using depth electrodes or cortical strips and grids can be made by highly specialized epilepsy surgery centers. Recordings based on the stereotactic surgical placement of intracranial electrodes (stereo-EEG, SEEG) [3, 5, 6], associated with long-term video recordings, have long been considered the gold standard for identifying the EZ. After invasive recordings, the EZ is currently identified by visually inspecting the video-SEEG recordings, a time-consuming procedure that requires the involvement of specifically trained neurophysiologists and is inevitably affected by the drawback of subjectivity. The rate of failure in epilepsy surgery in patients with seizures generated outside the temporal lobe [7] highlights the fact that identifying the EZ is still an unsolved problem and that more sophisticated diagnostic methods are required.

In most patients, the organization of the EZ corresponds to an intricately organized network (the epileptogenic network) of neuronal populations distributed in distinct and sometimes very large brain areas [8, 9], and this forms the basis of the definition adopted in the Report of the ILAE Commission on Classification and Terminology, 2005–2009 [10]: *focal epileptic seizures are conceptualised as originating within networks limited to one hemisphere and they may be discretely localised or more widely distributed. For each seizure type, ictal onset is consistent from one seizure to another, in some cases, however, there is more than one network, and more than one seizure type, but each individual seizure type has a consistent site of onset.*

It is widely assumed that epileptic phenomena are associated with abnormal changes in brain synchronization mechanisms, and a number of studies have shown that seizures are associated with the abnormal synchronization of distant structures [9, 11, 12].

Over the last few years, various methods have been developed with the aim of quantifying and characterizing the interactions and causal relationships of neuronal populations on the basis of EEG recordings, and initial evidence has shown that this can be a suitable approach to localizing the EZ. The aim of this chapter is to provide an overview of the different intracranial EEG signal processing methods used to identify the EZ, with particular attention being given to the methods aimed at characterizing effective brain connectivity using intracranial EEG recordings. We will then describe our studies of the connectivity pattern associated with type II focal cortical dysplasia (type II FCD, ILAE Classification 2010) [13], which are based on multivariate autoregressive parametric models [14] and measures derived from graph theory [15].

2 Intracranial EEG Signals: State-of-the-Art Applications and Analysis Methods to Localize the EZ

2.1 Intracranial EEG Recordings

As the correct identification of the EZ is required in order to define the strategy of epilepsy surgery, new technologies based on brain imaging have been developed with the aim of progressively increasing the number of patients who can be operated on after noninvasive investigations, and now fewer than 40 % of patients need invasive presurgical monitoring. However, despite these advances, when these diagnostic procedures are unable to identify the EZ unequivocally, or when the obtained information conflicts with or is insufficient to exclude the involvement of "eloquent cortical areas" in seizure generation (i.e., primary cortices that cannot be surgically removed without neurological sequelae), intracranial EEG monitoring must be carried out. Various intracranial EEG recording techniques are currently available, each of which has its advantages and drawbacks. Chronically implanted subdural electrodes allow recordings from large superficial areas of the neocortex but provide limited information concerning deep structures such as the hippocampus, insular region, or cortex within sulci, whereas intracerebral electrodes have the advantage of allowing excellent sampling from the mesial structures and the intrasulcal cortex and the disadvantage of providing information based on a limited volume of tissue.

The group of St. Anne's Hospital in Paris originally introduced intracranial SEEG at the beginning of the 1960s. Using this technique and a stereotactic frame, thin intracranial multi-contact tubular electrodes (rigid or flexible) made of biologically inert materials (e.g., platinum, iridium, or stainless steel) (Fig. 1) can be placed through small "burr holes" and accurately positioned in targeted brain areas. These thin electrodes make it possible to record the

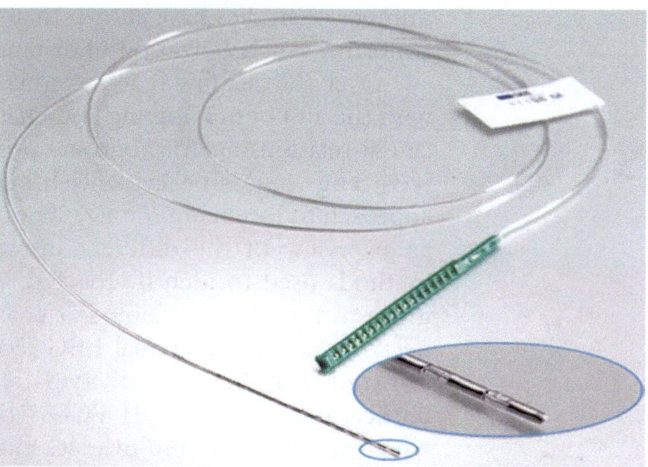

Fig. 1 An intracranial multi-contact platinum iridium electrode (Dixi, Beçancon, France)

electrical activity originating from the cortical and subcortical structures using a mean number of 12 electrodes in such a way that each crosses its trajectory between the initial and target points.

Selecting the brain structures to be explored, and the arrangement and locations of the electrodes, requires a careful analysis of all the data collected during the noninvasive presurgical work-up and the formulation of one or more hypotheses concerning the location of the EZ and the trajectories of the seizure discharges. After electrode implantation, a first visual analysis is made of all of the SEEG signals, one electrode at a time, and then a synthetic montage is created using bipolar derivations defined on the basis of the clinical characteristics of the ictal phenomena. The final montage shows all of the selected contacts and, together with MRI, allows the definition of different cortical zones depending on their particular patterns of electrical activity [16].

At least one spontaneous seizure needs to be recorded in order to make an accurate cortical neurophysiological map (see Fig. 2 for a representative example). In a subset of patients, electrical stimulation through intracranial electrodes can also be used to localize eloquent cortical areas, map functions, and, when necessary, elicit seizures. At the end of a SEEG recording, all of the clinical, anatomical, functional, and electrical data are compared in order to obtain the best definition of the EZ [17, 18].

2.2 Intracranial EEG Signal Analysis

There is a very wide range of methods of analyzing intracranial EEG signals [19, 20] that have been successfully used to study different questions in the field of epilepsy, such as seizure prediction, analysis of the source epileptic activity, and automatic recognition of epileptic transients or seizures [21–23].

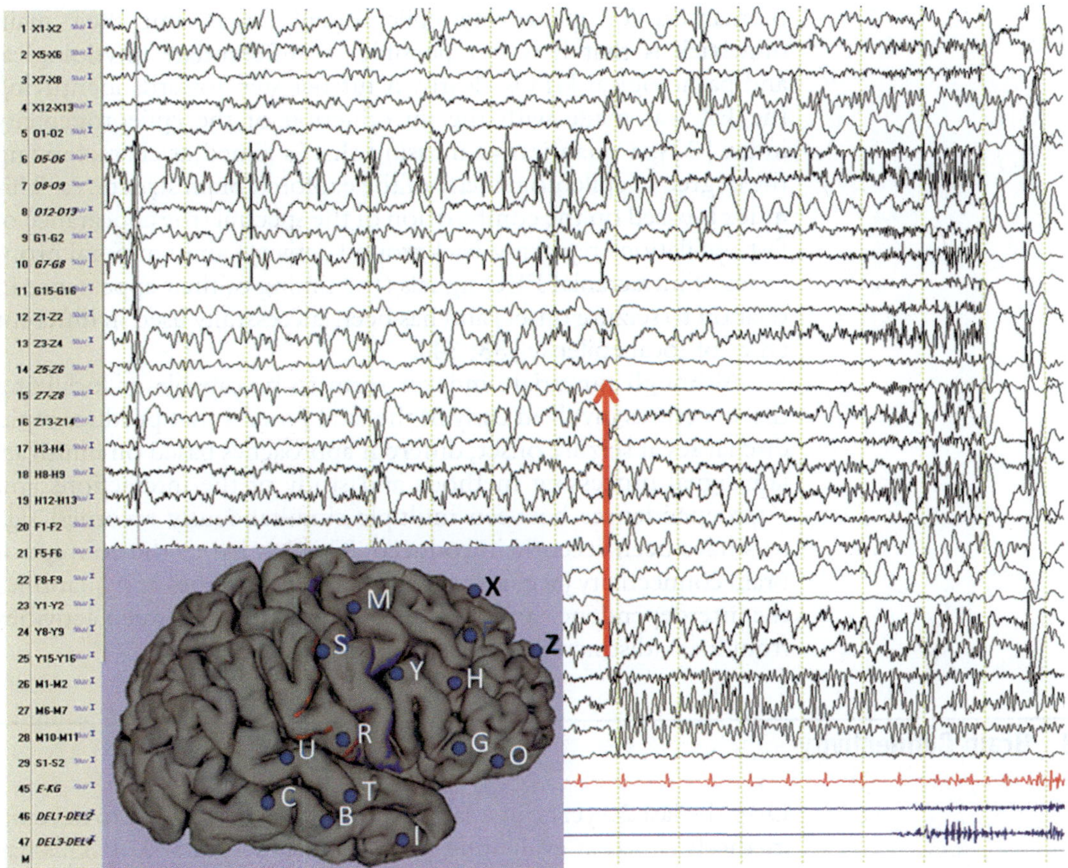

Fig. 2 A spontaneous seizure in *right* frontal lobe epilepsy (*red arrow*). The low-voltage fast activity is well located in electrodes O, G, and Z

In general terms, signal processing methods can be divided into the two main categories of univariate and multivariate methods. The most widely used univariate techniques are based on frequency domain analysis because obvious changes in EEG frequency content are very frequently associated with seizures. Over the last few years, considerable attention has been given to identifying (low-voltage) fast discharges or oscillations in the beta and/or gamma bands that could be a specific electrophysiological signature of epileptogenic areas [24–27]. More recently, technological improvements in digital EEG acquisition systems has made it possible to detect bursts of activity at very high frequencies, up to 500 Hz at the time of seizure onset [28]. On the basis of these observations, EEG activity above 80 Hz (high-frequency oscillations or HFOs) has been increasingly studied in subdural or intracranial EEG recordings with the aim of finding markers of epileptogenicity and identifying the area originating clinical seizures [29–31]. The findings of these studies suggest that HFOs may be potentially useful biomarkers for identifying epileptogenic areas, but their main drawback is that it is not possible to differentiate physiological and pathological HFOs clearly. A further step towards identifying the EZ was made with the introduction of the epileptogenicity index [32], a quantitative measure that is aimed at characterizing the degree of epileptogenicity of the explored brain structures. This index is based on the combination of the spectral (the appearance of fast oscillations replacing background activity) and temporal properties of intracranial EEG signals (the timing of their appearance in relation to seizure onset) and has been successfully used in different types of focal epilepsy [33, 34].

Although univariate indexes seem to be capable of localizing the structures involved early in the ictal process that produce rapid discharges at seizure onset, different approaches based on multivariate signal processing methods and study of the interactions and interdependences between multiple simultaneously recorded signals have been recently introduced. This has led to the concept of brain connectivity (see next section), which describes and quantifies the interdependence and degree of coordination between activities from local, distinct, or distant brain regions.

3   Brain Connectivity

Over the last 20 years, increasing efforts have been made to develop methods aimed at evaluating brain connectivity [35–37]. This has led to the development of various procedures that have different advantages and disadvantages and make different assumptions about the underlying relationship between the analyzed signals. However, the common hypothesis underlying these approaches is that the temporal evolution of the *correlation* between pairs of signals recorded in different brain regions reflects the connectivity between the regions themselves.

The concept of brain connectivity can be subdivided into three main categories: (1) anatomical (or structural) connectivity, which indicates the set of physical or structural connections linking neurons and may range from local circuits to large-scale pathways connecting distant brain regions; (2) functional connectivity, defined as the temporal correlation expressed in terms of the statistical dependence of spatially remote neuronal populations; and (3) effective connectivity, which refers to the influence that one neural system exerts over another, thus taking into account the direction of the information flow from one region towards another [38, 39]. Effective connectivity can be directly estimated on the basis of the signals (i.e., data driven) or be based on a model specifying causal links (i.e., model based) [14].

Most of the methods of estimating connectivity, particularly those focusing on functional connectivity, are bivariate (pair-wise) [40], which means that they are capable of evaluating the presence of interactions by considering only pairs of signals. However, this can lead to the identification of not only spurious interactions because it does not distinguish direct and indirect relationships but also spurious connections due to the effects of sources common to both signals [20]. Many recent papers have extensively described the advantages and disadvantages of bivariate measures in the fields of functional and effective connectivity [41, 42].

On the contrary, multivariate measures make it possible to explore a wide-ranging network by considering a whole set of signals in the same model. Interest in this field is growing because the quantitative analysis of complex networks is becoming a promising neuroscientific tool. Most of the estimators (mainly nonlinear approaches) belong to the family of methods measuring functional connectivity (see [20, 43] for a review). However, as previously stated, although they are useful in characterizing brain networks, they do not give any information concerning the direction of the interactions or the influence that one system has over another (causality).

In an attempt to overcome these limitations, directed connectivity measures have been developed with the aim of evaluating effective causality. The first definition of causality was given by Wiener in 1956 [44]: "If we can predict the first signal better by using the past information from the second one than by using the information without it, then we call the second signal causal to the first one." This general definition was formalized by Granger (1969) in terms of predictions in the context of stochastic processes and multivariate regression models of time series. Granger causality [45] was first defined in the field of econometrics for two processes $x(t)$ and $y(t)$. If the knowledge of the past of both $x(t)$ and $y(t)$ reduces the variance of the prediction error of $y(t)$ in comparison with the knowledge of the past of $y(t)$ alone, then the Granger signal $x(t)$ causes the signal $y(t)$. The lack of reciprocity makes it possible to evaluate the direction of the information flow between the signals. Granger's definition of causality can be easily implemented by means of autoregressive (AR) models and extended to the case of multivariate systems [14, 46].

A number of directed connectivity measures have been developed on the basis of the concept of Granger causality derived from the multivariate autoregressive (MVAR) modelling of multichannel EEG signals [14, 46] and successfully used to study the propagation of seizure activity in patients with focal epilepsies [13, 47–49]. Among them, directed transfer function (DTF) [50] and partial directed coherence (PDC) [51] are effective connectivity measures based on MVAR coefficients transformed into the frequency domain. PDC is particularly interesting because of its ability to distinguish direct and indirect causality flows in the estimated connectivity pattern.

4 Connectivity and EZ Localization

A number of studies of intracranial EEG recordings have developed and applied methods aimed at identifying and quantifying the coupling between the explored areas. These studies were based on the evidence showing that the structures involved in seizure generation are characterized by EEG abnormal transient coupling and changes in connectivity, particularly at the time of seizure onset [52–54]. Recent studies have found enhanced connectivity in the regions including the EZ, thus revealing the existence of highly interconnected nodes or "hubs" that may play a crucial role in the onset and propagation of ictal activity [13, 48, 55]. Moreover, it has been demonstrated that the activity recorded from distant regions shows altered connectivity patterns during seizures [56] and that network topology greatly changes during ictal events [13, 48, 57].

The earliest studies performed in the 1970s [58] used coherence (one of the most widely used bivariate measures) to localize epileptic foci in cats on the basis of the multi-electrode data obtained during a generalized seizure. Subsequently, Takigawa et al. [59] used directed coherence, an index that is capable of detecting the direction of information flow, to investigate bidirectional communication patterns between the frontal and occipital cortex in the scalp EEGs of epileptic patients.

Among the linear multivariate approaches, the *en bloc* or adapting version of DTF has been used to identify the epileptogenic focus during seizures [47, 49] and to investigate the dynamic changes in effective connectivity at the time of seizure onset [60–62]. Jung et al. [63] used DTF to estimate the ictal onset zone in patients with Lennox–Gastaut syndrome and demonstrated that areas with high average outflow values corresponded well with the surgical resection areas identified by means of conventional methods. Overall, these studies have shown that the multivariate approach is a suitable means of exploring the temporal dynamics of the mechanisms leading to ictal events [64, 65].

Other techniques based on nonlinear approaches have also been validated and used to study epileptic networks. One of these widely used in the field of intracranial EEG is the h^2 nonlinear regression index, which was introduced by Pijn and Lopes da Silva [66] and subsequently extended to SEEG by Wendling et al. [67]. This approach has mainly been used to investigate the coupling between the temporal neocortex and limbic structures in patients with temporal lobe epilepsy (TLE), and the results have shown that the regions involved in the EZ establish preferential functional links during seizures. Furthermore, the authors identified several subtypes of TLEs on the basis of the different interactions between temporal and neocortical structures (see [68, 69] for a review of the work done in this field).

Although all of these methods are based on measuring statistical coupling between pairs of signals, they have different underlying theoretical principles and there is no agreement about how they should be carried out. In a recent study, Wendling et al. [70] used simulations to evaluate the advantages and disadvantages of three different families of bivariate methods: linear and nonlinear regression, phase synchronization, and generalized synchronization. Their main conclusion was that there is no "universal" method insofar as none of the methods performed better in all situations, mainly because the results depend on the frequency content of the signals and some methods may be insensitive to the coupling parameters of the system.

Most of the studies of connectivity-based EZ localization have so far mainly concentrated on the ictal and pre-ictal phases, when abnormal critical patterns are more easily detectable; only a few were carried out with the aim of localizing the EZ on the basis of interictal intracranial EEG recordings. Bettus et al. [71] studied the interictal EEG patterns recorded in patients with drug-resistant mesial temporal lobe epilepsy (MTLE) using the h2 nonlinear correlation index and found that the EZ was characterized by decreased power in the theta band and a general increase in signal interdependences in comparison with the results obtained in a group of patients whose seizures originated in other areas. Wilke et al. [48] used adapting DTF, betweenness centrality, and indices derived from graph theory to identify critical network nodes during ictal and interictal electrocorticogram (EcoG) recordings of patients with intractable epilepsy. They observed patterns of altered network interactions (mainly in the gamma band) not only during seizures but also during interictal spike activity and random non-ictal periods.

5 Application in Focal Cortical Dysplasia

In our recent study [13], we evaluated the changes in dynamic connectivity patterns in the SEEG recordings of ten patients with Taylor-type focal cortical dysplasia (type II FCD) under interictal, pre-ictal, and ictal conditions using the PDC and indices derived from graph theory (in- and out-density, and betweenness centrality) in order to characterize the synchronization and connectivity properties of the SEEG signals recorded from inside the dysplasia (i.e., the lesional zone) and distinguish them from those of other regions involved in ictal activity or not.

Type II FCD is characterized by the presence of cortical dyslamination, and dysmorphic neurons and balloon cells. We selected this type of focal epilepsy in order to validate the appropriateness of our approach because this type of dysplasia is intrinsically epilepto-genic [72], the EZ normally overlaps the dysplasia, and the clear-cut borders of the anatomical lesion allow its precise exeresis and a good surgical outcome.

To estimate PDC an MVAR model has to be identified. An MVAR model describes a multivariate dataset as a linear combination of its own past plus uncorrelated white noise.

Given a set $S = \{x_n(k), \ 1 \leq m \leq M\}$ of M simultaneously observed stationary time series, the MVAR model with order p is defined as

$$\begin{bmatrix} x_1(k) \\ \vdots \\ x_M(k) \end{bmatrix} = \sum_{r=1}^{p} A_r \begin{bmatrix} x_1(k-r) \\ \vdots \\ x_M(k-r) \end{bmatrix} + \begin{bmatrix} w_1(k) \\ \vdots \\ w_M(k) \end{bmatrix},$$

where A_1, A_2, ... A_p are the coefficient matrices (dimensions $M \times M$), with the coefficients $a_{ij}(r)$ describing the linear interaction of $x_j(k-r)$ on $x_i(k)$, and $w_i(k)$ represents a random (Gaussian) white noise driving innovation.

The PDC, for each pair of channel i and j, are calculated from the Fourier transform of the MVAR coefficients as

$$\left| \pi_{ij}(f) \right| = \frac{\left| \overline{A}_{ij}(f) \right|}{\sqrt{\sum_{k=1} \left| \overline{A}_{kj}(f) \right|^2}},$$

where

$$\overline{A}(f) = I - \sum_{r=1}^{p} A_r \mathrm{e}^{-2i\pi \Delta t f r},$$

with Δt the sampling interval and I the identity matrix of dimension M.

The PDC $\pi_{ij}(f)$ describes the directional flow of information from signal $x_j(t)$ to $x_i(t)$, and its values range between 0 and 1.

Our methodological approach, from the SEEG signal acquisition to graph index measures, is described in Fig. 3. After SEEG signals were acquired and preprocessed, interactions between each pair of channels (nodes) were calculated by means of PDC in the frequency bands of interest, and a weighted adjacency matrix, whose elements represented the directed connectivity values between each pair of nodes, was generated by applying a threshold to each PDC values calculated by means of a bootstrap approach using phase randomization [73].Then each adjacency matrix was analyzed with measures derived from graph theory, namely, in- and out-density and betweenness centrality.

The main findings of our study showed that the epileptogenic area showed abnormal outgoing connectivity in comparison with all of the other regions examined by means of intracranial leads.

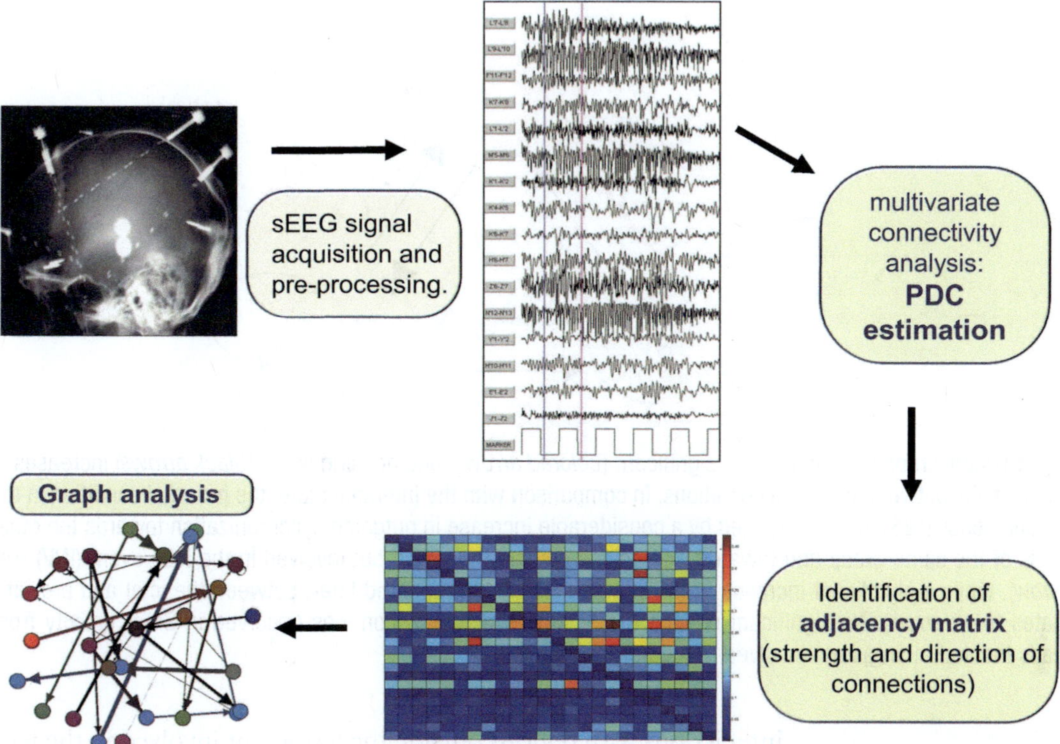

Fig. 3 Schematic representation of the analysis procedure used to study connectivity networks in SEEG signals

The main dynamic changes in connectivity occurred in the gamma band, and, most importantly, it was the frequency range that significantly differentiated the different brain structures in terms of synchronization patterns. The relationship between the role of gamma activity synchronization and HFOs in localizing the EZ is an important topic for further research.

The specific connectivity pattern characterizing the EZ was also present in the interictal state, even in the absence of obvious epileptiform activity. This is an important point and suggests that studying effective connectivity can add key information even when analyzing short segments of interictal activity, thus reducing the need for long-term presurgical monitoring (see also [69, 74–76]).

As a further step, we analyzed dynamic changes in the interactions within and between the different cortical regions in the gamma band and observed a global increase in the synchronization of not only the SEEG activity recorded from the dysplasia but also other regions showing epileptiform activity during seizures. However, the time course of the increase in gamma synchronization was different: a significant increase in outgoing synchronization between the electrodes within the EZ and those from the EZ towards other areas mainly occurred during the pre-ictal period, whereas increased

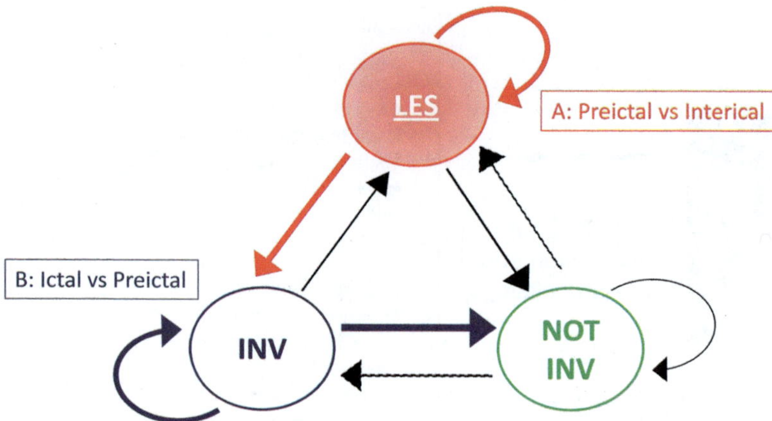

Fig. 4 Graphic representation of the significant (*colored arrows*) and nonsignificant (*black arrows*) increases in connectivity between temporal conditions. In comparison with the interictal state, the pre-ictal condition in the lesional leads (LES) is characterized by a considerable increase in outgoing synchronization towards the other leads of the same group and towards the regions outside the lesion but involved in the ictal event (INV) (*red arrow*), while a significant increase in the interactions of INV appeared later, between the ictal and pre-ictal states (*blue arrow*). No significant change in outgoing synchronization was observed in SEEG activity from leads located in areas not involved in the seizures (NOT-INV)

interactions with regions outside the lesion but involved in the ictal event occurred at the time of seizure onset (Fig. 4).

Our findings suggest that in patients with type II FCD, the region inside the dysplasia corresponds to the abnormal hub of the epileptic network that originates and sustains the seizures and plays a leading role in generating and propagating ictal EEG activity and in recruiting other distant areas to become involved in the seizure. This leading role of the dysplasia may account for the good postsurgical outcome of patients with type II FCD because the resection of dysplastic tissue removes the entire EZ responsible for seizure onset.

6   Conclusion and Future Developments

The aim of this review was to summarize state-of-the-art intracranial EEG signal processing methods in the field of presurgical epilepsy evaluation. Despite the large number of signal analysis approaches available, recent developments have shown that those aimed at investigating connectivity patterns offer a promising and efficient means of improving EZ localization.

Our data confirm that using advanced signal processing techniques to study synchronization can substantially improve the presurgical evaluation of patients with intractable focal epilepsy by providing quantitative information that is useful for localizing the EZ or at least greatly reduces the number of leads requiring further examination. However, further efforts are needed to clarify which measures of connectivity work best in localizing the epileptogenic network and validate them in larger and possibly multicenter patient populations that include patients with drug-resistant focal epilepsy of different (or unknown) etiology.

Acknowledgments

This work was supported by the EU Project Grant FP7-ICT-2009-6-270460 ACTIVE.

References

1. Hauser WA (2007) Incidence and prevalence. In: Engel J, Pedley TA (eds) Epilepsy: a comprehensive textbook. Lippincott-Raven Publishers, Philadelphia, pp 47–57

2. Beleza P (2009) Refractory epilepsy: a clinically oriented review. Eur Neurol 62(2):65–71

3. Munari C, Hoffmann D, Francione S et al (1994) Stereo-electroencephalography methodology: advantages and limits. Acta Neurol Scand Suppl 152:56–67

4. Rosenow F, Luders H (2001) Presurgical evaluation of epilepsy. Brain 124:1683–1700

5. Bancaud J, Talairach J (1973) Methodology of stereo EEG exploration and surgical intervention in epilepsy. Rev Otoneuroophtalmol 45(4):315–328

6. Cossu M, Chabardes S, Hoffmann D et al (2008) Presurgical evaluation of intractable epilepsy using stereo-electro-encephalography methodology: principles, technique and morbidity. Neurochirurgie 45(3):367–373

7. Tellez-Zenteno JF, Dhar R, Wiebe S (2005) Long-term seizure outcomes following epilepsy surgery: a systematic review and meta-analysis. Brain 128:1188–1198

8. Spencer SS (2002) Neural networks in human epilepsy: evidence of and implications for treatment. Epilepsia 43(3):219–227

9. Bartolomei F, Wendling F, Chauvel P (2008) The concept of an epileptogenic network in human partial epilepsies. Neurochirurgie 54(3):174–184

10. Berg AT, Berkovic SF, Brodie MJ et al (2010) Revised terminology and concepts for organization of seizures and epilepsies: report of the ILAE Commission on Classification and Terminology, 2005–2009. Epilepsia 51(4):676–685

11. Bartolomei F, Wendling F, Vignal JP et al (1999) Seizures of temporal lobe epilepsy: identification of subtypes by coherence analysis using stereo-electro-encephalography. Clin Neurophysiol 110(10):1741–1754

12. Le Van Quyen M, Soss J, Navarro R et al (2005) Preictal state identification by sychronization changes in long-term intracranial EEG recordings. Clin Neurophysiol 116(3):559–568

13. Varotto G, Tassi L, Franceschetti S et al (2012) Epileptogenic networks of type II focal cortical dysplasia: a stereo-EEG study. Neuroimage 61(3):591–598

14. Sakkalis V (2011) Review of advanced techniques for the estimation of brain connectivity measured with EEG/MEG. Comput Biol Med 41(12):1110–1117

15. Boccaletti S, Latora V, Moreno Y et al (2006) Complex networks: structure and dynamics. Phys Rep 424(4):175–308

16. Luders HO, Najm I, Nair D et al (2006) The epileptogenic zone: general principles. Epileptic Disord 8(Suppl 2):S1–S9

17. Centeno RS, Yacubian EM, Caboclo LO et al (2011) Intracranial depth electrodes implantation in the era of image-guided surgery. Arq Neuropsiquiatr 69(4):693–698

18. Cardinale F, Miserocchi A, Moscato A et al (2012) Talairach methodology in the multimodal imaging and robotics era. In: Scarabin J-M (ed) Stereotaxy and epilepsy surgery. John Libbey Eurotext, London, pp 245–272

19. Thakor NV, Tong S (2004) Advances in quantitative electroencephalogram analysis methods. Annu Rev Biomed Eng 6:453–495

20. Pereda E, Quiroga RQ, Bhattacharya J (2005) Nonlinear multivariate analysis of neurophysiological signals. Prog Neurobiol 77(1–2):1–37

21. Aarabi A, He B (2012) A rule-based seizure prediction method for focal neocortical epilepsy. Clin Neurophysiol 123(6):1111–1122

22. Benar CG, Grova C, Kobayashi E et al (2006) EEG-fMRI of epileptic spikes: concordance with EEG source localization and intracranial EEG. Neuroimage 30(4):1161–1170

23. Liu Y, Zhou W, Yuan Q et al (2012) Automatic seizure detection using wavelet transform and SVM in long-term intracranial EEG. IEEE Trans Neural Syst Rehabil Eng 20(6):749–755

24. Allen PJ, Fish DR, Smith SJ (1992) Very high-frequency rhythmic activity during SEEG suppression in frontal lobe epilepsy. Electroencephalogr Clin Neurophysiol 82(2):155–159

25. Fisher RS, Webber WR, Lesser RP et al (1992) High-frequency EEG activity at the start of seizures. J Clin Neurophysiol 9(3):441–448

26. Wendling F, Bartolomei F, Bellanger JJ et al (2003) Epileptic fast intracerebral EEG activity: evidence for spatial decorrelation at seizure onset. Brain 126:1449–1459

27. Worrell GA, Parish L, Cranstoun SD et al (2004) High-frequency oscillations and seizure generation in neocortical epilepsy. Brain 127:1496–1506

28. Jirsch JD, Urrestarazu E, LeVan P et al (2006) High-frequency oscillations during human focal seizures. Brain 129:1593–1608

29. Zijlmans M, Jacobs J, Kahn YU et al (2011) Ictal and interictal high frequency oscillations in patients with focal epilepsy. Clin Neurophysiol 122(4):664–671

30. Jacobs J, Staba R, Asano E et al (2012) High-frequency oscillations (HFOs) in clinical epilepsy. Prog Neurobiol 98(3):302–315

31. Brazdil M, Halamek J, Jurak P et al (2010) Interictal high-frequency oscillations indicate seizure onset zone in patients with focal cortical dysplasia. Epilepsy Res 90:28–32

32. Bartolomei F, Chauvel P, Wendling F (2008) Epileptogenicity of brain structures in human temporal lobe epilepsy: a quantified study from intracerebral EEG. Brain 131:1818–1830

33. Aubert S, Wendling F, Regis J et al (2009) Local and remote epileptogenicity in focal cortical dysplasias and neurodevelopmental tumours. Brain 132:3072–3086

34. Bartolomei F, Gavaret M, Hewett R et al (2011) Neural networks underlying parietal lobe seizures: a quantified study from intracerebral recordings. Epilepsy Res 93:164–176

35. Friston KJ (2004) Functional and effective connectivity in neuroimaging: a synthesis. Hum Brain Mapp 2:56–78

36. Jirsa VK, McIntosh AR (2007) Handbook of brain connectivity, vol 1. Springer, Berlin

37. Rubinov M, Sporns O (2010) Complex network measures of brain connectivity: uses and interpretations. Neuroimage 52(3):1059–1069

38. Sporns O, Chialvo DR, Kaiser M et al (2004) Organization, development and function of complex brain networks. Trends Cogn Sci 8(9):418–425

39. Lehnertz K (2011) Assessing directed interactions from neurophysiological signals—an overview. Physiol Meas 32(11):1715–1724

40. Lehnertz K, Bialonski S, Horstmann MT et al (2009) Synchronization phenomena in human epileptic brain networks. J Neurosci Methods 183(1):42–48

41. Ansari-Asl K, Senhadji L, Bellanger JJ et al (2006) Quantitative evaluation of linear and nonlinear methods characterizing interdependencies between brain signals. Phys Rev E Stat Nonlin Soft Matter Phys 74:031916

42. Osterhage H, Mormann F, Wagner T et al (2008) Detecting directional coupling in the human epileptic brain: limitations and potential pitfalls. Phys Rev E Stat Nonlin Soft Matter Phys 77:011914

43. Gourevitch B, Bouquin-Jeannes RL, Faucon G (2006) Linear and nonlinear causality between signals: methods, examples and neurophysiological applications. Biol Cybern 95(4):349–369

44. Wiener N (1956) Nonlinear prediction and dynamics. In: Proceedings of third Berkeley symposium, pp 247–252

45. Granger CWJ (1969) Investigating causal relations by econometric models and cross-spectral methods. Econometrica 37:424–438

46. Kaminski M, Liang H (2005) Causal influence: advances in neurosignal analysis. Crit Rev Biomed Eng 33(4):347

47. Franaszczuk PJ, Bergey GK, Kaminski MJ (1994) Analysis of mesial temporal seizure onset and propagation using the directed transfer function method. Electroencephalogr Clin Neurophysiol 91(6):413–427

48. Wilke C, Worrell G, He B (2011) Graph analysis of epileptogenic networks in human partial epilepsy. Epilepsia 52(1):84–93

49. Wilke C, Van Drongelen W, Kohrman M et al (2010) Neocortical seizure foci localization by means of a directed transfer function method. Epilepsia 51(4):564–572

50. Kaminski M, Blinowska K (1991) A new method of the description of the information flow in the brain structures. Biol Cybern 65(3):203–210

51. Baccalà LA, Sameshima K (2001) Partial directed coherence: a new concept in neural structure determination. Biol Cybern 84(6):463–474

52. Le Van Quyen M, Adam C, Baulac M et al (1998) Nonlinear interdependencies of EEG signals in human intracranially recorded temporal lobe seizures. Brain Res 792(1):24–40

53. Gotman J, Levtova V (1996) Amygdala-hippocampus relationships in temporal lobe seizures: a phase-coherence study. Epilepsy Res 25(1):51–57

54. Bartolomei F, Wendling F, Bellanger JJ et al (2001) Neural networks involving the medial temporal structures in temporal lobe epilepsy. Clin Neurophysiol 112(9):1746–1760

55. Morgan RJ, Soltesz I (2008) Nonrandom connectivity of the epileptic dentate gyrus predicts a major role for neuronal hubs in seizures. Proc Natl Acad Sci U S A 105(16):6179–6184

56. Guye M, Regis J, Tamura M et al (2006) The role of corticothalamic coupling in human temporal lobe epilepsy. Brain 129:1917–1928

57. Ponten SC, Bartolomei F, Stam CJ (2007) Small-world networks and epilepsy: graph theoretical analysis of intracerebrally recorded mesial temporal lobe seizures. Clin Neurophysiol 118(4):918–927

58. Gersch W, Goddard GV (1970) Epileptic focus location: spectral analysis method. Science 169(3946):701–702

59. Takigawa M, Wang G, Kawasaki H et al (1996) EEG analysis of epilepsy by directed coherence method. A data processing approach. Int J Psychophysiol 21(2–3):65–73

60. Wilke C, Ding L, He B (2008) Estimation of time-varying connectivity patterns through the use of an adaptive directed transfer function. IEEE Trans Biomed Eng 55(11):2557–2564

61. Astolfi L, Cincotti F, Mattia D et al (2008) Tracking the time-varying cortical connectivity patterns by adaptive multivariate estimators. IEEE Trans Biomed Eng 55(3):902–913

62. van Mierlo P, Carrette E, Hallez H et al (2011) Accurate epileptogenic focus localization through time-variant functional connectivity analysis of intracranial electroencephalographic signals. Neuroimage 56(3):1122–1133

63. Jung YJ, Kang HC, Choi KO et al (2011) Localization of ictal onset zones in Lennox-Gastaut syndrome using directional connectivity analysis of intracranial electroencephalography. Seizure 20(6):449–457

64. Schindler K, Leung H, Elger CE et al (2007) Assessing seizure dynamics by analysing the correlation structure of multichannel intracranial EEG. Brain 130:65–77

65. Schad A, Schindler K, Schelter B et al (2008) Application of a multivariate seizure detection and prediction method to non-invasive and intracranial long-term EEG recordings. Clin Neurophysiol 119(1):197–211

66. Pijn JP, Vijn PCM, Lopez da Silva FH (1989) The use of signal-analysis for the localization of an epileptogenic focus: a new approach. Adv Epileptol 17:272–276

67. Wendling F, Bartolomei F, Bellanger JJ et al (2001) Interpretation of interdependencies in epileptic signals using a macroscopic physical model of the EEG. Clin Neurophysiol 112(7):1201–1218

68. Wendling F, Bartolomei F, Senhadji L (2009) Spatial analysis of intracerebral electroencephalographic signals in the time and frequency domain: identification of epileptogenic networks in partial epilepsy. Philos Trans A Math Phys Eng Sci 367(1887):297–316

69. Wendling F, Chauvel P, Biraben A et al (2010) From intracerebral EEG signals to brain connectivity: identification of epileptogenic networks in partial epilepsy. Front Syst Neurosci 4:154

70. Wendling F, Ansari-Asl K, Bartolomei F et al (2009) From EEG signals to brain connectivity: a model-based evaluation of interdependence measures. J Neurosci Methods 183(1):9–18

71. Bettus G, Wendling F, Guye M et al (2008) Enhanced EEG functional connectivity in mesial temporal lobe epilepsy. Epilepsy Res 81(1):58–68

72. Palmini A (2010) Electrophysiology of the focal cortical dysplasias. Epilepsia 51(Suppl 1):23–26

73. Zoubir AM, Iskander DR (2004) Bootstrap techniques for signal processing. Cambridge University Press, UK

74. Schevon CA, Cappell J, Emerson R et al (2007) Cortical abnormalities in epilepsy revealed by local EEG synchrony. Neuroimage 35(1):140–148

75. Ortega GJ, Menendez de la Prida L, Sola RG et al (2008) Synchronization clusters of interictal activity in the lateral temporal cortex of epileptic patients: intraoperative electrocorticographic analysis. Epilepsia 49(2):269–280

76. Liao W, Zhang Z, Pan Z et al (2010) Altered functional connectivity and small-world in mesial temporal lobe epilepsy. PLoS One 5(1):e8525

Neuromethods (2015) 91: 103–130
DOI 10.1007/7657_2013_65
© Springer Science+Business Media New York 2013
Published online: 12 February 2014

On the Effect of Volume Conduction on Graph Theoretic Measures of Brain Networks in Epilepsy

Manolis Christodoulakis, Avgis Hadjipapas, Eleftherios S. Papathanasiou, Maria Anastasiadou, Savvas S. Papacostas, and Georgios D. Mitsis

Abstract

It is well established that both volume conduction and the choice of recording reference (montage) affect the connectivity measures obtained from scalp EEG, in the time and frequency domains. A number of measures have been proposed aiming to reduce this influence. Our purpose in this work is to establish the extent to which volume conduction and montage influence the graph theoretic measures of brain networks in epilepsy obtained from scalp EEG. We evaluate and compare two standard and most commonly used linear connectivity measures—cross-correlation in the time domain and coherence in the frequency domain—with measures that account for volume conduction, namely, corrected cross-correlation, imaginary coherence, phase lag index, and weighted phase lag index. We show that the graphs constructed with cross-correlation and coherence are affected by volume conduction and montage more markedly; however, they demonstrate the same trend—decreasing connectivity at seizure onset, which continues decreasing in the ictal and early postictal period, increasing again several minutes after the seizure has ended—with all other measures except imaginary coherence. In particular, networks constructed using cross-correlation yield better discrimination between the pre-ictal and ictal periods than the measures less sensitive to volume conduction such as the phase lag index and imaginary coherence. Thus, somewhat paradoxically, although removing the effects of volume conduction allows for a more accurate reconstruction of the true underlying networks this may come at the cost of discrimination ability with respect to brain state.

Key words Epilepsy, Volume conduction, Montage, Scalp EEG, Linear correlations, Graph theory, Complex networks

1 Introduction

Epilepsy is one of the most common neurological disorders of the brain, characterized by sudden and unpredictable seizures. It is a condition that affects 0.6–0.8 % of the world population, and in 15 % of the patients with epilepsy seizures cannot be controlled by antiepileptic drugs or surgery. It is essential for a patient to have a warning that a seizure is about to occur in order to avoid potentially dangerous situations. Moreover, reliable seizure prediction algorithms may enable the implementation of closed-loop therapeutic strategies [1].

The scientific community has continuously performed research towards improvement and development of automated seizure detection and prediction algorithms based on electroencephalographic (EEG) measurements in order to characterize the transition from the inter-ictal to the ictal state (pre-ictal phase) in quantitative terms. Most of these algorithms are based on linear and nonlinear time series analysis techniques of pre-seizure changes in the dynamics of either intracranial or scalp EEG recordings. However, the reported results are not always reliable and/or reproducible, often due to the large number of false positives [1].

In addition to the univariate measures that were employed initially, bivariate and/or multivariate methods have been used more recently in order to study the dynamics of the epileptic brain using EEG. In the last decade or so, many researchers have used complex network analysis—a methodology based on graph theory to investigate the brain. In this context the brain is considered as a complex network of distributed dynamical systems [2], and the underlying substrate of this network is signal correlations in functionally related brain areas. Graph theoretic measures provide a concise way to characterize functional connectivity in health and disease [3,4]; therefore, they may present a potentially useful set of markers for the epileptic brain as well. Different network types, such as random, small-world, and scale-free networks, are associated with different underlying brain properties. For instance, small-world networks that are thought to underpin anatomical and functional brain organization may support more efficient communication and synchronization compared to random networks [5]. Previous studies have provided evidence that epileptic seizures are characterized by changes in functional network features [6–8].

The effect of volume conduction on connectivity measures obtained from scalp EEG is well established; it has been shown that it may influence measures of correlation in the time and frequency domains considerably [9]. Therefore, alternative and/ or modified measures such as the odd part of the cross-correlation function, reduced/imaginary coherence, as well as the standard or the weighted phase lag index (WPLI) have been proposed in order to obtain correlation estimates between different scalp EEG channels that take into account volume conduction and reference effects [9–17]. Nunez et al. [9] introduced "reduced coherence," which is less affected by volume conduction effects, by subtracting the random coherence, i.e., the expected scalp potential coherence arising from uncorrelated neural sources from the measured coherence. This measure exploits the exponential decay of correlation due to volume conduction as a function of electrode separation, but it therefore requires detailed knowledge of all inter-electrode distances, which in turn may be impractical in the standard clinical setting. Another approach is the use of the imaginary part of coherency (IC), which is less sensitive to spurious connectivity due to volume conduction effects [18]. However, the IC was found to be less sensitive with respect to synchronization both in simulated coupled-oscillator models and absence seizure EEG data compared to the PLI, which is a measure of the distribution asymmetry for the phase difference between two signals [16]. An improved PLI, termed the WPLI, whereby the PLI is weighted by the magnitude of the cross-spectrum imaginary part, has been suggested more recently [13]. Moreover, methods that aim to reconstruct a suitable source space and then apply measures to determine functional interactions have been proposed [19–21]. However, the quality of source reconstruction strongly depends on the number of available recording electrodes, which in the clinical setting is typically low (around 21 electrodes).

The construction of brain networks (graphs) usually involves estimating the correlation value between different nodes (i.e., scalp or intracranial EEG electrodes), possibly applying a threshold in order to construct a binary graph, and consequently computing graph-theoretic measures of interest such as degree, and clustering coefficient. Correlation measures forming the substrate of graph theoretical analysis are known to be affected by the choice of reference and also by volume conducted currents from common sources [9,16]. Therefore, it is very likely that the estimated graph-theoretic measures may be influenced as well. This influence has been considered on a limited basis until now to our knowledge, using simulated data in a recent study [12]. This study compared several correlation metrics such as coherence, phase coherence, and PLI and concluded that graph-theoretic measures obtained from these measures were affected by volume conduction. The PLI was found to be affected considerably less by volume conduction than coherence and phase coherence, albeit it yielded networks that exhibited significantly different characteristics than random networks in the case of simulated independent sources. In the present study, we consider the effect of volume conduction and reference choice on graph-theoretic measures using long-duration scalp EEG data collected from patients with epilepsy. In this context, we employ several measures to assess correlation between electrodes, such as cross-correlation, the odd part of correlation, coherence and imaginary coherence, as well as the PLI and WPLI and we compare the graph-theoretic measures that result in each case during the pre-ictal, ictal, and postictal phases. Our study possibly provides a pragmatic account of the effects of volume conduction (and choice of reference) on graph-theoretic measures in the standard clinical setting of EEG recording in epilepsy, where only a limited number of scalp EEG electrodes is available. Preliminary results of this work have been presented in [22].

2   Methods

2.1   EEG Recordings

Long-term video-EEG recordings were collected from patients with epilepsy in the Neurology Ward of the Cyprus Institute of Neurology and Genetics, as part of a diagnostic or a presurgical evaluation. The XLTEK scalp EEG recording system was used. Twenty-one electrodes were placed according to the 10–20 international system (see electrode placement in Fig. 1) with two additional anterotemporal electrodes. Moreover, another four electrodes were used to record the electrooculogram (EOG) and electrocardiographic (ECG) signals, respectively. The data were recorded at a sampling rate of 200 Hz using a cephalic reference that was not part of the scalp derivations used to display the recorded channels. A bandpass filter 1–50 Hz was applied offline. We analyzed data from five patients that had at least one epileptic episode during the monitoring period, whereby seizures were identified and marked by specialized neurophysiologists (coauthors ESP and SSP). However, here we focus on one patient only in order to rigorously examine the effects of montage and volume conduction on the results instead of examining seizure-specific changes in the

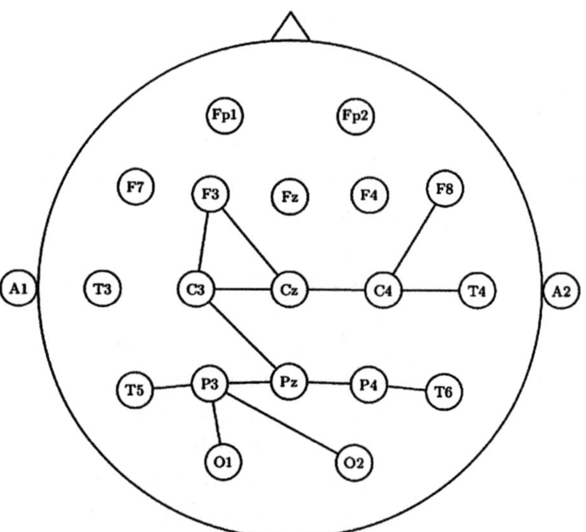

Fig. 1 Example of a functional brain network. If a connectivity measure between a pair of channel recordings (the nodes) exceeds the pre-specified threshold then a connection (i.e., an edge) is placed between the channel recordings. We refer to this set of connections as functional network. When applying graph theoretical measures to scalp EEG, we are faced with two considerable challenges: the fact that all channel recordings depend on a reference recording (the reference problem) and the fact that the electrodes will pick up a mixture of signals arising locally and signal volume conducted from other brain locations (the volume conduction problem)

employed measures. Specifically, in the following sections, we present results from one patient, whose seizure started focally on the EEG at P4-O2 and generalized afterwards. For this patient, we recorded 22 h of continuous scalp EEG signals. The duration of the seizure was 154 s, as marked by the neurophysiologists. Apart from the seizure itself, we analyzed a section of 15 min before and 15 min after the seizure onset.

2.2 Preprocessing

A 50 Hz Notch filter was applied to remove line noise, and subsequently the signals were bandpass filtered between 1 and 45 Hz. Eye artifacts were removed by applying independent component analysis (ICA) from the EEGLAB toolbox of Matlab. In order to assess the possible effects of muscle artifacts, we bandpass filtered the data between 1 and 20 Hz and the results were found to be similar overall; therefore, we present results when the data were preprocessed as mentioned above.

It is known that the montage (i.e., the choice of reference) affects connectivity measures [9] and as a consequence it may affect the corresponding graph-theoretic measures. For this reason, we mathematically converted the input data, which were originally recorded relative to the common cephalic reference, to three different montages: the common reference, the average reference, and the bipolar montage (see Section 2.2.1). We obtained results employing all three montages and compare the results below.

2.2.1 Recording Montages

Scalp EEG recording devices use differential amplifiers to compute the voltage of each EEG channel. A differential amplifier takes as input the measurements of two electrodes and produces the corresponding EEG channel as the difference between the two inputs, after it has been amplified. The choice of input electrodes to each amplifier is known as *montage*.

In the original recordings obtained with our system, each amplifier takes as input one of the 10–20 system electrodes (Fp1, Fp2, F7, F3, Fz, F4, F8, T3, C3, Cz, C4, T4, T5, P3, Pz, P4, T6, O1, O2, A1, A2) and one reference electrode (REF). This is an example of a *common reference montage*, since the reference electrode is common to all amplifiers. Additionally, we have mathematically re-referenced the data to Cz, which is often the reference electrode of choice. The *average reference montage* subtracts the average signal over all channels or a carefully chosen subset of them from the signal at each channel. In this work we used all 19 scalp channels to compute the average.

In the *bipolar montage*, contrary to the previous two montages, there is no input common to all the time series. Instead, pairs of electrodes placed in nearby locations of the scalp are used to obtain the time series by subtracting the corresponding measurements. In one example of such a montage, electrodes are taken in straight lines from the front to the back of the head, forming the pairs Fp1–F7, F7–T3, T3–T5, T5–O1, Fp2–F8, F8–T4, T4–T6, T6–O2, Fp1–F3, F3–C3, C3–P3, P3–O1, Fp2–F4, F4–C4, C4–P4, P4–O2, Fz–Cz, Cz–Pz.

2.3 Functional Network Construction

We calculated pairwise correlations between all pairs of time series (EEG data in common reference, average reference, or bipolar montage), using the connectivity measures described in Section 2.4. Edges were added between node pairs if the corresponding connectivity measure between each pair exceeded a pre-specified threshold, the value of which was dependent on the employed measure. For each measure we experimented with a range of thresholds, obtaining similar results overall in terms of the overall observed characteristics and the effect of volume conduction and reference. We will henceforth refer to the edges and connections obtained in this way as functional networks. Note that, following the above process, the obtained networks are binary and undirected, similar to the one shown in Fig. 1.

2.3.1 Challenges in Constructing Functional Networks from Scalp EEG Recordings

When applying graph theoretical measures to scalp EEG, we are faced with two considerable challenges in terms of obtaining and interpreting the correlation metrics that are used to construct the network edges (Fig. 1): the fact that all channel recordings depend on a reference recording (the *reference problem*) and the fact that the electrodes will pick up a mixture of signals arising locally as well as signal volume conducted from other brain regions (the *volume conduction problem*). With respect to the reference problem, note that any channel recording always reflects a potential *difference* between a recording electrode and a reference electrode. The latter could take various forms such as a common reference electrode (for instance Cz), a mathematically constructed average reference of all electrode signals, or a nearby recording electrode, as in the case of bipolar recordings. We should note that there is no distinction between recording and reference electrode, in the sense that the reference will never be silent and will contain some electrical brain activity [23]; thus, channel recordings will always be sensitive to signals picked up at *both* the recording and reference sites. Further, each channel will contain a mixture of signals arising in the vicinity of the recording and reference electrode(s) and signals from far-away sources that conducted through the head volume to the site of the recording and reference electrodes. To sum up, signals that are used to obtain the network edges will reflect a mixture of activity in the vicinity of recording and reference electrode locations and signal volume conducted from elsewhere.

2.4 Connectivity Measures

Common measures for estimating the correlation between pairs of time series include cross-correlation, coherence, synchronization likelihood, Granger causality, directed coherence, mutual information, PLI, and many more; see, e.g., [24] for a review. In this work we compare the performance of six measures of signal correlation. These include *cross-correlation* and *coherence*, which quantify linear correlations in the time and frequency domain respectively, and are the most commonly used measures for estimating EEG signal correlations. The remaining four measures were the *corrected cross-correlation* (i.e., the odd part of the cross-correlation function), *imaginary coherence*, *PLI*, and *WPLI*. These also reflect linear correlations but are less sensitive to the effects of volume conduction than standard correlation and coherence.

Cross-correlation
Given two time series $x(t)$ and $y(t)$, with $t \in 1 \ldots n$, the normalized cross-correlation function between x and y as a function of the lag, τ is given by

$$C_{xy}(\tau) = \frac{1}{(n-\tau)} \sum_{t=1}^{n-\tau} \left(\frac{x(t)}{\sigma_x} \right) \left(\frac{y(t+\tau)}{\sigma_y} \right),$$

where σ_x and σ_y are the standard deviations of x and y, respectively. C_{xy} is computed for a range of values of the lag τ, which depends on the sampling frequency; e.g., at 200 Hz as in our case, for a required range of $[-100\ 100]$ msec, τ lies within $[-20\ 20]$. Note that when the mean value of both signals is not subtracted beforehand (as done here), the normalized cross-covariance function should be used instead:

$$\sum_{t=1}^{n-\tau} \left(\frac{x(t) - \overline{x}}{\sigma_x} \right) \left(\frac{y(t+\tau) - \overline{y}}{\sigma_y} \right).$$

C_{xy} takes values between -1 and 1, with 1 indicating the largest positive correlation, -1 the largest negative correlation, and 0 no correlation. For the purposes of constructing the graphs and obtaining the graph theoretical measures, the absolute value of the cross-correlation is computed and the correlation between the two signals is computed as $\max_{\tau} |C_{xy}|$, over the desired range of τ.

Corrected cross-correlation
Cross-correlation often takes its maximum at zero lag in the case of scalp EEG measurements (e.g., see Figs. 3 and 4 below). Consistent zero-lag correlations could be due to volume conduction effects: this is because currents from underlying sources are conducted instantaneously through the head volume to the EEG sensors (i.e., assuming that scalp potentials have no delays compared to their underlying sources (quasi-static approximation)). Thus signals arising from a common source will be simultaneously picked up by different electrodes effecting a spurious zero-lag correlation between the two electrode signals. It should be noted though that zero-lag correlations could also be due to a third common source or even true direct physiological interactions. In principle, true direct interactions between any two physiological sources will

typically incur a nonzero delay due to transmission speed, provided that the sampling frequency is high enough to capture such delays. However, consistent nonzero lag correlations are unlikely to be due to the effects of common sources (subsuming volume conduction and reference effects [18]). In order to measure true interactions not occurring at zero lag, we calculate the odd part of cross-correlation, which is a measure of its asymmetry, as defined in [17], by subtracting the negative-lag part of $C_{xy}(\tau)$ from its positive-lag counterpart:

$$\tilde{C}_{xy}(\tau) = C_{xy}(\tau) - C_{xy}(-\tau) \quad \text{for } \tau > 0.$$

Note that $\tilde{C}_{xy}$ provides a lower bound estimate of the nonzero-lag cross-correlations and is notably smaller than C_{xy}.

Coherence
Coherency may be viewed as the equivalent measure of cross-correlation in the frequency domain; it measures the linear correlation between two signals x and y as a function of the frequency f. It is defined as the cross-spectral density between x and y normalized by the auto-spectral densities of x and y:

$$\Gamma_{xy}(f) = \frac{S_{xy}(f)}{\sqrt{S_{xx}(f) S_{yy}(f)}},$$

where S_{xy} is the cross-spectral density and S_{xx} and S_{yy} are the auto-spectral densities of x and y, respectively.

Coherency is a complex number, as the cross-spectral density is complex, whereas the auto-spectral density function is real. Therefore, in many cases coherence (or the squared coherence), which is defined as the magnitude of coherency (or its square), is employed as a measure of correlation in the frequency domain, i.e.:

$$\kappa_{xy}(f) = \frac{\left|\langle S_{xy}(f)\rangle\right|}{\sqrt{\left|\langle S_{xx}(f)\rangle\right|\left|\langle S_{yy}(f)\rangle\right|}}.$$

For finite data sets, coherency is estimated by using standard power spectral density estimation methods, such as the Welch method, i.e., by dividing the EEG signals into segments of equal length (eight segments with 50 % overlap in our case) and averaging the individual spectral estimates. In the above equation, $<\cdot>$ denotes the average over all segments [24].

The value of $\kappa_{xy}(f)$ ranges between 0 and 1, with 1 indicating perfect linear correlation and 0 no correlation between x and y at frequency f. We calculated the maximum coherence value both for the broadband EEG signals (1–45 Hz) as well as for each EEG frequency band (delta (1–4 Hz), theta (4–8 Hz), alpha (8–13 Hz), beta (13–30 Hz), and gamma (30–45 Hz)) separately, in order to assess the degree of correlation between the two signals in these bands.

Imaginary coherence

As stated above, coherency is a complex number. Nolte et al. [18] observed that the imaginary part of coherency is insensitive to volume conduction, while the real part is strongly affected. This is due to the fact that signals arising from the same electrical source in the brain, under the quasi-static approximation, will be volume conducted to any two recording locations with no time delay, thus influencing only the real part of the cross-spectrum. Hence, *imaginary coherence* is defined as the imaginary part of coherency:

$$\mathrm{IC}_{xy}(f) = \mathrm{Imag}\big(\Gamma_{xy}(f)\big).$$

As in the case of coherence, the maximum absolute value in each frequency band quantifies the correlation between the two signals in that band.

Phase lag index

Stam, Nolte, and Daffertshofer [16] proposed the PLI, which is defined as a measure of asymmetry of the phase difference distribution between two signals (in this case, EEG recordings from different electrodes). Following [16], the instantaneous phases were obtained by first bandpass filtering the signals in the frequency bands of interest and then using the Hilbert transform to obtain the phase of the analytic signal. Phase differences $(\Delta\phi)$ between a given pair of channels were wrapped in the interval $-\pi \leq \Delta\phi \leq \pi$:

$$\mathrm{PLI}_{xy} = |\langle \mathrm{sgn}(\Delta\phi(\tau)) \rangle|,$$

where $\Delta\phi$ is the instantaneous phase difference between x and y.

PLI ranges between 0 and 1, with 0 indicating no correlation and 1 maximal correlation.

WPLI

Vinck et al. [13] argued that PLI's sensitivity to noise and volume conduction is hindered by the discontinuity of the measure, which is caused by small perturbations turning phase lags into leads and vice versa. To overcome this problem, they defined the WPLI, which modifies PLI by weighting the contribution of observed phase leads and lags by the magnitude of the imaginary component of the cross-spectrum:

$$
\begin{aligned}
\mathrm{WPLI} &= \frac{\left|\langle \mathrm{Imag}\big(S_{xy}(f)\big)\rangle\right|}{\left|\langle \mathrm{Imag}\big(S_{xy}(f)\big)\rangle\right|} \\
&= \frac{\left|\langle |\mathrm{Imag}\big(S_{xy}(f)\big)| \cdot \mathrm{sgn}\big(\mathrm{Imag}\big(S_{xy}(f)\big)\big)\rangle\right|}{\left|\langle \mathrm{Imag}\big(S_{xy}(f)\big)\rangle\right|}.
\end{aligned}
$$

Similar to PLI, WPLI ranges between 0 and 1, with 0 indicating no correlation and 1 maximal correlation.

2.4.1 Network Properties

Once a functional network is constructed, properties of either the individual nodes or the individual edges or the network as a whole are examined, aiming to identify consistent changes in the pre-ictal, ictal, and postictal periods. In this section we define a number of network properties that characterize a network as a whole rather than examining individual nodes or edges. In the following, let n denote the number of nodes of the network and N the set of all nodes.

Average degree

The *degree*, k_i, of a node i is defined as the number of *neighboring nodes* in the network, i.e., the nodes to which node i is connected (equivalently, the number of edges incident to i). A summary of the degrees of a network is given by the *average degree*, which quantifies how well connected the graph is:

$$K = \frac{1}{n} \sum_{i \in N} k_i.$$

Characteristic path length and global efficiency

Although the average degree reflects the average number of connections any network node may have, it does not provide any information regarding the actual distribution of the edges and, hence, how easy it is for the information to flow in the network. Such information can be captured by the *shortest* (or geodesic) *path length*, $d_{i,j}$, between a pair of nodes, i and j. It is defined as the minimum number of edges that have to be traversed to get from node i to j. Then, the *characteristic path length* is defined as the average shortest path length over all pairs of nodes in the network:

$$L = \frac{1}{n(n-1)} \sum_{\substack{i,j \,\in\, N, \\ i \neq j}} d_{i,j}.$$

Unfortunately, the characteristic path length is only reliable when the network is fully connected; note that for any pair of nodes not connected through a path, the shortest path length $d_{i,j} = \infty$. A workaround for this is to consider only connected pairs of nodes, but this does not reflect the connectivity of the whole network.

To overcome the above problem, Latora and Marchiori [25] defined the *efficiency* between a pair of nodes as the inverse of the shortest distance between the nodes. When such a path does not exist, the efficiency is zero. *Global efficiency* is the average efficiency over all pairs of nodes:

$$E = \frac{1}{n(n-1)} \sum_{\substack{i,j \,\in\, N, \\ i \neq j}} \frac{1}{d_{i,j}}.$$

Clustering coefficient

A *cluster* in a network is a group of nodes that are highly interconnected. The measure of *clustering coefficient*, C_i, [26] of a node i is defined as the fraction of existing edges between neighbors of i over the maximal number of such possible connections:

$$C_i = \frac{2\,t_i}{k_i(k_i - 1)},$$

where k_i is the degree (i.e., the number of neighbors) of node i, and t_i denotes the number of edges between neighbors of i. Then the *global clustering coefficient*, C, of the network is defined as the mean clustering coefficient among all nodes:

$$C = \frac{1}{n} \sum_{i \in N} C_i.$$

3   Results

In this section, we examine the degree to which each connectivity measure is affected by the choice of reference and volume conduction effects in scalp EEG recordings. We first investigate the extent to which each montage is affected by volume conduction and subsequently use the montage that is less affected for further analyses.

3.1   The Effect of Montage on Connectivity Measures

Ideally one wants to estimate the activity of local cortical generators from EEG recordings and then study their interactions. The choice of reference channel will affect both the local cortical estimates and their interactions. It is known that using a common reference can substantially inflate coherence estimates particularly at smaller distances [23] as a common signal is subtracted from all channels. On the other hand, the average reference is known to produce estimates of coherence that are close to coherence estimates obtained from reference-independent potentials [23], possibly because the average reference montage is thought to approximate the local reference free potentials [23]. However, this advantage of the average reference only comes into play with a large number of electrodes and extensive coverage of the head; in the case of a standard 10–20 system as used here it may provide a poor approximation of the reference-free potentials. In the setting of a very limited number of electrodes, as in the case of standard 10–20 recordings routinely used in the clinical setting, the most pragmatic solution for obtaining estimates of local superficial cortical generators is to use a bipolar montage [23], as—in such cases the use of more sophisticated methods such as source reconstruction or spline Laplacians is not possible. What is clear from these considerations is that the choice of the reference electrode will influence the estimate of the local activity (electrode signal) and the estimate of interactions between pairs of such electrodes. Thus, it is also expected to influence the graph theoretic measures of interest. Below we study the effects of choice of reference electrode (montage) on these measures in our data and compare between such montages.

For each 5-s window in the 30-min period that we analyzed, we construct a functional network using one of the connectivity measures described in Section 2.4 (cross-correlation, corrected cross-correlation, coherence, imaginary coherence, PLI, and WPLI) using data from all three montages (bipolar, common reference, and average reference). The threshold we applied was chosen empirically and fixed for a given connectivity measure—though we show in the following subsections that this choice, unless too high or too low, does not affect the overall characteristics of the network properties and the corresponding seizure-related changes. For each connectivity measure, we maintained a constant threshold, regardless of montage, so as to compare the different montages. Figure 2 illustrates the average degree of the network as a function of time (in seconds), whereby each point in the plot corresponds to the average degree of the 5-s window that starts at the second which corresponds to the respective abscissa value (as stated above, windows are taken every 5 s without overlap). The vertical dashed line shows the beginning of the seizure, while the dotted line indicates its end, as marked by the expert neurophysiologists. The average network degree is essentially a measure of the average connectivity in the network as it is a function of the connections that survive the threshold.

It is evident in all measures and all montages that changes in the functional brain networks occur several seconds after the seizure onset (but before the seizure ends) and continue for several minutes. The network generally does not return to its pre-seizure state according to most measures. Specifically, all measures, except the imaginary coherence, indicate a gradual decrease in the network average degree after the seizure onset, which persists for approximately 6 min, followed by a gradual increase, albeit to lower levels than the pre-seizure period. The aforementioned decrease is more pronounced when the network was constructed by standard cross-correlation. The average degree obtained from IC on the other hand increases slightly after the seizure onset and remains higher throughout the remaining period.

PLI, WPLI, imaginary coherence, and corrected cross-correlation are less affected by the choice of reference than cross-correlation and coherence. In the graphs corresponding to cross-correlation and coherence we also observe that the average degree (a measure of the overall level of network connectivity) is largest for the common reference montage (green line), followed by the average reference (red

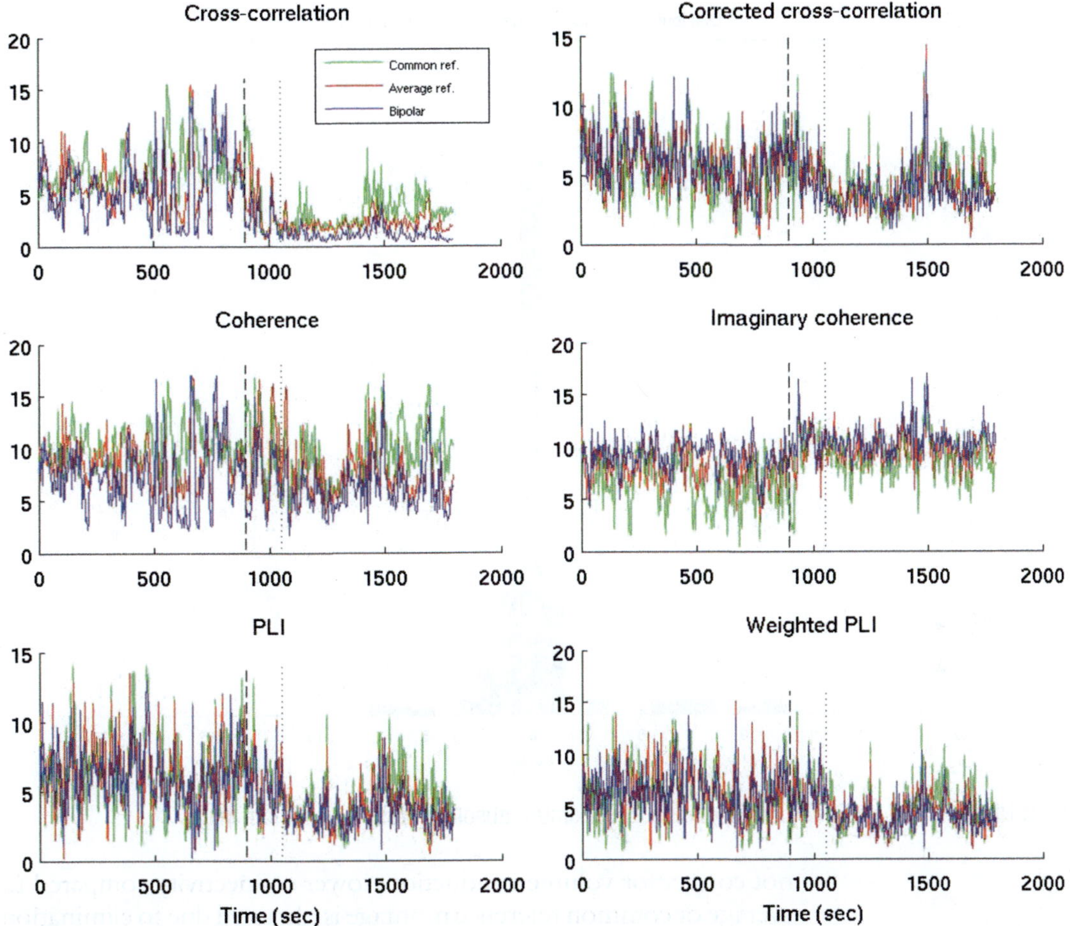

Fig. 2 The average degree of the functional brain networks as a function of time (s); comparison of montages for all connectivity measures

line) and bipolar montages (blue line). This again is consistent with the notion of more pronounced instantaneous spurious correlations at zero lag in the two former montages. Interestingly, imaginary coherence is reversely affected by the montage of choice, with the bipolar montage yielding higher connectivity, followed by the common reference and average reference montages.

We should note that the average degree obtained when using measures sensitive to zero-lag correlations is affected not only by the reference problem but also volume conduction. The bipolar montage yields better estimates of the local gradient of the potential along the scalp surface than a fixed reference at a remote distance [23]. This increases sensitivity and spatial resolution for superficial generators but reduces the sensitivity to distant sources. Thus, we expect that the bipolar montage will be less sensitive to volume-conducted currents. This is consistent with what we observe in the case of the bipolar montage, particularly for cross-correlation and coherence which do

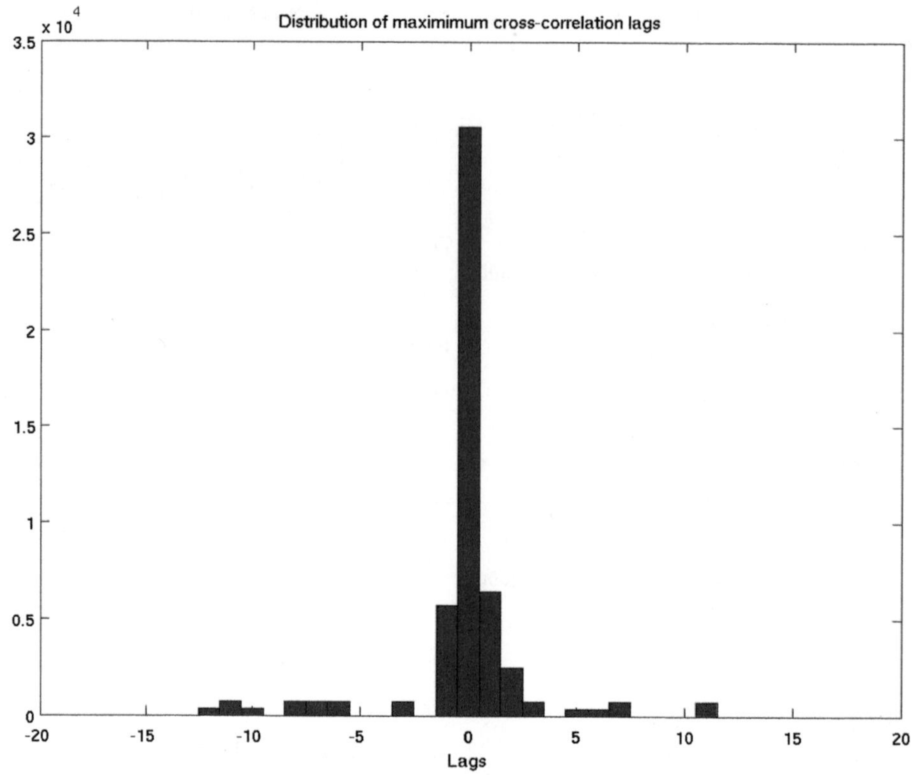

Fig. 3 Histogram of time lag corresponding to maximum absolute cross-correlation values

not correct for volume conduction; lower connectivity compared to average or common reference montage is obtained due to elimination of (a number of) spurious connections. Regarding IC, whereby the bipolar montage yields maximum connectivity in comparison to other montages, this could be attributed to the fact that IC tends to favor significant connections between more remote electrode locations [18] which may subsequently lead to a possible "overcorrection" of volume conduction effects, as discussed further in Section 3.3.1.

In the following sections functional brain networks have been constructed using bipolar montage time series. However we note again that despite the differences, the overall dynamic changes of the connectivity measures as a function of brain state (inter-ictal, ictal, postictal) retained the same basic characteristics regardless of montage.

3.2 Time-Domain Connectivity Measures

In this section we examine the properties of functional brain networks based on time-domain measures, namely, standard and corrected cross-correlation. Pairwise associations between channels were calculated by taking the maximal absolute cross-correlation value among all time lags between -250 and 250 ms (50 samples lead and lag, respectively). In order to illustrate the possible extent

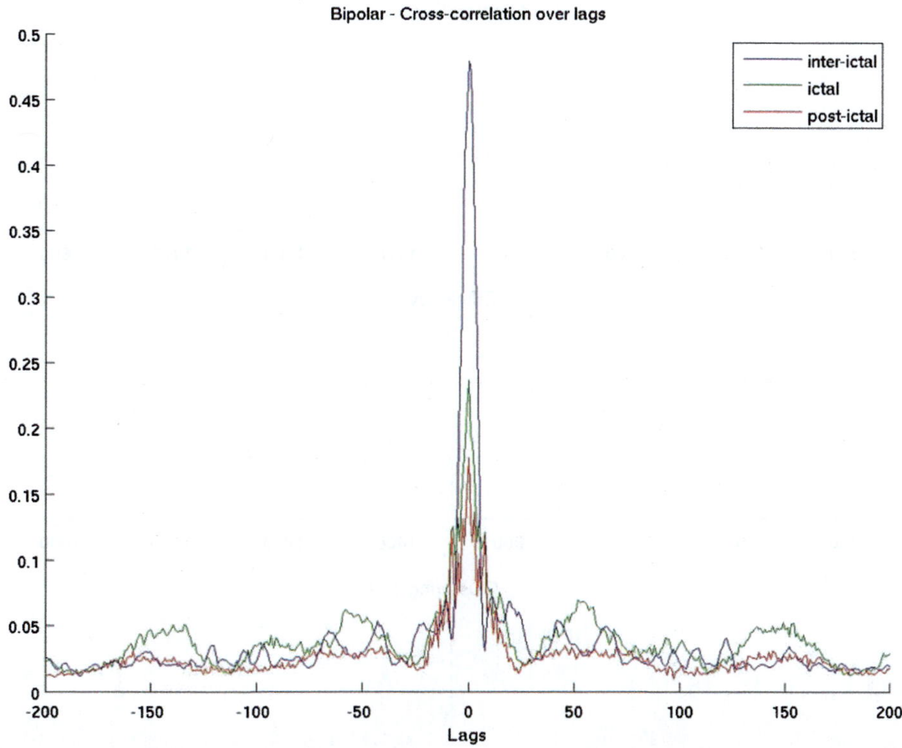

Fig. 4 Absolute cross-correlation as a function of time lag, in the inter-ictal, ictal, and postictal periods

of volume conduction effects, we first investigated the characteristics of the obtained cross-correlation functions by identifying the time lag corresponding to the maximum cross-correlation absolute value. The resulting histogram is shown in Fig. 3.

The majority (59 %) of the maximal values occur at zero lag, followed by those at lags ± 1. Note that when the zero lag was excluded when obtaining the maximum value, the maximum occurred mostly at lags ± 1 (i.e., ± 5 ms) in the vast majority of cases (results not shown). As discussed in Section 2.4, the zero-lag cross-correlations in the setting of scalp EEG are likely to be due to volume conduction; this in turn implies that cross-correlation is affected by the latter.

This effect is also evident in Fig. 4, where the absolute cross-correlation is shown as a function of the time lag, in the inter-ictal, ictal, and postictal periods, averaged over all pairs of channels. We note that the cross-correlation function is mostly symmetric and that the magnitude of the zero-lag component is roughly an order of magnitude larger than the largest nonzero lag components, consistent with observations reported in [17] using resting MEG of normal controls.

Figure 5 illustrates the network properties defined in Section 2.4.1 as a function of time, for the standard and corrected

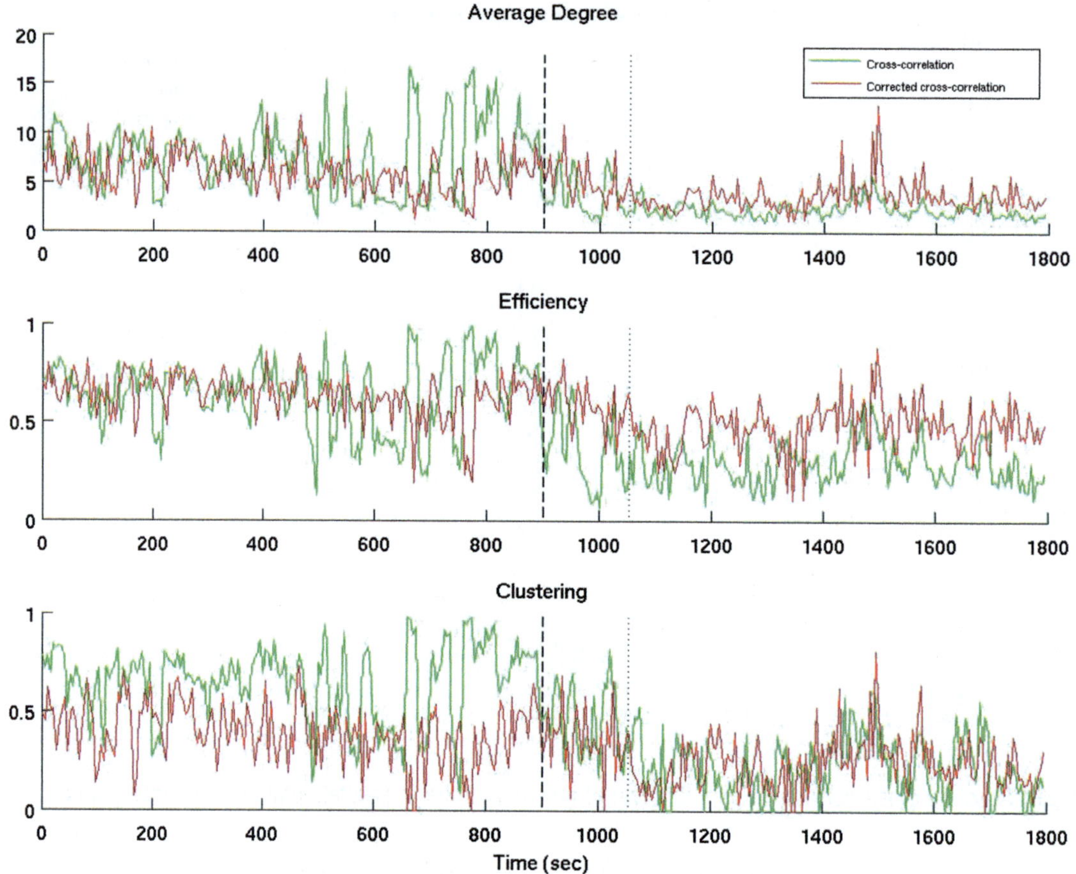

Fig. 5 Network properties of the functional brain networks constructed using cross-correlation (*green line*) and corrected cross-correlation (*red line*)

cross-correlation functional brain networks. Whereas the results from both measures exhibit similar trends before, during, and after the seizure, the seizure-related changes are more pronounced for standard cross-correlation—the drops in average degree, efficiency, and clustering coefficient during the ictal and postictal period are overall steeper. For corrected cross-correlation, the average clustering coefficient is the measure that exhibits the clearest seizure-related changes; whereas the average degree remains relatively constant, clustering drops in the postictal period, suggesting a different topology (less clusters) even though the average network connectivity does not change much. Generally, for the graph theoretical indices we examined, the discrimination ability in terms of modulation by brain state is more pronounced for cross-correlation than for corrected cross-correlation. Note that the thresholds used had different numerical values; the maximum values for the corrected cross-correlation were considerably smaller than their counterparts obtained from standard cross-correlation. We selected these values such that the average degree was comparable in both

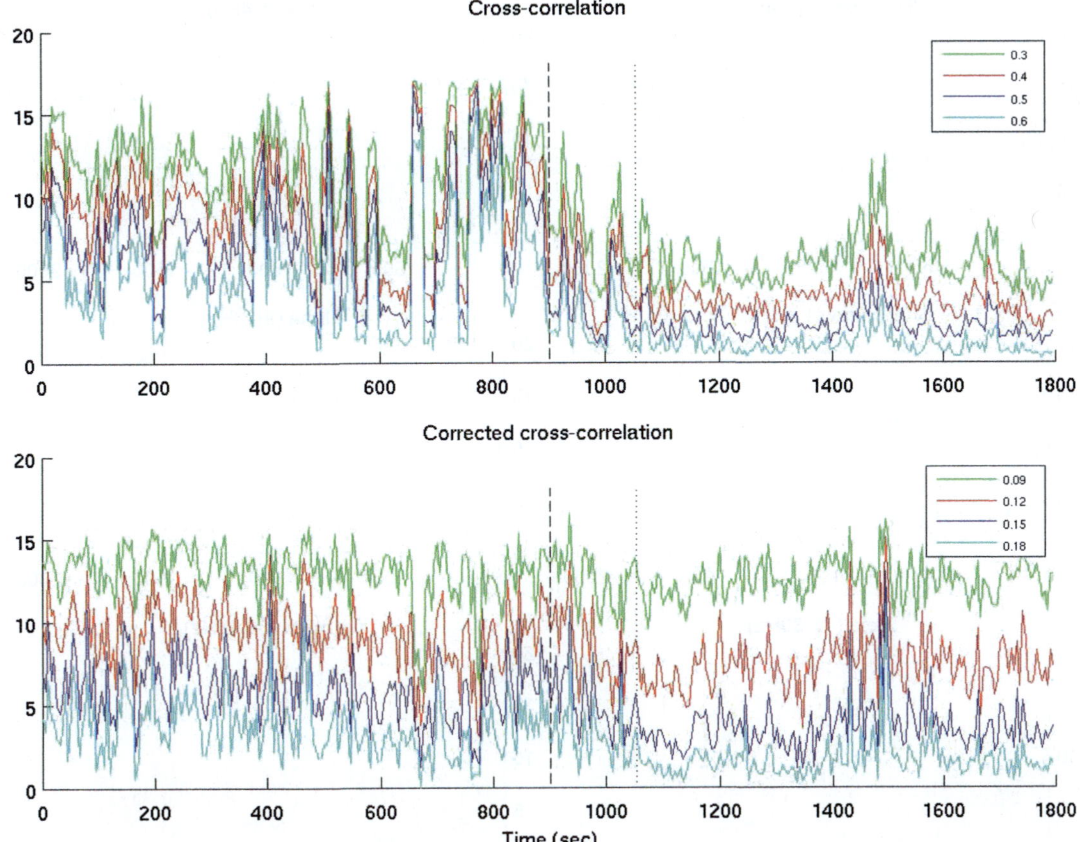

Fig. 6 Average degree of networks constructed with standard and corrected cross-correlation, for various thresholds

cases. As mentioned above, we assessed the possible effects of muscle artifacts by repeating the analysis when the data were bandpass filtered between 1 and 20; the results were qualitatively the same, and hence we do not show these separately.

In order to show that the above results are relatively independent of the threshold choice, we repeated the process for various thresholds. As we can see in Fig. 6 for the average degree, the threshold choice does not affect the general manner in which the functional brain networks evolve in time, with the effect being a vertical shift due to the change in threshold value—e.g., a lower threshold yields more significant connections and hence a higher average degree. Similarly, neither the efficiency nor the clustering measures were affected by the threshold choice, other than being shifted downwards as the threshold increased (results not shown separately).

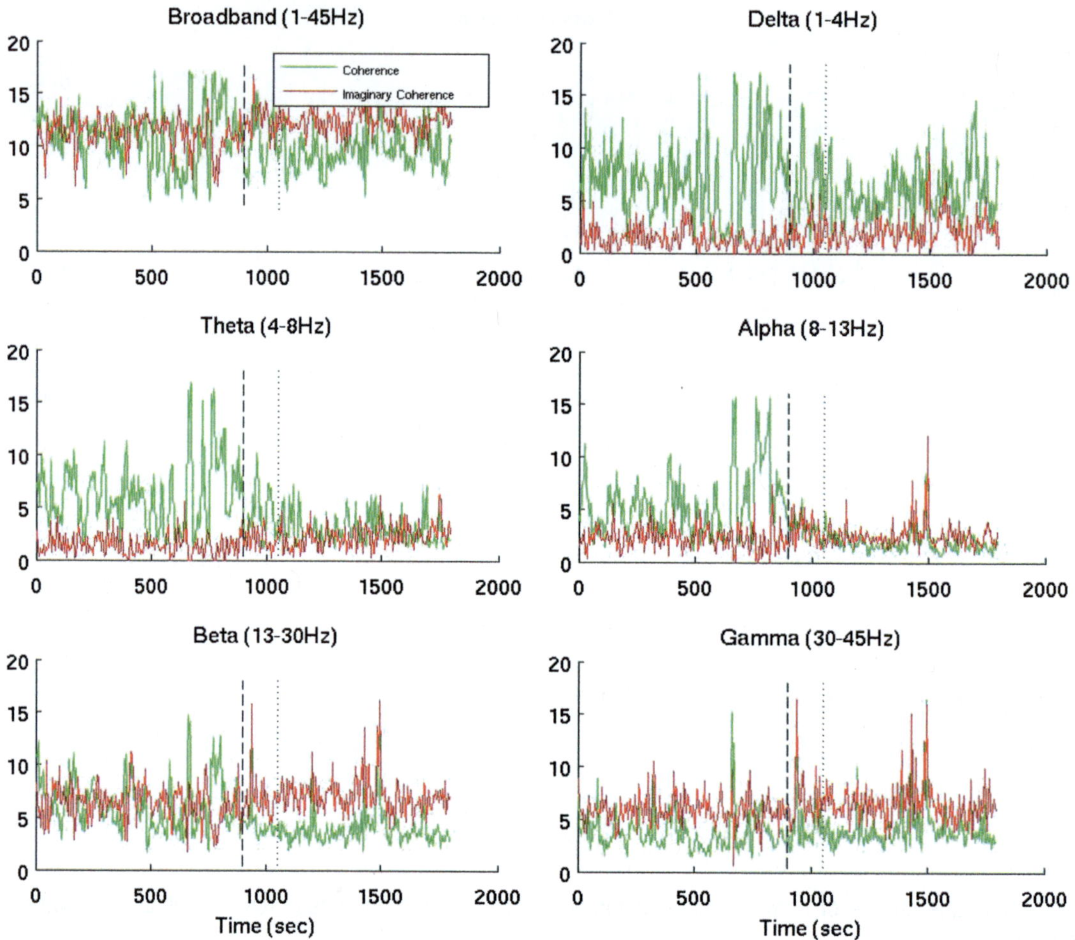

Fig. 7 Average network degree of networks constructed with coherence (green lines) and imaginary coherence (red lines) in different frequency bands

3.3 Frequency-Domain Connectivity Measures

3.3.1 Coherence

Figure 7 illustrates the average degree of the networks constructed with coherence (green line) and imaginary coherence (red line) obtained in the frequency bands of interest for the EEG signal. It can be observed that the clearer drop in the average degree, when the latter was obtained by standard coherence, occurred in the theta and alpha bands and to a lesser degree in the beta band. Also, the average degree for the broadband signal was mostly determined by coherence values in the delta band, as the latter yielded the larger degree values among all bands but did not exhibit such a clear drop. On the other hand, IC exhibits a different behavior overall, with the larger degree values obtained in the higher frequency bands (beta, gamma), which in turn were reflected in the broadband signal. Contrary to coherence, a slight increase is observed in the theta and alpha bands, suggesting that the two measures reflect different aspects of the underlying

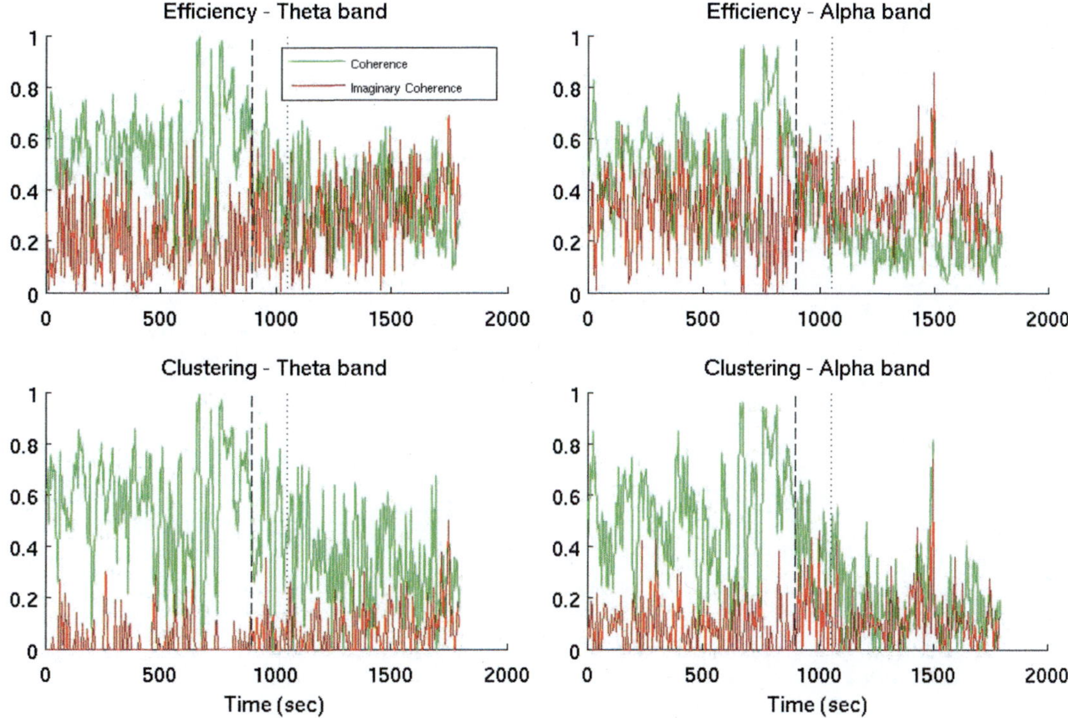

Fig. 8 Efficiency and clustering for the theta (4–8 Hz) and alpha (8–13 Hz) bands

functional interactions. As the theta and alpha bands yielded the clearest drop in average degree, we show the corresponding efficiency and clustering coefficient values in Fig. 8. Similar to the results shown above for cross-correlation, their values obtained by coherence drop markedly in the ictal and postictal period (green lines), whereas no clear patterns emerge when their values were obtained by imaginary coherence (red lines). Again this suggests that imaginary coherence reflects different underlying interactions.

In order to further demonstrate this, we show the top 20 % of most significant connections of representative inter-ictal, ictal, and postictal functional brain networks constructed using coherence and imaginary coherence in Fig. 9, in each of the frequency bands of interest. It can be seen that the two types of networks have only a small fraction of edges in common. Furthermore, it is evident that networks constructed using the coherence measure favor short-range connections, which suggests that coherence may be affected by volume conduction at any stage of the seizure progress (before, during, or after seizure onset). On the contrary, this is not the case with imaginary coherence, in which long-range connections seem to occur most often. Moreover, at low frequencies, we observe some structure in the imaginary coherence networks, from occipital left to frontal right during both the inter-ictal and postictal periods but not during the ictal period.

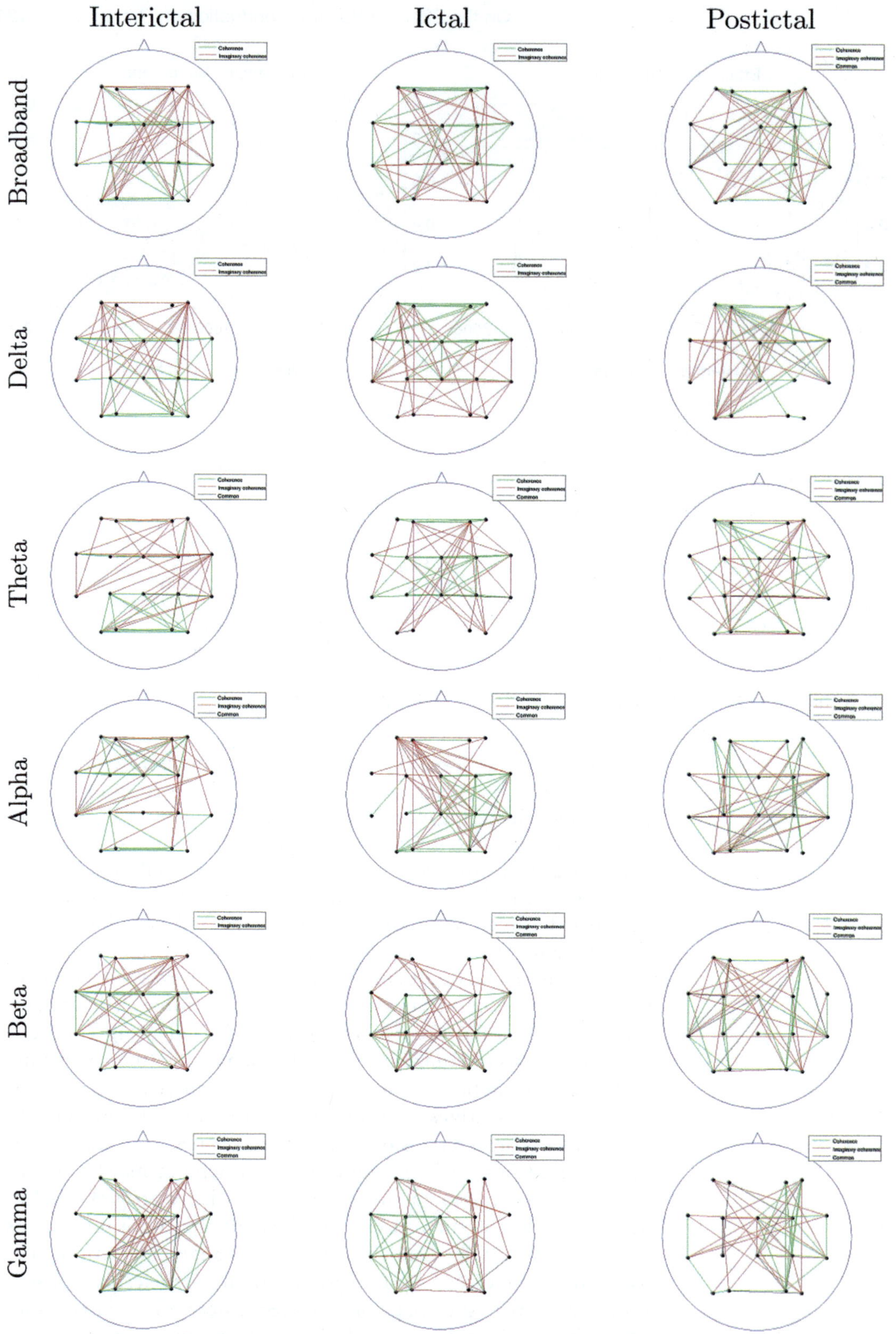

Fig. 9 The top 20 % of connections (in terms of strength) for representative brain networks constructed using coherence and imaginary coherence. *Green* edges correspond to coherence only, *red* to imaginary coherence only, and *black* to both. Note that in most cases coherence and imaginary coherence do not yield the same edges

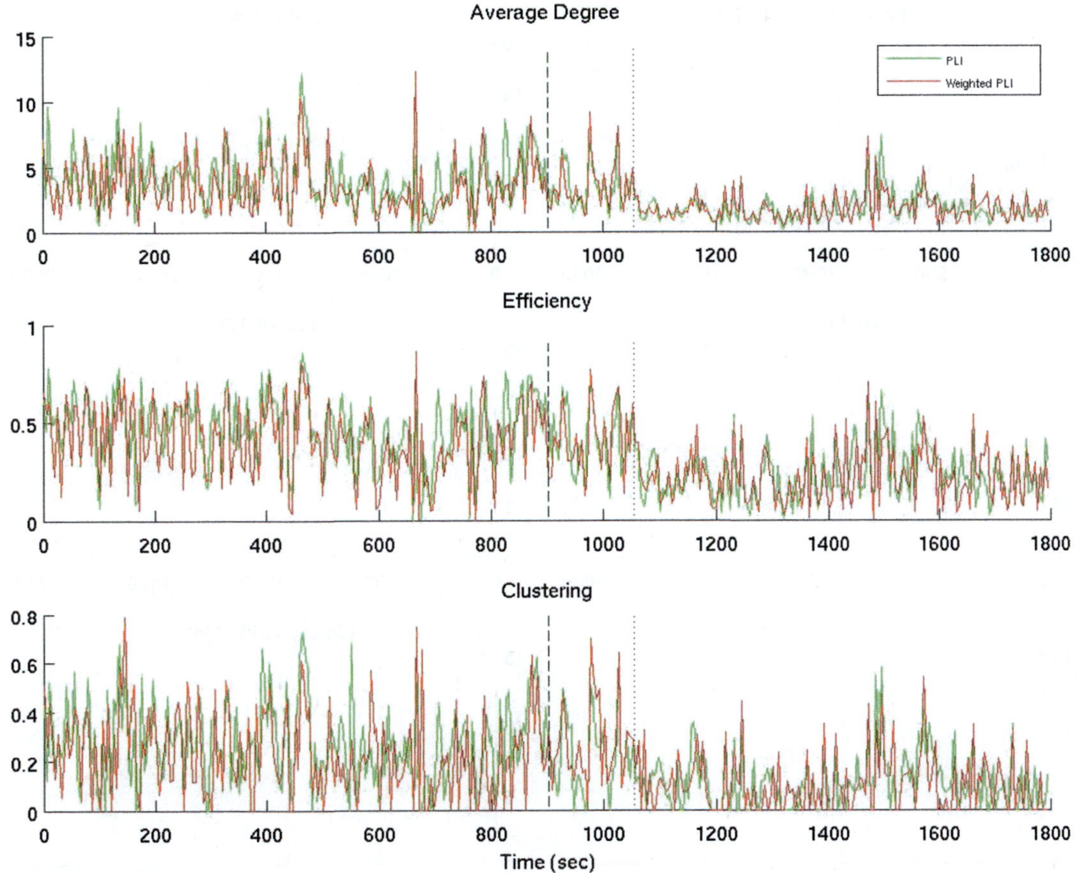

Fig. 10 Functional network properties for PLI vs. weighted PLI as a function of time

3.3.2 PLI and WPLI

Both PLI and WPLI were designed to mitigate the effects of volume conduction [13,16]. As shown in Section 3.1, both the PLI and WPLI were mostly unaffected by the choice of montage. Figure 10 depicts the network properties of PLI and WPLI as a function of time. The same trend is observed as with correlation-based measures and coherence (Figs. 4 and 6), i.e., all three graph-theoretic measures decrease gradually starting at several seconds after the seizure onset until several minutes later, and they increase again after that. The main drops in the three network theoretical measures occur during the postictal period.

In Fig. 11 we demonstrate the contribution of the different frequency bands to the average degree of the networks constructed with PLI and WPLI. We notice that the decrease in the network connectivity that is observed for the broadband signal at seizure onset is mainly due to the beta band. In the remaining bands the

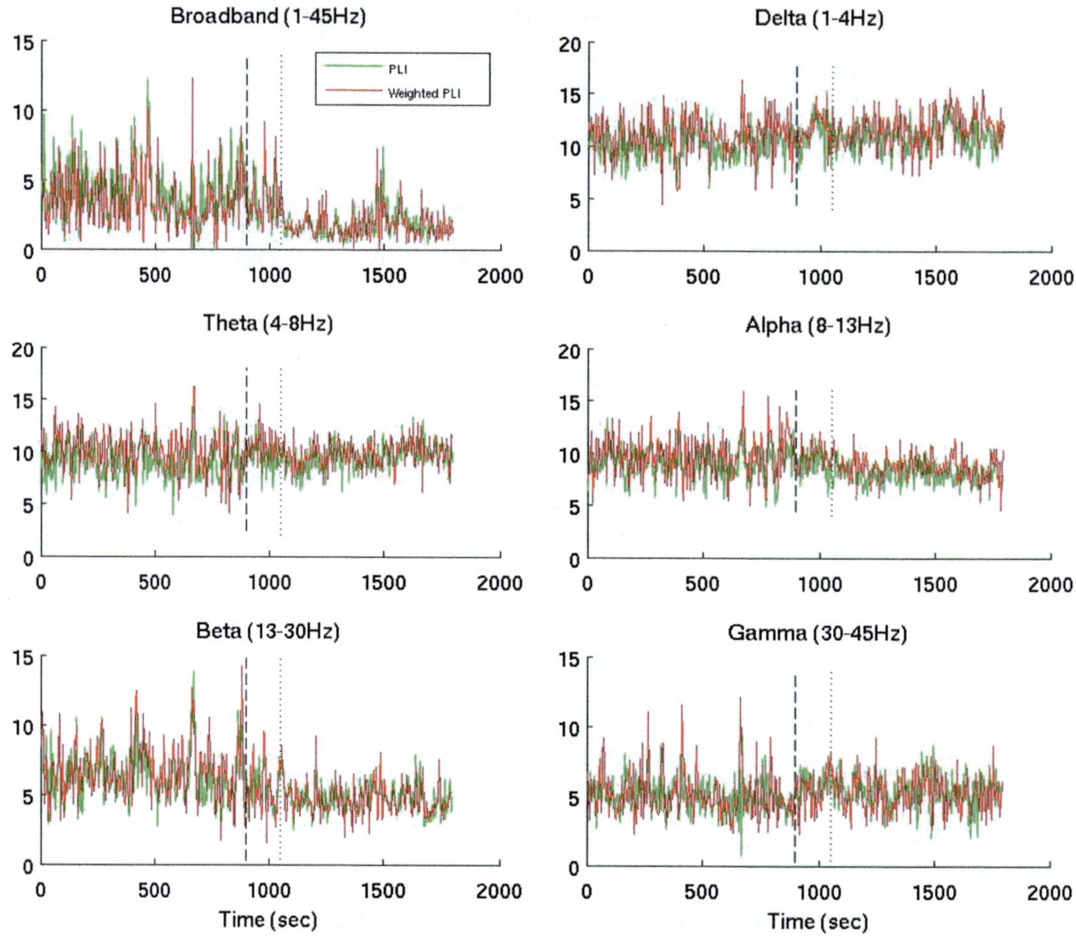

Fig. 11 Average network degree of networks constructed with PLI and weighted PLI, in the EEG frequency bands of interest

PLI- and WPLI-based network average degrees show very little modulation with brain state similar to what is observed with imaginary coherence.

4 Conclusions

During the last decade, bivariate and multivariate methods have been widely applied, in conjunction with complex network analysis, to study the dynamics of the epileptic brain using EEG. In the standard clinical setting of long-term EEG recordings in patients with epilepsy, only scalp EEG recorded from a limited number of electrodes is typically available. However, network analyses based on signal correlations of scalp EEG recordings are affected by the choice of reference electrode (montage) and volume conduction and more generally the fact that a common signal is picked up by two electrodes leading to spurious correlation at zero lag. Here, we investigated the choice of reference and effects of volume conduction on six bivariate signal correlation measures—cross-correlation, corrected cross-correlation, coherence, imaginary coherence, PLI, and WPLI. The results show that both montage and volume conduction have considerable effect on the obtained results, though in our data they do not appear to change the overall nature of the seizure-related changes in the obtained graph-theoretic measures (average degree, efficiency, clustering coefficient). Interestingly, our data also suggest that in some cases, measures that are more prone to the effects of volume conduction (e.g., standard cross-correlation) may yield results that are more sensitive in detecting seizure-related changes in brain state (inter-ictal, ictal, postictal, see Fig. 2). That said, clustering and average degree measures of corrected-cross-correlation (Fig. 5) and PLI/WPLI (Fig. 10) also show changes when moving from the inter-ictal to the ictal and postictal periods.

Choice of reference and volume conduction: Of the six measures we investigated, those mostly affected by the choice of recording reference (montage) were the standard cross-correlation and, to a lesser degree, coherence. In these two measures, the average reference and the common reference montages resulted in higher connectivity (as reflected in the average degree, see Fig. 2) than the bipolar montage. Corrected cross-correlation, PLI, and WPLI were the least affected measures, whereby all three montages yielded similar connectivity patterns. Interestingly, imaginary coherence, contrary to all other measures, was reversely affected by the montage, with the bipolar montage resulting in higher connectivity than the other two montages. The subtraction of a common reference potential in the case of the common reference montage is instantaneous; therefore, it may produce spurious correlations at zero lag. In the case of average reference, this would only produce a good approximation of the reference-free potentials, given sufficient electrode coverage of the head [23]; thus, in our case it may still result in considerable common signal being subtracted out instantaneously from each electrode derivation. In case of the bipolar recordings, different signals are used as references in each electrode derivation. Therefore, measures that are sensitive to zero-lag correlations are expected to (1) be more sensitive to montage change and (2) show inflated correlations due to common signal in the common and average reference montage but less so in the bipolar montage. Our results (Fig. 2) are consistent with this notion.

Network connectivity during succession of inter-ictal, ictal, and postictal states: In terms of network connectivity, all methods but imaginary coherence produced the same trend in the network evolution before, during, and after the seizure onset. Namely, all three connectivity measures we used to analyze the networks—average degree, efficiency, and clustering coefficient—decreased a few seconds after the seizure onset and remained lower than their pre-seizure values in the postictal period (Figs. 5, 7, 8, and 10). Imaginary coherence was the only measure that did not yield this decreasing trend but remained relatively unchanged (with possibly a slight increasing trend—Figs. 7 and 8).

The average degree (overall level of connectivity) obtained by coherence (Fig. 7) is modulated strongly by brain state in similar fashion as when based on cross-correlation (Fig. 5). In particular in the delta, theta, and alpha band marked decreases are observed in the ictal and postictal periods. In contrast, the average degree when based on imaginary coherence shows little variation with brain state. The possible exceptions to this are the alpha and gamma band where, contrary to the case of coherence, the average degree increases in the ictal and postictal period. This different behavior may be explained by the fact that coherence (i.e., the squared magnitude of coherency) was mostly determined by the real part of the cross-spectrum between different electrodes, which was found to be significantly larger than the cross-spectrum imaginary part, which in turn determines IC. Hence, the observed changes in the coherence-related changes reflect seizure-related changes in the real part of the cross-spectrum. Moreover, coherence and IC were found to yield different patterns of significant connections, with the first favoring short-range connections and the second long-range connections (Fig. 9).

The observation of high clustering coefficient and efficiency values pre-ictally for cross-correlation and coherence (Figs. 5 and 8) may be explained by the presence of a high number of short-range (local) connections due to the effects of volume conduction on these measures. During the seizure and even more so in the postictal period, this topology seems to be disrupted and the clustering coefficient and efficiency drop steeply, along with a simultaneous steep decrease in the average network degree. For the corrected cross-correlation (Fig. 5), whereas the average degree remains relatively constant, the clustering drops and is actually the measure that exhibits the clearest seizure-related changes, suggesting a change in the network topology.

Our data is consistent with a substantial decrease in overall network correlation that starts in the ictal period but is most pronounced in the postictal period. The most marked changes occur in the alpha and theta bands (Fig. 7). This is consistent with the observations reported by Muller et al. [27] using scalp EEG, who found a decrease in correlation in the lower frequencies (predominantly in the alpha and theta bands), whereby the most pronounced decrease also occurred in the postictal state. In addition to the changes in correlation, we also observe a decrease in the clustering coefficient and the efficiency (i.e., an increase in path length), which is again most pronounced in the postictal state (Figs. 5 and 8). Thus our data is consistent with a more regular network during the inter-ictal period, which mainly becomes disconnected during the ictal and postictal periods.

Previous studies applying graph-theoretic measures to intracranial recordings have reported global and local changes in network topology during the pre-ictal and inter-ictal phases [6]. During seizures, it has been observed that these networks develop through an apparent topological progression [28], whereby the topology changed severely in organization. Specifically, Kramer et al. [28] observed that at seizure onset, at seizure termination, and in the postictal interval, the number of edges of the brain network increased compared to the middle of the seizure and the pre-ictal interval using ECoG recordings. In terms of topology, they observed that during the pre-ictal and initial ictal intervals the largest subnetworks exhibited small-world topologies while just before seizure termination these subnetworks evolved toward a more random configuration, in agreement with previous observations [29].

Our results agree only partially with these observations. However, it is probably not instructive to directly compare our data to these studies because of the marked biophysical differences of the two sets of signals. Intracranially measured signals are generated locally and at a markedly different spatial scale. At different spatial scales, varying (and even opposing) seizure-related changes in synchronization may ensue [28]. In fact, in concurrent recordings of intracranial and scalp EEG it has been shown that ictal patterns as observed at the scalp are not a direct reflection of the underlying intracranial ictal activity but can be a manifestation of subsequent processes such as seizure development, synchronization, and propagation [30].

Also, it should be noted that our results were found to be dependent on seizure type. Specifically, the overall pattern of seizure-related changes presented here was observed in three out of the five subjects that we analyzed. This in turn may relate to modulation of the overall connectivity in different directions between generalized seizures, where synchronization increases (e.g., [15]) and focal-onset seizures such as the main data in our paper and in [27], where decreased synchronization was observed. However, one should be cautious in making generalizations even in this case, as in one of the focal epilepsy patients we examined, an overall reverse trend of what is shown above was observed. This suggests that seizure type and possibly even focus location may influence the results to a significant degree. For instance, it is likely that in the case of absence seizures such as the ones examined in [16], in addition to short-range connections, long-range connections may contribute to the pronounced increase in network connectivity (global synchronization), which does not seem to be the case with the data presented here. Furthermore, one should be cautious in directly comparing our results with respect to efficiency and clustering coefficient, which seem to agree more with the studies of Kramer et al. [6, 26], as the overall changes observed in network connectivity as reflected by average degree were markedly different. Note however that despite the seizure type-related differences that we observed in our data with respect to the modulation of graph-theoretic measures by brain state, the effects of reference choice and volume conduction were found to be similar in all subjects.

The seizure-related modulations (i.e., the differences between inter-ictal, ictal, and postictal states) were more profound in the case of the simple cross-correlation and coherence, the graph measures of which are dominated by zero-lag correlations and thus most likely affected by volume conduction. Thus we arrive at a seeming paradox, whereby network measures sensitive to volume conduction offer better discriminability than network measures that are designed to mitigate volume conduction in order to provide a more accurate account of the underlying network connectivity. This is in contrast with observations by [16], where it was reported that PLI was better discriminating than measures sensitive to volume conduction. The main difference in our data lies in the direction of the modulation in synchronization, which in turn may be related to seizure type: Stam et al. examined absence seizures in [15], where they show an increased level of synchronization; our main results here in contrast correspond to focal-onset seizures where synchronization decreases.

How can we understand the seemingly paradoxical observation of lower discriminability of measures corrected for common signals? One possibility is this: Our data suggest that during the inter-ictal period, connectivity is high but mostly dominated by short-range zero-lag components, particularly in the alpha and theta bands, i.e., consistent with resting state activity. During the seizure and postictally, this activity is disrupted, and thus the zero-lag correlations decrease. However, the modulation of this zero-lag component may add to seizure-state discriminability. The measures corrected for zero-lag correlations also exhibit seizure state-related changes (corrected cross-correlation and PLI), but they are insensitive to the aforementioned zero-lag component and thus appear less discriminating in terms of seizure state. In other words, our data seems to indicate that if the primary aim was to discriminate seizure state, rather than characterize the true underlying network, measures sensitive to volume conduction could be more appropriate, at least in the case presented here. This remains to be verified by further work in more subjects.

In conclusion, we examined the effects of volume conduction and montage on graph-theoretic measures of epileptic brain networks obtained from scalp EEG. We compared classical connectivity measures with measures that account for common signals due to reference and volume conduction. We found that despite the fact that common signals did influence graph theoretical measures, the overall direction and pattern of seizure state-related changes were similar between the two sets of measures, with the exception of imaginary coherence. The type of seizure and overall network connectivity level play an important role in the results obtained from graph theoretical measures of epileptic networks.

References

1. Mormann F, Andrzejak RG, Elger CE, Lehnertz K (2007) Seizure prediction: the long and winding road. Brain 130(Pt 2):314–333
2. Stam CJ, Reijneveld JC (2007) Graph theoretical analysis of complex networks in the brain. Nonlinear Biomed Phys 1(1):3
3. Bullmore ET, Sporns O (2009) Complex brain networks: graph theoretical analysis of structural and functional systems. Nat Rev Neurosci 10(3):186–198
4. Rubinov M, Sporns O (2010) Complex network measures of brain connectivity: uses and interpretations. Neuroimage 52 (3):1059–1069
5. Achard S, Bullmore ET (2007) Efficiency and cost of economical brain functional networks. PLoS Comput Biol 3(2):e17
6. Kramer MA, Kolaczyk ED, Kirsch HE (2008) Emergent network topology at seizure onset in humans. Epilepsy Res 79(2–3):173–186
7. Ponten SC, Bartolomei F, Stam CJ (2007) Small-world networks and epilepsy: graph theoretical analysis of intracerebrally recorded mesial temporal lobe seizures. Clin Neurophysiol 118(4):918–927
8. Christodoulakis M, Anastasiadou M, Papacostas SS, Papathanasiou ES, Mitsis GD (2012) Investigation of network brain dynamics from EEG measurements in patients with epilepsy using graph-theoretic approaches. In: 2012 IEEE 12th international conference on bioinformatics & bioengineering (BIBE), 11–13 November 2012, pp 303–308. doi: 10.1109/BIBE.2012.6399693
9. Nunez PL, Srinivasan R, Westdorp F, Wijesinghe RS, Tucker DM, Silberstein RB, Cadusch PJ (1997) EEG coherency I: statistics, reference electrode, volume conduction, Laplacians, cortical imaging, and interpretation at multiple scales. Electroencephalogr Clin Neurophysiol 103(5):499–515
10. Nunez PL, Silberstein RB, Shi Z, Carpenter MR, Srinivasan R, Tucker DM, Doran SM, Cadusch PJ, Wijesinghe RS (1999) EEG coherency II: experimental comparisons of multiple measures. Clin Neurophysiol 110 (3):469–486
11. Guevara R, Luis J, Velazquez P, Nenadovic V, Wennberg R, Senjanovi G, Velazquez JLP, Senjanovic G, Dominguez LG (2005) Phase synchronization measurements using electroencephalographic recordings: what can we really say about neuronal synchrony? Neuroinformatics 3(4):301–314
12. Peraza LR, Asghar AUR, Green G, Halliday DM (2012) Volume conduction effects in brain network inference from electroencephalographic recordings using phase lag index. J Neurosci Methods 207(2):189–199
13. Vinck M, Oostenveld R, van Wingerden M, Battaglia F, Pennartz CM (2011) An improved index of phase-synchronization for electrophysiological data in the presence of volume-conduction, noise and sample-size bias. Neuroimage 55 (4):1548–1565
14. Thatcher RW (2012) Coherence, phase differences, phase shift, and phase lock in EEG/ERP analyses. Dev Neuropsychol 37(6):476–496
15. Haufe S, Tomioka R, Nolte G, Müller K-R, Kawanabe M (2010) Modeling sparse connectivity between underlying brain sources for EEG/MEG. IEEE Trans Biomed Eng 57 (8):1954–1963
16. Stam CJ, Nolte G, Daffertshofer A (2007) Phase lag index: assessment of functional connectivity from multi channel EEG and MEG with diminished bias from common sources. Hum Brain Mapp 28(11):1178–1193
17. Nevado A, Hadjipapas A, Kinsey K, Moratti S, Barnes GR, Holliday IE, Green GG (2012) Estimation of functional connectivity from electromagnetic signals and the amount of empirical data required. Neurosci Lett 513 (1):57–61
18. Nolte G, Bai O, Wheaton L, Mari Z, Vorbach S, Hallett M (2004) Identifying true brain interaction from EEG data using the imaginary part of coherency. Clin Neurophysiol 115 (10):2292–2307

19. David O, Friston KJ (2003) A neural mass model for MEG/EEG: coupling and neuronal dynamics. Neuroimage 20(3):1743–1755

20. Gross J, Kujala J, Hämäläinen M, Timmermann L, Schnitzler A, Salmelin R (2001) Dynamic imaging of coherent sources: studying neural interactions in the human brain. Proc Natl Acad Sci 98(2):694–699

21. Lehmann D, Faber PL, Gianotti LRR, Kochi K, Pascual-Marqui RD (2006) Coherence and phase locking in the scalp EEG and between LORETA model sources, and microstates as putative mechanisms of brain temporo-spatial functional organization. J Physiol Paris 99(1):29–36

22. Christodoulakis M, Hadjipapas A, Papathanasiou ES, Anastasiadou M, Papacostas SS, Mitsis GD (2013) Graph theoretic analysis of scalp EEG brain networks in epilepsy – the influence of montage and volume conduction. In: 2013 IEEE 13th international conference on bioinformatics & bioengineering (BIBE), 10–13 November 2013

23. Nunez PL, Srinivasan R (2006) Electric fields of the brain: the neurophysics of EEG. Oxford University Press, New York

24. Pereda E, Quiroga RQ, Bhattacharya J (2005) Nonlinear multivariate analysis of neurophysiological signals. Prog Neurobiol 77(1–2):1–37

25. Latora V, Marchiori M (2001) Efficient behavior of small-world networks. Phys Rev Lett 87 (19):198701

26. Watts DJ, Strogatz SH (1998) Collective dynamics of 'small-world' networks. Nature 393(6684):440–442

27. Müller MF, Baier G, Jiménez YL, Marín García AO, Rummel C, Schindler KA (2011) Evolution of genuine cross-correlation strength of focal onset seizures. J Clin Neurophysiol 28(5):450–462

28. Kramer MA, Eden UT, Kolaczyk ED, Zepeda R, Eskandar EN, Cash SS (2010) Coalescence and fragmentation of cortical networks during focal seizures. J Neurosci 30(30):10076–10085

29. Schindler KA, Bialonski S, Horstmann M-T, Elger CE, Lehnertz K (2008) Evolving functional network properties and synchronizability during human epileptic seizures. Chaos 18(3):033119

30. Pacia SV, Ebersole JS (1997) Intracranial EEG substrates of scalp ictal patterns from temporal lobe foci. Epilepsia 38(6):642–654

Neuromethods (2015) 91: 131–157
DOI 10.1007/7657_2014_68
© Springer Science+Business Media New York 2014
Published online: 2 August 2014

Methods for Seizure Detection and Prediction: An Overview

**Giorgos Giannakakis, Vangelis Sakkalis,
Matthew Pediaditis, and Manolis Tsiknakis**

Abstract

Epilepsy is one of the most common neurological diseases and the most common neurological chronic disease in childhood. Electroencephalography (EEG) still remains one of the most useful and effective tools in understanding and treatment of epilepsy. To this end, many computational methods have been developed for both the detection and prediction of epileptic seizures. Techniques derived from linear/nonlinear analysis, chaos, information theory, morphological analysis, model-based analysis, all present different advantages, disadvantages, and limitations. Recently, there is the notion of selecting and combining the most robust features from different methods for revealing various signals' characteristics and making more reliable assumptions. Finally, intelligent classifiers are employed in order to distinguish epileptic state out of normal states. This chapter reviews the most widely adopted algorithms for the detection and prediction of epileptic seizures, emphasizing on information theory based and entropy indices. Each method's accuracy has been evaluated through performance measures, assessing the ability of automatic seizure detection/ prediction.

Keywords EEG, Epilepsy, Seizure detection, Seizure prediction, Entropy, Nonlinear analysis, Morphological analysis, Seizure modeling, Classification

---

1 Introduction

Epilepsy is a chronic neurological disorder characterized by neurological brain dysfunctions resulting in epileptic seizures. An epileptic seizure is a transient occurrence of signs and/or symptoms due to abnormal, excessive, and synchronous neuronal activity in the brain [1]. It is estimated that in 2012 between 0.4 and 1 % of the world's population (around 50 million people) have active epilepsy [2]. Among many diagnostic and imaging methods, the electroencephalogram (EEG) is by far the most used and effective technique in the daily clinical treatment. Given that it is noninvasive, is relatively accurate, and has low cost, it has been established as a necessary tool for clinicians and people who deal with epilepsy.

Until recently, seizures were identified only visually by an expert neurologist. However, this procedure constitutes a laborious task especially in the case of long-term EEG recordings.

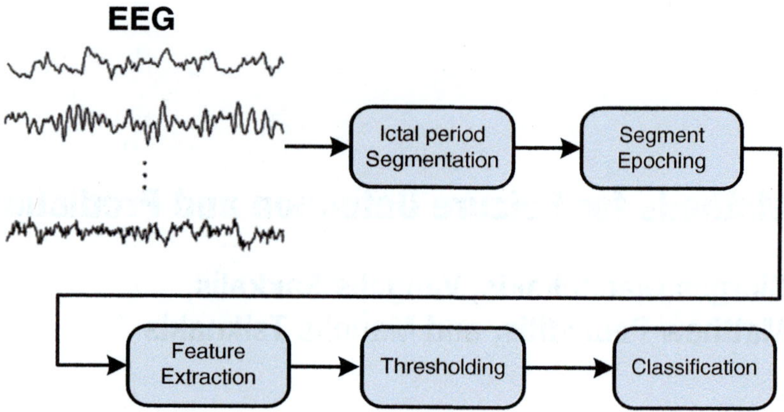

Fig. 1 Schematic representation of seizure detection system

Therefore, automatic computer aided algorithms have evolved in order to shorten and automate this procedure and many seizure detection methods are reported in the international literature [3, 4]. Figure 1 shows a schematic representation of a seizure detection system.

Most studies present their solution to the problem of seizure detection in the context of a decision support system for the neurologist expert. As there are many types of seizures, this is sometimes a difficult task, taking into account the nature, temporal length, and singularities of each seizure type. When a patient experiences seizures of different types one needs to categorize EEG ictal periods into a specific type, although some epileptic syndromes are difficult to be characterized as being of specific category. A more demanding task, which is still considered an open scientific question, is the prediction of a seizure [5], which profoundly will improve the quality of life of people suffering from severe seizures. Besides, the understanding of underlying mechanisms leading in seizures and the origin of a seizure is in each case are still under question.

Towards this direction, many EEG analysis algorithms have been proposed. Linear analysis has been widely used based mainly on synchronization features as a primer and straightforward approach. Although these methods can reveal in some cases the existence of epileptic seizures, they have their limits if someone takes the nature of real human EEG data into account [6, 7]. Under this prism, EEG signals can be interpreted as the result of a system containing highly nonlinear elements. The study of nonlinear EEG dynamics can reveal hidden information and provide a more complete picture about underlying brain processes [8, 9]. Nonlinear analysis has been used with increased accuracy over the last decade in the area of seizure detection and prediction.

2   Methods

In this section, linear methods (highlighting time–frequency methods), nonlinear methods (highlighting measures derived from information theory), methods based on signal's morphological characteristics, and vision-based methods are presented.

2.1   Linear Methods

Linear methods have been widely used in the area of epilepsy detection mainly due to their simplicity and versatility. One of the simplest linear statistic metrics is the variance of the signal. It offers an insight into dynamics underlying the EEG and is usually calculated in consecutive windows. A further linear method is based on the autocorrelation function, exploiting the periodic nature of seizures. Liu et al. [10], using Scored Autocorrelation Moment (SAM) analysis, distinguished EEG epochs containing seizures with an accuracy of 91.4 % although signals did not present differences in their spectral properties.

Furthermore, seizure onset and offset determination may be succeeded using linear prediction filters (LPF) [11]. An LPF estimates the spectral characteristics of a signal, with its accuracy depending on the stationarity of the latter. When there are spikes, sharp waves or rapid changes, the filter's prediction error increases, leading to an identification of a possible seizure.

Discrete wavelet transform (DWT), which is a transformation extracting scale-frequency components from data (each component with resolution matched to its scale) may also be applied in seizure detection. In [12], normal and seizure signals were classified with an accuracy of 99.5 % using DWT and a linear classifier. Another linear feature, the relative fluctuation index [13], can measure the intensity of the fluctuation of EEG signals, which is defined as:

$$F_i = \sum_{j=1}^{M-1} \left| a_i(j+1) - a_i(j) \right|,$$

where a_i denotes the amplitude of the filtered EEG signal at the ith band with length M. During a seizure, there is a larger fluctuation in the EEG signals than during an ictal-free period. Therefore, values of fluctuation index during a seizure are usually larger than during rest EEG. Using this index, along with other features, a study by Yuan et al. [13] achieved 94.90 % mean accuracy in a segment-based aspect and 93.85 % mean accuracy in an event-based aspect.

2.1.1   Time–Frequency Methods

Various studies that employ time–frequency features have also been used in the area of seizure detection [14]. Hassanpour et al. [15] used time–frequency patterns as signatures in order to detect seizures.

Tzallas et al. [16–18] performed an extensive investigation of well-known time–frequency distributions, extracting features from the Power Spectral Density (PSD) time–frequency grid followed by artificial neural network (ANN) classification. With this methodology, an accuracy between 89 and 100 % for three different datasets was achieved using reduced interference time–frequency distribution and ANNs. A time–frequency matched filter was introduced in [19, 20] in order to reveal seizure patterns. Rankine et al. [21] proposed a related methodology analyzing changes in preictal, ictal, and postictal states. Moreover, an improved time–frequency dictionary in terms of reconstruction accuracy and discrimination between seizure and non-seizure states is presented in [22].

2.2 Nonlinear Methods

Epileptic seizures can be seen as manifestations of intermittent spatiotemporal transitions of the human brain from chaos to order [23]. Nonlinear analysis of EEG has attracted increasing interest by many research groups mainly because it incorporates the non-stationary nature of a signal. It perceives brain mechanisms as part of a macroscopic system in a way to understand its spatio-temporal dynamic properties. The revealed underlying information of ongoing EEG leads to promising results not only in the detection but also in the prediction of upcoming seizures [24].

2.2.1 Fractal Dimension

Fractal dimension is a nonlinear time domain measure characterizing the complexity of a time series. The degree of complexity increases if the fractal dimension increases. Various algorithms have been developed [25] in order to calculate the fractal dimension such as box counting [26], Katz's algorithm [27], Petrosian's algorithm [28], and Higuchi's algorithm [29]. According to the last, the time series $x(i), i = 1, 2, \ldots, N$ formulates the vector

$$X_m^k = \left\{ x(m), x(m+k), \ldots, x\left(m + \left\lfloor \frac{N-m}{k} \right\rfloor \cdot k \right) \right\},$$

where k is the time lag, $m = 1, 2, \ldots, k$ and $\lfloor y \rfloor$ the round down integer of argument y. For each X_m^k, the average is formed:

$$L_m(k) = \frac{\sum_{i=1}^{\left\lfloor \frac{(N-m)}{k} \right\rfloor} \left| x(m+ik) - x(m+(i-1)k) \right| (n-1)}{\left\lfloor \frac{N-m}{k} \right\rfloor \cdot k}$$

Finally, the sum of averages is calculated as

$$L(k) = \sum_{m=1}^{k} L_m(k)$$

The linear estimation of the slope of the curve $\ln(L(k))$ versus $\ln(1/k)$ is an estimate of the fractal dimension.

2.2.2 Lyapunov Exponent

Lyapunov exponent (λ) is a nonlinear metric measuring the exponential divergence of two time series trajectories in phase space. Considering the m-dimensional time vector of a time series $X = \{x(t), x(t + 1), \ldots, x(t + m - 1)\}$ and two neighboring points X_{t_0} and X_t in phase space at time t_0 and t respectively, the distances of the points in the ith direction are $dX_i|_{t_0}$ and $dX_i|_t$ respectively. Given the following equation,

$$dX_i|_t \approx e^{\lambda_i t} dX_i|_{t_0}$$

the Lyapunov exponents are λ_i.

Finally, the maximal Lyapunov exponent can be defined as

$$\lambda_{\max} = \lim_{t \to \infty} \lim_{dX_i|_{t_0} \to 0} \lim \frac{1}{t} \ln \frac{dX_i|_t}{dX_i|_{t_0}}$$

and measures the biggest increase rate of the error in the initial conditions. Lyapunov exponents characterize the chaotic nature of a time series, i.e., a slight shift in initial conditions can lead to a non-deterministic difference in the phase space trajectory. Using Lyapunov exponents and recurrent neural networks, Guler et al. [30] achieved 96.79 % accuracy rate in the detection of epileptic seizures.

On the other hand, Kannathal et al. [31] tested nonlinear measures including the correlation dimension (CD), maximal Lyapunov exponent, Hurst exponent (H), and Kolmogorov–Sinai entropy (K–S entropy) in order to distinguish epileptic from normal EEG activity. All measures showed high discriminating ability, with slightly better results being reported for the CD (p-value: 0.0001) and K–S entropy (p-value: 0.0001).

2.3 Information Theory Based Analysis and Entropy

Entropy is a physical measure derived from thermodynamics, describing the order or disorder of a physical system. High entropy values equal to high levels of disorder of a system, whereas low values describe a more ordered system, capable of producing more work. Signal processing and analysis research disciplines borrowed entropy from information theory in order to address and describe the irregularity, complexity, or unpredictability characteristics of a signal. Given these properties, entropy has been widely used towards automatic seizure detection [3, 32, 33].

2.3.1 Shannon Entropy

After some initial approaches by H. Nyquist and R. Hartley, research leaders at Bell Labs, Shannon established in 1948 quantitatively the foundations of information theory [34]. According to these, a signal is divided into J non-overlapping value bins and the ratios of samples falling into j^{th} bin to the total samples N are calculated

$$H = -\sum_{j=1}^{J} p_j \log_2\left(p_j\right) \quad \text{where} \quad p_j = \frac{N(x_j)}{N},$$

where $N(x_j)$ is the amount of samples that fall into bin j of total J bins to the total samples N. EEG Shannon entropy has been correlated with desflurane effect compartment concentrations [35]. It has also been used in order analyze long term EEG coming from patients with frontal lobe epilepsy [36].

2.3.2 Spectral Entropy

Spectral Entropy was introduced by Inouye [37, 38] measuring the proportional contribution of each spectral component to the total spectral distribution [39].

$$H = -\sum_{j=1}^{J} p_j \log_2\left(p_j\right) \quad \text{where} \quad p_j = \frac{S_j}{S},$$

where S is the total spectrum and S_j is the spectrum at frequency bin j of total J bins.

A traditional approach to estimate spectral entropy is through Fourier power spectrum [40, 41], which is applicable mainly where a signal's stationarity conditions are satisfied, e.g., the resting EEG. However, many clinical applications are highly non-stationary with transient and rapid changes in their spectra distributions. In addition to that, a time-varying entropy index is necessary in some cases [42]. This can be partially dealt with the short time Fourier transform (STFT) revealing spectral distributions over successive windows [40]. However, this approach faces the intrinsic problem of window size selection that arises from the Heisenberg Fourier Uncertainty Principle [43].

$$\Delta t \Delta f \geq \frac{1}{4 \cdot \pi}$$

Due to this, a small window size increases temporal resolution but makes spectral resolution poor whereas a wide window size achieves the opposite effect. It is considered that the optimal distribution is a Gaussian that minimizes the product of time–frequency variances [44].

To overcome these limitations, Quiroga et al. [45] introduced wavelet entropy (WE) which is based on multi-resolution decomposition by means of the wavelet transform (WT). This technique has already been applied in EEG/ERP signal analysis [46–48]. The problem with this approach is that the results are strongly dependent on the selection of the mother wavelet function. WE was efficiently applied in order to discriminate between EEG signals of controls and epileptic patients [49–51]. Rosso et al. [52] compares the Gabor transform and the wavelet transform claiming the superiority of the second because a variable window is used for the analysis. Subsequently, the time evolution of wavelet entropy and relative wavelet entropy was investigated, showing significant decrease during ictal periods. However, different wavelet basis functions can produce different results, making their interpretation sometimes ambiguous.

In order to yield an optimal time–frequency distribution and subsequently time–frequency spectral entropy, adaptive algorithms are used. Adaptive Optimal Kernel (AOK) time–frequency representation [53] is an effective method in representing signals in the time frequency plane. The main advantage of having an adaptive signal-dependent method is that in each case the kernel is selected according to how well it is suited to signal's characteristics. The method is adjusted by the choice of the kernel which involves a compromise between cross term reduction, loss of time–frequency resolution, and maintenance of certain properties of distribution [44, 54]. Spectral entropy by using this method was presented in [55].

2.3.3 Approximate Entropy

Approximate entropy (ApEn) was introduced by Pincus [56] to quantify the regularity and predictability of a time series data of physiological signals. Being a modification of the Kolmogorov–Sinai entropy [57], it was especially developed for determination of the regularity of biologic signals in the presence of white noise [58].

Given a time series $X(n) = \{x(n)\} = \{x(1), x(2), \ldots, x(N)\}$ of N samples, the ApEn value is calculated through the following steps:

1. The vector sequences $X_m(i) = \{x(i), x(i+1), \ldots, x(i+m-1)\}$ which represent m consecutive values commencing with the ith point are formed.

2. The distance between $x_m(i)$ and $x_m(j)$ is calculated, defined by

$$d[X_m(i), X_m(j)] = \max_{1 \leq k \leq m} \{|x(i+k-1) - x(j+k-1)|\}$$

3. For each $x_m(i)$ the number $N_i^m(r)$ of vectors is calculated

$$d[X_m(i), X_m(j)] \leq r$$

with r representing the noise filter level.

Then, the parameters $C_i^m(r)$ are estimated as,

$$C_i^m(r) = \frac{N_i^m(r)}{N - m + 1}$$

4. $\phi^m(r)$ is defined as the mean value of the parameters C_i^m:

$$\phi^m(r) = \frac{\sum_{i=1}^{N=m+1} \ln C_i^m(r)}{N - m + 1}$$

5. $\mathrm{ApEn}(m,\ r,\ N)$ is calculated using $\phi^m(r)$ and $\phi^{m+1}(r)$ as

$$\mathrm{ApEn}(m, r, N) = \phi^m(r) - \phi^{m+1}(r)$$

ApEn has already been used in many applications such as analysis of heart rate variability [59–62], analysis of the endocrine hormone release pulsatility [63], and detection of epilepsy [64, 65]. The majority of studies indicate that during ictal periods ApEn presents a significant decrease in comparison with EEG during normal periods [3, 65, 66]. ApEn was also used in order to classify EEG signals among five different states (including an ictal state) with an increased accuracy [67].

The calculation of ApEn depends on the parameters embedding dimension (m), noise filter level (r), and data length (N). Besides, it is arguable whether the standard deviation used at the noise filter level would be calculated from the original data series or from the individual selected EEG segments. However, there is no specific guideline for their optimal determination even though most research studies use the parameters described in [56, 61] as a rule of thumb. But ApEn statistics do not present relative consistency [61] leading to problems in hypothesis formulation and testing. As signals of different source and pathologies can have quite different properties, these parameters should be determined, based on the specific use. The need for a consistent determination of parameters was studied in a recent work [68] where a preliminary analysis of these parameters was established.

2.3.4 Sample Entropy

Sample entropy (SampEn), which is presented in [61], also estimates complexity in time series providing an unbiased measure regarding the length of time series.

$$H = \ln\left(\frac{A^m(r)}{B^m(r)}\right)$$

The calculation of sample entropy starts with the steps 1 and 2 already described for the ApEn calculation. The following steps are given below:

3. For each $X_m(i)$ the number $N_i^m(r)$ of vectors is calculated

$$d[X_m(i), X_m(j)] \leq r$$

with r representing the noise filter level.

Then, the parameters $B_i^m(r)$ and $B^m(r)$ are defined as,

$$B_i^m(r) = \frac{N_i^m(r)}{N - m - 1}$$

$$B^m(r) = \frac{1}{N - m} \sum_{i=1}^{N-m} B_i^m(r)$$

4. The dimension is incremented to $m = m + 1$ and the number $N_i^{m+1}(r)$ is calculated so that

$$d[X_{m+1}(i), X_{m+1}(j)] \leq r$$

Then, the parameters $A_i^{m}(r)$ and $A^{m}(r)$ are defined as

$$A_i^{m}(r) = \frac{N_i^{m+1}(r)}{N - m - 1}$$

$$A^{m}(r) = \frac{1}{N - m} \sum_{i=1}^{N-m} A_i^{m}(r)$$

5. Finally, sample entropy is defined as

$$\mathrm{SampEn}(m, r) = \ln\left[\frac{B^{m}(r)}{A^{m}(r)}\right]$$

The advantage of SampEn is that its calculation is independent of time series size as it restricts self-matches and uses a simpler calculation algorithm, reducing execution time [61]. However, despite its advantages, SampEn is not widely used [69]. Sample entropy was used as feature for automatic seizure detection in [70]. It was also applied in [71] combined with Lempel–Ziv as indicators to discriminate focal myoclonic events and localize myoclonic focus.

2.3.5 Kullback–Leibler Entropy

Kullback–Leibler entropy (K–L entropy) measures the degree of similarity between two probability distributions and can be interpreted as a method quantifying differences in information content [72]. K–L entropy was applied to intracranial multichannel EEG recordings and indicates its ability to detect seizure onset based on spectral distribution properties [73].

2.3.6 Lempel–Ziv Complexity

The Lempel–Ziv measure estimates the rate of recurrence of patterns along a time series, reflecting a signal's complexity. Lempel–Ziv has been applied to epileptic EEG signal showing increased values during ictal periods [74]. Another work has applied LZ complexity on E-ICA and ST-ICA transformed signals in an attempt to isolate seizure activity [75]. Both of these studies have been applied to limited datasets, pointing out the need of a more thorough evaluation.

2.3.7 Permutation Entropy

Permutation entropy is a measure of complexity introduced by Bandt [76]. Its application to absence epilepsy on rats indicated superiority on prediction of epileptic seizures and identification of preictal periods (54 % detection rate) comparing with sample entropy [77]. The same study achieved 98.6 % correct identification of interictal periods.

2.3.8 Order Index

Order index is another nonlinear feature that was proposed by Ouyang [78] measuring the irregularity of non-stationary time series. In a recent work [79], a comparative analysis of order index along with other linear and nonlinear features was performed.

2.4 Morphological Analysis

Most algorithms in the literature select features based on amplitude, spectral properties, and synchronization of ongoing EEG in order to identify a seizure. However, little progress has been made in order to incorporate a neurologist's experience in analyzing a waveform morphology and shape for making a decision on optimal epilepsy treatment. Some studies working towards shape analysis of epileptic seizures give quite promising results not only by their techniques themselves but also by the prospect of integrating the present and future perception of neurologists' expertise. In this way, Deburchgraeve et al. [80] extracts segments that morphologically resemble seizures by a combined nonlinear energy operator and a spikiness index. Then a detector is applied exploiting the repetitive nature of a seizure is applied. The spike and wave complexes of epileptic syndromes can also be extracted by a two-stage algorithm, the first enhancing the existence of spikes and the second applying a patient-specific template matching [81]. Interictal spikes have also been detected using Walsh transform in addition to the fulfillment of clinical criteria establishing a simulated epileptic spike [82, 83].

2.5 Vision-Based Analysis

In some cases, epilepsy monitoring is performed with synchronized video and EEG recordings. Epileptic syndromes are evaluated based not only on scalp recordings but also on human motion features extracted from video sequences. Analysis involves mainly detection of the myoclonic jerks, eye motion (eyeball doze, eyeball upwards roll, eyelid movements), head jerks and movements, facial expressions (mouth, lips malformations). However, it can be understood that a seizure specific organization and combination of motion features should be applied in order to provide better results [84]. This promising area of research helps neurologist experts have a more complete picture preventing them from false alarms and leading to decision support with increased accuracy. A thorough review can be found in [85]. Vision-based analysis in epilepsy can be divided into two categories, marker-based and marker-free techniques. Marker-based techniques track objects/markers placed in representative parts of the human body that convey information related to the epileptic manifestation. On the other hand, marker-free techniques use advanced image processing and computer vision tools to extract motion-related information directly from the image sequences in the video. Both techniques return time-varying signals, which form the basis for further feature extraction in the time- and frequency domain. The extracted features finally feed a classifier such as an ANN or a decision tree with the aim to detect epileptic seizures.

3   Seizure Prediction

Nowadays, seizure detection is considered practically an issue that has been solved with satisfactory accuracy. On the other hand, seizure prediction remains an open scientific problem in a way that there is no consistent approach for predicting a seizure accurately within a significant amount of time before it occurs. However, many algorithms have been tested in their ability to forecast seizures.

3.1   Early Approaches

The notion of seizure prediction was firstly mentioned in 1975 [86] based on spectral analysis of EEG data collected from two electrodes. In 1981, Rogowski et al. [87] investigated preictal periods using pole trajectories of an autoregressive model. Gotman et al. [88] investigated rates of interictal spiking as indicators of upcoming seizures.

3.2   Linear Methods

3.2.1   Statistical Measures

Among other measures Mormann et al. [89] investigated the statistical moment of the EEG amplitudes in order to detect the preictal state. Other linear measures like power have been used in [90] and signal variance has been used in [91] to predict seizure onset.

3.2.2   Hjorth Parameters

Hjorth parameters, namely, activity, mobility, and complexity, are time domain parameters useful for the quantitative evaluation of EEG [92]. The parameter of activity represents the variance of signal's amplitude, the mobility represents the square root of the ratio between the variances of the first derivative and the amplitude, and the complexity is derived as the ratio between the mobility of the first derivative of the EEG and the mobility of the EEG itself. Mormann et al. used Hjorth parameters among others as features for seizure prediction [89]. Mobility has been also used followed by SVM classification achieving better false positive rates (fpr) in comparison with plain spectral analysis [93].

3.2.3   Accumulated Energy

The accumulated energy is computed from the average energy across all values of the signal x of a window k

$$E_k = \frac{1}{N_\mathrm{w}} \sum_{i=1}^{N_\mathrm{w}} x_k^2(i),$$

where N_w is the window length.

Then, the average of ten values of average energies are added to the running accumulated energy.

$$\mathrm{AE}_m = \frac{1}{10} \sum_{k=10m-9}^{10m} E_k + \mathrm{AE}_{m-1},$$

where $m = 1,2,\ldots,N/N_\mathrm{w}$ and $\mathrm{AE}_0 = 0$.

This measure can be considered as the running average of average window energies. Accumulated energy has been used in various studies of seizure prediction [94–96].

3.2.4 AR Modelling

In [97], a feature extraction and classification system was proposed based on Autoregressive Models, SVM and Kalman Filtering. Its performance regarding false positives rates per hour is quite promising with a mean prediction time ranging from 5 to 92 min.

3.3 Nonlinear Methods

Most of the nonlinear methods exploit the reconstruction of a time series $x(i)$, $i = 1, 2, \ldots, N$ in phase space domain forming the m-dimensional time delayed vectors

$$X_m(i) = \{x(i), x(i + 1 \cdot \tau), \ldots, x(i + (m - 1) \cdot \tau)\},$$

where m is the embedding dimension and τ is the time delay.

This reconstruction conveys important information about the nonlinear dynamics of a signal and it is used to many methods some of them described below.

3.3.1 Lyapunov Exponent

The calculation method of Lyapunov exponents was analyzed in the previous section of this chapter. Iasemidis et al. [98–100] applied for the first time nonlinear analysis to seizure prediction. The idea behind this approach is that the transition from normal to epileptic EEG is reflected by a transition from chaotic to a more ordered state, and therefore, the spatiotemporal dynamical properties of the epileptic brain are different for different clinical states.

3.3.2 Dynamical Similarity Index

Dynamical similarity index is a method introduced in [101] which calculates brain state dynamics through phase state reconstruction and compares a running window state against a reference window with the use of the cross correlation integral. Various studies using this index have shown promising results even in the detection of preictal states of temporal lobe epilepsy [102, 103] and neocortical partial epilepsy [104].

3.3.3 Correlation Dimension

In order to calculate the correlation dimension, the correlation integral defined by Grassberger and Procaccia [105] is needed

$$C_m(\varepsilon, N_m) = \frac{2}{N_m(N_m - 1)} \sum_{i=1}^{N_m} \sum_{j=i+1}^{N_m} \Theta\big(\varepsilon - \|x_i - x_j\|\big),$$

$$N_m = N - (m - 1) \cdot \tau,$$

where Θ is the Heaviside function. This integral counts the pairs (x_i, x_j) whose distance is smaller than ε. Then, the correlation dimension D is defined by

$$D = \lim_{\substack{\varepsilon \to 0 \\ N \to \infty}} \frac{\partial \ln C_m(\varepsilon, N_m)}{\partial \ln \varepsilon}$$

In preictal states, drops in correlation dimension were observed making this measure able to identify states preceding seizures [95, 106, 107].

3.3.4 Entropies

Zandi et al. [108] used entropic measure of positive zero-crossing intervals achieving 0.28 false positive rate and average prediction time 25 min.

Kolmogorov entropy (KE) is a nonlinear measure of the rate at which information about the state of a system is lost [109] and defined as

$$KE = \lim_{\substack{\varepsilon \to 0 \\ m \to \infty}} \frac{1}{\tau} \ln \frac{C_m(\varepsilon, N_m)}{C_{m+1}(\varepsilon, N_m)}$$

where C_m, C_{m+1} are the correlation integrals of embedding dimensions m, $m + 1$, respectively and τ the time delay.

In time-varying EEG signal the temporal evolution of KE is required. Time series can be divided in N_w windows, and the mean KE [110] can be calculated as

$$KE = \frac{1}{N_w} \sum_{k=1}^{N_w} KE(k)$$

Higher positive values of KE suggest a chaotic behavior of the system, whereas lower values suggest a more ordered system. KE has been applied to seizure anticipation in infant epilepsy [90].

3.3.5 Phase Synchronization

Phase synchronization is a nonlinear bivariate measure that has been applied widely in the field of seizure prediction. It is represented through mean phase coherence

$$R = \left| \frac{1}{N} \sum_{k=1}^{N} e^{i[\varphi_a(t_k) - \varphi_b(t_k)]} \right|$$

In [111, 112] a detection of preictal state based on phase synchronization decrease is presented. This decrease can be attributed to interactions between channels that are located to epileptogenic focus (pathological synchronization) and others located out of seizure focus [113, 114].

4 Combination of Methods and Feature Selection

Detection methods can produce one or more features which are representative measures describing information needed to each problem establishment. Different features can reveal different aspects of information a signal contains. Due to this, different features can achieve high sensitivity, some others high specificity and vice versa.

A combination of methods is necessary in order to exploit most of available information. In this case, an efficient feature selection procedure is required in order to highlight the most robust features capable to produce best results. It is reported that when using SVM as a classifier, which is considered an efficient classifier, the selected features affect the performance more than the used classifier itself [115]. Towards the same direction, omitting good features may be more detrimental for SVMs than including bad ones [116].

A study of EEG-based neonatal seizures [117] used total 55 features derived from time domain, frequency domain and information theory achieving remarkable results (good detection rate (GDR): 89 % at 1 false detection per hour (FD/h)).

In a recent study [79], nonlinear features and features derived from information theory were evaluated in surface EEG recordings. Signals were recorded from 8 epileptic patients whose seizures identified and classified by a neurologist expert as absence like seizures. The recordings were collected from 21 scalp loci of the international 10–20 system with all electrodes referenced to the earlobe. An electrode placed in the middle of the distance between Fp1 and Fp2 on the subject's forehead served as ground. For this study, 75 seizures were selected within artifact free EEG time series (without eye blinks, spikes, head movements, chewing, general discharges), as other related studies [3] indicate. Figure 2 shows an example of that study including ApEn, Order Index, and Multiscale Variance Index (MVI).

An important factor in the effectiveness of methods is the proper determination of their intrinsic parameters. In a recent study [68], a preliminary analysis of parameters used on ApEn was performed in order to ensure an improved detection rate and accuracy.

5   Decision Making and Classification

After the feature extraction, seizure detection/prediction can be evaluated using either threshold based methods or trained classifiers. The stage of thresholding can be optional or the usage of classifiers should be applied instead.

5.1   Threshold Based Analysis

Threshold based analysis focus on the determination of thresholds that can categorize feature values to ictal or non-ictal states. This can be the statistical evaluation of the variability of features' values according to the desired significance level. The most common approach is the threshold to be determined as a product of a constant and the standard deviation of the feature space distribution. The three sigma rule [118] for any unimodal probability

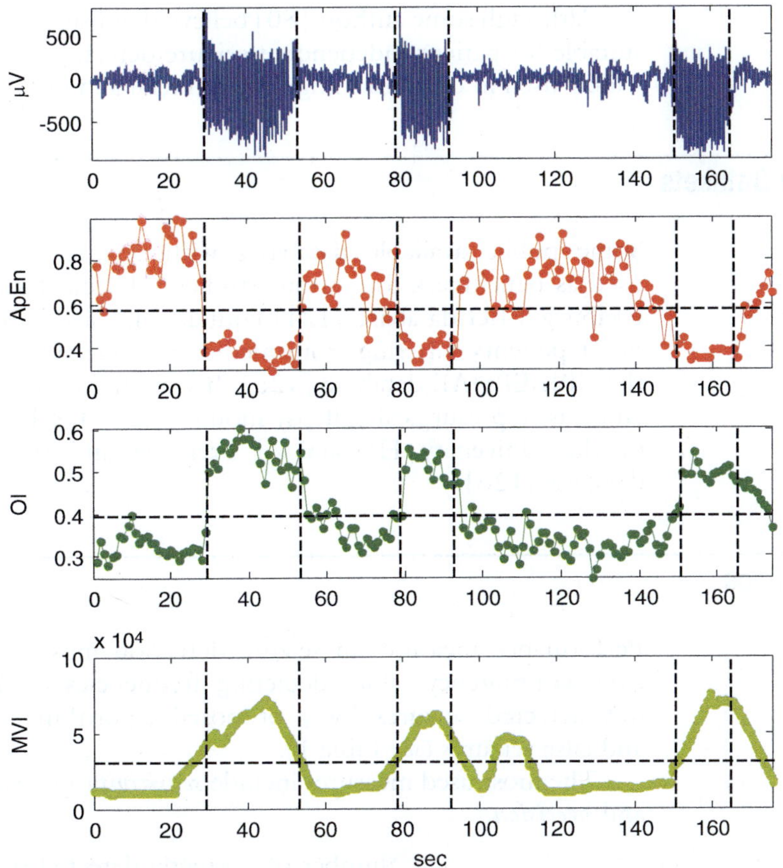

Fig. 2 EEG recording containing and the corresponding variation of approximate entropy (ApEn), Order Index (OI), and Multiscale Variance Index (MVI). The *vertical dashed lines* denote the start and end of a seizure, while the *horizontal dashed line* represents the detection threshold

density is a rule of thumb that has been applied to many problems of threshold determination. A more complex determination of threshold could be achieved using Chebyshev inequality [68].

$$P\{|\text{feature} - \mu| \geq k\sigma\} \leq \frac{1}{k^2},$$

where μ, σ are the mean value, standard deviation of the selected feature distribution and k the chosen statistical threshold.

5.2 Usage of Classifiers

Various classifiers have been employed in order to categorize data and features into classes and make conclusions about the methods performance. Classifiers like Expert Systems [118], Decision trees [119], Artificial Neural Networks (ANN) [17, 18, 30, 32, 120–122], and Support Vector Machines (SVM) [123] have been employed in order to increase a method's detection accuracy.

Although some authors [80] believe that using a classifier is not suitable for patient-independent seizure detection, they have been widely used in distinguishing between ictal and non-ictal periods.

6 Epilepsy Datasets

Some public available databases with seizure data have been used as benchmark for various studies. The most known are the Freiburg EEG database [124] contains invasive EEG recordings of 21 patients suffering from medically intractable focal epilepsy, the EPILEPSIAE database [125] that contains recordings during an invasive presurgical epilepsy monitoring at the Epilepsy Center of the University Hospital of Freiburg and the Bonn EEG database [126].

7 Performance

Performance measures in seizure detection mostly are extracted from contingency tables depicting frequencies of detected and not detected seizures by a proposed algorithm against actual and false seizures (*see* Table 1).

The most used measures include *sensitivity*, *specificity*, *accuracy*, and *precision*.

$$\text{Sensitivity} = \frac{\text{Number of correctly detected seizures}}{\text{Total number of algorithm positive outcomes}}$$

$$= \frac{TP}{TP + FN}$$

$$\text{Specificity} = \frac{\text{Number of correctly normal states}}{\text{Total number of algorithm negative outcomes}}$$

$$= \frac{TN}{TN + FP}$$

Table 1

Contingency table used in order to evaluate performance of a detection algorithm against experiment outcome

		Detection algorithm		
		Seizure	No seizure (control)	Total
Experiment outcome	Positive (P)	TP	FP	nP = TP + FP
	Negative (N)	FN	TN	nN = FN + TN
Total		nS = TP + FN	nNS = FP + TN	N

TP represents true positives, FP represents false positives, FN represents false negatives, and TN represents true negatives

$$\text{Accuracy} = \frac{\text{Number of correctly detected seizures} + \text{correctly normal states}}{\text{Total number of cases}}$$

$$= \frac{\text{TP} + \text{TN}}{\text{TP} + \text{FP} + \text{FN} + \text{TN}}$$

$$\text{Precision} = \frac{\text{Number of correctly detected seizures}}{\text{Total number of seizures}} = \frac{\text{TP}}{\text{TP} + \text{FP}}$$

In addition, there are also positive predictive value (PPV) and negative predictive value (NPV) defined as

$$\text{PPV} = \frac{\text{Number of correctly detected seizures}}{\text{Total number of detected incidents}} = \frac{\text{TP}}{\text{TP} + \text{FP}}$$

$$\text{NPV} = \frac{\text{Number of correctly detected normal states}}{\text{Total number of normal states}} = \frac{\text{TN}}{\text{TN} + \text{FN}}$$

which address the probability of a detected seizure to be an actual seizure and the probability of a detected normal state EEG to be actual normal state EEG, respectively.

Finally, tests can be evaluated by the likelihood ratio which is the probability a segment characterized as ictal classified as ictal divided by the probability a segment characterized as ictal classified as not ictal. This indicates how much likely it is a segment which tests positive is ictal compared with one who tests negative.

$$\text{Likelihood ratio} = \frac{\text{Sensitivity}}{1 - \text{Specificity}}$$

7.1 Detection Performance of Algorithms

It is understood that datasets from different sources can significantly affect a method's behavior and accuracy depending on seizure type, level of noise, patient movements, etc. Although many researchers use their own datasets, however, a public available EEG dataset [126] consisting of signals from five different states: (a) normal EEG with eyes open, (b) EEG with eyes closed, (c) EEG from interictal period (seizure free) within the epileptogenic zone, (d) EEG of interictal period from the hippocampal formation of the opposite hemisphere of the brain, (e) EEG of seizure activity, was created and used as benchmark for many researchers that evaluated their methods. The best achieved detection accuracies of these methods using this dataset are presented in Table 2.

Other studies using their own datasets have also evaluated their methods by means of detection accuracy and they are presented in Table 3.

As these datasets are heterogeneous referring to different seizure types, no safe conclusion for their comparative accuracy can be performed.

Table 2
Seizure detection accuracies for selected studies using data described in [126]

Authors	Year	Method	Dataset	Best achieved accuracy rate (%)
Gautama et al. [127]	2003	Delay vector variance	Bonn EEG database [126]	86.2
VP Nigam, D Graupe [128]	2004	Nonlinear filtering	Bonn EEG database [126]	97.2
I. Güler, E.D. Ubeyli [129]	2005	Wavelet coefficient metrics	Bonn EEG database [126]	98.68
Srinivasan et al. [120]	2005	Time and frequency features, ANN	Bonn EEG database [126]	99.6
Güler et al. [30]	2005	Lyapunov exponent, Recurrent neural networks	Bonn EEG database [126]	96.79
Kannathal et al. [32]	2005	Entropy measures, ANFIS classifier	Bonn EEG database [126]	92.22
Güler and Übeyli [130]	2006	PSD features, modified mixture of experts	Bonn EEG database [126]	98.6
Srinivasan et al. [64]	2007	ApEn, Elman network	Bonn EEG database [126]	100
Polat and Güneş [119]	2007	PSD, decision trees	Bonn EEG database [126]	98.72
Subasi [131]	2007	Wavelet coefficients metrics, mixture of experts	Bonn EEG database [126]	94.5
Tzallas et al. [17]	2007	Time–frequency analysis, ANN	Bonn EEG database [126]	96.3 (average accuracy)
Ghosh-Dastidar et al. [121]	2007	Wavelet-chaos, neural network	Bonn EEG database [126]	96.7
Polat and Güneş [132]	2008	PCA-FFT, AIRS classifier	Bonn EEG database [126]	100
Ghosh-Dastidar et al. [133]	2008	PCA, TBFNN	Bonn EEG database [126] (subset)	96.6
Ocak [65]	2009	Wavelet transform, ApEn	Bonn EEG database [126]	94.85
Tzallas et al. [18]	2009	Time–frequency analysis, ANN	Bonn EEG database [126]	100
Chandaka et al. [123]	2009	Cross-correlation, support vector machines	Bonn EEG database [126]	95.96

(continued)

Table 2
(continued)

Authors	Year	Method	Dataset	Best achieved accuracy rate (%)
Altunay et al. [11]	2010	Linear prediction filter	Bonn EEG database [126]	93.33
Guo et al. [134]	2010	Discrete wavelet transform, line length feature, MLPNN	Bonn EEG database [126]	97.77
Kumar et al. [33]	2010	Entropy measures, Recurrent Elman network (REN)	Bonn EEG database [126]	99.75
Übeyli [122]	2010	Lyapunov exponents, probabilistic neural networks	Bonn EEG database [126]	98.05
Fathima et al. [12]	2011	Discrete wavelet transform	Bonn EEG database [126]	99.5
Guo et al. [135]	2011	Genetic programming features, K-nearest neighbor classifier	Bonn EEG database [126]	99.2
Martis et al. [136]	2012	EMD features	Bonn EEG database [126]	95.33

Table 3
Seizure detection accuracies for selected studies using their own datasets

Authors	Year	Seizure type	Method	Dataset	Best achieved accuracy rate (%)
Liu et al. [10]	1992	Neonatal seizures	Scored autocorrelation Moment	EEG segments (58 with seizures, 59 without seizures)	91.4
Alkan et al. [137]	2005	Absence seizures (petit mal)	MUSIC, autoregressive Spectrum, MPLNN	EEG of 11 subjects (6 normal, 5 epileptic) with 20 absence seizures (petit mal) total	92
Greene et al. [138]	2007	Neonatal seizures	Combination of EEG and ECG features, LD classifier	10 neonates, 633 seizures	86.32
Vukkadala et al. [139]	2009	No specific type	Approximate entropy, Elman neural network	Intracranial EEG of 21 subjects (12 normal, 9 epileptic) containing 30 ictal and 30 non-ictal periods	93.33

(continued)

Table 3
(continued)

Authors	Year	Seizure type	Method	Dataset	Best achieved accuracy rate (%)
Yuan et al. [13]	2012	Intractable focal seizures	Linear/nonlinear features, extreme learning machine	21 patients EEG data (81 seizures)	94.9
Zhou et al. [140]	2013	Intractable focal seizures	Lacunarity, Bayesian LDA	21 patients EEG data (81 seizures)	96.67

7.2 Prediction Performance of Algorithms

In order to evaluate the predictability and the performance of different methods, a framework called the seizure prediction characteristic [5, 141] was introduced. According to this, in a seizure prediction scheme four parameters should be taken into account: the seizure occurrence period (SOP) which is the time period during which the seizure is to be expected, the seizure prediction horizon (SPH) which is a minimum window of time between the alarm raised by the prediction method and the beginning of SOP, the false prediction rate (FPR) which is the number of false predictions per time interval and the sensitivity which is the fraction of correctly predicted seizures within total seizures. These characteristics can evaluate the effectiveness of a prediction scheme.

8   Discussion

This chapter reviews and presents a comparative presentation of the state-of-the-art methods in the area of seizure detection and prediction. Most methods are EEG-based as EEG is by far the most widely used tool for seizure detection and prediction. Besides that, the area of vision-based detection and its combination with biosignal analysis is also briefly presented as a very promising tool in clinical and research practice.

Each method presents its advantages and limitations achieving its maximum effect regarding the application and seizure type. In many cases seizure characteristics are notably varying even among different seizures of the same patient and this phenomenon is more common in small ages [144]. Therefore, a combination of these methods is, in many cases, substantial exploiting the advantages of each one maximizing the effectiveness.

An objective comparative assessment would be ideal in order to be able to choose the optimal methods for specific applications. However, in the related literature, it is difficult to define such a framework because of different datasets used in different studies. Seizure types, sample size, dataset precision (number of electrodes, sampling frequency, etc.) and adopted experiment setup are the main factors that prevent objectivity in methods' evaluation. In addition, as there are no standard guidelines with reliable standardized data, most studies apply their methods on small datasets or the collection of data is driven exclusively by their application. Hence, they often demonstrate good accuracy for selected EEG segments but it is not safe to make a generalized assumption about their performance.

Despite the intrinsic difficulties, this review aims to introduce and foster the understanding of available methods in the area of seizure detection and prediction. Its contribution is towards the construction of a consistent framework for the area of epilepsy detection/prediction that can balance computational complexity and accuracy.

References

1. Fisher RS, van Emde Boas W, Blume W, Elger C, Genton P, Lee P, Engel J Jr (2005) Epileptic seizures and epilepsy: definitions proposed by the International League Against Epilepsy (ILAE) and the International Bureau for Epilepsy (IBE). Epilepsia 46(4):470–472

2. World Health Organization (WHO) (2012) Report, fact sheet N° 999, http://www.who.int/mediacentre/factsheets/fs999/en/index.html

3. Acharya UR, Molinari F, Sree SV, Chattopadhyay S, Ng KH, Suri JS (2012) Automated diagnosis of epileptic EEG using entropies. Biomed Signal Process Contr 7(4):401–408

4. Sakkalis V, Giurcaneanu CD, Xanthopoulos P, Zervakis ME, Tsiaras V, Yang Y, Karakonstantaki E, Micheloyannis S (2009) Assessment of linear and nonlinear synchronization measures for analyzing EEG in a mild epileptic paradigm. IEEE Trans Inform Tech Biomed 13(4):433–441

5. Winterhalder M, Maiwald T, Voss HU, Aschenbrenner-Scheibe R, Timmer J, Schulze-Bonhage A (2003) The seizure prediction characteristic: a general framework to assess and compare seizure prediction methods. Epilepsy Behav 4(3):318–325

6. Kantz H, Schreiber T (2004) Nonlinear time series analysis. Cambridge University Press, Cambridge

7. Paivinen N, Lammi S, Pitkanen A, Nissinen J, Penttonen M, Gronfors T (2005) Epileptic seizure detection: a nonlinear viewpoint. Comput Methods Programs Biomed 79 (2):151–159

8. Faure P, Korn H (2001) Is there chaos in the brain? I. Concepts of nonlinear dynamics and methods of investigation. C R Acad Sci III 324(9):773–793

9. Korn H, Faure P (2003) Is there chaos in the brain? II. Experimental evidence and related models. C R Biol 326(9):787–840

10. Liu A, Hahn JS, Heldt GP, Coen RW (1992) Detection of neonatal seizures through computerized EEG analysis. Electroencephalogr Clin Neurophysiol 82(1):30–37

11. Altunay S, Telatar Z, Erogul O (2010) Epileptic EEG detection using the linear prediction error energy. Expert Syst Appl 37 (8):5661–5665

12. Fathima T, Bedeeuzzaman M, Farooq O, Khan Y (2011) Wavelet based features for epileptic seizure detection. MES J Tech Manag 2(1):108–112

13. Yuan Q, Zhou W, Liu Y, Wang J (2012) Epileptic seizure detection with linear and nonlinear features. Epilepsy Behav 24 (4):415–421

14. Musselman M, Djurdjanovic D (2012) Time-frequency distributions in the classification of

epilepsy from EEG signals. Expert Syst Appl 39(13):11413–11422

15. Hassanpour H, Mesbah M, Boashash B (2004) Time-frequency based newborn EEG seizure detection using low and high frequency signatures. Physiol Meas 25 (4):935–944

16. Tzallas AT, Tsipouras MG, Fotiadis DI (2007) Automatic seizure detection based on time-frequency analysis and artificial neural networks. Comput Intell Neurosci 2007:80510

17. Tzallas AT, Tsipouras MG, Fotiadis DI (2007) The use of time-frequency distributions for epileptic seizure detection in EEG recordings. 29th annual international conference of the IEEE engineering in medicine and biology society, 23–26 Aug 2007, pp 3–6

18. Tzallas AT, Tsipouras MG, Fotiadis DI (2009) Epileptic seizure detection in EEGs using time-frequency analysis. IEEE Trans Inf Technol Biomed 13(5):703–710

19. Khlif MS, Mesbah M, Boashash B, Colditz P (2007) Multichannel-based newborn EEG seizure detection using time-frequency matched filter. 29th annual international conference of the IEEE engineering in medicine and biology society, pp 1265–1268

20. Boashash B, Mesbah M, Golditz P (2003) Time-frequency detection of EEG abnormalities. Elsevier, Amsterdam, pp 663–669

21. Rankine L, Mesbah M, Boashash B (2007) A matching pursuit-based signal complexity measure for the analysis of newborn EEG. Med Biol Eng Comput 45(3):251–260

22. Nagaraj SB, Stevenson N, Marnane W, Boylan G, Lightbody G (2012) A novel dictionary for neonatal EEG seizure detection using atomic decomposition. 34th annual international conference of the IEEE engineering in medicine and biology society, pp 1073–1076

23. Sabesan S, Chakravarthy N, Tsakalis K, Pardalos P, Iasemidis L (2009) Measuring resetting of brain dynamics at epileptic seizures: application of global optimization and spatial synchronization techniques. J Combin Optim 17 (1):74–97

24. Stam CJ (2005) Nonlinear dynamical analysis of EEG and MEG: review of an emerging field. Clin Neurophysiol 116(10):2266–2301

25. Esteller R, Vachtsevanos G, Echauz J, Litt B (2001) A comparison of waveform fractal dimension algorithms. IEEE Trans Circ Syst Fund Theor Appl 48(2):177–183

26. Sarkar N, Chaudhuri BB (1994) An efficient differential box-counting approach to compute fractal dimension of image. IEEE Trans Syst Man Cybern 24(1):115–120

27. Katz MJ (1988) Fractals and the analysis of waveforms. Comput Biol Med 18 (3):145–156

28. Petrosian A (1995) Kolmogorov complexity of finite sequences and recognition of different preictal EEG patterns. Proceedings of the eighth IEEE symposium on computer-based medical systems, Lubbock, TX, pp 212–217

29. Higuchi T (1988) Approach to an irregular time-series on the basis of the fractal theory. Physica D 31(2):277–283

30. Güler NF, Übeyli ED, Güler İ (2005) Recurrent neural networks employing Lyapunov exponents for EEG signals classification. Expert Syst Appl 29(3):506–514

31. Kannathal N, Acharya UR, Lim CM, Sadasivan PK (2005) Characterization of EEG—a comparative study. Comput Methods Programs Biomed 80(1):17–23

32. Kannathal N, Choo ML, Acharya UR, Sadasivan PK (2005) Entropies for detection of epilepsy in EEG. Comput Methods Programs Biomed 80(3):187–194

33. Kumar SP, Sriraam N, Benakop PG, Jinaga BC (2010) Entropies based detection of epileptic seizures with artificial neural network classifiers. Expert Syst Appl 37(4):3284–3291

34. Shannon CE (1948) A mathematical theory of communication. Bell Syst Tech J 27:379–423

35. Bruhn J, Lehmann LE, Ropcke H, Bouillon TW, Hoeft A (2001) Shannon entropy applied to the measurement of the electroencephalographic effects of desflurane. Anesthesiology 95(1):30–35

36. Steuer R, Ebeling W, Bengner T, Dehnicke C, Hattig H, Meencke H-J (2004) Entropy and complexity analysis of intracranially recorded EEG. Int J Bifurcat Chaos 14(02):815–823

37. Inouye T, Shinosaki K, Sakamoto H, Toi S, Ukai S, Iyama A, Katsuda Y, Hirano M (1991) Quantification of EEG irregularity by use of the entropy of the power spectrum. Electroencephalogr Clin Neurophysiol 79 (3):204–210

38. Inouye T, Shinosaki K, Iyama A, Matsumoto Y (1993) Localization of activated areas and directional EEG patterns during mental arithmetic. Electroencephalogr Clin Neurophysiol 86(4):224–230

39. He Z-y, Chen X, Luo G (2006) Wavelet entropy measure definition and its application for transmission line fault detection and identification (part I: definition and methodology). International conference on power system technology, PowerCon 2006, Chongqing, 22–26 Oct 2006, pp 1–6

40. Powell GE, Percival IC (1979) A spectral entropy method for distinguishing regular and irregular motion of Hamiltonian systems. J Phys Math Gen 12(11):2053–2071

41. Blanco S, Garay A, Coulombie D (2013) Comparison of frequency bands using spectral entropy for epileptic seizure prediction. ISRN Neurol 2013:5

42. Bezerianos A, Tong S, Thakor N (2003) Time-dependent entropy estimation of EEG rhythm changes following brain ischemia. Ann Biomed Eng 31(2):221–232

43. Zygierewicz J, Durka PJ, Klekowicz H, Franaszczuk PJ, Crone NE (2005) Computationally efficient approaches to calculating significant ERD/ERS changes in the time-frequency plane. J Neurosci Methods 145 (1–2):267–276

44. Cohen L (1995) Time–frequency analysis. Prentice Hall, New York, NY

45. Quian Quiroga R, Rosso OA, Basar E (1999) Wavelet entropy: a measure of order in evoked potentials. Electroencephalogr Clin Neurophysiol Suppl 49:298–302

46. Rosso OA, Blanco S, Yordanova J, Kolev V, Figliola A, Schurmann M, Basar E (2001) Wavelet entropy: a new tool for analysis of short duration brain electrical signals. J Neurosci Methods 105(1):65–75

47. Quiroga RQ, Rosso OA, Basar E, Schurmann M (2001) Wavelet entropy in event-related potentials: a new method shows ordering of EEG oscillations. Biol Cybern 84 (4):291–299

48. Giannakakis GA, Tsiaparas NN, Xenikou MFS, Papageorgiou C, Nikita KS (2008) Wavelet entropy differentiations of event related potentials in dyslexia. 8th IEEE international conference on bioinformatics and bioengineering, Athens, 8–10 Oct 2008, pp 1–6

49. Hornero R, Abasolo DE, Espino P (2003) Use of wavelet entropy to compare the EEG background activity of epileptic patients and control subjects. The proceedings seventh international symposium on signal processing and its applications, vol 2, 1–4 July 2003, pp 5–8

50. Li X (2006) Wavelet spectral entropy for indication of epileptic seizure in extracranial EEG. In: King I, Wang J, Chan L-W, Wang D (eds) Neural information processing. Springer, Berlin, pp 66–73

51. Mirzaei A, Ayatollahi A, Gifani P, Salehi L (2010) EEG analysis based on wavelet-spectral entropy for epileptic seizures detection. 3rd international conference on proceedings of biomedical engineering and informatics (BMEI), Yantai, 16–18 Oct 2010, pp 878–882

52. Rosso OA, Blanco S, Rabinowicz A (2003) Wavelet analysis of generalized tonic-clonic epileptic seizures. Signal Process 83 (6):1275–1289

53. Jones DL, Baraniuk RG (1995) An adaptive optimal-kernel time-frequency representation. IEEE Trans Signal Process 43 (10):2361–2371

54. Cohen L (1989) Time frequency-distributions—a review. Proc IEEE 77(7):941–981

55. Giannakakis GA, Tsiaparas NN, Papageorgiou C, Nikita KS (2009) Spectral entropy of dyslexic ERP signal by means of adaptive optimal kernel. 16th international conference on digital signal processing, Santorini, Greece, 5–7 July 2009, pp 1–6

56. Pincus SM (1991) Approximate entropy as a measure of system complexity. Proc Natl Acad Sci U S A 88(6):2297–2301

57. Kolmogorov AN (1958) New metric invariant of transitive dynamical systems and endomorphisms of lebesgue spaces. Doklady Russ Acad Sci Phys 119(N5):861–864

58. Bein B (2006) Entropy. Best Pract Res Clin Anaesthesiol 20(1):101–109

59. Pincus SM, Cummins TR, Haddad GG (1993) Heart-rate control in normal and aborted sids infants. Am J Physiol 264(3): R638–R646

60. Pincus SM, Goldberger AL (1994) Physiological time-series analysis—what does regularity quantify. Am J Physiol 266(4):H1643–H1656

61. Richman JS, Moorman JR (2000) Physiological time-series analysis using approximate entropy and sample entropy. Am J Physiol Heart Circ Physiol 278(6):H2039–H2049

62. Kim WS, Yoon YZ, Bae JH, Soh KS (2005) Nonlinear characteristics of heart rate time series: influence of three recumbent positions in patients with mild or severe coronary artery disease. Physiol Meas 26(4):517–529

63. Pincus SM, Mulligan T, Iranmanesh A, Gheorghiu S, Godschalk M, Veldhuis JD (1996) Older males secrete luteinizing hormone and testosterone more irregularly, and jointly more asynchronously, than younger males. Proc Natl Acad Sci U S A 93 (24):14100–14105

64. Srinivasan V, Eswaran C, Sriraam N (2007) Approximate entropy-based epileptic EEG detection using artificial neural network's. IEEE Trans Inf Technol Biomed 11 (3):288–295

65. Ocak H (2009) Automatic detection of epileptic seizures in EEG using discrete wavelet transform and approximate entropy. Expert Syst Appl 36(2):2027–2036

66. Zhou Y, Huang RM, Chen ZY, Chang X, Chen JL, Xie LL (2012) Application of approximate entropy on dynamic characteristics of epileptic absence seizure. Neural Regen Res 7(8):572–577

67. Guo L, Rivero D, Pazos A (2010) Epileptic seizure detection using multiwavelet transform based approximate entropy and artificial neural networks. J Neurosci Methods 193(1):156–163

68. Giannakakis G, Sakkalis V, Pediaditis M, Farmaki C, Vorgia P, Tsiknakis M (2013) An approach to absence epileptic seizures detection using approximate entropy. 35th annual international conference of the IEEE engineering in medicine and biology society (EMBC), Osaka, Japan, 3–7 July 2013, pp 413–416

69. Abásolo D, Hornero R, Espino P, Álvarez D, Poza J (2006) Entropy analysis of the EEG background activity in Alzheimer's disease patients. Physiol Meas 27(3):241

70. Song Y, Crowcroft J, Zhang J (2012) Automatic epileptic seizure detection in EEGs based on optimized sample entropy and extreme learning machine. J Neurosci Methods 210(2):132–146

71. Molteni E, Perego P, Zanotta N, Reni G (2008) Entropy analysis on EEG signal in a case study of focal myoclonus. Conf Proc IEEE Eng Med Biol Soc 2008:4724–4727

72. Abásolo D, Muñoz D, Espino P (2012) Kullback-leibler entropy analysis of the electroencephalogram background activity in Alzheimer's disease patients. In: The proceedings of 9th IASTED international conference on biomedical engineering, BioMed 2012, pp 43–46

73. Quiroga RQ, Arnhold J, Lehnertz K, Grassberger P (2000) Kulback-Leibler and renormalized entropies: applications to electroencephalograms of epilepsy patients. Phys Rev E Stat Phys Plasmas Fluids Relat Interdiscip Topics 62(6):8380–8386

74. Abasolo D, James CJ, Hornero R (2007) Non-linear analysis of intracranial electroencephalogram recordings with approximate entropy and Lempel-Ziv complexity for epileptic seizure detection. Conf Proc IEEE Eng Med Biol Soc 2007:1953–1956

75. James CJ, Abasolo D, Gupta D (2007) Space-time ICA versus Ensemble ICA for ictal EEG analysis with component differentiation via Lempel-Ziv complexity. Conf Proc IEEE Eng Med Biol Soc 2007:5473–5476

76. Bandt C, Pompe B (2002) Permutation entropy: a natural complexity measure for time series. Phys Rev Lett 88(17):174102

77. Li XL, Ouyang GX, Richards DA (2007) Predictability analysis of absence seizures with permutation entropy. Epilepsy Res 77(1):70–74

78. Ouyang G, Dang C, Richards DA, Li X (2010) Ordinal pattern based similarity analysis for EEG recordings. Clin Neurophysiol 121(5):694–703

79. Sakkalis V, Giannakakis G, Farmaki C, Mousas A, Pediaditis M, Vorgia P, Tsiknakis M (2013) Absence seizure epilepsy detection using linear and nonlinear EEG analysis methods. 35th annual international conference of the IEEE engineering in medicine and biology society (EMBC), Osaka, Japan, 3–7 July 2013, pp 6333–6336

80. Deburchgraeve W, Cherian PJ, De Vos M, Swarte RM, Blok JH, Visser GH, Govaert P, Van Huffel S (2008) Automated neonatal seizure detection mimicking a human observer reading EEG. Clin Neurophysiol 119(11):2447–2454

81. Nonclercq A, Foulon M, Verheulpen D, De Cock C, Buzatu M, Mathys P, Van Bogaert P (2009) Spike detection algorithm automatically adapted to individual patients applied to spike-and-wave percentage quantification. Neurophysiol Clin 39(2):123–131

82. Adjouadi M, Sanchez D, Cabrerizo M, Ayala M, Jayakar P, Yaylali I, Barreto A (2004) Interictal spike detection using the Walsh transform. IEEE Trans Biomed Eng 51(5):868–872

83. Adjouadi M, Cabrerizo M, Ayala M, Sanchez D, Yaylali I, Jayakar P, Barreto A (2005) Detection of interictal spikes and artifactual data through orthogonal transformations. J Clin Neurophysiol 22(1):53–64

84. Pediaditis M, Tsiknakis M, Vorgia P, Kafetzopoulos D, Danilatou V, Fotiadis D (2010) Vision-based human motion analysis in epilepsy—methods and challenges. 10th IEEE international conference on information technology and applications in biomedicine (ITAB), Corfu, 3–5 Nov 2010, pp 1–5

85. Pediaditis M, Tsiknakis M, Leitgeb N (2012) Vision-based motion detection, analysis and recognition of epileptic seizures-A systematic review. Comput Methods Programs Biomed 108(3):1133–1148

86. Viglione SS, Walsh GO (1975) Epileptic seizure prediction. Electroencephalogr Clin Neurophysiol 39:435–436

87. Rogowski Z, Gath I, Bental E (1981) On the prediction of epileptic seizures. Biol Cybern 42(1):9–15

88. Gotman J, Ives JR, Gloor P, Olivier A, Quesney LF (1982) Changes in interictal eeg spiking and seizure occurrence in humans. Epilepsia 23(4):432–433

89. Mormann F, Kreuz T, Rieke C, Andrzejak RG, Kraskov A, David P, Elger CE, Lehnertz K (2005) On the predictability of epileptic seizures. Clin Neurophysiol 116 (3):569–587

90. van Drongelen W, Nayak S, Frim DM, Kohrman MH, Towle VL, Lee HC, McGee AB, Chico MS, Hecox KE (2003) Seizure anticipation in pediatric epilepsy: use of Kolmogorov entropy. Pediatr Neurol 29 (3):207–213

91. McSharry PE, Smith LA, Tarassenko L (2003) Comparison of predictability of epileptic seizures by a linear and a nonlinear method. IEEE Trans Biomed Eng 50 (5):628–633

92. Hjorth B (1970) EEG analysis based on time domain properties. Electroencephalogr Clin Neurophysiol 29(3):306–310

93. Park Y, Luo L, Parhi KK, Netoff T (2011) Seizure prediction with spectral power of EEG using cost-sensitive support vector machines. Epilepsia 52(10):1761–1770

94. Litt B, Esteller R, Echauz J, D'Alessandro M, Shor R, Henry T, Pennell P, Epstein C, Bakay R, Dichter M, Vachtsevanos G (2001) Epileptic seizures may begin hours in advance of clinical onset: a report of five patients. Neuron 30(1):51–64

95. Maiwald T, Winterhalder M, Aschenbrenner-Scheibe R, Voss HU, Schulze-Bonhage A, Timmer J (2004) Comparison of three nonlinear seizure prediction methods by means of the seizure prediction characteristic. Physica D-Nonlinear Phenomena 194(3–4):357–368

96. Gigola S, Ortiz F, D'Attellis CE, Silva W, Kochen S (2004) Prediction of epileptic seizures using accumulated energy in a multiresolution framework. J Neurosci Methods 138(1–2):107–111

97. Chisci L, Mavino A, Perferi G, Sciandrone M, Anile C, Colicchio G, Fuggetta F (2010) Real-time epileptic seizure prediction using AR models and support vector machines. IEEE Trans Biomed Eng 57(5):1124–1132

98. Iasemidis LD, Shiau DS, Chaovalitwongse W, Sackellares JC, Pardalos PM, Principe JC, Carney PR, Prasad A, Veeramani B, Tsakalis K (2003) Adaptive epileptic seizure prediction system. IEEE Trans Biomed Eng 50 (5):616–627

99. Iasemidis LD, Shiau DS, Sackellares JC, Pardalos PA, Prasad A (2004) Dynamical resetting of the human brain at epileptic seizures: application of nonlinear dynamics and global optimization techniques. IEEE Trans Biomed Eng 51(3):493–506

100. Iasemidis LD, Sackellares JC, Zaveri HP, Williams WJ (1990) Phase space topography and the Lyapunov exponent of electrocorticograms in partial seizures. Brain Topogr 2 (3):187–201

101. Van Quyen ML, Martinerie J, Baulac M, Varela F (1999) Anticipating epileptic seizures in real time by a non-linear analysis of similarity between EEG recordings. Neuroreport 10 (10):2149–2155

102. Le Van Quyen M, Adam C, Martinerie J, Baulac M, Clemenceau S, Varela F (2000) Spatio-temporal characterizations of nonlinear changes in intracranial activities prior to human temporal lobe seizures. Eur J Neurosci 12(6):2124–2134

103. Le Van Quyen M, Martinerie J, Navarro V, Boon P, D'Have M, Adam C, Renault B, Varela F, Baulac M (2001) Anticipation of epileptic seizures from standard EEG recordings. Lancet 357(9251):183–188

104. Navarro V, Martinerie J, Quyen MLV, Clemenceau S, Adam C, Baulac M, Varela F (2002) Seizure anticipation in human neocortical partial epilepsy. Brain 125(3):640–655

105. Grassberger P, Procaccia I (1983) Characterization of strange attractors. Phys Rev Lett 50 (5):346–349

106. Lehnertz K, Andrzejak RG, Arnhold J, Kreuz T, Mormann F, Rieke C, Widman G, Elger CE (2001) Nonlinear EEG analysis in epilepsy: Its possible use for interictal focus localization, seizure anticipation, and prevention. J Clin Neurophysiol 18(3):209–222

107. Aschenbrenner-Scheibe R, Maiwald T, Winterhalder M, Voss HU, Timmer J, Schulze-Bonhage A (2003) How well can epileptic seizures be predicted? An evaluation of a nonlinear method. Brain 126:2616–2626

108. Zandi AS, Dumont GA, Javidan M, Tafreshi R (2009) An entropy-based approach to predict seizures in temporal lobe epilepsy using scalp EEG. Conf Proc IEEE Eng Med Biol Soc 2009:228–231

109. Lee H, Kohrman M, Hecox K, Drongelen W (2013) Seizure prediction. In: He B (ed) Neural engineering. Springer, New York, NY

110. Zhang L-Y (2009) Kolmogorov entropy changes and cortical lateralization during complex problem solving task measured with EEG. J Biomed Sci Eng 2(8):661–664

111. Mormann F, Andrzejak RG, Kreuz T, Rieke C, David P, Elger CE, Lehnertz K (2003) Automated detection of a preseizure state based on a decrease in synchronization in intracranial electroencephalogram recordings from epilepsy patients. Phys Rev E Stat Nonlin Soft Matter Phys 67(2):021912

112. Mormann F, Kreuz T, Andrzejak RG, David P, Lehnertz K, Elger CE (2003) Epileptic seizures are preceded by a decrease in synchronization. Epilepsy Res 53(3):173–185

113. Lehnertz K, Mormann F, Kreuz T, Andrzejak RG, Rieke C, David P, Elger CE (2003) Seizure prediction by nonlinear EEG analysis. IEEE Eng Med Biol Mag 22(1):57–63

114. Mormann F, Lehnertz K, David P, Elger CE (2000) Mean phase coherence as a measure for phase synchronization and its application to the EEG of epilepsy patients. Physica D 144(3–4):358–369

115. Guyon I, Weston J, Barnhill S, Vapnik V (2002) Gene selection for cancer classification using support vector machines. Mach Learn 46(1–3):389–422

116. Guyon I, Aliferis C, Cooper G, Elisseeff A, Pellet J-P, Spirtes P, Statnikov A (2008) Design and analysis of the causation and prediction challenge. WCCI workshop on causality

117. Temko A, Thomas E, Marnane W, Lightbody G, Boylan G (2011) EEG-based neonatal seizure detection with support vector machines. Clin Neurophysiol 122(3):464–473

118. Pukelsheim F (1994) The three sigma rule. Am Stat 48(2):88–91

119. Polat K, Güneş S (2007) Classification of epileptiform EEG using a hybrid system based on decision tree classifier and fast Fourier transform. Appl Math Comput 187 (2):1017–1026

120. Srinivasan V, Eswaran C, Sriraam N (2005) Artificial neural network based epileptic detection using time-domain and frequency-domain features. J Med Syst 29(6):647–660

121. Ghosh-Dastidar S, Adeli H, Dadmehr N (2007) Mixed-band wavelet-chaos-neural network methodology for epilepsy and epileptic seizure detection. IEEE Trans Biomed Eng 54(9):1545–1551

122. Übeyli ED (2010) Lyapunov exponents/probabilistic neural networks for analysis of EEG signals. Expert Syst Appl 37 (2):985–992

123. Chandaka S, Chatterjee A, Munshi S (2009) Cross-correlation aided support vector machine classifier for classification of EEG signals. Expert Syst Appl 36(2, Part 1):1329–1336

124. Freiburg EEG database. https://epilepsy.uni-freiburg.de/freiburg-seizure-prediction-proj ect/eeg-database

125. The European Epilepsy Database. http://epi lepsy-database.eu

126. Andrzejak RG, Lehnertz K, Mormann F, Rieke C, David P, Elger CE (2001) Indications of nonlinear deterministic and finite-dimensional structures in time series of brain electrical activity: dependence on recording region and brain state. Phys Rev E Stat Nonlin Soft Matter Phys 64(6):061907

127. Gautama T, Mandic DP, Van Hulle MM (2003) Indications of nonlinear structures in brain electrical activity. Phys Rev E Stat Nonlin Soft Matter Phys 67(4 Pt 2): 046204

128. Nigam VP, Graupe D (2004) A neural-network-based detection of epilepsy. Neurol Res 26(1):55–60

129. Guler I, Ubeyli ED (2005) Adaptive neuro-fuzzy inference system for classification of EEG signals using wavelet coefficients. J Neurosci Methods 148(2):113–121

130. Güler İ, Übeyli ED (2007) Expert systems for time-varying biomedical signals using eigenvector methods. Expert Syst Appl 32 (4):1045–1058

131. Subasi A (2007) EEG signal classification using wavelet feature extraction and a mixture of expert model. Expert Syst Appl 32 (4):1084–1093

132. Polat K, Gunes S (2008) Artificial immune recognition system with fuzzy resource allocation mechanism classifier, principal component analysis and FFT method based new hybrid automated identification system for classification of EEG signals. Expert Syst Appl 34(3):2039–2048

133. Ghosh-Dastidar S, Adeli H, Dadmehr N (2008) Principal component analysis-enhanced cosine radial basis function neural network for robust epilepsy and seizure detection. IEEE Trans Biomed Eng 55 (2):512–518

134. Guo L, Rivero D, Dorado J, Rabunal JR, Pazos A (2010) Automatic epileptic seizure detection in EEGs based on line length feature and artificial neural networks. J Neurosci Methods 191(1):101–109

135. Guo L, Rivero D, Dorado J, Munteanu CR, Pazos A (2011) Automatic feature extraction

using genetic programming: an application to epileptic EEG classification. Expert Syst Appl 38(8):10425–10436

136. Martis RJ, Acharya UR, Tan JH, Petznick A, Yanti R, Chua CK, Ng EY, Tong L (2012) Application of empirical mode decomposition (emd) for automated detection of epilepsy using EEG signals. Int J Neural Syst 22 (6):1250027

137. Alkan A, Koklukaya E, Subasi A (2005) Automatic seizure detection in EEG using logistic regression and artificial neural network. J Neurosci Methods 148(2):167–176

138. Greene BR, Boylan GB, Reilly RB, de Chazal P, Connolly S (2007) Combination of EEG and ECG for improved automatic neonatal seizure detection. Clin Neurophysiol 118 (6):1348–1359

139. Vukkadala S, Vijayalakshmi S, Vijayapriya S (2009) Automated detection of epileptic EEG using approximate entropy in elman networks. Int J Recent Trends Eng 1 (1):307–312

140. Zhou W, Liu Y, Yuan Q, Li X (2013) Epileptic seizure detection using lacunarity and Bayesian linear discriminant analysis in intracranial EEG. IEEE Trans Biomed Eng 60 (12):3375–3381

141. Yacoob Y, Davis LS (1996) Recognizing human facial expressions from long image sequences using optical flow. IEEE Trans Pattern Anal Mach Intell 18(6):636–642

142. Schad A, Schindler K, Schelter B, Maiwald T, Brandt A, Timmer J, Schulze-Bonhage A (2008) Application of a multivariate seizure detection and prediction method to non-invasive and intracranial long-term EEG recordings. Clin Neurophysiol 119 (1):197–211

143. James CJ, Gupta D (2009) Seizure prediction for epilepsy using a multi-stage phase synchrony based system. Conf Proc IEEE Eng Med Biol Soc 2009:25–28. doi:10.1109/ IEMBS.2009.5334898

144. Lombroso CT (1996) Neonatal seizures: a clinician's overview. Brain Dev 18(1):1–28

Neuromethods (2015) 91: 159–169
DOI 10.1007/7657_2013_62
© Springer Science+Business Media New York 2013
Published online: 12 February 2014

Graph-theoretic Indices of Evaluating Brain Network Synchronization: Application in an Alcoholism Paradigm

Vangelis Sakkalis, Theofanis Oikonomou, Vassilis Tsiaras, and Ioannis Tollis

Abstract

Electroencephalographic signal synchronization studies have been a topic of increasing interest lately, using both linear and nonlinear formulations. In this direction a graph-theoretic approach devised to study and stress the coupling dynamics of task-performing dynamical networks is proposed in this communication. Graph theoretical measures and visualization provide the tools to visualize and characterize the topology of a brain network as in an alcoholism paradigm during mental rehearsal of pictures, which is known to reflect synchronization impairment. More specifically, in this chapter the synchronization between all pairs of channels is calculated using (a) the magnitude squared coherence, (b) an estimation of phase synchronization, and (c) a robust nonlinear state-space generalized synchronization assessment method. Synchronization matrices define graphs whose topological structure and properties are characterized using measures for graphs and weighted networks. The results are in accordance with previous psychophysiology studies suggesting that an alcoholic subject has impaired synchronization of brain activity and loss of lateralization during the rehearsal process, most prominently in alpha (8–12 Hz) band, as compared to a control subject. Lower beta (13–30 Hz) synchronization was also evident in an alcoholic subject.

Key words EEG, Synchronization, Coherence, Scalp EEG, Brain networks, Graph theory, Complex networks

---

1   Introduction

Impaired cognitive functioning has repeatedly been reported in alcohol-dependent individuals [1]. The fact that alcoholics have cognitive deficits in performing complex coordinated tasks suggests some related differentiation in brain functional connectivity as expressed by synchronization between different neural assemblies [2]. Synchronous oscillations of certain types of such assemblies in different frequency bands captured by both linear and nonlinear methods have been successfully used in the past as indices of cerebral engagement in cognitive tasks or brain pathologies [3].

Magnitude squared coherence (MSC) has been a well-established and traditionally used tool in alcohol research to investigate the linear relation between two signals or electroencephalographic (EEG) channels. However, related studies on different pathologies, like controlled epilepsy, using similar EEG analysis approaches suggest that nonlinear interdependencies should also be investigated [3, 4]. Phase synchronization presents a different approach in analyzing the possible nonlinear interdependencies of the EEG signal and focuses on the phases of the signals. A robust phase coupling measure is the phase locking value (PLV) [5]. Finally, another group of synchronization measures are based on the assumption that neurons are highly nonlinear devices, which in some cases show chaotic behavior. Such measures belong to the generalized synchronization (GS) concept and are based on analyzing the interdependence between the amplitudes of the signals in a state-space reconstructed domain [6]. The above measures are used to construct interdependence matrices, based on which network visualization is attempted.

The focus of this chapter is on visualizing and quantifying statistically significant differences in coupling of EEG channels in alcohol addicted subjects vs. controls, using their graph theoretic signatures. Special interest in using graph theory to study neural and functional networks has been in focus recently, since it offers a unique perspective of studying local and distributed brain interactions [7, 8]. In this study, network topology and functional connectivity are evaluated by measuring several properties of the emerged graphs, i.e., mean vertex degree, clustering coefficient index, assortativity coefficient, shortest path length, and efficiency.

2 Methods

2.1 Data Acquisition and Analysis

The EEG signals used in this work arise from 30 right-handed subjects (had no personal or family history of any neurological disease, no age difference, and normal vision or corrected normal vision) that were recorded in an electrically shielded, sound- and light-attenuated room. Participants were sitting in a reclined chair and fixated a point in the center of a computer display located 1 m away from participants' eyes. Each subject was fitted with a 61-lead electrode cap (ECI, Electrocap International) according to the entire 10/20 International montage along with an additional 41 sites as follows: FPz, AFz, AF1, AF2, AFz, AF8, F1, F2, F5, F6, FCz, FC2, FC3, FC4, FC5, FC6, FC7, FC8, C1, C2, C5, C6, CPz, CP1, CP2, CP3, CP4, CP5, CP6, TP7, TP8, PI, P2, P5, P6, POz, PO1, PO2, PO7, and PO8 (Standard Electrode Position Nomenclature, American EEG Association 1990). All scalp electrodes were referred to Cz. Midline electrodes were excluded. Subjects were grounded with a nose electrode, and the electrode impedance was always below 5 kΩ. Two additional bipolar deviations were used to record the vertical and horizontal EOG. The signals were amplified with a gain of 10,000 by Ep-A2 amplifiers (Sensorium, Inc., Charlotte, VT) with a bandpass between 0.02 and 50 Hz. The amplified signals were sampled at a rate of 256 Hz, and ten artifact-free epochs of 1 s were selected while each subject performs a picture rehearsal task as described below. Trials with excessive eye and body movements (>73.3 μV) were rejected online. All synchronization measures were averaged over the epochs.

In this experiment, each subject was exposed to pictures of objects chosen from the 1980 Snodgrass and Vanderwart picture set. These stimuli were randomized (but not repeated) and presented on a white background at the center of a computer monitor and were approximately 5–10 cm × 5–10 cm, thus subtending a visual angle of 0.05°–0.1°. Ten trials were shown. The interval between each trial was fixed to 3.2 s. The participants were instructed to memorize the pictures in order to be able to identify them later.

2.2 Magnitude Squared Coherence

Let us suppose that we have two simultaneously measured discrete time series and , $n = 1, \ldots, N$. MSC or simply coherence is the cross spectral density function S_{xy}, which is simply derived via the FFT of the cross-correlation, normalized by their individual auto-spectral density functions. Hence, MSC is calculated using the Welch's method, as

$$\gamma_{xy}(f) = \frac{\left| \langle S_{xy}(f) \rangle \right|^2}{\left| \langle S_{xx}(f) \rangle \right| \left| \langle S_{yy}(f) \rangle \right|} \tag{1}$$

where $<.>$ indicates window averaging. The estimated MSC for a given frequency f ranges between 0 (no coupling) and 1 (maximum linear interdependence).

2.3 Phase Locking Value

One of the mostly used phase synchronization measures is the PLV approach. It assumes that two dynamic systems may have their phases synchronized even if their amplitudes are zero correlated [5]. The PLV is defined as the locking of the phases associated to each signal, such as

$$\left| \phi_x(t) - \phi_y(t) \right| = \text{const} \tag{2}$$

In order to estimate the instantaneous phase of our signal, we transform it using the Hilbert transform (HT), whereby the analytical signal $H(t)$ is computed as

$$H(t) = x(t) + i\tilde{x}(t) \tag{3}$$

where is the HT of x(t), defined as

$$\tilde{x}(t) = \frac{1}{\pi} PV \int_{-\infty}^{\infty} \frac{x(t')}{t - t'} \, dt', \tag{4}$$

where PV denotes the Gauchy principal value. The analytical signal phase is defined as

$$\phi(t) = \arctan(\tilde{x}(t)/x(t)) \tag{5}$$

Therefore for the two signals $x(t)$ and $y(t)$ of equal time length with instantaneous phases $\phi_X(t)$, $\phi_Y(t)$, respectively, the PLV bivariate metric is defined given by

$$\text{PLV} = \left| \frac{1}{N} \sum_{j=0}^{N-1} \exp(i(\phi_X(j\Delta t) - \phi_Y(j\Delta t))) \right| \tag{6}$$

where Δt is the sampling period and N is the sample number of each signal. PLV takes values within the $[0, 1]$ interval, where 1 indicates perfect phase synchronization and 0 indicates lack of synchronization.

2.4 Nonlinear Synchronization (State-Space Approach)

However, a physiological time series such as the EEG appears to have more than the single degree of freedom represented just by plotting the voltage as a function of time. To free up some of these unknown parameters, a standard technique is to map the scalar time series to a vector-valued one in a higher dimensional space $\mathbb{R}^m$, thereby giving it an extension in space as well as time, forming dynamical evolving trajectories, known as attractors [9]. Hence, one may measure how neighborhoods (i.e., recurrences) in state space located in one attractor maps into the other. This idea turned out to be the most robust and reliable way of assessing the extent of GS [6]. First, we reconstruct delay vectors out of our time series:

$$\mathbf{x}_n = x_n, x_{n-\tau}, \ldots, x_{n-(m-1)\tau} \text{ and}$$

$$\mathbf{y}_n = y_n, y_{n-\tau}, \ldots, y_{n-(m-1)\tau}, \tag{7}$$

where $n = 1 \ldots N$, and m and τ are the embedding dimension and time lag, respectively. Let $r_{n,j}$ and $s_{n,j}$, $j = 1, \ldots, k$ denote the time indices of the k nearest neighbors of $\mathbf{x}_n$ and $\mathbf{y}_n$, respectively. For each $\mathbf{x}_n$ the squared mean Euclidean distance to its k neighbors is defined as

$$R_n^{(k)}(X) = \frac{1}{k} \sum_{j=1}^{k} \left(\mathbf{x}_n - \mathbf{x}_{r_{n,j}} \right)^2 \tag{8}$$

And the Y-conditioned squared mean Euclidean distance $R_n^{(k)}(X|Y)$ is defined by replacing the nearest neighbors by the equal time partners of the closest neighbors of $\mathbf{y}_n$.

If the set of reconstructed vectors (point cloud $\mathbf{x}_n$) has an average squared radius $R(X) = (1/N)\sum_{n=1}^{N} R_n^{(N-1)}(X)$, then $R_n^{(k)}(X|Y) \approx R_n^{(k)}(X) \ll R(X)$ if the systems are strongly correlated, while if they are independent. Hence, an interdependence measure is defined as

$$S^{(k)}(X|Y) = \frac{1}{N} \sum_{n=1}^{N} \frac{R_n^{(k)}(X)}{R_n^{(k)}(X|Y)} \tag{9}$$

Since by construction, it is clear that S ranges between 0 (indicating independence) and 1 (indicating maximum synchronization).

2.5 Network Construction (Graph Topology)

For every subject, run, band, and synchronization measure the interdependence for each channel pair (there are $61(61-1)/2 = 1,830$ channel pairs since the number of active EEG channels is 61) is calculated. The results were stored to 61×61 interdependence matrices (IM) with elements ranging from 0 to 1. In order to obtain a graph from an IM we need to convert it into an $N \times N$ adjacency matrix, A. The easiest way of achieving that is to define a *threshold* variable T, such that $0 \leq T \leq 1$. The value $A(i, j)$ is either 1 or 0, indicating the presence or the absence of an edge between nodes i and j, respectively. Namely, $A(i, j)=1$ if $IM(i, j) \geq T$; otherwise, $A(i, j) = 0$. An adjacency matrix defines a graph. Thus given an IM we may define a graph for each value of T (i.e., if the threshold takes the values $T = 0.001, 0.002, \ldots, 0.999, 1$ then 1,000 such graphs may be defined; one for every thousandth of T) [10]. For each edge of a graph we may define its value as $W(i, j) = IM(i, j)$ when $IM(i, j) \geq T$.

After constructing A, we visualize the network edges as straight line segments (Fig. 2). Additionally we can visualize the edge values using the heat map color scheme and width of edge segments; for high edge values that correspond to strong interdependence we draw red-shaded thick line, while for low edge values we draw blue-shaded thin line (Fig. 3). Next, we compute various properties of the resulting graph. These include the average degree K, the assortativity coefficient r [11], the clustering coefficient C, the average shortest path length L, and the efficiency E_f [14].

Before we describe the above network statistics we should define a graph and a few graph-theoretic concepts. A graph $G = (V, E)$ consists of a set of n nodes $V = \{v_1, v_2, \ldots, v_n\}$ and a set of m edges E, where e_{ij} denotes an edge between nodes v_i and v_j. The neighborhood N_i of a node v_i is defined as the set of vertices that have an edge to v_i, namely, $N_i = \{v_j \mid (v_i, v_j) \in E\}$.

2.5.1 Average Vertex Degree

The degree k_i of a node is the number of vertices in its neighborhood, i.e., $|N_i|$. The average degree of a graph is the average of the degrees of all nodes, as

$$K = \sum_{i \in V} k_i / n \tag{10}$$

A network is called assortative when high-degree vertices connect preferentially to each other and low-degree vertices to each other [11, 12]. A network is called disassortative when high-degree vertices preferentially connect to low-degree vertices. Anatomical brain networks tend to be disassortative, while functional networks tend to be assortative [13]. The assortativity coefficient is the (sample) Pearson product–moment correlation coefficient between the degrees of the end vertices of edges E of graph $G = (V, E)$.

For unweighted and undirected graphs this coefficient can be written as [7]

$$ r = \frac{4m \sum\limits_{\{u,v\}\in E} k_u k_v - \left[\sum\limits_{\{u,v\}\in E} (k_u + K_v)\right]^2}{2m \sum\limits_{\{u,v\}\in E} \left(k_u^2 + k_v^2\right) - \left[\sum\limits_{\{u,v\}\in E} (k_u + K_v)\right]^2} \tag{11} $$

where k_u and k_v are the degrees of the end vertices u and v, respectively, of edge $\{u,v\} \in E$. For directed unweighted graphs the above measure can still be used if we ignore the direction of edges [7]. The assortativity coefficient r lies in the range $-1 \leq r \leq 1$, where -1 indicates totally disassortative network while 1 indicates a totally assortative network.

2.5.2 Clustering Coefficient

The clustering coefficient (CC) C_i of a node v_i is the fraction of the existing edges between the nodes in v_i's neighborhood over the number of all possible edges between them. For an undirected graph, if a node v_i has k_i neighbors, then $k_i(k_i - 1)/2$ is the number of all possible edges in its neighborhood. Hence

$$ C_i = 2\left|\{e_{jk}\}\right|/k_i(k_i - 1) : v_j, v_k \in N_i \tag{12} $$

This measure equals to 1 if every neighbor connected to v_i is also connected to every other node within the neighborhood and 0 if no node connected to v_i connects to any other node connected to v_i. The clustering coefficient for the whole graph is the average of C_i for each node, $C = \sum_{i=1}^{n} C_i/n$, and is a measure of the tendency of graph nodes to form local clusters [7]. For unweighted graphs, the local CC associated with a node i having k_i connections is defined as the ratio of the triangles that contain this node divided by the maximum possible number of such triangles $k_i(k_i - 1)/2$.

2.5.3 Average Shortest Path Length

The shortest path (distance) d_{ij} between two nodes v_i and v_j is the minimum number of edges we need to traverse in order to go from node v_i to node v_j. The average shortest path length is a measure of interconnectedness of the graph and is defined as

$$ L = \sum\nolimits_{i,j \in V, i \neq j} d_{ij}/n(n-1) \tag{13} $$

One serious drawback of L is that it can only be used for connected graphs.

2.5.4 Graph Efficiency

A measure that is also based on shortest paths but works for disconnected graphs as well is the graph efficiency [14]. The efficiency ε_{ij} in the communication between nodes v_i and v_j is inversely proportional to the shortest distance: $\varepsilon_{ij} = \frac{1}{d_{ij}}$. When there is no path in the graph between node v_i and node v_j, $d_{ij} = +\infty$ and consistently $\varepsilon_{ij} = 0$. The average efficiency of graph G can be defined as

$$Ef = \frac{1}{n(n-1)} \sum_{v_i \neq v_j \in V} \varepsilon_{ij} = \frac{1}{n(n-1)} \sum_{v_i \neq v_j \in V} \frac{1}{d_{ij}} \qquad (14)$$

2.5.5 Threshold Selection Approaches

Most EEG visualization methods require manual threshold selection [10], and the important question is how to choose a threshold value T. One intuitive way to choose a global threshold T or a range of thresholds is described in [10], where the threshold was chosen according to differentiation in network statistics between healthy and schizophrenic patients. For each one of threshold values $T \in \{0, 0.001, 0.002, \ldots, 0.009, 1\}$ and for fixed interdependence measure and frequency band a graph was created for each patient. Then for each threshold the mean value and the standard deviation (across networks of the same group) of the average degree K, the clustering coefficient C, and the average shortest path length L were calculated. Then using t-tests with significance level $0.001 \leq p \leq 0.02$, threshold values that lead to statistical differentiation in K, L, and C between the two groups were identified. In this chapter we repeat the approach of [10] in the threshold selection using data from alcoholic and control patients. Figure 1 depicts the average degree as a function of threshold. The networks were calculated using the nonlinear synchronization method (Section 2.4) at beta band. There is statistically significant difference between the average degrees of graphs of alcoholic and control patients when the threshold is in interval [0.36, 0.45].

Another way of setting threshold values is by using statistical hypothesis testing and randomization techniques (surrogate data) [15] to estimate a threshold above which the values of an interdependency measure are statistically significant [16]. In this case each interdependency link $IM(i, j)$ has its own threshold T_{ij}.

After identifying threshold regions where there is statistically important differentiation between the two groups, we can choose a threshold and define the networks. Then one can get information about the network structure using network statistics. For example consider the case where the graph of the average IM of control group has greater C compared to the graph of the average IM of the alcoholic group for some band, state, and threshold. This means that the first graph has denser local organization than the second. However, we do not get any details as to where this denser organization lies in the brain, which nodes makes this happen, in what way these nodes are connected to the rest of the graph, and generally how the two networks look like. In order to get that kind of information and understand the topology of the graph we visualize the graph.

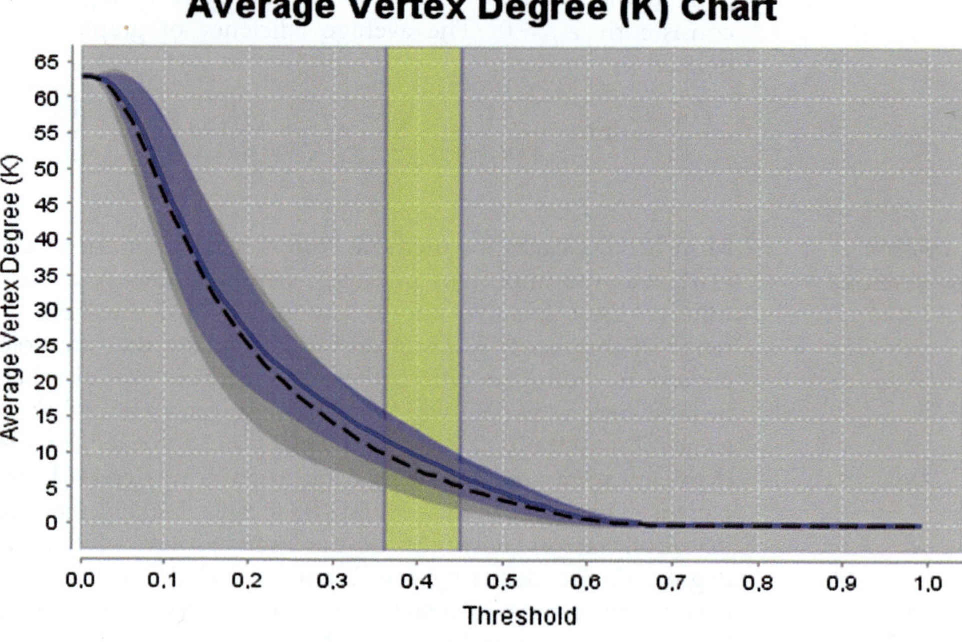

Fig. 1 Average degree as a function of threshold. The networks were calculated using the nonlinear synchronization method (Section 2.4) at beta band. With *black dashed line* is drawn the mean value (across patients) of the average degree of networks derived from alcoholic patients, while with blue color line is shown the mean value (across control subjects) of average degree of networks derived from control subjects. There is statistically significant difference between the average degrees of graphs of alcoholic and control patients when the threshold is in interval [0.36, 0.45]

3   Results

Our study was able to visualize the established brain networks as identified by the significant coupling calculated using coherence, PLV, and the nonlinear GS method, assuming no pre-specified threshold. The networks corresponding to the alcoholic brain tend to retain a hemispheric symmetry, whereas the networks corresponding to the control brain display a prominent lateralization (Fig. 2), when the broadband signals are analyzed. Restricting our analysis on beta band this effect is further enhanced. Figure 2 illustrates the results obtained using the GS method and the broadband signals, while Fig. 3 shows the results obtained using mscoherence and the beta band signal. From these two figures we conclude that the network topologies differ between the two groups, but it is not significantly affected by the choice of the synchronization method and/or the band. However, focusing on

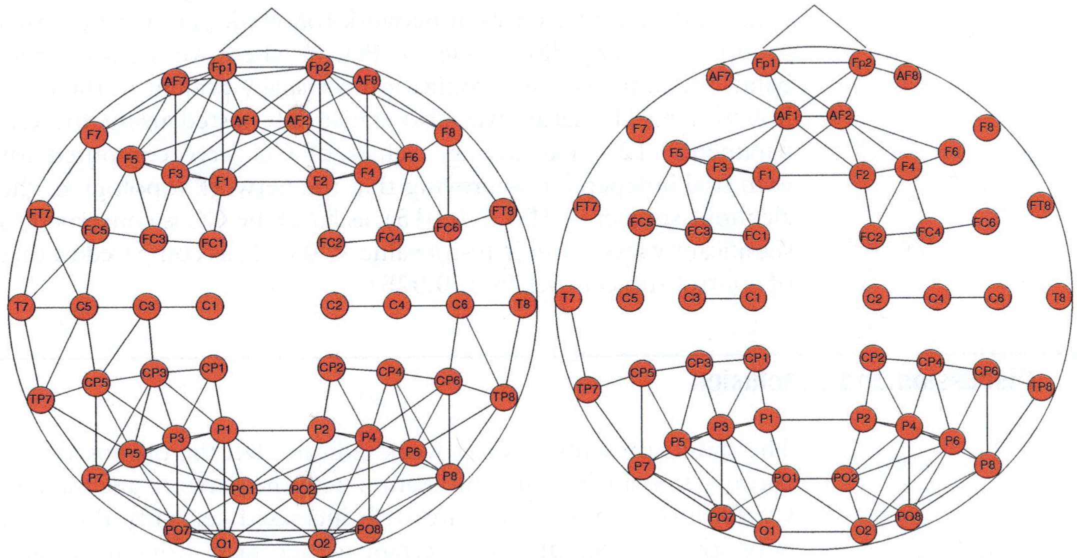

Fig. 2 The average (across control patients and runs) network (*left graph*) appears to exhibit lateralization compared to the average (across alcoholic patients and runs) network (*right graph*) which exhibits interhemispheric symmetry, when the broadband signals are analyzed using the interdependence measure proposed by Arnhold et al. [20]. The threshold is $T = 0.42$ (see Fig. 1)

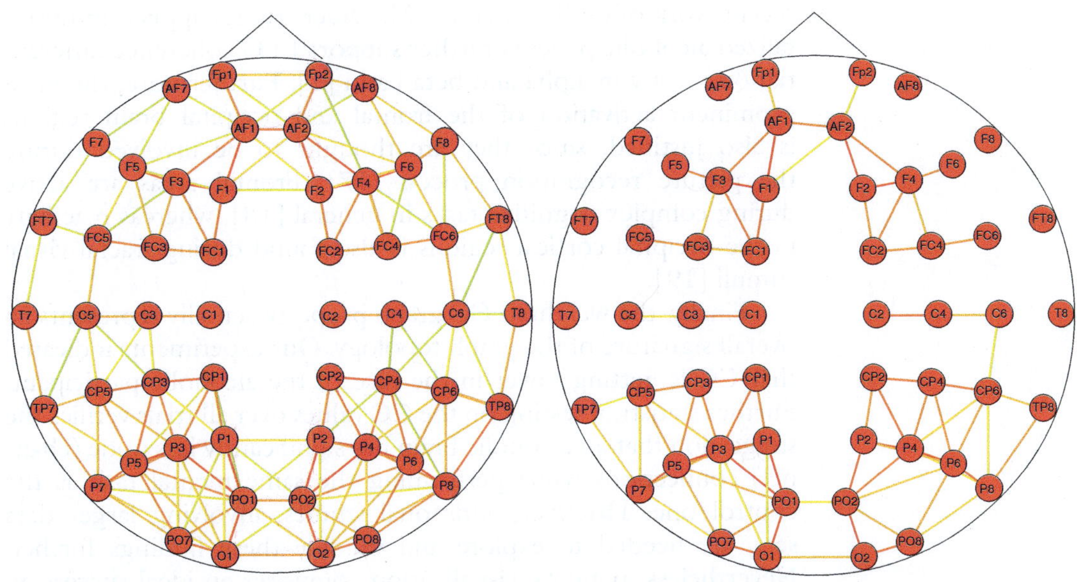

Fig. 3 Average across patient (control and alcoholic) and run networks. For each patient and run the interdependence was calculated using mscoherence in beta band. Higher synchronization is evident in the control subjects (*left network*) compared to the alcoholic ones

beta band, the differences in network topologies between the two groups are clearly discernible. In Fig. 3, where we focus on beta band, the impaired synchronization is visually evident in the alcoholic brain. The latter was also statistically tested using the CC parameter (12). The average values of CC were computed for each trial independently proving that the network topology of the alcoholic subject (0.106 ± 0.015), as far as the CC is concerned, is significantly less dense (t-test p-value < 0.001) as compared to that of control subject (0.175 ± 0.025).

4 Discussion and Conclusion

The results presented in this work indicate that the proposed synchronization analysis in combination with the network analysis and visualization is able to picture with increased certainty the brain network topology during a certain mental task. Both linear and nonlinear interdependence patterns may be revealed and easily illustrated. Using such a framework the alcoholic individual was found to have impaired synchronization of brain activity and loss of lateralization during the rehearsal process (Figs. 2 and 3), most prominently in alpha band, as compared to a control.

Lower beta synchronization was also evident in the alcoholic subject (Fig. 2). Similar findings were reported during moderate-to-heavy-alcohol intake during rest and mental rehearsal in the recent work of de Bruin et al. [2]. Older studies applied in hospitalized alcoholic patients further support EEG coherence differentiation mostly in alpha and beta band [1]. Furthermore, the most prominent activation of the frontal and occipital brain regions is also justified, since they are thought to be involved during the picture recognition process [17]. Frontal areas are active during complex cognitive tasks in general [18], whereas reactivity in the occipital cortical regions is also found during fractal visual stimuli [19].

Finally, the weighted CC graph property actually represents an overall signature of the graph topology. Our experiments indicated that CC is getting lower in the case of the alcoholic participant. Further statistical testing on the CC values over all trials verifies the suggestion that an alcoholic brain has significantly less node (channel) connections while performing the same mental task as the control one. However, additional studies, involving larger data sets, are needed to explore and validate these findings further. Nevertheless, network visualization comprises an ideal succession and supplement of statistical synchronization brain studies, as described in this exemplar case.

Acknowledgment

The authors would like to thank Henri Begleiter at the Neurodynamics Laboratory, State University of NY Health Center at Brooklyn, for providing the EEG data set.

References

1. Winterer G, Enoch MA, White KV, Saylan M, Coppola R, Goldman D (2003) EEG phenotype in alcoholism: increased coherence in the depressive subtype. Acta Psychiatr Scand 108 (1):51–60

2. de Bruin EA, Stam CJ, Bijl S, Verbaten MN, Kenemans JL (2006) Moderate-to-heavy alcohol intake is associated with differences in synchronization of brain activity during rest and mental rehearsal. Int J Psychophysiol 60:304–314

3. Sakkalis V (2011) Review of advanced techniques for the estimation of brain connectivity measured with EEG/MEG. Comput Biol Med 41(12):1110–1117

4. Sakkalis V (2011) Applied strategies towards EEG/MEG biomarker identification in clinical and cognitive research. Biomark Med 5 (1):93–105

5. Mormann F, Lehnertz K, David P, Elger C (2000) Mean phase coherence as a measure for phase synchronization and its application to the EEG of epilepsy patients. Phys D 144:358–369

6. Quian Quiroga R, Kraskov A, Kreuz T, Grassberger P (2002) Performance of different synchronization measures in real data: a case study on electroencephalographic signals. Phys Rev E 65:041903

7. Sakkalis V, Tsiaras V, Tollis I (2010) Graph analysis and visualization for brain function characterization using EEG data. J Healthc Eng 1(3):435–460

8. Fingelkurts AA, Kähkönen S (2004) Functional connectivity in the brain—is it an elusive concept? Neurosci Biobehav Rev 28:827–836

9. Takens F, Rand DA, Young LS (1981) Detecting strange attractors in turbulence. Dynamical Syst Turbulence Lecture Notes Mathematics 898:366–381

10. Sakkalis V, Oikonomou T, Pachou E, Tollis I, Micheloyannis S, Zervakis M (2006) Time-significant wavelet coherence for the evaluation of schizophrenic brain activity using a graph theory approach. In: Proceedings of the 28th annual international conference of the IEEE engineering in medicine and biology society, New York, USA, August 2006

11. Newman MEJ (2002) Assortative mixing in networks. Phys Rev Lett 89(20):208701

12. Newman MEJ (2003) Mixing patterns in networks. Phys Rev E 67(2):026126

13. Park CH, Kim SY, Kim Y-H, Kim K (2008) Comparison of the small-world topology between anatomical and functional connectivity in the human brain. J Phys A 387 (23):5958–5962

14. Latora V, Marchiori M (2001) Efficient behavior of small-world networks. Phys Rev Lett 87 (19):198701

15. Andrzejak RG, Kraskov A, Stogbauer H, Mormann F, Kreuz T (2003) Bivariate surrogate techniques: necessity, strengths, and caveats. Phys Rev E 68(6):066202

16. Sakkalis V, Tsiaras V, Zervakis M, Tollis IG (2007) Optimal brain network synchrony visualization: Application in an alcoholism paradigm. Conf Proc IEEE Eng Med Biol Soc 2007:4285–4288

17. Zhang XL, Begleiter H, Porjesz B, Wang W, Litke A (1995) Event related potentials during object recognition tasks. Brain Res Bulletin 38 (6):531–538

18. Sakkalis V, Zervakis M, Micheloyannis S (2006) Significant EEG feature selection for the evaluation of cognitive task complexity functions. Brain Topogr 19(1–2):53–60

19. Sakkalis V, Giurcăneanu CD, Xanthopoulos P, Zervakis M, Tsiaras V, Yang Y, Micheloyannis S (2009) Assessment of linear and nonlinear synchronization measures for analyzing EEG in a mild epileptic paradigm. IEEE Trans Inf Technol Biomed 13(4):433–441. doi:10.1109/TITB.2008.923141

20. Arnhold J, Grassberger P, Lehnertz K, Elger CE (1999) A robust method for detecting interdependences: application to intracranially recorded EEG. Physica D 134(4):419–430

Neuromethods (2015) 91: 171–204
DOI 10.1007/7657_2014_69
© Springer Science+Business Media New York 2014
Published online: 18 May 2014

Time-Varying Effective Connectivity for Investigating the Neurophysiological Basis of Cognitive Processes

Jlenia Toppi, Manuela Petti, Donatella Mattia, Fabio Babiloni, and Laura Astolfi

Abstract

This chapter describes the methodological advancements developed during the last 20 years in the field of effective connectivity based on Granger causality and linear autoregressive modeling. At first we introduce the concept of Granger causality and its application to the connectivity field. Then, a detailed description of both stationary and time-varying versions of Partial Directed Coherence (PDC) estimator for effective connectivity will be given. The General Linear Kalman Filter (GLKF) approach is described an algorithm, recently introduced for estimating the temporal evolution of the parameters of adaptive multivariate model, able to overcome the limits of existing time-varying approaches. Then a detailed description of the graph theory approach and of possible indexes which could be defined is given. At the end, the potentiality of the described methodologies is demonstrated in an application aiming at investigating the neurophysiological basis of motor imagery processes.

Keywords Electroencephalography (EEG), Effective connectivity, Non-stationarity, Graph theory, Statistical assessment, Multiple comparisons, Motor imagery

> "Learn from yesterday, live for today,
> hope for tomorrow. The important thing is to not
> stop questioning. Curiosity has its
> own reason for existing."
>
> Albert Einstein

---

1 Introduction

The development of a methodology for the estimation of the information flows between different and differently specialized cerebral areas, starting from noninvasive measurements of the neuroelectrical brain activity, has gained more and more importance in the field of Neuroscience, as an instrument to investigate the neural basis of cerebral processes. In fact, the knowledge of the complex cerebral dynamics underlying human cognitive processes cannot be fully achieved by the reconstruction of temporal/spectral activations of different parts of the brain. The description of the brain circuits involved in the execution of a task is a crucial point for understanding the mechanisms at the basis of the specific function under examination, as well as for the development of applications in the clinical and rehabilitation fields.

In the last two decades, several approaches were developed and applied to neuroelectrical signals in order to estimate the connectivity patterns elicited among cerebral areas. Several studies investigated the properties of all the available methods providing different solutions for different application fields and highlighting the best approaches able to reproduce the brain circuits related to noninvasive EEG measurements.

The theory of Granger Causality [1], based on the statistical studies on causality by Norbert Wiener [2], is one of the most generally adopted in EEG based connectivity studies, due to its many advantages in terms of generality, easiness, and possibility to provide a fully multivariate analysis of brain circuits in terms of existence, strength and direction of the functional links. At the same time, this method avoids the necessity of a priori information on the brain circuits under investigation. Since it provides not a mere description of a synchronicity between distant brain structures, but an hypothesis on the brain circuits, including the influence exerted by a neural system to the others, it can be considered an estimator of the so-called effective connectivity [3–5].

The functional connectivity between cortical areas is then defined as the temporal correlation between spatially remote neurophysiologic events and it could be estimated by using different methods both in time as well as in frequency domain based on bivariate or multivariate autoregressive (MVAR) models [6–9]. Several studies have proved the higher efficiency in estimating functional connectivity of methods, such as Directed Transfer Function (DTF) [8] or Partial Directed Coherence (PDC) [10], which are defined in frequency domain and based on the use of MVAR models built on original time-series [11]. In fact the bivariate approach is affected by a high number of false positives due to the impossibility of the method in discarding a common effect on a couple of signals of a third one acquired simultaneously [12]. Moreover the bivariate methods give rise to very dense patterns of propagation, thus it is impossible to find the sources of propagation [6, 13]. The PDC technique has been demonstrated [10] to rely on the concept of Granger causality between time series [1], according to which an observed time series $x(n)$ causes another series $y(n)$ if the knowledge of $x(n)$'s past significantly improves prediction of $y(n)$. Moreover, the PDC is also of particular interest because it can distinguish between direct and indirect connectivity flows in the estimated connectivity pattern better than DTF and its direct modified version, the directed DTF [14].

All the MVAR based methodologies for the functional connectivity estimation require the hypothesis of stationarity of the signals included. Thus, the temporal dynamics of the influences between cerebral areas are completely loss with evident disadvantages in describing causal networks at the basis of time-varying complex neurophysiological mechanisms. To overcome this limitation, different algorithms for the estimation of MVAR models with time dependent coefficients were recently developed [15]. In particular, the first proposed methodology was based on a short-time window approach, assuming the stationarity of signals in short time intervals [16].

In 2001 the multi-trial Recursive Least Square (RLS) method with Forgetting Factor was introduced. It consisted in the adaptive estimation of the MVAR model by means of a recursive algorithm involving a weighted influence of the past of the investigated signals [17, 18]. However, even if the RLS overcomes the limits of the short time approach, providing a more accurate estimation of the dynamics of non-stationary data, due to its computational complexity it presents a limitation in the number of signals to be considered at the same time in the estimation [19, 20]. Since few years ago, the problem of the model dimension has been solved by reducing the number of electrodes time series to be included in the model [19–21] or by using cortical waveforms derived for some regions of interest from high resolution EEG data [17]. However, the need to reduce the model dimension introduces a significant error due to the "hidden source dilemma." In fact, each time a relevant source of information of the problem is removed from the autoregressive model, spurious connectivity links are introduced in the estimation, degrading the reconstruction of the connectivity network. Moreover, the a priori selection of sources, both at scalp or cortical levels, could not be applied every time the dynamics at the basis of the investigated cerebral processes are not completely known and the selection of a subgroup of areas mostly involved in the investigated task is not so obvious. In 2010, a new method based on the Kalman filter, the General Linear Kalman Filter (GLKF), was provided as good alternative to RLS in the description of temporal evolution of information flows between the nodes of a network and as a solution to the limitation in the number of signals to be considered simultaneously in estimation process [19].

Several studies demonstrated that the processes at the basis of attentive or memory processes involve not isolate and specific cerebral areas but groups of brain areas strictly connected to each other. The communication between such areas is characterized by a specific timing and is subjected to a temporal evolution strictly linked to the explicated cognitive function. For this reason, the study of complex cerebral mechanisms such as those at the basis of cognitive processes require methodologies able to describe phenomena evolving in time and which globally involve the brain in terms of functional networks. The methodologies available since few years ago did not allow to follow the temporal evolution of cerebral networks with sufficient accuracy also taking into account all the sources involved in the processes [18]. The methodological advancements achieved with GLKF approach provided an important contribution to the study of complex mechanisms at the basis of cognitive processes such as attention and memory, without applying any a priori sources selection which could affect the results [22].

In the present chapter, we firstly introduce the theoretical background highlighting the state of the art and the open issues of the methodologies applied in the field. In particular, we provide a detailed description of the PDC estimator in its stationary and time-varying versions and then we introduce the concepts of statistical assessment of connectivity patterns to be used as a preliminary step in graph theory approach needed for the extraction of relevant properties of investigated networks. Secondly, we show an application of the described methodologies with the aim to demonstrate that the advanced approaches for time-varying connectivity estimation and for the extraction of salient indexes describing the most important features of the investigated networks could be used for the study of cerebral mechanisms at the basis of motor imagery processes. A pilot testing study was conducted in a healthy subject. The study provided high density EEG data measured during the execution of a motor imagery task, the imagination of prolonged grasping of both hands. On the basis of scientific evidence derived from literature, descriptors of the cognitive functions based on connectivity networks elicited by the specific task were provided.

2   Theoretical Background: Kalman Filter for Estimating Time-Varying Connectivity

2.1   Multivariate Methods for the Estimation of Connectivity

Let Υ be a set of signals, obtained from noninvasive EEG recordings or from the reconstruction of neuroelectrical cortical activities based on scalp measures:

$$\Upsilon = \left[y_1(t), y_2(t), \ldots, y_N(t) \right]^T \tag{1}$$

where t refers to time and N is the number of electrodes or cortical areas considered.

Supposing that the following MVAR process is an adequate description of the dataset Υ:

$$\sum_{k=0}^{p} \Lambda(k)\, \Upsilon(t-k) = E(t) \quad \text{with } \Lambda(0) = I, \tag{2}$$

where $\Upsilon(t)$ is the data vector in time, $E(t) = [e1(t), \ldots, en(t)]^T$ is a vector of multivariate zero-mean uncorrelated white noise processes, $\Lambda(1)$, $\Lambda(2)$, $\ldots$, $\Lambda(p)$ are the $N \times N$ matrices of model coefficients and p is the model order, usually chosen by means of the Akaike Information Criteria (AIC) for MVAR processes [23].

Once an MVAR model is adequately estimated, it becomes the basis for subsequent spectral analysis. To investigate the spectral properties of the examined process, Eq. 2 is transformed to the frequency domain:

$$\Lambda(f)\Upsilon(f) = E(f) \tag{3}$$

where:

$$\Lambda(f) = \sum_{k=0}^{p} \Lambda(k)\mathrm{e}^{-j2\pi f \Delta tk} \tag{4}$$

and Δt is the temporal interval between two samples.

2.2 Partial Directed Coherence

The PDC [10] is a full multivariate spectral measure, used to determine the directed influences between pairs of signals in a multivariate data set. PDC is a frequency domain representation of the existing multivariate relationships between simultaneously analyzed time series that allows the inference of functional relationships between them. This estimator was demonstrated to be a frequency version of the concept of Granger causality [1], according to which a time series $x[n]$ can be said to have an influence on another time series $y[n]$ if the knowledge of past samples of x significantly reduces the prediction error for the present sample of y.

It is possible to define PDC as

$$\pi_{ij}(f) = \frac{\Lambda_{ij}(f)}{\sqrt{\displaystyle\sum_{k=1}^{N}\Lambda_{kj}(f)\Lambda_{kj}^{*}(f)}} \tag{5}$$

which bears its name by its relation to the well-known concept of partial coherence [24].

The PDC from node j to node i, $\pi_{ij}(f)$, describes the directional flow of information from the signal $y_j[n]$ to $y_i[n]$, whereupon common effects produced by other signals $y_k[n]$ on the latter are subtracted leaving only a description that is exclusive from $y_j[n]$ to $y_i[n]$. PDC squared values are in the interval $[0:1]$ and the normalization condition

$$\sum_{n=1}^{N} \left|\pi_{nj}(f)\right|^{2} = 1 \tag{6}$$

is verified. According to this normalization, $\pi_{ij}(f)$ represents the fraction of the information flow of node j directed to node i, as compared to all the j's interactions with other nodes.

Even if this formulation derived directly from information theory, the original definition was modified in order to give a better physiological interpretation to the estimate results achieved on electrophysiological data. In particular, a new type of normalization, already used for another connectivity estimator such as DTF [8] was introduced. Such normalization consisted in dividing each estimated value of PDC by the root squared sum of all the elements of the relative row, obtaining the following definition:

$$\pi_{ij}^{\text{row}}(f) = \frac{\Lambda_{ij}(f)}{\sqrt{\sum_{k=1}^{N} \Lambda_{ik}(f)\Lambda_{ik}^{*}(f)}} \tag{7}$$

Even in this formulation PDC squared values are in the range [0:1], but the normalization condition is as follows:

$$\sum_{n=1}^{N} \left| \pi_{\text{in}}^{\text{row}}(f) \right|^2 = 1 \tag{8}$$

Moreover, a squared formulation of PDC has been introduced and can be defined as follows for the two types of normalization:

$$\text{sPDC}_{ij}^{\text{col}}(f) = \frac{\left|\Lambda_{ij}(f)\right|^2}{\sum_{k=1}^{N} \left|\Lambda_{kj}(f)\right|^2} \tag{9}$$

$$\text{sPDC}_{ij}^{\text{row}}(f) = \frac{\left|\Lambda_{ij}(f)\right|^2}{\sum_{k=1}^{N} \left|\Lambda_{ik}(f)\right|^2} \tag{10}$$

The main difference with respect to the original formulation is in the interpretation of these estimators. Squared PDC can be put in relationship with the power density of the investigated signals and can be interpreted as the fraction of ith signal power density due to the jth measure. The higher performance of squared methods in respect to simple PDC has been demonstrated in a simulation study [25]. This study revealed higher accuracy for the methods based on squared formulation of PDC (i) in the estimation estimation of connectivity patterns on data characterized by different lengths and signal-to-noise ratios (SNRs) and (ii) in distinction between direct and indirect pathways.

2.3 Adaptive Partial Directed Coherence

The original formulation of PDC is based on the hypothesis of stationarity of the signals included in the estimation process. Such hypothesis leads to a complete loss of the information about the temporal evolution of estimated information flows.

For overcoming this limitation, a time-varying adaptation of squared PDC was introduced [25]. The adaptation consisted in modifying the original formulation of PDC by including time dependence in the MVAR coefficients. Thus, the adaptive squared PDC estimator can be defined as follows:

$$\mathrm{sPDC}_{ij}^{\mathrm{row}}(f,t) = \frac{\left|\Lambda_{ij}(f,t)\right|^2}{\sum\limits_{k=1}^{N}\left|\Lambda_{ik}(f,t)\right|^2} \tag{11}$$

$$\mathrm{sPDC}_{ij}^{\mathrm{col}}(f,t) = \frac{\left|\Lambda_{ij}(f,t)\right|^2}{\sum\limits_{k=1}^{N}\left|\Lambda_{kj}(f,t)\right|^2}, \tag{12}$$

where t refers to a time dependence of the MVAR coefficients and $\Lambda_{ij}(f,t)$ represents the ij entry of the matrix of model coefficients Λ at frequency f and time t.

The estimation of time-varying MVAR parameters can be performed by means of two different approaches available at the moment, the RLS and the GLKF. The GLKF, whose higher accuracy in following temporal dynamics of investigated connectivity patterns, also in presence of a high number of sources, has been already demonstrated [22], will be described below.

2.4 The General Linear Kalman Filter

In the GLKF an adaptation of the Kalman filter to the case of multi-trial time series is provided. In particular:

$$\begin{aligned} Q_t &= G_{t-1}Q_{t-1} + V_t \\ O_t &= H_t Q_t + W_t \end{aligned} \tag{13}$$

where O_t represents the observation, Q_t is the state process, H_t and G_t are the transition matrices, and V_t and W_t are the additive noises. To obtain the connection with the time-varying MVAR, it is necessary to make the following associations:

$$Q_t = \begin{bmatrix} \Lambda_1(t)^N \\ \vdots \\ \Lambda_p(t)^N \end{bmatrix},$$

$$O_t = \begin{pmatrix} y_1^{(1)}(t) & \cdots & y_M^{(1)}(t) \\ \vdots & \ddots & \vdots \\ y_1^{(N)}(t) & \cdots & y_M^{(N)}(t) \end{pmatrix} = \Upsilon_t \tag{14}$$

$$G_{t-1} = I_{dp}, \quad H_t = \left(O_{t-1}, \quad \cdots, \quad O_{t-p} \right), \tag{15}$$

where N denotes the number of trials, whereas M is the dimension of the measured process [19]. In particular, for $t = [p + 1:T]$ the following steps are repeated:

$$Q_{t-1}^{+} = Q_{t-1} + K_t\left(O_t - H_t Q_{t-1}\right) \tag{16}$$

$$Q_t = G_{t-1}Q_{t-1}^{+} \tag{17}$$

$$K_t = P_{t-1}H_t^T/S_t \tag{18}$$

$$S_t = H_t P_{t-1} H_t^T + tr\left(\overline{W}_t\right) I_k \quad \textit{where} \ \ \overline{W}_t = E\left[W_t W_t^T\right] \tag{19}$$

$$P_{t-1}^+ = \left(I_{dp} - K_t H_t\right) P_{t-1} \tag{20}$$

$$P_t = G_{t-1} P_{t-1}^+ G_{t-1}^T + \overline{V}_t \quad \textit{where} \ \ \overline{V}_t = E\left[V_t V_t^T\right] \tag{21}$$

The rationale behind this set of equations is the demand for a linear and recursive estimator for the state $Q_t\ (Q_{(t-1)}, O_t)$ as a function of the previous state $Q_{(t-1)}$ and of the actual observation matrix O_t. In particular starting from the coefficient matrix at time $t-1$, its estimation is updated to $Q_{(t-1)}^+$ by means of Eq. 16 and used for estimating the values corresponding to the following time sample Q_t by means of Eq. 17. K_t is called the Kalman gain matrix and it weights the prediction error. Its formulation is reported in Eq. 18 where $P_{(t-1)}$ is the covariance matrix whose evolution is estimated by means of Eqs. 20 and 21.

The quality of estimation is related to the definition of two parameters, c1 and c2, regulating the compromise between the estimation accuracy and the speed of adaptation to transitions. A description of the application range of c1 and c2 and a suggestion about their typical values for EEG signals under different conditions of SNR and amount of trials is available in [22].

2.5 Statistical Validation of Connectivity Patterns

Random correlations between signals induced by environmental noise or by chance can lead to the presence of spurious links in the connectivity estimation process. In order to assess the significance of estimated patterns, the value of functional connectivity for a given pair of signals and for each frequency sample, obtained by computing PDC, has to be statistically compared with a threshold level which is related to the lack of transmission between considered signals. Such threshold could be inferred by means of two different approaches. The first approach, mainly used in stationary applications, extracts the statistical threshold from an empirical distribution representing the null-case for PDC estimator. Such threshold represents thus the inferior limit above which the estimated link is not due to chance. On the contrary, the second approach, mainly used in time-varying applications, infers the statistical threshold directly from the baseline condition, a period in which the subject is exposed, without performing the task, to all the external stimuli due to environmental noise and to the paradigm administration. In this way, all PDC values estimated during baseline period represent the reference distribution to be compared with the values inferred during task condition. In particular, once achieved the reference distribution, the threshold is computed by applying on it a percentile with a significance level of 5 %. A threshold value is thus achieved for each pair, direction and frequency band. A schematic representation of the process is reported in Fig. 1.

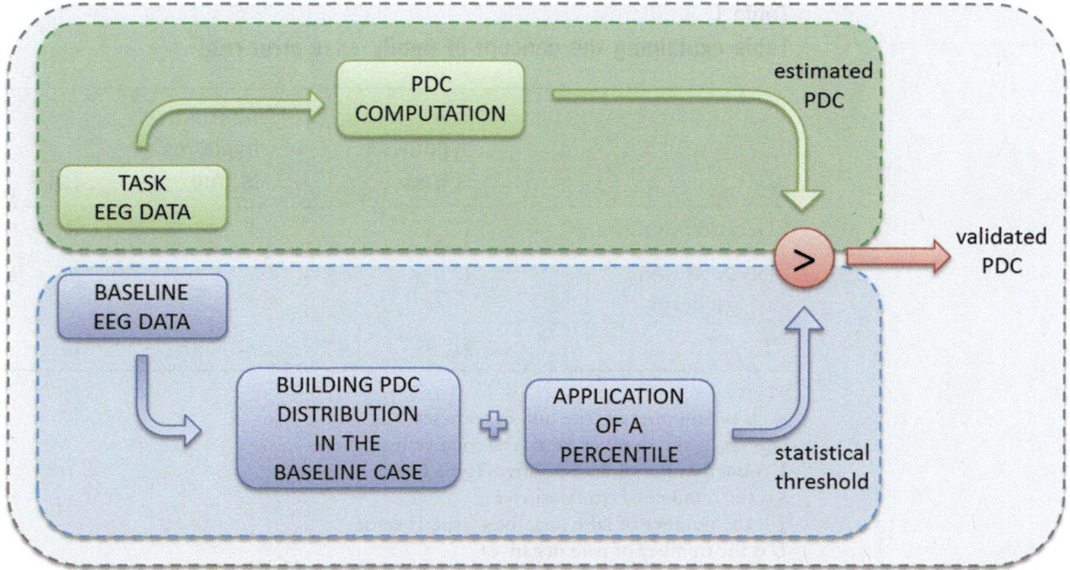

Fig. 1 Schematic representation of validation process for time-varying connectivity

2.6 Reducing the Occurrence of Type I Errors in Assessment of Connectivity Patterns

The statistical validation process has to be applied on each couple of signals for each direction and for each frequency sample. This necessity leads to the execution of a high number of simultaneous univariate statistical tests with evident consequences in the occurrence of type I errors. The statistical theory provides several techniques that could be usefully applied in the context of the assessment of connectivity patterns in order to avoid the occurrence of false positives.

The family-wise error rate represents the probability of observing one or more false positives after carrying out simultaneous univariate tests. Supposing to have m null hypotheses H_1, H_2, ..., H_m. Each hypothesis could be declared significant or not-significant by means of a statistical test. Table 1 summarizes the situation after multiple significance tests are simultaneously applied.

The FWER is the probability of making even one type I error in the family:

$$\text{FWER} = \Pr(V \geq 1) \tag{22}$$

Many methodologies are available for preventing type I errors [26], but in the following sections we limit the discussion to False Discovery Rate (FDR) and Bonferroni adjustments, which are the most used methodologies in neuroscience field.

2.6.1 Bonferroni Adjustment

The Bonferroni adjustment [27] starts from the consideration that if we perform N univariate tests, each one of them with an unknown significant probability α, the probability p that at least one of the test is significant is given by [28]:

Table 1
Table explaining the concept of family-wise error rate

	Null hypothesis is true	Alternative hypothesis is true	Total
Declared significant	V	S	R
Declared non-significant	U	T	$m - R$
Total	m_0	$m - m_0$	m

where:
m_0 is the number of true null hypotheses
$m - m_0$ is the number of true alternative hypotheses
V is the number of false positives (type I error)
S is the number of true positives
T is the number of false negatives (type II error)
U is the number of true negatives
R is the number of rejected null hypotheses

$$p < N\alpha \tag{23}$$

In other words this means that if $N = 20$ tests are performed with the usual probability $\alpha = 0.05$, at least one of them will become significant statistically by chance alone. However, the Bonferroni adjustment requires that the probability p for which this event could occur (i.e., one result will be statistically significant by chance alone) could be equal to α. By using the Eq. 23, the single test will be performed at a probability.

$$\beta^* = \alpha/N \tag{24}$$

This β^* is the actual probability at which the statistical tests are performed in order to conclude that all of the tests are performed at α level of statistical significance, Bonferroni adjusted for multiple comparisons. The Bonferroni adjustment is quite flexible since it does not require the hypothesis of independence of the data to be applied but it's really conservative. In fact, it allows to highly reduce the number of false positives but at the same time, it introduces a lot of false negatives. For this reason, in 1995 a new approach, called the False Discovery Rate and less conservative of Bonferroni method, was introduced. Its capability to prevent both type I and type II errors was demonstrated in [29].

2.6.2 False Discovery Rate

The false discovery rate (FDR), suggested by Benjamini and Hochberg is the expected proportion of erroneous rejections among all rejections [30]. Considering V as the number of false positives and S as the number of true positives, the FDR is given by:

$$\mathrm{FDR} = E\left[\frac{V}{V + S}\right] \tag{25}$$

where $E[\,]$ is the symbol for expected value.

In the following we report the False Discovery Rate controlling procedure described by Benjamini and Hochberg in 1995. Let $H_1, H_2, \ldots, H_m$ be the null hypotheses, with m as the number of univariate tests to be performed, and $p_1, p_2, \ldots, p_m$ their corresponding p-values. Let's order in ascending order the p-values as $p(1) \leq p(2) \leq \ldots \leq p(m)$ and then select the largest i for which the condition

$$P_{(i)} \leq \frac{i}{m}\alpha \qquad (26)$$

is verified. Only the first i null hypotheses ($H_1, \ldots, H_i$), corresponding to the first order i p-value will be rejected.

In the case of independent tests, an approximation for evaluating the corrected significance level has been introduced [30, 31]:

$$\beta^* = \frac{(m + 1)}{2m}\alpha \qquad (27)$$

In this case the new level of significance is β^*. Such value guarantees that each test is performed with the imposed significance α.

2.7 Graph Theory Approach

The methodological advancement in the functional connectivity field has been leading to the description of neurological mechanisms at the basis of complex cerebral processes involving a high number of sources. Once the connectivity pattern, that was achieved for the investigated condition, was qualitatively described, a quantitative characterization of its main properties is necessary in order to synthetize the huge amount of information derived from the application of such advanced methodologies. The extraction of indexes describing global and local properties of the investigated networks could open the way to several different applications in neuroscience field.

A graph is a mathematical object consisting of a set of vertices (or nodes) and edges (or connections) indicating the presence of some sort of interaction between the vertices. The adjacency matrix A contains the information about the connectivity structure of the graph and can be derived directly from the investigated network. When a directed edge exists from the node j to the node i, the corresponding entry of the adjacency matrix is $A_{ij} = 1$ in binary graphs or $A_{ij} = v$ (where v is the value achieved by the estimator) in weighted graphs otherwise $A_{ij} = 0$. The adjacency matrix can be used for the extraction of salient information about the characteristic of the investigated network by defining several indexes based on its elements.

In relation to the estimator used for building the network, the associated graph could be:

undirected → if the estimator is able to extract only the value of the information flows and not its direction. In this case the adjacency matrix is symmetric ($A_{ij} = A_{ji}$).

directed → if the estimator allows to reconstruct not only the magnitude but also the direction of the connection. In this case the adjacency matrix is asymmetric ($A_{ij} \neq A_{ji}$).

If the estimator used for the analysis is based on multivariate approach, as in the case of PDC [10], the corresponding graph is binary/weighted and directed.

2.7.1 Adjacency Matrix Extraction

Once the functional connectivity pattern is estimated, it is necessary to define an associated adjacency matrix for each network, on which salient indexes able to characterize the network properties can be extracted. The generic ijth entry of a directed binary adjacency matrix is equal to 1 if there is a functional link directed from the jth to the ith signal and to 0 if no link exists. The construction of an adjacency matrix can be performed by comparing each estimated connectivity value to its correspondent threshold value. In particular:

$$G_{ij} = \begin{cases} 1 \to A_{ij} \geq \tau_{ij} \\ 0 \to A_{ij} < \tau_{ij} \end{cases} \tag{28}$$

where G_{ij} and A_{ij} represent the entry (i, j) of the adjacency matrix G and the connectivity matrix A, respectively, and τ_{ij} is the corresponding threshold.

Different approaches have been developed for evaluating the threshold values, most of them based on qualitative assumptions aiming at fixing the number of edges or the degree of some nodes or at maximizing some properties of the investigated networks. The selection of the threshold to be used for the extraction of adjacency matrices is a crucial step of the graph theory approach; in fact the type of threshold used in the process might affect the structure and the topological properties of the investigated networks. Recently, it was demonstrated that the use of statistical thresholds computed on null-case distribution or on baseline condition allows to prevent erroneous description of network main properties [32].

2.7.2 Graph Theory Indexes

Different indexes can be defined on the basis of the adjacency matrix extracted from a given connectivity pattern. The most commonly used, will be described in the following:

Global Efficiency. The global efficiency is the average of the inverse of the geodesic length and represents the efficiency of the communication between all the nodes in the network [33]. It can be defined as follows

$$E_{\mathrm{g}} = \frac{1}{N(N-1)} \sum_{i \neq j} \frac{1}{d_{ij}} \tag{29}$$

where N represents the number of nodes in the graph and d_{ij} the geodesic distance between i and j (defined as the length of the shortest path between i and j).

Local Efficiency. The local efficiency is the average of the global efficiencies computed on each sub-graph G_i belonging to the network and represents the efficiency of the communication between all the nodes around the node i in the network [33]. It can be defined as follows

$$E_1 = \frac{1}{N} \sum_{i=1}^{N} E_{\mathrm{g}}(G_i), \tag{30}$$

where N represents the number of nodes in the graph and G_i the sub-graph achieved deleting the ith row and the ith column from the original graph.

Degree. The degree of a node consists in the number of links connected directly to it. In directed networks, the indegree is the number of inward links and the outdegree is the number of outward links. Connections weight is ignored in calculations [34]. Degree of node k can be defined as follows

$$\deg_k = \sum_{\substack{i \in N \\ i \neq k}} a_{ij} + \sum_{\substack{j \in N \\ j \neq k}} a_{ij} \tag{31}$$

where a_{ij} represents the entry ij of the adjacency matrix A. It is possible to compute the degree for incoming (in-degree) and outcoming (out-degree) connections separately. In particular the formulation for these two indexes is:

$$\deg_k^{\mathrm{in}} = \sum_{\substack{i \in N \\ i \neq k}} a_{ij} \tag{32}$$

$$\deg_k^{\mathrm{out}} = \sum_{\substack{j \in N \\ j \neq k}} a_{ij} \tag{33}$$

Characteristic Path Length. The characteristic path length is the average shortest path length in the network, where the shortest path length between two nodes is the minimum number of edges that must be traversed to get from one node to another. It can be defined as follows

$$L = \frac{1}{n} \sum_{i \in N} L_i = \frac{1}{n} \sum_{i \in N} \frac{\sum_{j \in N, j \neq i} d_{ij}}{n - 1}, \tag{34}$$

where L_i is the average distance between node i and all other nodes and d_{ij} is the geodesic distance between node i and node j [34].

Clustering Coefficient. The clustering coefficient describes the intensity of interconnections between the neighbors of a node [35]. It is defined as the fraction of triangles around a node or the fraction of node's neighbors that are neighbors of each other. The binary directed version of Clustering Coefficient is defined as follows [36]

$$C = \frac{1}{n}\sum_{i \in N} C_i = \frac{1}{n}\sum_{i \in N} \frac{t_i}{\left(k_i^{out} + k_i^{in}\right)\left(k_i^{out} + k_i^{in} - 1\right) - 2\sum_{j \in N} g_{ij} g_{ji}},$$

(35)

where t_i represents the number of triangles involving node i, k_i^{in} and k_i^{out} are the number of incoming and outcoming edges of nodes i, respectively, and g_{ij} is the entry ij of adjacency matrix G.

Small-Worldness. A network G is defined as small-world network if $L_G \cong L_{rand}$ and $C_G \gg C_{rand}$ where L_G and C_G represent the characteristic path length and the clustering coefficient of a generic graph, respectively, and L_{rand} and C_{rand} represent the corresponding quantities for a random graph [35]. On the basis of this definition, a measure of small-worldness of a network can be introduced as follows

$$S = \frac{C_G \big/ C_{rand}}{L_G \big/ L_{rand}}$$

(36)

So a network is said to be a small world network if $S > 1$ [37].

The indexes described above do not allow the characterization of the network in terms of symmetries or influences between two different areas of the scalp. For this purpose, new indexes able to describe the topological properties of the investigated networks were defined [38]. For the evaluation of such indexes it is important to modify the adjacency matrix A in A' by disposing in the first N_1 rows and N_1 columns the connectivity values related to the first cerebral area and in the second N_2 rows and N_2 columns the connectivity values related to the second cerebral area.

Asymmetry. The asymmetry index is the percentage difference in the number of internal connections between two different spatial regions. It can be defined as follows

$$S = \frac{\sum_{i=1}^{N_1}\sum_{j=1}^{N_1} A'_{ij}}{N_1^2} - \frac{\sum_{i=N_1+1}^{N_2}\sum_{j=N_1+1}^{N_2} A'_{ij}}{N_2^2}$$

(37)

where N_1 and N_2 indicate the number of electrodes belonging to the two scalp regions respectively and A'_{ij} represents the adjacency matrix modified according to the scalp's regions to be compared. It could assume values in the range $[-1:1]$, where 0 means same number of internal connections within the two regions, 1 means connections only within the first region and -1 means only connections within the second region.

Influence. The influence index is the percentage difference in the number of interconnections between two different spatial regions.

$$I = \frac{\sum_{i=1}^{N_1}\sum_{j=N_1+1}^{N_2} A'_{ij} - \sum_{i=N_1+1}^{N_2}\sum_{j=1}^{N_1} A'_{ij}}{N_1 \cdot N_2} \tag{38}$$

where N_1 and N_2 indicate the number of electrodes belonging to the two scalp regions respectively and A'_{ij} represents the adjacency matrix modified according to the scalp's regions to be compared. It could assume values in the range $[-1{:}1]$, where 0 means no connections or same number of interconnections between the two regions, 1 means connections only from the second region to the first and -1 means only connections from the first region to the second.

It has been demonstrated that Asymmetry and Influence indexes could be used in neuroscience field for investigating the symmetries and influences between the two hemispheres or between frontal and parietal areas [38].

3   Application of TV Connectivity to the Investigation of Motor Imagery Processes

The methodological advancements provided during the last 10 years in the field of effective connectivity allowed to have available a set of state-of-the-art tools able (1) to describe with high accuracy the causal information exchanged between spatially distinct cerebral areas, (2) to follow with high speed their temporal evolution, and (3) to extract with high consistence graph theory indexes able to describe their main local and global properties. Such advancements allowed the use of effective connectivity in all the challenging applications in which open issues still remain. In order to demonstrate the potentiality of the tools actually available we proposed their application for the description of neurophysiological processes at the basis of cognitive functions.

In particular, in the following, we applied the proposed methodology to the data of a subject performing a motor imagery paradigm. The ambitious aim is to demonstrate that the estimated connectivity networks and the corresponding graph theory indexes extracted from these data are able to selectively describe the neurophysiological basis of motor imagery processes. The study could be considered as a preliminary step for the investigation of salient descriptors of motor imagery neurological processes allowing to provide information to be used in Brain Computer Interface field, where motor imagery is extensively employed.

The data used for testing the proposed methodology were recorded on a healthy subject who took part in a motor imagery experiment. Subject was comfortably seated in front of a computer screen and was asked to perform, according to the position of a red target on the screen, one of the following tasks: prolonged grasping of both hands (G) for the whole task length or just to relax (R). Regarding hands grasping imagery, subject was encouraged to perform a kinesthetic imagery rather than visually "seeing" themselves moving hands. The experiment was divided into 6 runs of 24 trials each (randomly ordered). The task length was set to 15 s and the inter-trial interval to 2 s. EEG potentials were recorded by a 61-channel system by means of an electrode cap (BrainAmp, Brainproducts GmbH, Germany). Sampling rate was 200 Hz. Surface EMG was recorded to ensure that participant refrained from contracting their arm muscles during motor imagery.

3.1 Effective Connectivity Analysis

EEG data were down-sampled at 100 Hz to optimize the following connectivity analysis and band-pass filtered in the range [1–45] Hz. Ocular artifacts were removed by means of Independent Component Analysis (ICA), whose higher accuracy in respect to the regression approaches has been already demonstrated [39]. EEG signals were then segmented in the interval [−500:1,000]ms according to the onset of the red target on the screen and classified according to the executed tasks. Residual artifacts were then removed by applying a semi-automatic procedure based on threshold criterion. In fact, all the trials containing signals whose absolute value exceeded 100 μV were considered as artifacts and excluded from the analysis. Artifacts-free trials, extracted for each condition, were subjected to time-varying functional connectivity analysis by means of GLKF. We chose $c1 = c2 = 0.01$ as adaptation constants to be used in the estimation process. In fact a simulation study highlighted them as valid values for obtaining accurate estimates with high speed of adaptation to temporal evolutions of EEG signals [22]. The connectivity patterns estimated for each time sample were averaged in six time intervals ([−500:−250] ms; [−250:0] ms; [0:250] ms; [250:500] ms; [500:750] ms; [750:1,000] ms) defined according to the "GO" stimulus and in four frequency bands, defined according to the Individual Alpha Frequency (IAF), determined from the Fast Fourier Transform spectra over posterior leads (parietal, parieto-occipital, and occipital) [40]. Ranges of the individually defined bands were: Theta (IAF-6 ÷ IAF-2), Alpha (IAF-2 ÷ IAF + 2), Beta (IAF + 2 ÷ IAF + 14), and Gamma (IAF + 15 ÷ IAF + 30). In order to discard all the information flows due to environmental noise or to the instrumentation used for administering the paradigm, a statistical comparison between connectivity patterns elicited during task execution and rest condition was computed for a significance level of 5 %, FDR corrected for multiple comparisons. The statistical threshold was also used for deriving the corresponding adjacency matrices on which different graph theory indexes were computed [32,41]. A set of graph indexes was extracted for each task, time interval and frequency band. In order to compare, at single subject level, the indexes achieved in each task along the six time intervals, a bootstrap method was applied. The resampling process consisted in the computation of the investigated index on a new adjacency matrix achieved by randomly deleting, at each computation, 10 % of the connections in the original adjacency matrix. 50 iterations were performed for bootstrap method.

3.2 Results

3.2.1 Effective Connectivity

The methodologies proposed in the Methods section were applied to the EEG data recorded during Grasping imagery task. The results achieved for the considered subject performing Grasping task in Theta, Alpha, and Beta bands were reported in Figs. 2, 3, and 4, respectively.

In the first column of Figs. 2, 3, and 4 we reported effective connectivity patterns elicited along the six considered time intervals by a representative subject performing Grasping imagery task in Theta, Alpha, and Beta frequency bands respectively. Connectivity patterns are represented on a 2D scalp model, seen from above with the nose pointing to the top of the page. Each electrode is represented by means of a white circle and the causal connections between different electrodes are represented by arrows linking the electrodes. Color and size of the arrows code the corresponding PDC value. In the second column of Figs. 2, 3, and 4 we reported the corresponding graphical representation of the role of each electrode involved in the connectivity pattern. Each electrode is represented by means of a sphere. The color of each sphere codes for the role of the electrode within the network: green if the electrode is a hub in the network, meaning that it manages the information flow in both directions (in and out): red if the electrode is a sink in the network, blue if the electrode is a source for the network. If the sphere is white, the electrode has no significant role in the network. The diameter of each sphere codes for the magnitude of the degree index computed for the corresponding electrode.

From a visual inspection of connectivity patterns elicited by the representative subject during Grasping task in Theta band (Fig. 2), it is possible to follow the temporal evolution along the six investigated time intervals of effective cerebral networks sustaining the motor imagery processes. In the two intervals preceding the "GO" stimulus no connections significantly different from rest condition could be appreciated. After the GO onset, the connectivity pattern started its temporal evolution highlighting, since 250 ms after the beginning of imagery task execution, a significant involvement of left centro-parietal electrodes (C3, CP3, C5) and then of the frontal areas of the scalp (F5, F3, F1, Fz, F2, F4). The involvement of these two scalp regions increased along time and was at its maximum at the end of the analyzed window. The graphical representation of the electrodes role confirmed what was shown by the time-varying

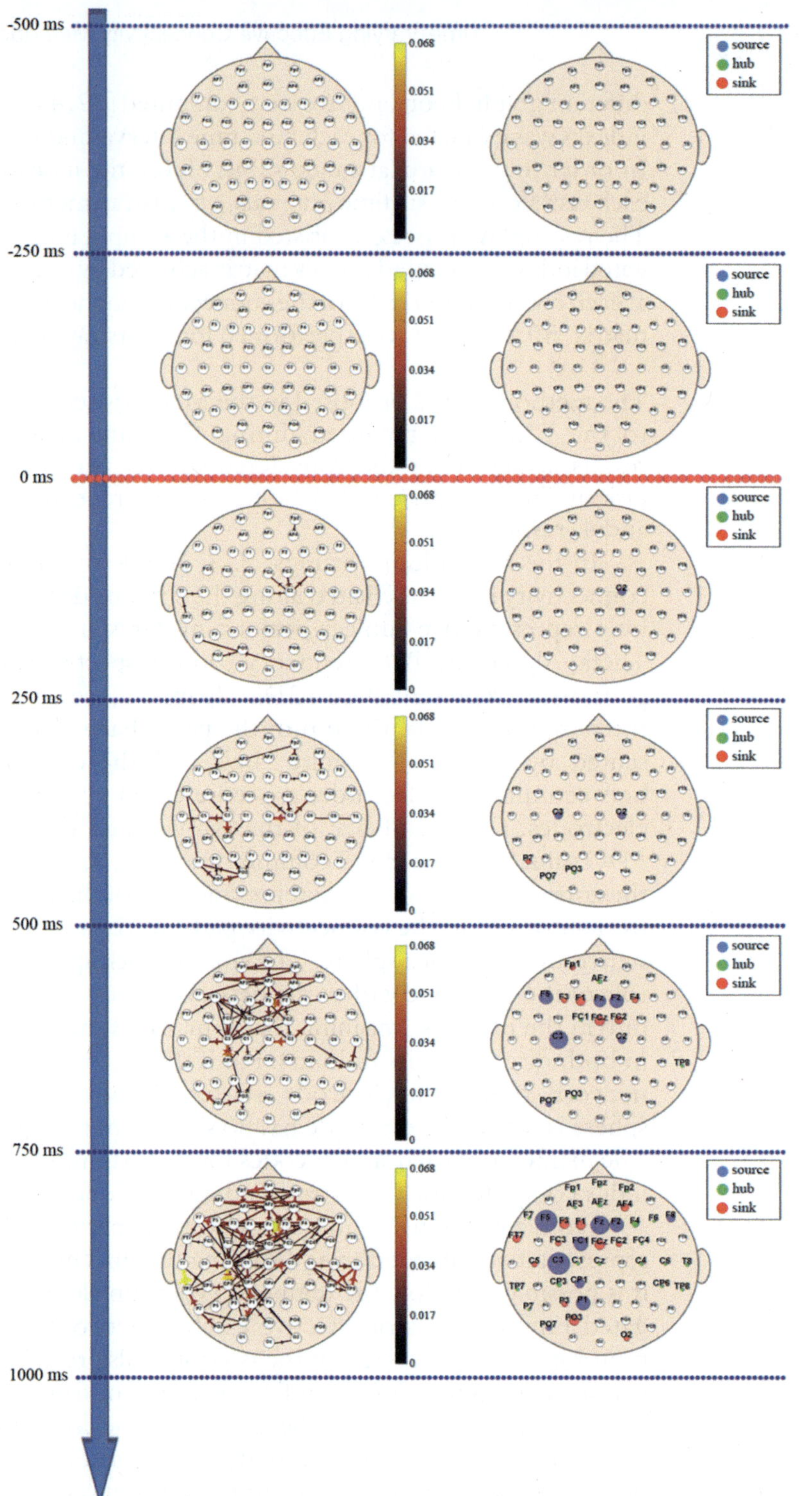

Fig. 2 *First column*. Effective connectivity patterns elicited along the six considered time intervals by a subject performing Grasping imagery task in Theta band. Connectivity patterns are represented on a 2D scalp model, seen from above with the nose pointing to the top of the page. Color and size of the *arrows* code the corresponding PDC value. *Second column*. Graphical representation of the role of each electrode involved in the connectivity patterns elicited during Grasping imagery task in Theta band. The color and the diameter of each sphere code for the role (hub in *green*; sink in *red*; source in *blue*) and the magnitude of the degree of the corresponding electrode, respectively

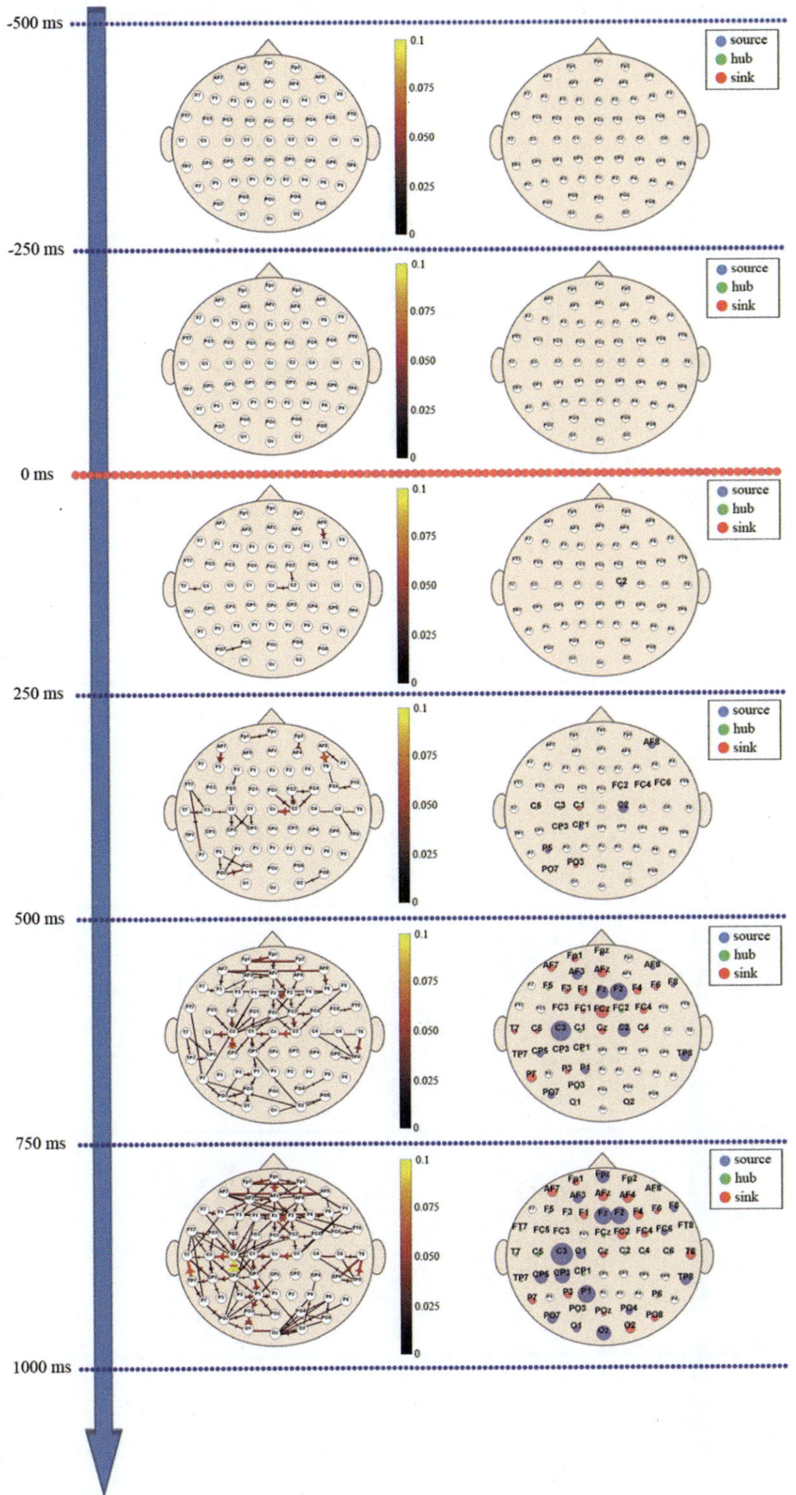

Fig. 3 *First column.* Effective connectivity patterns elicited along the six considered time intervals by a subject performing Grasping imagery task in Alpha band. Connectivity patterns are represented on a 2D scalp model, seen from above with the nose pointing to the top of the page. Color and size of the *arrows* code the corresponding PDC value. *Second column.* Graphical representation of the role of each electrode involved in the connectivity patterns elicited during Grasping imagery task in Alpha band. The color and the diameter of each sphere code for the role (hub in *green*; sink in *red*; source in *blue*) and the magnitude of the degree of the corresponding electrode, respectively

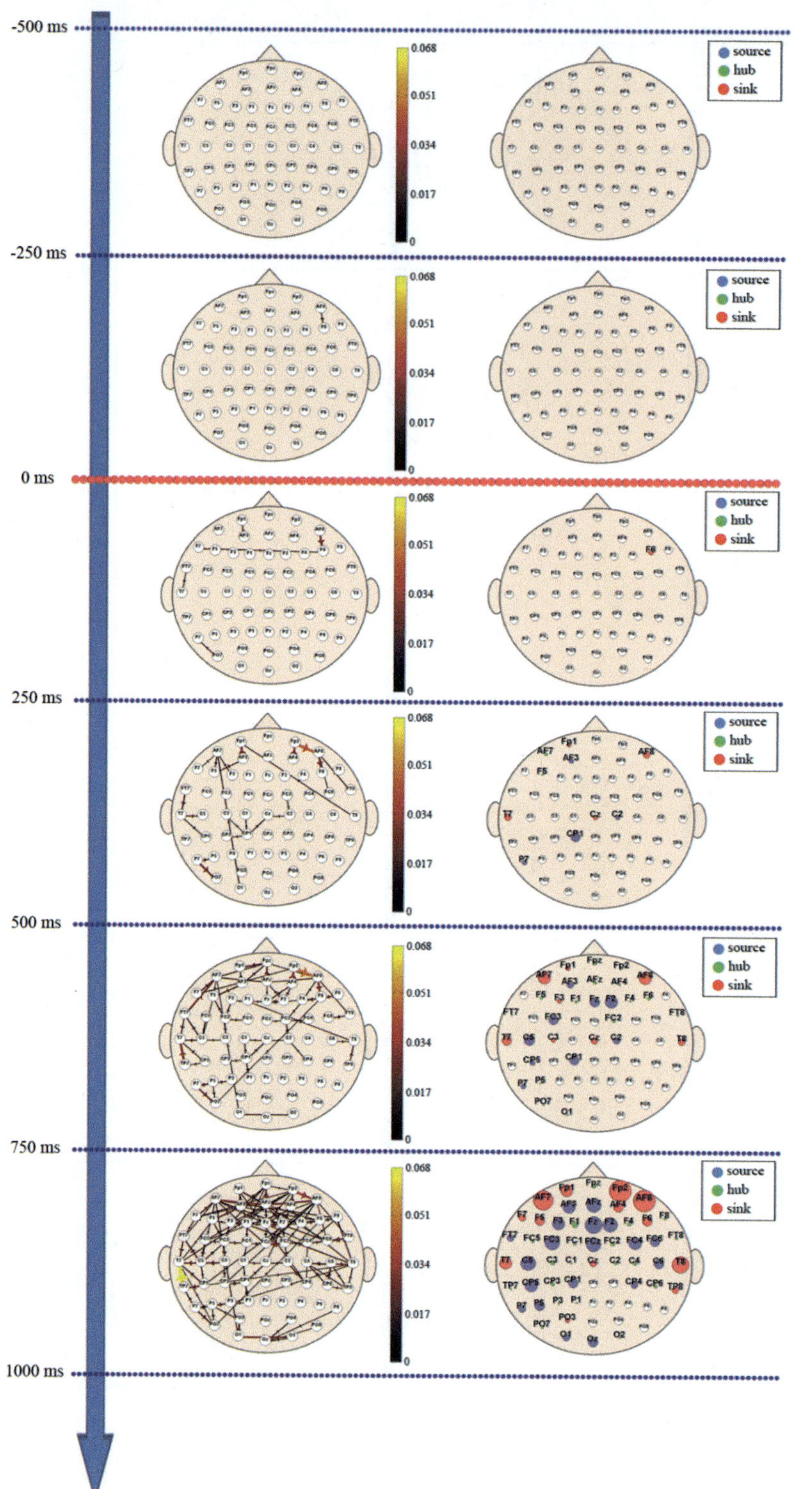

Fig. 4 *First column*. Effective connectivity patterns elicited along the six considered time intervals by a subject performing Grasping imagery task in Beta band. Connectivity patterns are represented on a 2D scalp model, seen from above with the nose pointing to the top of the page. Color and size of the *arrows* code the corresponding PDC value. *Second column*. Graphical representation of the role of each electrode involved in the connectivity patterns elicited during Grasping imagery task in Beta band. The color and the diameter of each sphere code for the role (hub in *green*; sink in *red*; source in *blue*) and the magnitude of the degree of the corresponding electrode, respectively

connectivity patterns. In particular, after 250 ms from the GO stimulus, C3 and C2 electrodes started to acquire a dominant role as source of information in the network. After 500 ms from the beginning of the task execution the role of C3 electrode as source of information gained importance and was associated to an increase of the involvement of frontal electrodes as source/sink of information in the network.

A similar behavior, even with small differences, could be found in Alpha (Fig. 3) and Beta (Fig. 4) frequency bands. In particular, in Alpha band after 500 ms from the GO onset a higher involvement of left hemisphere could be highlighted if compared with the results achieved in Theta band. Since 750 ms from the beginning of the task execution, the connectivity pattern is clearly divided in two subnetworks, one over the left motor areas (around C3 electrode) and the other one over the frontal cerebral regions. The role of C3 as source of information was confirmed also in Alpha band, while the involvement of frontal part of the brain is less evident than those achieved in Theta band. In Beta band, after 500 ms from the GO stimulus, we found an involvement of the frontal areas and then after 750 ms from the beginning of the task execution an engagement of the left hemisphere over the motor areas. Such involvement was lower than the one described in the other two frequency bands.

The observation of connectivity patterns achieved for each frequency band along the six considered time intervals revealed a high involvement of the left centro-parietal areas located around C3 electrode and a small participation of the right hemisphere (C4) to the task execution. In order to better investigate the causal flows involving the two electrodes placed over the sensory motor areas, C3 and C4, we reported the time–frequency connectivity distributions for causal links spreading out from C3 and C4 electrodes and directed to their nearest neighbors, in Figs. 5 and 6, respectively. In each matrix we reported the PDC values for the corresponding causal link achieved along all the time samples considered in the window [−500:1,000] ms and over all the included frequency samples in the range [1–45] Hz. In particular, Fig. 5 shows the time–frequency connectivity distribution for causal links spreading out from C3 electrode and directed to its nearest neighbors (C3 → C1 (right), C3 → CP3 (down), C3 → C5 (left), C3 → FC3 (up)). The results reported in Fig. 5 described more extensively and precisely what was found in Figs. 2, 3, and 4 over sensory motor areas along the six considered time intervals. In particular, no significant activations resulted before the "GO" onset between Task and Rest conditions. Then, since 250 ms after the beginning of imagery task a weak connection directed from C3 to C5 was found in the frequency range 2–6 Hz (Theta Band). After 250 ms from the "GO" onset stronger causal links directed from C3 to C5 and CP3 electrodes resulted in the frequency range of Theta and Alpha bands. Significant connections in Beta band were instead found after 500 ms from the beginning of the imagery task, spreading out from C3 and directed to FC3, C5, CP3 and very weakly to C1.

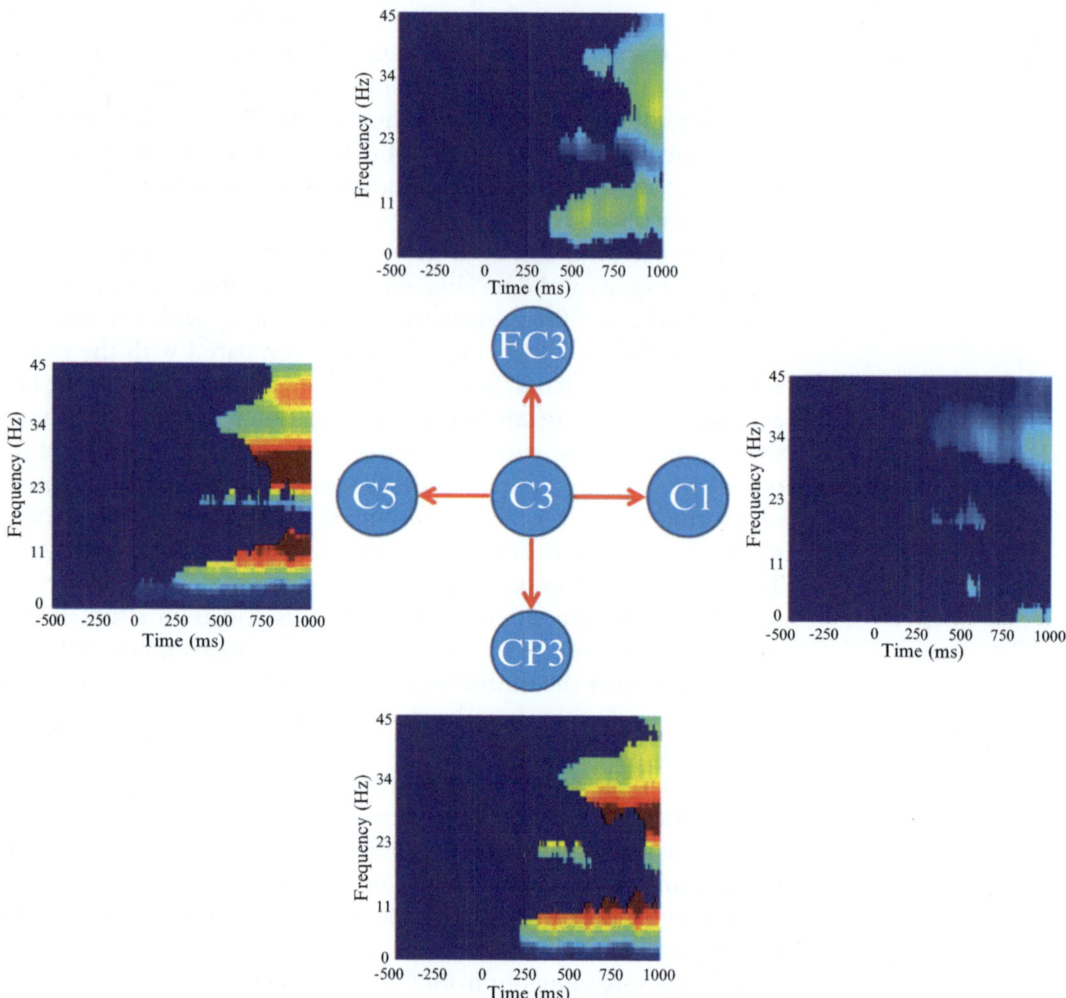

Fig. 5 Time–frequency connectivity distribution for causal links spreading out from C3 electrode and directed to its nearest neighbors (C3 → C1, C3 → CP3, C3 → C5, C3 → FC3). In each matrix we reported the PDC values for the corresponding causal link achieved along all the time samples in [−500:1,000] ms considered window and over all the included frequency samples in the range [1–45]Hz

Figure 6 shows the time–frequency connectivity distribution for causal links spreading out from C4 electrode and directed to its nearest neighbors (C4 → C6 (right), C4 → CP4 (down), C4 → C2 (left), C4 → FC4 (up)). The strength of the significant connections was weaker than the one achieved for causal links spreading out from C3, confirming the important role of the left hemisphere (dominant hemisphere because the subject is right-handed) in motor imagery tasks. Moreover, the weaker connections outcoming from C4 were also delayed if compared to those achieved for C3 electrode. In fact, they appeared after 500 ms from the "GO" stimulus onset.

This type of representation, reported in Figs. 5 and 6, allowed to follow the temporal evolution of connectivity patterns precisely

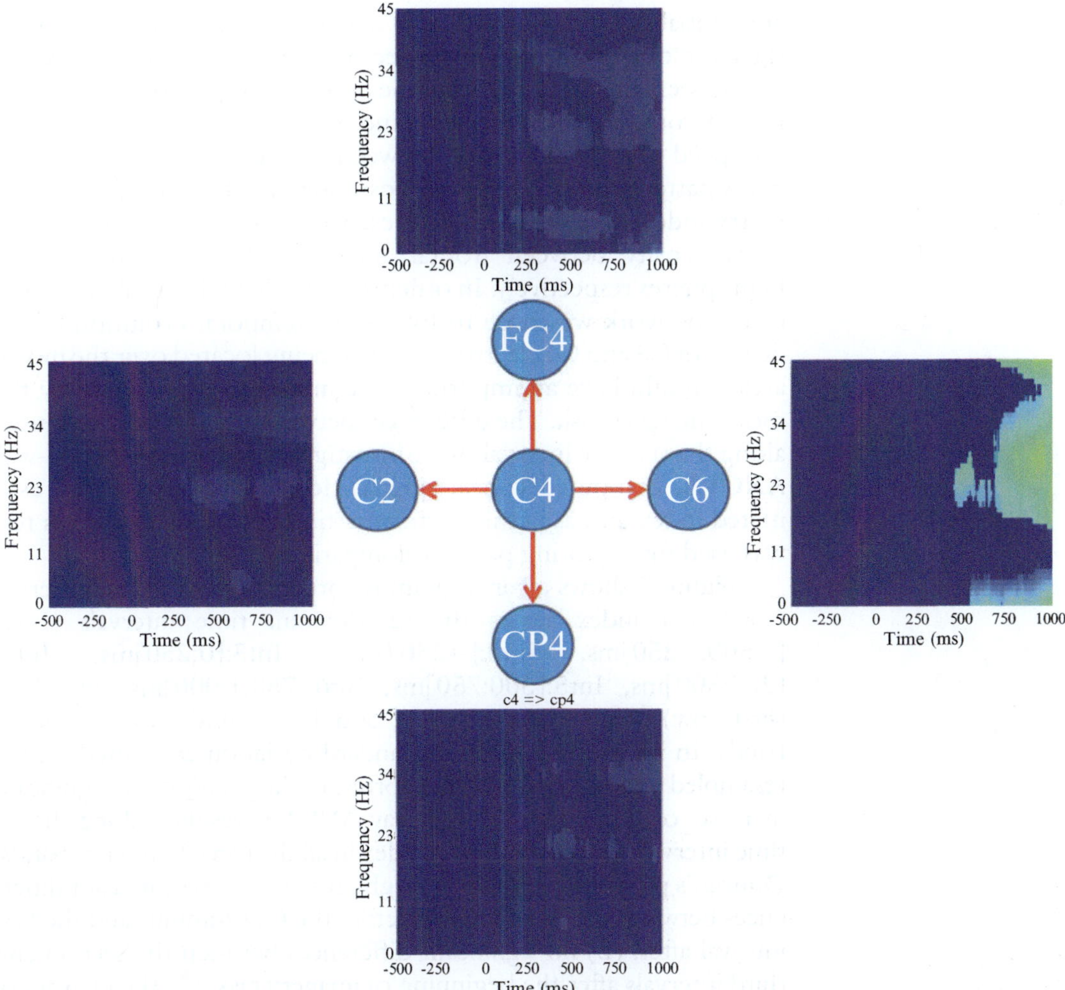

Fig. 6 Time–frequency connectivity distribution for causal links spreading out from C4 electrode and directed to its nearest neighbors (C4 → C6, C4 → CP4, C4 → C2, C4 → FC4). In each matrix we reported the PDC values for the corresponding causal link achieved along all the time samples in [−500:1,000] ms considered window and over all the included frequency samples in the range [1–45]Hz

for each time sample in the investigated observation window and each frequency sample in the predefined spectral range. However, such representation modality didn't allow to provide a concise representation of the whole estimated pattern. In order to get such more synthetic representation, we averaged PDC distributions in six temporal intervals and four frequency bands, in order to display the whole pattern as in Figs. 2, 3, and 4.

3.2.2 Graph Theory Approach

The properties of the observed connectivity patterns elicited in the different phases of motor imagery task were quantitatively described by means of several graph theory indexes characterizing both local and global aspects of investigated networks. In particular, we investigated global properties of elicited networks by means of (1) Clustering coefficient, highlighting the presence of electrodes clusters in the network, (2) Local Efficiency index, describing the efficiency and the speed of communication between different electrodes involved in the pattern and (3) Anterior/Posterior and (4) Left/Right Asymmetry indexes, describing the presence of asymmetries in connections density between frontal and parietal areas and the two hemispheres respectively. In order to investigate the local properties of the network we chose to follow the temporal evolution of the degree of C3 and C4 electrodes, which being located over the motor areas, should have an important role in the pattern elicited by the motor imagery task. The differences between the indexes computed along the six time intervals were investigated by means of a one-way ANOVA, computed considering as dependent variable the estimated index and as within factor the time intervals. Duncan's test was used for exploring pairwise comparisons.

Figure 7 shows a bar diagram reporting the trend of Clustering Coefficient index across the six different time intervals (Int1: [−500:−250]ms, Int2:[−250:0]ms, Int3:[0:250]ms, Int4: [250:500]ms, Int5:[500:750]ms, Int6:[750:1,000]ms) in Theta (first row), Alpha (second row), and Beta (third row) frequency bands. In red we reported the standard deviation computed on the resampled distribution of the corresponding index. A significant increase, confirmed by the one-way ANOVA, resulted along the six time intervals for the clustering index in all the three frequency bands. Duncan's pairwise comparisons highlighted: (1) no significant differences between the two intervals before the GO stimulus and the first interval after, (2) no significant differences between the second and third intervals after the beginning of imagery task, (3) the increase of clustering coefficient in the last interval was significantly different from all the others intervals.

Figure 8 shows a bar diagram reporting the trend of Local Efficiency index across the six different time intervals (Int1: [−500:−250]ms, Int2:[−250:0]ms, Int3:[0:250]ms, Int4: [250:500]ms, Int5:[500:750]ms, Int6:[750:1,000]ms) in Theta (first row), Alpha (second row), and Beta (third row) frequency bands. In red we reported the standard deviation computed on the resampled distribution of the corresponding index. One-way ANOVA test confirmed a statistical difference between the Local Efficiency values achieved for the six time intervals in all the three frequency bands. Duncan's pairwise comparisons stated: (1) no significant differences between the two intervals before the GO stimulus and the first interval after (the two after in Beta band), (2) a significant increase along the last three intervals after the GO stimulus in theta and alpha bands and along the two last intervals in Beta band.

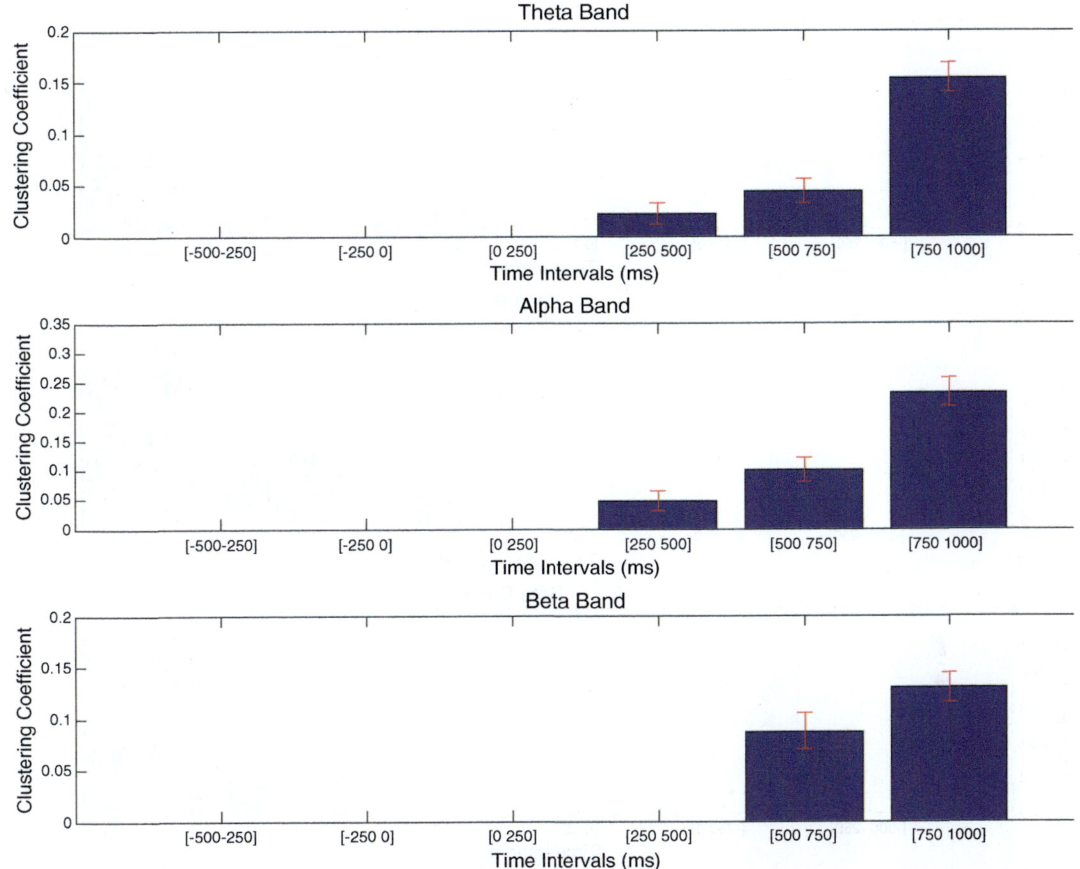

Fig. 7 Bar diagrams reporting the trend of clustering coefficient index across the six different time intervals (Int1: [−500:−250]ms, Int2:[−250:0]ms, Int3:[0:250]ms, Int4:[250:500]ms, Int5:[500:750]ms, Int6:[750:1,000]ms) in Theta (*first row*), Alpha (*second row*), and Beta (*third row*) frequency bands

Figure 9 shows a bar diagram reporting the trend of Left/Right Asymmetry index across the six different time intervals (Int1: [−500:−250]ms, Int2:[−250:0]ms, Int3:[0:250]ms, Int4: [250:500]ms, Int5:[500:750]ms, Int6:[750:1,000]ms) in Theta (first row), Alpha (second row), and Beta (third row) frequency bands. In red we reported the standard deviation computed on the resampled distribution of the corresponding index. A significant increase, confirmed by the one-way ANOVA, resulted along the six time intervals for the Left/Right Asymmetry index in all the three frequency bands. The positive values of asymmetry index confirmed the dominance of left hemisphere in grasping imagery task. Moreover, such asymmetry is higher in Alpha band than in Theta and Beta bands. Duncan's pairwise comparisons highlighted: (1) no significant differences between the two intervals before the GO stimulus and the first interval after, where the index remained

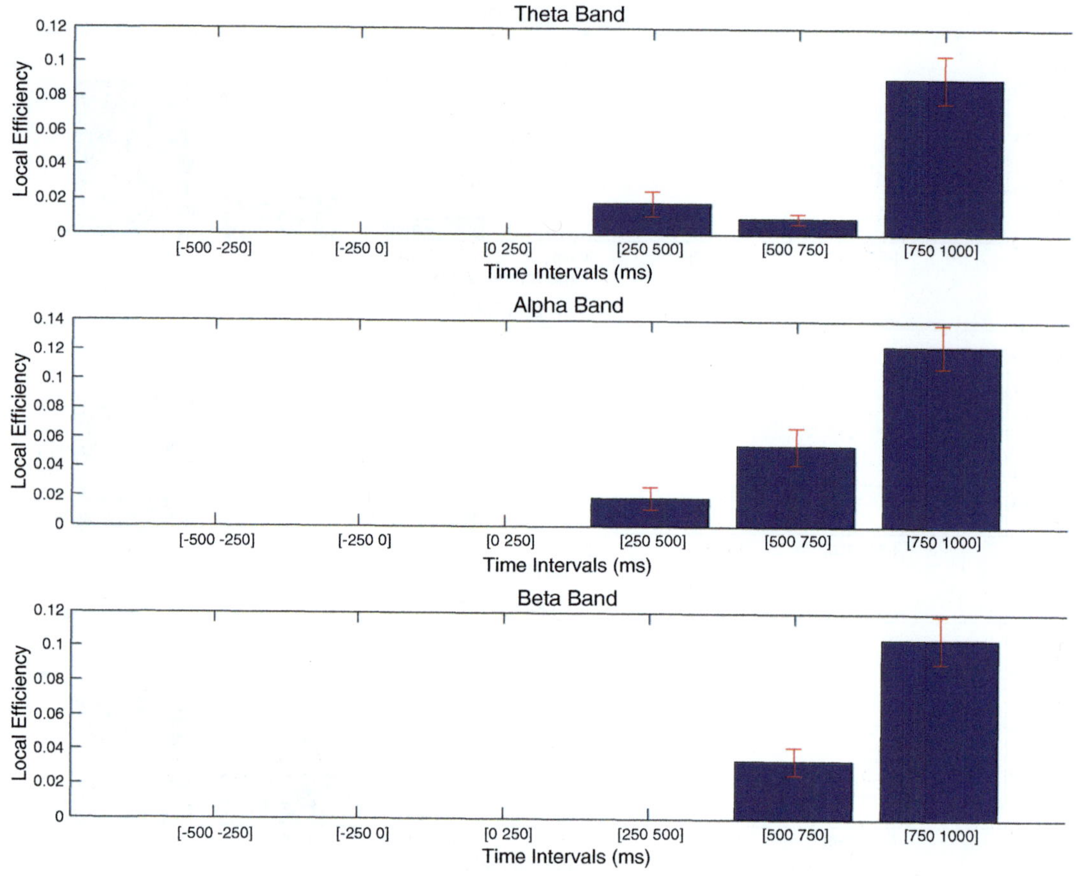

Fig. 8 Bar diagrams reporting the trend of local efficiency index across the six different time intervals (Int1: [−500:−250]ms, Int2:[−250:0]ms, Int3:[0:250]ms, Int4:[250:500]ms, Int5:[500:750]ms, Int6:[750:1,000] ms) in Theta (*first row*), Alpha (*second row*), and Beta (*third row*) frequency bands

around zero, (2) a significant increase of the index along the four intervals after GO stimulus in all the three frequency bands.

Figure 10 shows a bar diagram reporting the trend of Anterior/Posterior Asymmetry index across the six different time intervals (Int1:[−500:−250]ms, Int2:[−250:0]ms, Int3:[0:250]ms, Int4:[250:500]ms, Int5:[500:750]ms, Int6:[750:1,000]ms) in Theta (first row), Alpha (second row), and Beta (third row) frequency bands. In red we reported the standard deviation computed on the resampled distribution of the corresponding index. Significant differences resulted between the Anterior/Posterior asymmetry values computed for the six time intervals in all the three frequency bands. In particular, a significant higher involvement of the frontal areas in respect to the parietal ones resulted. The asymmetry was mainly evident in Theta and Beta band Duncan's pairwise comparisons highlighted: (1) no significant differences between the two intervals before the GO stimulus and the one

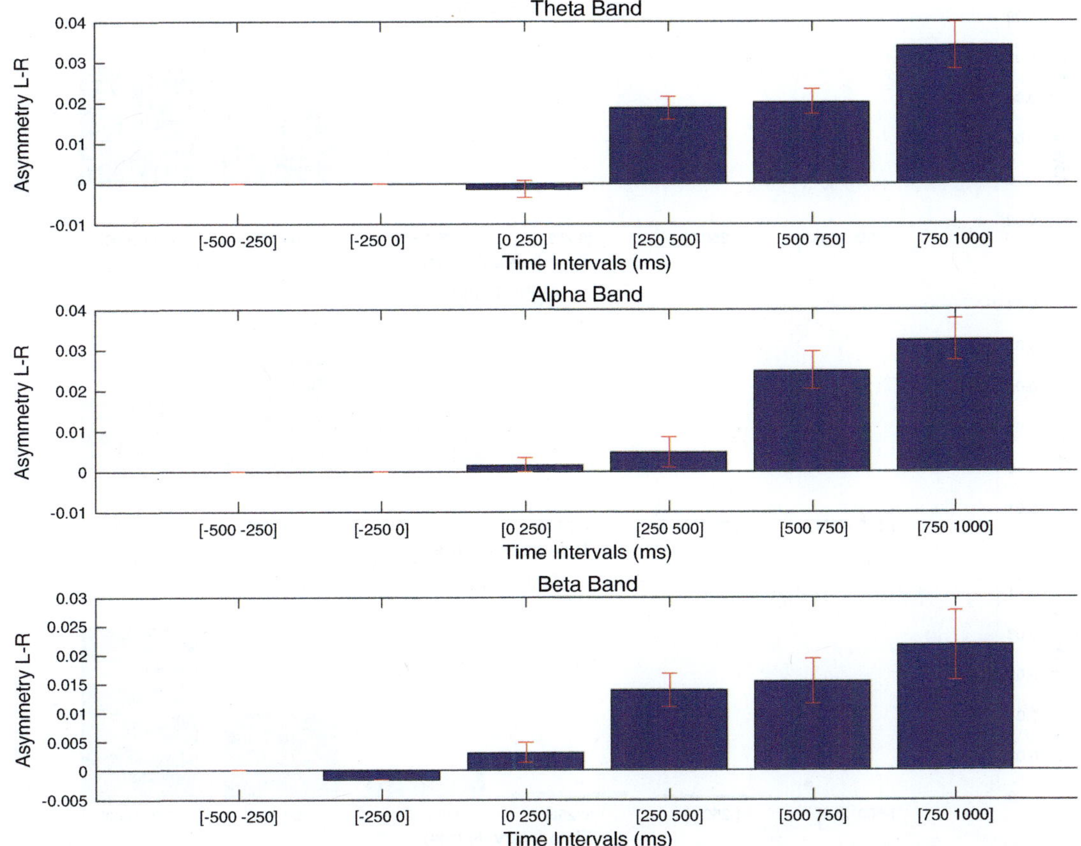

Fig. 9 Bar diagrams reporting the trend of left/right Asymmetry index across the six different time intervals (Int1:[−500:−250]ms, Int2:[−250:0]ms, Int3:[0:250]ms, Int4:[250:500]ms, Int5:[500:750]ms, Int6: [750:1,000]ms) in Theta (*first row*), Alpha (*second row*), and Beta (*third row*) frequency bands

after in Alpha and Beta bands and the two after in Theta band, (2) significant increase of the index along the last three intervals after the beginning of the motor imagery execution for all the three frequency bands, except for alpha band where in the last interval a significant reduction was found.

Figure 11 shows a bar diagram reporting the trend of C3 (blue bars) and C4 (red bars) degree index across the six different time intervals (Int1:[−500:−250]ms, Int2:[−250:0]ms, Int3:[0:250] ms, Int4:[250:500]ms, Int5:[500:750]ms, Int6:[750:1,000]ms) in Theta (first row), Alpha (second row), and Beta (third row) frequency bands. One-way ANOVA analysis revealed a significant increase of the degree index computed for C3 and C4 electrodes along the six time intervals in all the three frequency bands. Duncan's pairwise comparisons highlighted: (1) no significant differences in C3 and C4 degree index between the two intervals before

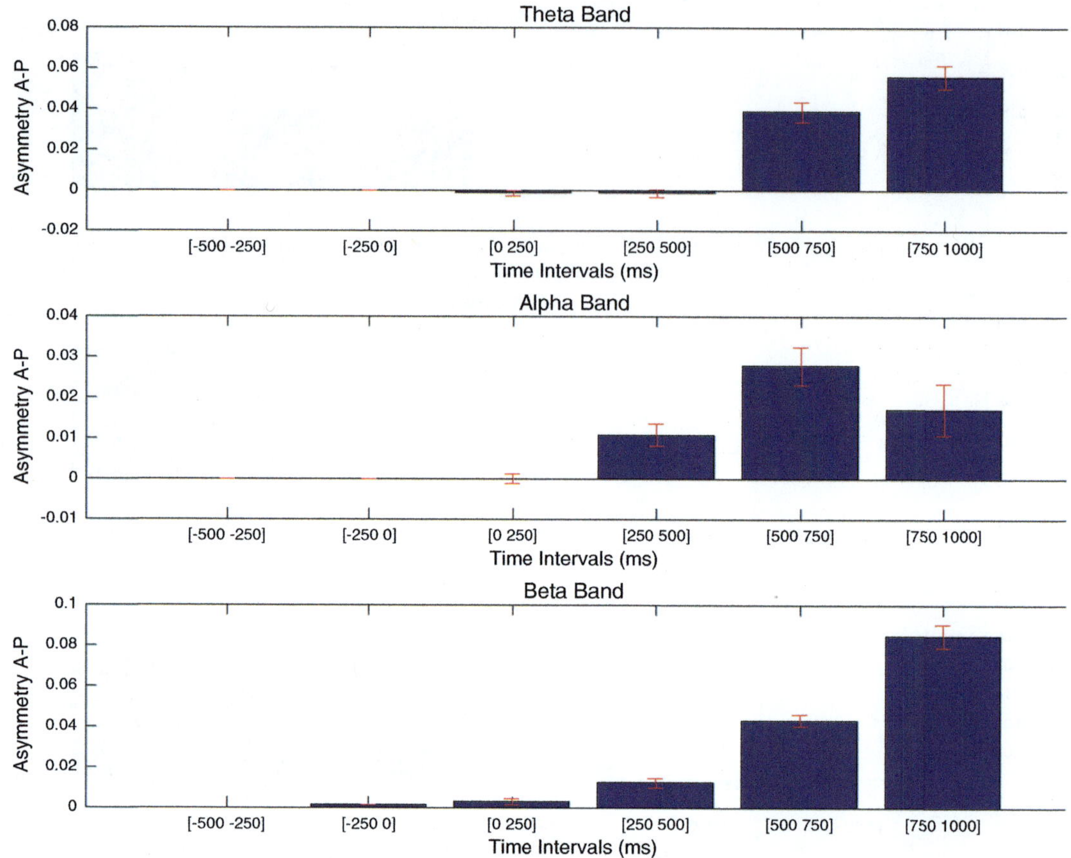

Fig. 10 Bar diagrams reporting the trend of anterior/posterior asymmetry index across the six different time intervals (Int1:[−500:−250]ms, Int2:[−250:0]ms, Int3:[0:250]ms, Int4:[250:500]ms, Int5:[500:750]ms, Int6: [750:1,000]ms) in Theta (*first row*), Alpha (*second row*), and Beta (*third row*) frequency bands

the GO stimuls and the one after, (2) significant increase of the two indexes along the last three intervals after the beginning of the motor imagery execution for all the three frequency bands, (3) significant differences between the degree of the two electrodes in the last three time intervals after the GO stimulus. In particular, the degree of C3 resulted higher than the one computed for C4.

On the whole, the results reported in this section highlighted the temporal evolution of salient properties of networks elicited by motor imagery tasks. The motor imagery execution led to an increase of the global properties of the elicited networks highlighted by an increase of efficiency and speed of communication between electrodes, associated to a higher number of clusters combining different electrodes. The time-varying analysis revealed also an increase, along the six considered time intervals, of the involvement of the left hemisphere (higher than the one achieved for the right hemisphere) and of the frontal areas. The analysis of local

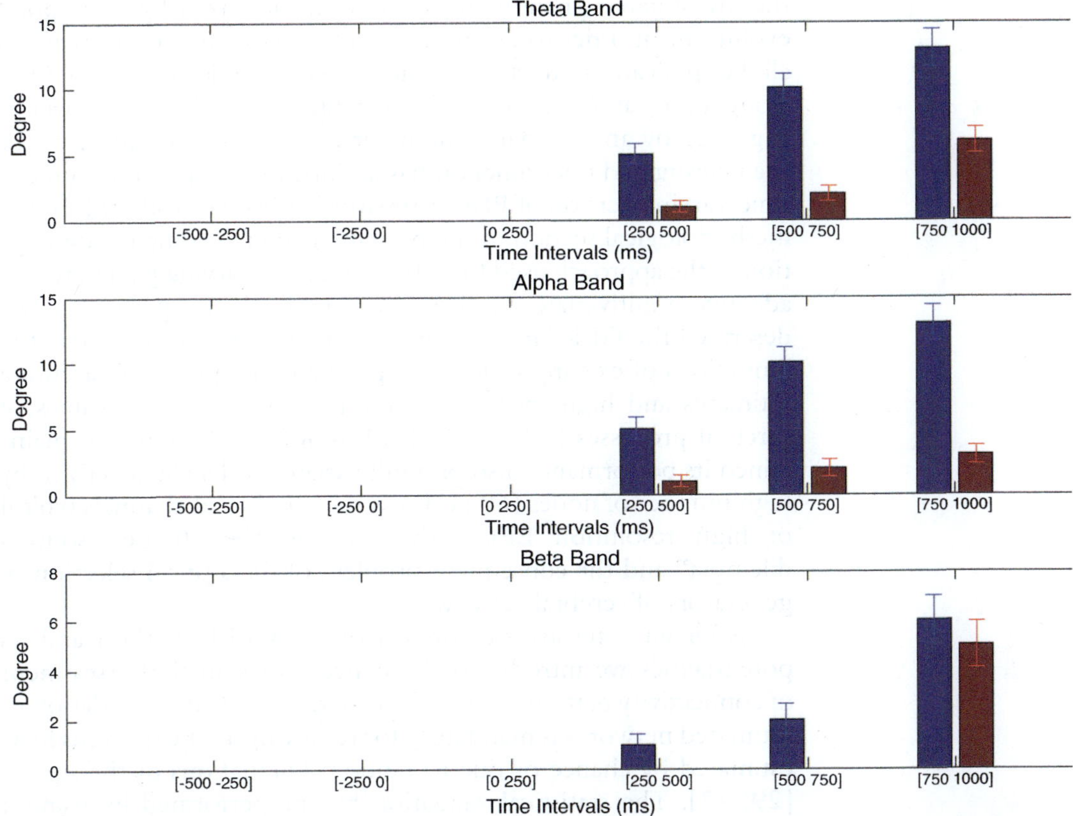

Fig. 11 Bar diagrams reporting the trend of C3 (*blue bars*) and C4 (*red bars*) degree index across the six different time intervals (Int1:[−500:−250]ms, Int2:[−250:0]ms, Int3:[0:250]ms, Int4:[250:500]ms, Int5: [500:750]ms, Int6:[750:1,000]ms) in Theta (*first row*), Alpha (*second row*), and Beta (*third row*) frequency bands

properties of investigated networks confirmed the dominance of the left hemisphere through the higher degree of C3 electrode than the one achieved for C4.

4   Conclusion

In the present chapter we provided an accurate description of the state-of-the-art methodologies for time-varying connectivity estimation and their possible application for investigating the neurophysiological basis of cognitive functions.

First we introduced the concept of Granger causality as theoretical basis of multivariate connectivity estimators extensively used in neuroscience field [1]. Then a detailed description of PDC estimator was provided [10]. Using its classic formulation based on the hypothesis of stationarity of data included in the multivariate model, its application is limited to all the situations in which the investigated phenomena are not characterized by a temporal evolution. In order to extend the study of effective connectivity to all the applications in which the networks are subjected to evolution along time, as in all the tasks in which the subject's answer is triggered by an external stimulus or in all the paradigm in which the investigated phenomenon has an intrinsic temporal dynamic, a time-varying version of PDC estimator has been introduced. In the methodological section of the present chapter, a complete description of the approach used for estimating time-varying parameters of adaptive multivariate models was provided. In particular, we described the GLKF method as an approach able to overcome the limitations of existing time-varying approaches providing accurate estimates and high speed in adaptations to the fast dynamics of cerebral processes [18, 19, 22]. Moreover, such approach maintained its performance also on multivariate model characterized by high number of nodes (60 nodes as the number of channels typical of high resolution EEG), thus solving the "hidden sources dilemma" and the consequent problem of the a priori selection of generators of cerebral activity.

After an extensive description of the GLKF method and its potentialities we introduced the concept of statistical assessment of connectivity patterns [19, 22]. In fact, the statistical validation of estimated networks is mandatory for removing all the spurious links estimated by chance or due to random fluctuations of the signals [29, 42]. The statistical validation can be performed by using a reference distribution representing the null case of PDC estimator. Such distribution is built starting from baseline EEG data and its 95th percentile corresponds to the statistical threshold to be used for discarding spurious links. The use of such statistical threshold is also important for the extraction of an adjacency matrix used for computing graph theory indexes describing the main properties of investigated networks. In fact, such procedure allowed the extraction of stable consistent and repeatable graph indexes to be used as accurate descriptors of investigated phenomena.

The methodological advancement provided during the last 10 years and described in this chapter, allowed to study complex mechanisms at the basis of cognitive functions in humen.

In particular, we proposed an application of the examined methodologies to the investigation of neurophysiological basis of motor imagery processes. We showed the results achieved for a healthy subject executing the imagery of prolonged hands grasping. The application of a time-varying approach for connectivity estimation allowed to follow with high accuracy the temporal evolution of the elicited cerebral processes. The results also demonstrated that the speed of adaptation to transitions of the PDC estimator was high. In fact, a significant connectivity pattern, associated to the motor imagery task, started appearing 250 ms after the request was detected in the case of the subject under investigation.

This temporal delay in the connectivity estimates could be defined as acceptable especially if we considered that it included also the reaction time of the subject in decoding the request and starting the task execution.

Moreover, the achieved results described in this section were in agreement with the literature related to the paradigm used in the experiment but at the same time they provided new information for the description of cerebral networks at the basis of such motor imagery task. The investigation of simple (both hands grasping) motor imagery tasks revealed time-varying connectivity patterns involving mainly the frontal areas (Fpz, Fz, FCz) and the left central and parietal regions (C3, CP3, and P3) of the scalp in Theta, Alpha, and Beta frequency bands. The engagement of cerebral areas mainly in the hemisphere contralateral to the hand used during the imagination (left hemisphere considering that the subject was right-handed) and the prevalent involvement of Alpha and Beta bands in the tasks execution were in agreement with the literature related to the execution of motor tasks. In fact it was demonstrated that the execution of voluntary movements results in a circumscribed desynchronization in the upper Alpha and lower Beta bands localized close to contralateral sensorimotor area [43–45]. Such results can be extended also to the imagination of voluntary movements, in fact, several studies based on human brain imaging using hemodynamic markers (PET and fMRI), extra-cerebral magnetic sources (MEG), and electric field (EEG) have shown that motor imagery activates many of the same cortical areas as those involved in planning and execution of motor movements (e.g., medial supplemental motor area, premotor cortex, dorsolateral prefrontal cortex, and posterior parietal cortex) [46–52].

The application of advanced methods for time-varying connectivity allowed to selectively describe the distinct phases of motor imagery processes, without any a priori selection of the phenomenon sources but including all the neurophysiological generators in the connectivity estimation process (all the 61 acquired channels). Overcoming the necessity to arbitrarily select the channels to be included in the analysis allowed to achieved consistent networks whose structure and properties were in agreement with the literature related to cognitive processes. Moreover the use of a statistical approach for validating connectivity patterns and extracting adjacency matrices on which graph indexes were computed allowed the extraction of quantifiable indexes characterizing the investigated networks. Such indexes selectively described the properties of the networks elicited during different phases for the administered cognitive task. These results are completely new and can be seen as a confirmation of the quality of the methodology developed.

Acknowledgments

This work was partly supported by the European ICT Program FP7-ICT-2009-4 Grant Agreement 287320 CONTRAST and by the grant provided by the Minister of Foreign Affair, "Direzione Generale Sistema Paese".

References

1. Granger CWJ (1969) Investigating causal relations by econometric models and cross-spectral methods. J Econ 37(3):424–438

2. Wiener N (1956) The theory of prediction. In: Beckenbach EF (ed) Modern mathematics for engineers, vol 1. McGraw-Hill, New York

3. Horwitz B (2003) The elusive concept of brain connectivity. Neuroimage 19(2):466–470

4. Lee L, Harrison LM, Mechelli A (2003) A report of the functional connectivity workshop. Neuroimage 19(2):457–465

5. Aertsen A, Preissl H (1991) Dynamics of activity and connectivity in physiological neuronal networks. In: Schuster HG (ed) Non linear dynamics and neuronal networks. VCH, New York, NY, 281–302 p

6. Blinowska KJ, Kaminski M, Kaminski J, Brzezicka A (2010) Information processing in brain and dynamic patterns of transmission during working memory task by the SDTF function. Conf Proc IEEE Eng Med Biol Soc 2010:1722–1725

7. Sameshima K, Baccalá LA (1999) Using partial directed coherence to describe neuronal ensemble interactions. J Neurosci Methods 94(1):93–103

8. Kaminski MJ, Blinowska KJ (1991) A new method of the description of the information flow in the brain structures. Biol Cybern 65(3):203–210

9. Turbes CC, Schneider GT, Morgan RJ (1983) Partial coherence estimates of brain rhythms. Biomed Sci Instrum 19:97–102

10. Baccalá LA, Sameshima K (2001) Partial directed coherence: a new concept in neural structure determination. Biol Cybern 84:463–474

11. Kus R, Kaminski M, Blinowska KJ (2004) Determination of EEG activity propagation: pair-wise versus multichannel estimate. IEEE Trans Biomed Eng 51(9):1501–1510

12. Blinowska KJ, Kuś R, Kamiński M (2004) Granger causality and information flow in multivariate processes. Phys Rev E Stat Nonlin Soft Matter Phys 70(5 Pt 1):050902

13. David O, Cosmelli D, Friston KJ (2004) Evaluation of different measures of functional connectivity using a neural mass model. Neuroimage 21(2):659–673

14. Astolfi L, Cincotti F, Mattia D, Marciani MG, Baccala LA, de Vico Fallani F et al (2007) Comparison of different cortical connectivity estimators for high-resolution EEG recordings. Hum Brain Mapp 28(2):143–157

15. Blinowska KJ (2011) Review of the methods of determination of directed connectivity from multichannel data. Med Biol Eng Comput 49(5):521–529

16. Ding M, Bressler SL, Yang W, Liang H (2000) Short-window spectral analysis of cortical event-related potentials by adaptive multivariate autoregressive modeling: data preprocessing, model validation, and variability assessment. Biol Cybern 83(1):35–45

17. Astolfi L, Cincotti F, Mattia D, De Vico Fallani F, Tocci A, Colosimo A et al (2008) Tracking the time-varying cortical connectivity patterns by adaptive multivariate estimators. IEEE Trans Biomed Eng 55(3):902–913

18. Möller E, Schack B, Arnold M, Witte H (2001) Instantaneous multivariate EEG coherence analysis by means of adaptive high-dimensional autoregressive models. J Neurosci Methods 105(2):143–158

19. Milde T, Leistritz L, Astolfi L, Miltner WHR, Weiss T, Babiloni F et al (2010) A new Kalman filter approach for the estimation of high-dimensional time-variant multivariate AR models and its application in analysis of laser-evoked brain potentials. Neuroimage 50(3):960–969

20. Weiss T, Hesse W, Ungureanu M, Hecht H, Leistritz L, Witte H et al (2008) How do brain areas communicate during the processing of noxious stimuli? An analysis of laser-evoked event-related potentials using the Granger causality index. J Neurophysiol 99(5):2220–2231

21. Zhu C, Guo X, Jin Z, Sun J, Qiu Y, Zhu Y et al (2011) Influences of brain development and ageing on cortical interactive networks. Clin Neurophysiol 122(2):278–283

22. Toppi J, Babiloni F, Vecchiato G, De Vico Fallani F, Mattia D, Salinari S et al (2012) Towards the time varying estimation of complex brain connectivity networks by means of a general linear Kalman filter approach. Conf Proc IEEE Eng Med Biol Soc 2012:6192–6195

23. Akaike H (1974) A new look at statistical model identification. IEEE Trans Autom Control 19:716–723

24. Baccalá LA (2001) On the efficient computation of partial coherence from multivariate autoregressive model. Biol Cybern 84:463–474

25. Astolfi L, Cincotti F, Mattia D, Marciani MG, Baccalà LA, de Vico Fallani F et al (2006) Assessing cortical functional connectivity by partial directed coherence: simulations and application to real data. IEEE Trans Biomed Eng 53(9):1802–1812

26. Nichols T, Hayasaka S (2003) Controlling the familywise error rate in functional neuroimaging: a comparative review. Stat Methods Med Res 12(5):419–446

27. Bonferroni C (1936) Teoria statistica delle classi e calcolo delle probabilità, vol 8. Libreria Internazionale Seeber, Florence

28. Zar JH (2010) Biostatistical analysis. Prentice Hall, Upper Saddle River, 964 p

29. Toppi J, Babiloni F, Vecchiato G, Cincotti F, De Vico Fallani F, Mattia D et al (2011) Testing the asymptotic statistic for the assessment of the significance of partial directed coherence connectivity patterns. Conf Proc IEEE Eng Med Biol Soc 2011:5016–5019

30. Benjamini Y, Hochberg Y (1995) Controlling the false discovery rate: a practical and powerful approach to multiple testing. J R Stat Soc 57(1):289–300

31. Benjamini Y, Yekutieli D (2001) The control of the false discovery rate in multiple testing under dependency. Ann Stat 29(4):1165–1188

32. Toppi J, De Vico Fallani F, Vecchiato G, Maglione AG, Cincotti F, Mattia D et al (2012) How the statistical validation of functional connectivity patterns can prevent erroneous definition of small-world properties of a brain connectivity network. Comput Math Methods Med 2012:130985

33. Latora V, Marchiori M (2001) Efficient behavior of small-world networks. Phys Rev Lett 87(19):198701

34. Sporns O, Chialvo DR, Kaiser M, Hilgetag CC (2004) Organization, development and function of complex brain networks. Trends Cogn Sci 8(9):418–425

35. Watts DJ, Strogatz SH (1998) Collective dynamics of "small-world" networks. Nature 393(6684):440–442

36. Fagiolo G (2007) Clustering in complex directed networks. Phys Rev E Stat Nonlin Soft Matter Phys 76(2 Pt 2):026107

37. Humphries MD, Gurney K (2008) Network "small-world-ness": a quantitative method for determining canonical network equivalence. PLoS ONE 3(4):e0002051

38. Toppi J, Petti M, De Vico Fallani F, Vecchiato G, Maglione AG, Cincotti F et al (2012) Describing relevant indices from the resting state electrophysiological networks. Conf Proc IEEE Eng Med Biol Soc 2012:2547–2550

39. Hoffmann S, Falkenstein M (2008) The correction of eye blink artefacts in the EEG: a comparison of two prominent methods. PLoS ONE 3(8):e3004

40. Klimesch W (1999) EEG alpha and theta oscillations reflect cognitive and memory performance: a review and analysis. Brain Res Brain Res Rev 29(2–3):169–195

41. De Vico Fallani F, Astolfi L, Cincotti F, Mattia D, Tocci A, Salinari S et al (2008) Brain network analysis from high-resolution EEG recordings by the application of theoretical graph indexes. IEEE Trans Neural Syst Rehabil Eng 16(5):442–452

42. Takahashi DY, Baccalà LA, Sameshima K (2007) Connectivity inference between neural structures via partial directed coherence. J Appl Stat 34(10):1259–1273

43. Pfurtscheller G, Aranibar A (1979) Evaluation of event-related desynchronization (ERD) preceding and following voluntary self-paced movement. Electroencephalogr Clin Neurophysiol 46(2):138–146

44. Pfurtscheller G, Berghold A (1989) Patterns of cortical activation during planning of voluntary movement. Electroencephalogr Clin Neurophysiol 72(3):250–258

45. Pfurtscheller G, Lopes da Silva FH (1999) Event-related EEG/MEG synchronization and desynchronization: basic principles. Clin Neurophysiol 110(11):1842–1857

46. Porro CA, Francescato MP, Cettolo V, Diamond ME, Baraldi P, Zuiani C et al (1996) Primary motor and sensory cortex activation during motor performance and motor imagery: a functional magnetic resonance imaging study. J Neurosci 16(23):7688–7698

47. Schnitzler A, Salenius S, Salmelin R, Jousmäki V, Hari R (1997) Involvement of primary motor cortex in motor imagery: a neuromagnetic study. Neuroimage 6(3):201–208

48. Crone NE, Miglioretti DL, Gordon B, Sieracki JM, Wilson MT, Uematsu S et al (1998) Functional mapping of human sensorimotor cortex with electrocorticographic spectral analysis. I.

Alpha and beta event-related desynchronization. Brain 121(Pt 12):2271–2299

49. Cochin S, Barthelemy C, Roux S, Martineau J (1999) Observation and execution of movement: similarities demonstrated by quantified electroencephalography. Eur J Neurosci 11 (5):1839–1842

50. Jeannerod M, Frak V (1999) Mental imaging of motor activity in humans. Curr Opin Neurobiol 9(6):735–739

51. Miller KJ, Schalk G, Fetz EE, den Nijs M, Ojemann JG, Rao RPN (2010) Cortical activity during motor execution, motor imagery, and imagery-based online feedback. Proc Natl Acad Sci U S A 107(9):4430–4435

52. Zapparoli L, Invernizzi P, Gandola M, Verardi M, Berlingeri M, Sberna M et al (2012) Mental images across the adult lifespan: a behavioural and fMRI investigation of motor execution and motor imagery. Exp Brain Res 224(4):519–540

Neuromethods (2015) 91: 205–229
DOI 10.1007/7657_2014_71
© Springer Science+Business Media New York 2014
Published online: 5 June 2014

Assessment of Sensory Gating Deficit in Schizophrenia Using a Wavelet Transform Methodology on Auditory Paired-Click Evoked Potentials

Cristina Farmaki, Vangelis Sakkalis, Klevest Gjini, Nash N. Boutros, and George Zouridakis

Abstract

Schizophrenia is a chronic mental disorder that results in various neurocognitive and emotional deficits. Among them, the auditory sensory gating deficit, which refers to the neurophysiological processes of filtering out redundant or unnecessary auditory stimuli, has been considered a leading endophenotype for the illness. Sensory gating is efficiently probed using electroencephalography (EEG) recordings in a simple paired-click auditory evoked potential paradigm. This chapter will discuss the use of EEG in the search for biomarkers of schizophrenia and more specifically, the assessment of the sensory gating process in patients with schizophrenia via time-frequency decomposition of the EEG signals. In order to study the functional abnormalities associated with schizophrenia we will focus on the oscillatory changes, spanning a wide range of frequency bands, during an auditory paired-click experiment. To that end, we will employ a combination of the wavelet transform and a nonparametric permutation test based on Monte Carlo randomization. Results from the application of this methodology on real EEG data from both patients with schizophrenia and healthy control participants will be discussed in detail.

Keywords Schizophrenia, Sensory gating deficit, Electroencephalography (EEG), Auditory paired-click stimuli, Time-frequency analysis, Wavelet transform, Nonparametric permutation test

---

1   Introduction

1.1 Electrophysiological Abnormalities in Schizophrenia

Schizophrenia is a complex neuropsychiatric disorder characterized by persistent neurocognitive deficits (memory, concentration, and learning), difficulty with reality testing, and by core emotional impairments. It is often described in terms of positive and negative symptoms. Positive symptoms are those that most healthy individuals do not normally experience, but are present in a large proportion of schizophrenia patients, including delusions and tactile, auditory, visual, olfactory, and/or gustatory hallucinations. Positive

---

G.Z. is Visiting Professor at BCBL. His permanent appointment is with the Departments of Engineering Technology, Computer Science, and Electrical and Computer Engineering, University of Houston, Houston, TX 77204, USA.

symptoms usually show good response to approved medications, from which typicals act by blocking D2 receptors while atypicals have more actions on other DA and 5HT effects. On the other hand, negative symptoms are deficits of normal emotional responses or of other thought processes, and they commonly include blunted affect, poverty of speech (alogia), inability to experience pleasure (anhedonia), lack of motivation (avolition), and lack of desire to form relationships (asociality). People with prominent negative symptoms often respond poorly to medication, and therefore the symptoms persist even in treated patients. Developing new therapies to target treatment-resistant symptoms requires better understanding of the pathophysiology of schizophrenia, a challenging task, as schizophrenia is a heterogeneous syndrome with many different biological subtypes [1, 2], each showing preferential response to different treatments [3].

Neurophysiological biomarkers may be objective indices of prominent features in schizophrenia patients such as cognitive dysfunction [4]. The search for relevant biomarkers of the disease and its subtypes has involved biochemical, genetic, and technological approaches. The latter includes all the usual brain imaging technologies, such as electroencephalography (EEG), magnetic resonance imaging (MRI), magnetoencephalography (MEG), positron emission tomography (PET), and near-infrared spectroscopy. Among these technologies, EEG and MEG have the significant advantage of greater temporal resolution, since changes in neuronal activity occurring over a few milliseconds can be detected, versus, e.g., a scale of several seconds for functional MRI. Additionally, EEG and MEG are noninvasive, with no radiation exposure, while EEG also offers greater accessibility with much lower cost. Although EEG suffers from poor spatial resolution, due to dispersion or smearing of the electrical signal by the cerebrospinal fluid, skull, and scalp, sophisticated signal processing techniques, such as event-related potentials (ERP) analysis, machine learning, and source localization techniques (including forward modeling [5] using a proper head model), can be useful in analyzing and detecting EEG abnormalities, and even in diagnostic classification [6], or treatment response prediction [7].

There are numerous studies of the abnormalities associated with schizophrenia, based on various forms of EEG analysis, including resting-state EEG, studies of EEG during cognitive task performance, and studies using ERPs. Resting-state EEG has been used in various studies, in order to detect potentially important abnormalities in schizophrenia [8, 9]. Niculin et al. [10] point out that long-term temporal interactions, in the scale of tens of seconds, may be necessary to examine cognitive processes, such as logical reasoning and working memory, known to be disturbed in schizophrenia. They evaluated long-range temporal correlations (auto-correlations)—LRTC—in the amplitude dynamics of neuronal oscillations, over 5–50 s, in patients with schizophrenia and normal controls, during the resting state. LRTC values in both alpha and beta bands were lower across most EEG leads in patients compared to healthy controls, suggesting increased variability of the neuronal activity. The authors speculated that this may be associated with thought disorder; nevertheless no correlations with measures of psychopathology were reported. Resting EEG was also used in [11], where the authors observed increased theta-alpha power in both patients of schizophrenia and their unaffected relatives, compared to healthy controls, which may indicate an elementary neurophysiological problem associated with genetic liability of schizophrenia. Resting-state EEG activity, in the form of quantitative-EEG oscillations, was studied by Koutsoukos et al. [12], before and during auditory verbal hallucinations in patients with treatment-resistant schizophrenia. They detected increased phase coupling of theta and gamma bands in left frontotemporal regions during the hallucinations, which suggests a theta-gamma interaction involved in the production and experience of auditory verbal hallucinations in schizophrenia patients.

The inability of patients with schizophrenia to modulate electrophysiological activity in response to a cognitive task has been demonstrated in several studies that use EEG analysis. Carlino et al. [13] recorded EEG of 17 schizophrenia patients and 17 matched healthy controls during eyes closed, eyes open, and forward and backward counting. They calculated the correlation dimension (D2) of the EEG signals, a nonlinear measure of complexity, and confirmed increased D2 values during the cognitive tasks compared with the resting state in healthy controls, while the patients showed increased D2 values at rest but they could not increase them further during the counting tasks. On the other hand, McCormick et al. [14] investigated the hypothesis that empathy arises from experiencing others' sensory or motor experiences as though they were one's own. They evaluated the mirror neuron system via measurement of the Mu rhythm[1] of schizophrenia patients and matched healthy subjects, and, unexpectedly, concluded that actively psychotic patients showed greater Mu suppression after actual and observed hand movements than did healthy controls and nonpsychotic patients. They suggested the existence of mirror neuron activity among patients with schizophrenia during the psychotic phase of the illness, and its correlation with severity of psychosis.

---

[1] The mirror neuron system is a set of specialized neurons that become active both during motor actions and during the observation of another individual's motor actions. It can be assessed through the Mu rhythm, a frequency rhythm between 8 and 13 Hz, most prominent over the sensorimotor cortex. Mu rhythm in healthy individuals is suppressed when the subject performs a motor task or observes another performing a motor task.

The ERPs constitute another EEG technique that has served several different purposes in the study of schizophrenia. In the first place, ERPs have been extensively used to characterize abnormal cognition, as in the case of [15], where an ERP index of working memory (WM) capacity (i.e., the number of objects that can be held in memory), the contralateral delay activity (CDA), was used in order to examine whether WM deficits in schizophrenia are qualitatively similar to the individual differences in WM capacity among healthy subjects. Their findings demonstrated that WM impairment in schizophrenia patients is not associated with the same patterns of neural activity that characterize low WM capacity in healthy individuals; instead, it reflects a specific impairment in the ability to distribute attention broadly. EEG data from a working memory (WM) experiment on stabilized schizophrenia patients and healthy controls were also analyzed by Sakkalis et al. in [16]. In contrast to the aforementioned report, the authors in [16] used the wavelet coherence, a measure particularly helpful in studying nonstationary time-varying brain dynamics, in order to study and characterize any disturbances present in the functional connectivity network of schizophrenia patients. The proposed analysis indicated significant task differentiation in gamma band more prominent in frontal, frontal-central, and temporal regions, whereas the results were in accordance with the disturbance of connections between the neurons giving additional information related to the localization of most prominent disconnection.

ERP indices have also been used as biomarkers, associated with particular symptoms of the disorder [17, 18]. Identification of such biomarkers would improve our understanding of the neurophysiological abnormalities underlying these symptoms. An example of a putative schizophrenia biomarker is the suppression of the P50 component of ERP response to repeated identical auditory stimuli. The P50 suppression is associated with impairment of sensory gating, a neural process that is explained in detail in the next section.

1.2 Evaluation of Sensory Gating in Schizophrenia Using EEG

Sensory gating [19, 20] denotes the ability of the central nervous system (CNS) to modulate the sensitivity to incoming stimuli, in order to focus on task-relevant input and avoid redundant information overflow. A widely used electrophysiological procedure to assess sensory gating in humans is the auditory paired-stimulus paradigm (PSP) [21, 22], during which a pair of identical auditory stimuli ("clicks or brief tones"), S1 and S2, is presented at an interval of 500 ms. In addition, stimulus pairs are separated by long inter-stimulus intervals (8–10 s) to assure full neuronal recovery prior to repeating stimulation and, thus, avoid interference of the effects of consecutive trials [23]. Three evoked potential components are commonly used to investigate the sensory gating: P50, N100, and P200. The quality of the sensory gating mechanism is represented by the ratio of the corresponding component amplitude of the response to the second stimulus to the component amplitude of the response to the first stimulus (S2/S1). Normally, in healthy subjects the amplitudes of P50, N100, and P200 to S2 are significantly attenuated in respect with that to S1 [24, 25], hence yielding a low ratio S2/S1, which indicates a high sensory gating capability due to a strong inhibition of irrelevant input.

Schizophrenia patients, on the other hand, consistently exhibit insufficient inhibitory processing of repetitive, irrelevant auditory stimuli, reflected in gating deficit of each one of the aforementioned components. Many researchers report that schizophrenia patients have a diminished gating of the auditory P50 [21, 25–28], while reduced gating of the N100 response to repeated stimulation has also been demonstrated [29, 30]. Finally, numerous publications have confirmed that both amplitude and gating of the P200 component are reduced in schizophrenia [20, 24, 31, 32]. The increased ratio S2/S1 that reveals the gating deficit could result from either an abnormally attenuated response to S1 stimulus [29, 33] when coupled with inability to further attenuate the weakened response to S1, i.e., difficulty in registering novel information and then in attenuating the abnormal response, or a larger evoked response to S2 stimulus [34, 35], i.e., difficulty in gating out redundant information. As proof exists supporting both theories [36], whether the abnormal gating ratios in schizophrenia are associated with gating in or gating out is an important issue to be cleared.

Most of the auditory sensory gating studies have used time-domain analyses to study the sensory gating changes in schizophrenia patients. While the earliest attempts focused on the amplitude, gating ratio, and latency of the P50 component [34, 35, 37], there is an increasing body of publications studying the implication of the N100 and P200 components in the sensory gating procedure [20, 24, 27, 38, 39]. In these studies the authors use the averaged response in order to cancel out background noise and they extract time-domain measures, such as the amplitudes of the corresponding components, their latencies, the gating ratios S2/S1, as well as the absolute difference S2-S1. Arnfred in [40] explored the application of difference waves (i.e., point-to-point subtraction of the S2 waveform from the S1 waveform) on auditory P50 gating data, inferring that the patients had an attenuated difference P50, which was most prominent in the subsample of patients with severe negative symptoms. Besides the averaged-trials techniques, single-trial analysis studies have also been conducted in order to improve classification between healthy and schizophrenia subjects. Iyer et al. [41] used an iterative independent component analysis (iICA) procedure to measure single-trial evoked potentials in the context of an auditory paired-stimuli paradigm, and achieved 100 % classification rates using amplitude and latency of the N100 component. The authors in [42] also employed a single-trial analysis to study the trial-to-trial temporal variability of the P50 component, concluding that patients with schizophrenia had significantly more temporal variability than healthy controls.

The traditional view of ERPs, sometimes referred to as the additive ERP model, assumes that ERP components reflect transient bursts of neuronal activity, time locked to the eliciting stimulus, that arise from one or more neural generators subserving specific sensory and cognitive operations during information processing. In this view, ERPs are superimposed on ongoing background EEG "noise" with amplitude and phase distributions that are completely unrelated to processing of task events. On the contrary, time-frequency analysis of single-trial EEG epochs, which has been recently employed in a number of studies of the sensory gating procedure, reveals that EEG does not simply reflect random background noise, but there are event-related changes in the magnitude and phase of EEG oscillatory activity at specific frequencies that contribute to the event processing [43]. A time-frequency representation provides a more refined and detailed account of the brain's event-related neurooscillatory activity, relative to the more static view provided by traditional ERP approaches [44]. In accordance with that, Sakkalis et al. [45] demonstrated the superiority of a novel wavelet-based methodology, compared to the traditional Fourier transform analysis, in the evaluation of different brain regions' activations of healthy subjects, during a mathematical thinking experiment.

In this context, the authors in [46] demonstrated that beta band responses to S1 may predict subsequent P50 suppression to S2, thus suggesting that beta activity reflects stimulus encoding while in [47] beta activity synchronization was shown to reflect widespread network involvement. Schizophrenia patients have exhibited reduced low frequency, as well as beta band, response to S1 stimulus in several studies [30, 48]. On the contrary, although evoked gamma response contributes morphologically to the auditory P50 ERPs [30] and some studies have reported reduced gamma response to S1 [33] or reduced gamma band gating deficits [49] in schizophrenia patients, there are studies reporting no significant differences in the gamma response between healthy controls and schizophrenia patients [30]. Therefore, it is necessary to study more in depth the functional significance of oscillatory activity within these bands, in order to extract relevant information about the neurophysiological and cognitive deficits of individuals suffering from schizophrenia.

In the present study, a time-frequency analysis based on wavelets, across a broad frequency range (8–60 Hz), was applied on event-related EEG data of both healthy subjects and schizophrenia patients. The data were derived from an auditory paired-click paradigm, in order to investigate the sensory gating changes in subjects suffering from schizophrenia compared to healthy controls, in different frequency bands. The extracted time-frequency representations underwent a nonparametric statistical analysis based on Monte Carlo randomization, and the features that were found to significantly differentiate the two groups of subjects were fed into three classification schemes, based on linear discriminant analysis, support vector machines, and a radial basis function (RBF) network. The following sections include the basic theoretical issues behind the methodology used, details about the participants and the experimental setup, and the results of the statistical analysis and the classification outcome.

2 Time-Frequency Analysis: Background Theory

2.1 Continuous Wavelet Transform

Wavelets were first introduced by Gossman and Morlet [50] in order to overcome the inability of Fourier and Gabor transforms to analyze signals involving different range of frequencies, due to their fixed window size. Wavelets can be literally defined as "small" waves that have limited duration and 0 average values. They are mathematical functions capable of localizing a function or a set of data both in time and frequency. Their main advantage is that they have a varying window size, being wide for slow frequencies and narrow for the fast ones, thus leading to an optimal time-frequency resolution in all the frequency ranges. Furthermore, owing to the fact that windows are adapted to the transients of each scale, stationarity is not a requirement for signal analysis using wavelets. Therefore, the wavelet transform is considered as one of the superior techniques in analyzing nonstationary signals, such as EEG.

In Continuous Wavelet Transform (CWT), the signal to be analyzed is matched and convolved with the wavelet basis function at continuous time and frequency increments. A *wavelet family* $\psi_{\alpha\beta}$ is a set of elemental functions generated by dilations and translations of a unique admissible[2] *mother wavelet* $\psi(t)$:

$$\psi_{a,b}(t) = |a|^{-1/2} \psi\left(\frac{t-b}{a}\right)$$

where $a, b \in R$, $a \neq 0$ are the scale and translation parameters, respectively. As a increases the wavelet becomes more narrow and by varying b, the mother wavelet is displaced in time. Thus, the

---

[2] For the continuous wavelet transform to be invertible, the mother wavelet $\psi(t)$ must satisfy the admissibility condition: $\int_{-\infty}^{\infty} \frac{|\widehat{\psi}(\omega)|}{\omega} d\omega < \infty$, where $\widehat{\psi}(\omega)$ is the Fourier transform of the wavelet. The admissibility condition implies that the mother wavelet has no DC component, that is, $\widehat{\psi}(0) = 0$. For this to occur, the mother wavelet must contain oscillations, i.e., it must have sufficient negative area to cancel out the positive area.

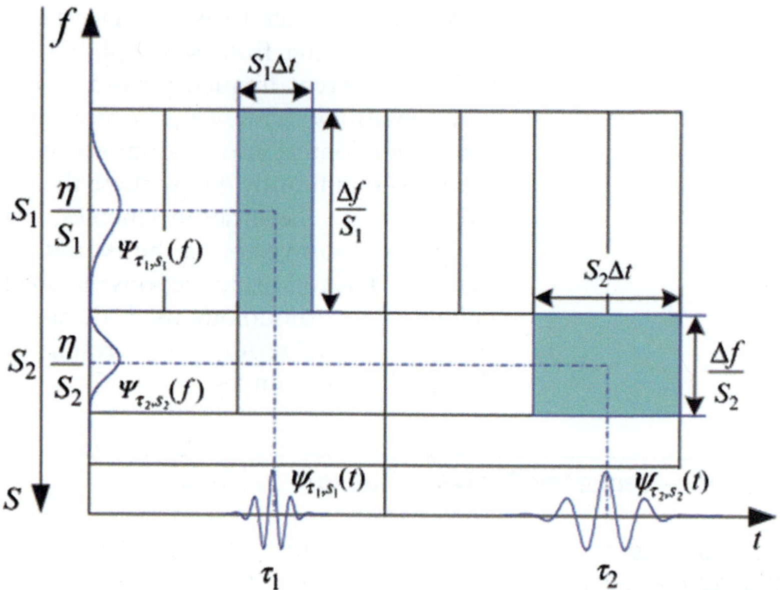

Fig. 1 Variations of time and frequency resolution of the Morlet wavelet at the locations $\left(\tau_1, \frac{\eta}{S_1}\right)$ and $\left(\tau_2, \frac{\eta}{S_2}\right)$ on the time-frequency plane. The horizontal axis τ represents the time shifts of the mother wavelet, while the axis S represents the different scales of the mother wavelet

wavelet family gives a unique pattern and its replicas at different scales and with variable localization in time.

Therefore, the CWT of a signal (t) at time b and scale a is defined as:

$$C(a, b) = \int_{-\infty}^{\infty} x(t)|a|^{-1/2}\psi\left(\frac{t - b}{a}\right)\mathrm{d}t$$

Figure 1 illustrates an example of a commonly used mother wavelet, the Morlet wavelet on the time-frequency plane, and the effect of the different combinations of time shifts and scales of the wavelets on its shape.

2.2 Morlet Wavelet

The Morlet wavelet [51] is a complex wavelet, comprising real and imaginary sinusoidal oscillations, that is convolved with a Gaussian envelope so that the wavelet magnitude is largest at its center and tapered toward its edges (Fig. 2). The wavelet's Gaussian distribution around its center time point has a SD of σ_t. The wavelet also has a Gaussian shaped spectral bandwidth around its center frequency, f_0, that has a SD of σ_f. It is mathematically formed as follows:

$$w(t, f_0) = A\exp\left(-t^2/2\sigma_t^2\right)\exp\left(2i\pi f_0 t\right)$$

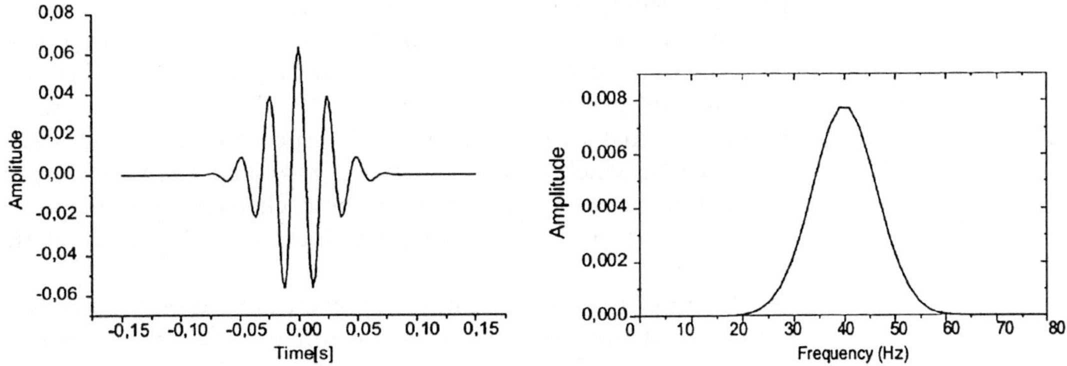

Fig. 2 Morlet wavelet in time and frequency domain

Wavelets are normalized so that their total energy is 1, the normalization factor A being equal to $(\sigma_t \sqrt{\pi})^{-1/2}$. The temporal SD, σ_t, is inversely proportional to σ_f (the exact relationship between them is defined by $\sigma_f = 1/2\pi\sigma_t$), consistent with the Heisenberg uncertainty principle that as temporal precision increases (i.e., shorter σ_t) frequency precision decreases (i.e., larger σ_f). Furthermore, a wavelet is defined by a ratio of the center frequency, f_0, to σ_f (i.e., $f_0/\sigma_f = c$), such that σ_f and σ_t vary with the center frequency, f_0. This ratio is usually constant [52], ensuring a constant number of cycles in the wavelet for all the frequencies used in the analysis. However, the open source Matlab-based software EEGLAB [53] offers the possibility of using different number of cycles throughout the range of frequencies, beginning with a small number of cycles for the low frequencies and gradually reaching a larger number of cycles for the highest frequencies. The advantage of this technique is the higher resolution for both low and high frequencies.

2.3 Event-Related Power Measures

Power reflects the magnitude of the neuroelectric oscillations at specific frequencies. When EEG oscillations are assumed to be stable (stationary) over time, the traditional FFT is often used to spectrally decompose this time-invariant EEG. Nevertheless, when EEG activity cannot be assumed to be stationary over the time period of interest, as in the case of ERPs, a time-frequency decomposition is necessary. When the magnitude values are squared for each time-frequency data point and then averaged over trials the result is a 2-dimensional matrix containing *total power* of the EEG at each frequency and time point. Total power captures the magnitude of the oscillations irrespective of their phase angles and it comprises two sources of event-related oscillatory power, evoked and induced power.

Evoked power expresses changes in EEG power that are phase-locked with respect to the stimuli onset across trials [44]. It is isolated by averaging the event-locked EEG epochs in time domain prior to decomposing the signal in both time and frequency spaces. The averaging process does not affect frequencies that are phase-locked with respect to event onset across repeated trials; therefore they are still present in the average ERP. This is not the case for oscillations that are out of phase with respect to the stimulus onset across trials, which cancel out toward zero after the time-domain averaging. *Induced power* reflects these changes in EEG power that are time-locked, but not phase-locked, with respect to the stimuli onset.

In order to calculate induced power, evoked power needs to be removed from the total power estimate. The authors in [54] propose a time-frequency decomposition applied to each single trial, followed by an averaging of the powers across trials, in order to identify non-phase-locked activity; however this technique yields the total oscillatory power of the EEG, without discarding the evoked activity. On the contrary, the authors in [55] claim to calculate the induced power by subtracting the estimated evoked responses from the corresponding single trial data, without providing any further information about how this subtraction is implemented. Zervakis et al. [56] introduced two novel measures, the Phase Intertrial Coherence (PIC) and the Phase-shift Inter-trial coherence (PsIC), in order to evaluate the phase-locked and non-phase-locked oscillatory activity, respectively. Nevertheless, the authors point out that these measures are useful in order to qualitatively evaluate the phase or non-phase-locked nature of oscillations, via the time-frequency maps of the measures, since the PsIC reflects mixed phase and non-phase-locked intertribal coherence. The focus of their study was not on the changes in energy with respect to a baseline or pre-stimulus period; therefore a quantitative evaluation of the extracted induced activity cannot be carried out. Moreover, in [57] it is pointed out that removing the mean ERP (i.e., the evoked power) from each epoch before time-frequency analysis would involve an implicit assumption that event-related brain dynamics can be modeled as a sum of a stable ERP and a reliable pattern of EEG amplitude modulation. Such a model would not take into account EEG phase-dependent inter-actions, whereas the actual effect of subtracting the evoked oscil-latory activity from each epoch prior to the wavelet transform would be relatively small, particularly at frequencies above 10 Hz, where auditory ERP amplitudes are 15 dB or more below mean EEG amplitudes. Apparently, there is great contro-versy in the field on whether or how this subtraction should be performed; therefore we chose not to include the induced oscilla-tory activity in our analysis. Figure 3 illustrates the properties of evoked and induced powers, where averaging clearly preserves the phase-locked evoked oscillations while it eliminates the non-phase-locked induced activity.

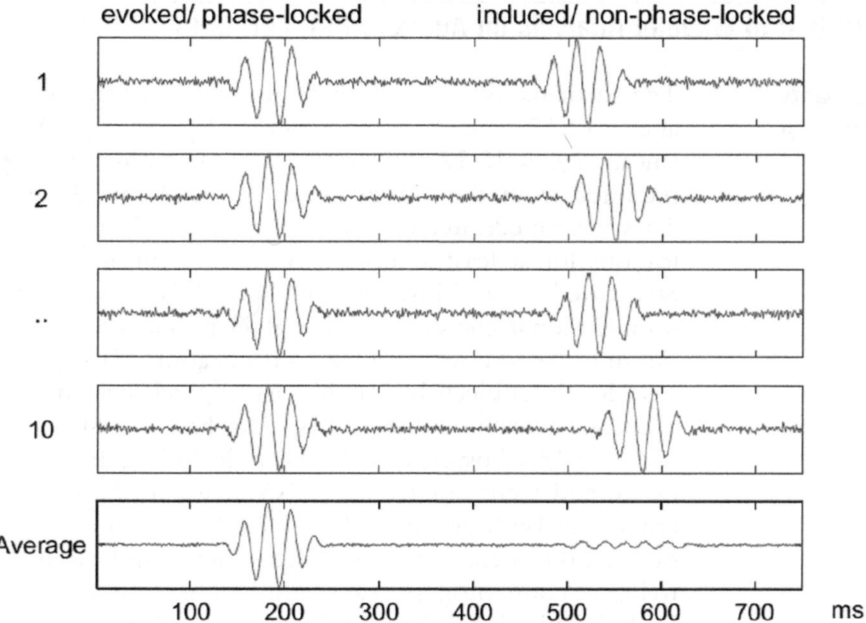

Fig. 3 Example of evoked (phase-locked) versus induced (non-phase-locked) EEG oscillations

Apart from the aforementioned power measures, event-related phase consistency across trials can be calculated, defined as the partial or exact synchronization of activity at a particular latency and frequency to a set of experimental events to which EEG data trials are time locked [53]. This measure was first introduced by Tallon-Baudry et al. in [58] as "Phase-locking factor"; however we use the term "Inter-Trial Phase Coherence" (ITPC) throughout this paper, in accordance with the definition in [53]. Typically, for n trials, if $F_k(f, t)$ is the spectral estimate of trial k, at frequency f and time t, then the ITPC is defined as:

$$\mathrm{ITCP}(f, t) = \frac{1}{n} \sum_{k=1}^{n} \frac{F_k(f, t)}{|F_k(f, t)|}$$

where $|\cdot|$ represents the complex norm.

The ITPC measure takes values between 0, which represents absence of synchronization between EEG data and time-locking events, and 1, which represents perfect synchronization. The time-frequency analysis yields complex vectors in the 2-D phase space, each of which is represented by the magnitude and phase of the spectral estimate. In order to calculate the ITPC, the lengths of each one of the trial activity vectors are normalized to 1, and then their complex average is computed. Thus, only the information about the phase of the spectral estimate of each trial is taken into account.

3 Application of Wavelet Analysis on Auditory PSP EEG Data

3.1 Participants and Experimental Setup

The participants in this study included 12 healthy controls (mean age = 42.17, SD = 7.16) and 12 schizophrenia patients (mean age = 42.17, SD = 10.50). Schizophrenia subjects were recruited from the outpatient clinics at Wayne State University. They were medicated and clinically stable without change in medications for at least 1 month. An agreement among chart and a Structured Clinical Interview for DSM-IV diagnosis was necessary for inclusion in the study. An initial telephone screening was carried out in order to establish a basic qualification of the participants.

Healthy subjects had no history of psychiatric, drug or alcohol abuse, or neurological disorders. They had no first-degree relatives with any Axis-I psychiatric diagnoses including drug use or dependence and were not on any CNS active medications. Exclusion criteria for both groups included: (a) history of a seizure disorder or any other neurological problems including head injury leading to loss of consciousness of any length of time; (b) currently meeting DSM IV criteria for drug or alcohol dependence; (c) positive urine test for any drugs of abuse; and (d) pregnancy in women. Written informed consents were obtained from all the subjects. This study was carried out following guidelines for proper human research conduct in accordance with the Declaration of Helsinki. The protocol and study procedures were approved by the Institutional Review Board at Wayne State University.

In the current study the PSP ("sensory gating" protocol) was used, which consists of the presentation of two consecutive stimuli, S1 and S2. The stimuli were identical clicks of duration of 4 ms each, intensity of 85 dB and frequency of 1,000 Hz, with time of rise/fall equal to 1 ms. The interstimulus and interpair intervals were set to vary randomly between 500 ± 25 ms and 8 ± 1 s, respectively. Subjects were instructed to relax, stay awake, and try to avoid too much eye movement during the recording session. For each subject 120 trials were recorded.

3.2 EEG Data Acquisition and Preprocessing

EEG data were recorded from 19 sites (Fp1, Fp2, Fz, F3, F4, F7, F8, Cz, C3, C4, T7, T8, Pz, P3, P4, P7, P8, O1, O2), according to the international 10–20 electrode placement system, and left and right mastoids. Vertical and horizontal electrooculograms were also recorded. A linked ears reference was chosen for acquisition of the data with an analog band-pass filter of 0.3–100 Hz and analog-to-digital sampling rate of 1,000 Hz. An electrode attached to the forehead was used as the ground electrode. Electrode impedances were kept below 5 kΩ.

No additional digital filtering was applied to the data offline. Eye movement and blink artifacts were removed using Independent

Component Analysis [59] and trials containing artifacts exceeding ±75 µV were rejected. After trial rejection the groups did not differ significantly on the number of final trials included: all subjects had at least 75 % of trials accepted (i.e., 90 out of 120 trials). The 200 ms before S1 stimulus were used for baseline correction over the remaining epoch. All preprocessing steps were performed using the open source Matlab software toolbox Fieldtrip [60].

3.3 Time-Frequency Analysis and Classification

For each subject and for all 19 channels that were recorded, all single trials were convolved with a complex Morlet wavelet. The decomposition was performed for a frequency range 8–60 Hz, using 4-Hz steps, and the wavelet length was increased linearly from 1 cycle at 8 Hz to 4 cycles at 60 Hz, yielding a time range from −130 ms (pre-stimulus) to 901 ms (post-stimulus), divided to 150 time points. The time-frequency decompositions were performed using the open source Matlab software EEGLAB [53]. Figures 4 and 5 illustrate two examples of time-frequency representations of relative-to-baseline changes of total power, ITPC, and evoked power, averaged over all sensors, for a representative healthy subject and a schizophrenia patient, respectively.

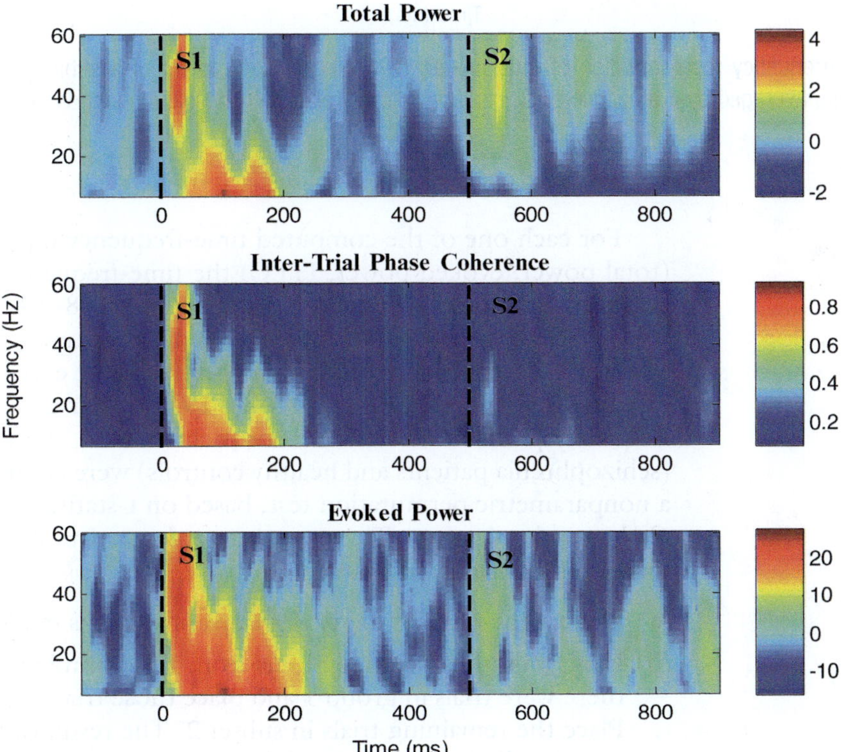

Fig. 4 Time-frequency representation of total power, ITPC, and evoked power of a healthy subject, averaged over all sensors. *Color bars* indicate relative-to-baseline changes in power

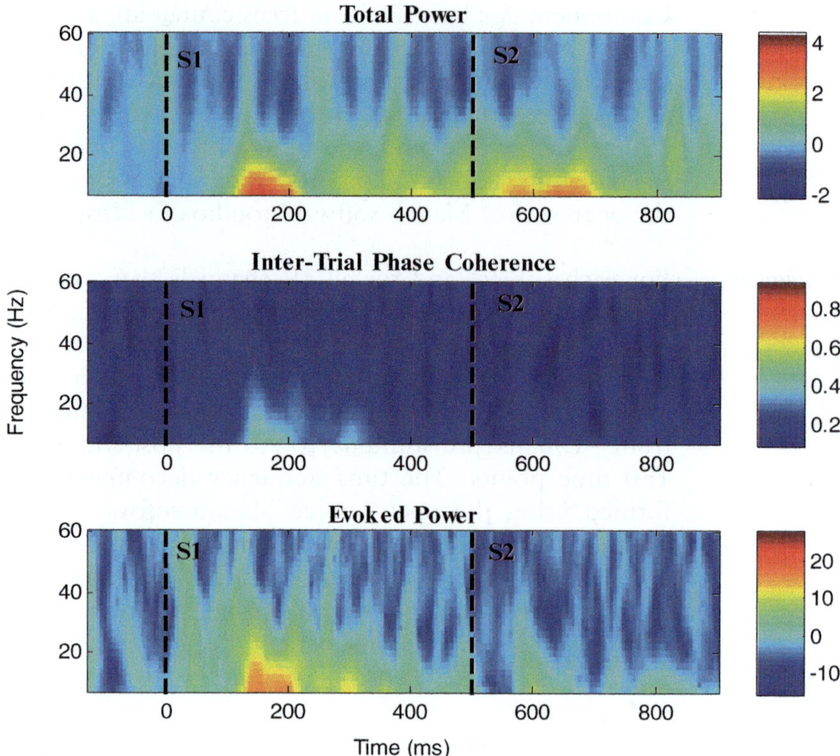

Fig. 5 Time-frequency representation of total power, ITPC, and evoked power of a subject suffering from schizophrenia, averaged over all sensors. *Color bars* indicate relative-to-baseline changes in power

For each one of the computed time-frequency representations (total power, evoked power, ITPC) the time-frequency bins were averaged over four frequency bands (alpha: 8–13 Hz, beta: 13–30 Hz, low gamma: 30–45 Hz, medium gamma: 45–60 Hz) and over 52 time epochs of 20 ms length (with the exception of the first one that lasted from -130 to -119 ms). For each new time-frequency point of interest and for each channel, the two groups (schizophrenia patients and healthy controls) were compared using a nonparametric permutation test, based on t-statistic with Monte Carlo randomization. The nonparametric statistical test is performed in the following way:

(a) Collect the trials of the two groups of subjects in a single set.

(b) Randomly draw as many trials from this combined dataset as there were trials in group 1 and place those trials into subset 1. Place the remaining trials in subset 2. The result of this procedure is called random partition.

(c) Calculate the test statistic (here, unpaired *t*-test) on this random partition.

(d) Repeat steps (b) and (c) a large number of times and construct a histogram of the test statistics.

(e) From the test statistic that was actually observed and the constructed histogram, calculate the proportion of random partitions that resulted in a larger test statistic than the observed one. This proportion is the Monte Carlo significance probability, which is also called the permutation *p-value*.

(f) If the *p-value* is smaller than a critical alpha value, then conclude that the data in the two groups are significantly different.

The accuracy of the Monte Carlo p-value increases with the number of draws from the permutation distribution. In the current study, the number of random partitions was set to 1,000 and the significance level to 5 %. To control for increased type I error rate due to multiple comparisons, Bonferroni correction was performed.

Furthermore, in order to compare groups in terms of sensory gating, the averaged time-frequency bins from 501 to 901 ms poststimulus (S2 response) were subtracted from the corresponding time-frequency bins from 1 to 401 ms (S1 response), for each one of the measures and the channels, yielding time-frequency gating representations of 4 frequency bands and 20 time epochs (Fig. 6). We chose S1–S2 amplitude difference as an expression of the gating measure because some subject's responses to S1 or S2 stimuli in time-frequency state-space were negative, making S2/S1 ratio difficult to interpret and impractical in statistical analysis.

The same statistical analysis as described above was performed for the gating representations. The features that were found to be statistically significant ($p < 0.05$) for certain channels and time-frequency bins, as a result of the aforementioned statistical analysis, were used as features in the classification schemes between schizophrenia patients and healthy controls.

The features were first normalized to the range [0,1] in order to avoid the ones in greater numeric ranges dominating those in smaller numeric ranges. Then, the normalized features were introduced into a wrapper for feature subset selection [61], in order to find an optimal feature subset in terms of classification results. An SVM classifier with a RBF kernel, a fivefold cross-validation scheme, and a best-first search engine were used for the estimation of the accuracy of the feature subsets.

The resulting optimal subset was used as a feature set in three different classifiers:

- A linear discriminant analysis (LDA) classifier.
- A support vector machine (SVM) classifier with a RBF kernel.
- A normalized Gaussian RBF network, which uses the k-means clustering algorithm to provide the basis functions and learns a linear regression on top of that.

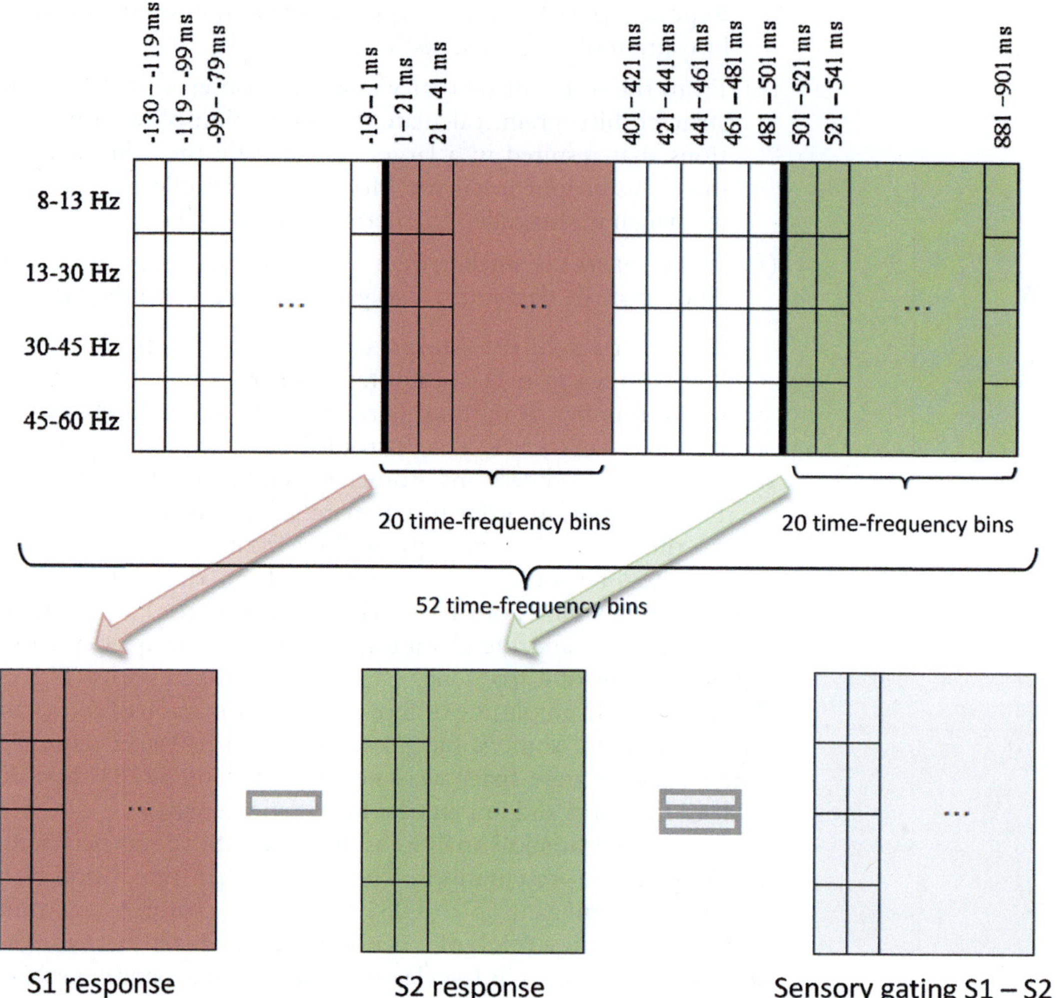

Fig. 6 Time-frequency representation of S1 response, S2 response, and sensory gating (S1–S2) after averaging over specific frequency bands and time ranges

For each one of the classifiers we used two different testing modes:

- A tenfold cross-validation
- Training on the 66 % of the data and testing on the remainder (percentage-split mode)

The wrapper for the feature subset selection, and the SVM and RBF network classifiers were applied using the open source data mining software Weka 3.6.7 [62]. For the LDA classifier the corresponding function of the Statistics Toolbox of Matlab R2010a was utilized.

4 Results

4.1 Statistical Analysis: Schizophrenia Patients–Healthy Subjects

The time bins that yielded significant differences ($p < 0.05$) between schizophrenia patients (SP) and healthy subjects (HS) in the statistical analysis are given below. In order to facilitate the comparison between S1 and S2 responses the time bins between 501 and 901 ms after S1 are converted to 1–501 ms after S2.

- *−19 to 1 ms:* SP showed significantly higher S1 total response than HS in low gamma band at F8 channel ($p = 0.044$).

- *1–21 ms:* SP showed significantly lower S1 evoked response than HS in beta band at the F3 channel ($p = 0.041$), and lower S1 ITPC response in alpha band at the F7 channel ($p = 0.038$), as well as lower gating of ITPC response in beta band, at the channels Fz and F3 ($p = 0.042$, $p = 0.037$, respectively).

- *61–81 ms:* SP showed significantly lower S1 evoked and ITPC responses ($p = 0.032$, $p = 0.027$, respectively) in medium gamma band at the C3 channel. Additionally, SP showed significantly higher S1 total response in medium gamma band, at T7 channel ($p = 0.041$).

- *81–101 ms:* SP showed significantly lower S2 evoked and ITPC responses ($p = 0.037$, $p = 0.033$, respectively) in medium gamma band at Fz.

- *101–121 ms:* SP showed significantly lower S1 evoked response in medium gamma band at Fp1 ($p < 0.02$), lower gating of evoked response in beta band at Fz ($p < 0.023$), and lower gating of ITPC response in beta band at T8 ($p = 0.03$) and in low gamma band, at P3 ($p < 0.029$).

- *141–161 ms:* SP showed significantly lower S1 evoked response in low gamma band at T8 ($p = 0.031$) and lower S2 ITPC response in medium gamma band at Fp2 ($p = 0.03$).

- *161–181 ms:* SP showed significantly higher S1 total response in low gamma band at C3 and Pz ($p = 0.025$, $p = 0.028$, respectively) and gating of total response in low gamma band at C3, Pz, and P3 ($p = 0.014$, $p = 0.017$, $p = 0.018$, respectively).

- *181–201 ms:* SP showed significantly higher S1 total response in beta band at P4 ($p = 0.026$) and gating of total response in beta at F4, C3, and P4 ($p = 0.013$, $p = 0.018$, $p = 0.02$, respectively) and in low gamma band at Cz, Pz, C3, C4, and P3 ($p = 0.014$, $p = 0.014$, $p = 0.02$, $p = 0.021$, $p = 0.018$, respectively). Additionally, SP showed significantly higher gating of ITPC response in medium gamma band at T8 ($p = 0.031$).

- *221–241 ms:* SP showed significantly higher S1 ITPC response in medium gamma band at C4 ($p = 0.04$).

- *241–261 ms:* SP showed significantly higher gating of total response in beta band at P4 ($p = 0.021$).

- *301–321 ms:* SP showed significantly lower S2 evoked response in beta band at O1 ($p = 0.033$) and in low gamma band at Pz ($p = 0.04$).

- *321–341 ms:* SP showed significantly lower S2 evoked response in beta band at O1 ($p = 0.018$) and lower S2 ITPC response in alpha and beta band at O1 ($p = 0.025$, $p = 0.026$, respectively), as well as significantly higher gating of evoked response in alpha band at O1 and O2 ($p = 0.024$ both) and in beta band at P3 and O1 ($p = 0.024$, $p = 0.02$, respectively) and higher gating of ITPC response in beta band at C3 ($p = 0.02$).

- *361–381 ms:* SP showed significantly higher gating of evoked response in alpha band at C3 and P7 ($p = 0.022$, $p = 0.023$, respectively) and in beta band at Cz, C3, and T7 ($p = 0.023$, $p = 0.019$, $p = 0.021$, respectively) and higher gating of ITPC response in alpha band at Cz, C3, and P7 ($p = 0.018$, $p = 0.016$, $p = 0.02$, respectively) and in beta band at C3 and T7 ($p < 0.0024$, $p = 0.021$, $p = 0.018$, respectively).

- *381–401 ms:* SP showed significantly higher S1 ITPC response in beta band at T7 and P8 ($p = 0.03$, $p = 0.031$, respectively), as well as higher gating of total response in low gamma band at Cz and C4 ($p = 0.026$, $p = 0.029$, respectively) and higher gating of evoked response in alpha band at C3 and P7 ($p = 0.021$, $p = 0.02$, respectively) and in beta band at T7 ($p = 0.028$).

- *441–461 ms:* SP showed significantly higher S1 ITPC response in alpha band at Cz, C3, and F3 ($p = 0.031$, $p = 0.03$, $p = 0.03$, respectively) and beta band at Fz ($p = 0.029$).

- *461–481 ms:* SP showed significantly higher S1 ITPC response in alpha band at Cz and C3 ($p = 0.023$, $p = 0.017$, respectively).

- *481–501 ms:* SP showed significantly higher S1 ITPC response in alpha and beta band at Cz ($p = 0.012$, $p = 0.029$, respectively), as well as higher S1 evoked response in medium gamma band at P8 ($p = 0.035$).

The following Tables 1, 2, and 3 illustrate the *p*-values of the statistically significant time-frequency bins, between schizophrenia patients and normal controls, in topographic maps, for each one of the power features: total power, evoked power and ITPC.

4.2 Classification

The feature selection, for the classification between schizophrenia patients and normal controls, yielded the following optimal subset of features:

- S1 total response in low gamma band at channel C3, in the time range 161–181 ms.

Table 1
_p_-Values of statistically significant time-frequency bins of total power between schizophrenia patients and normal controls

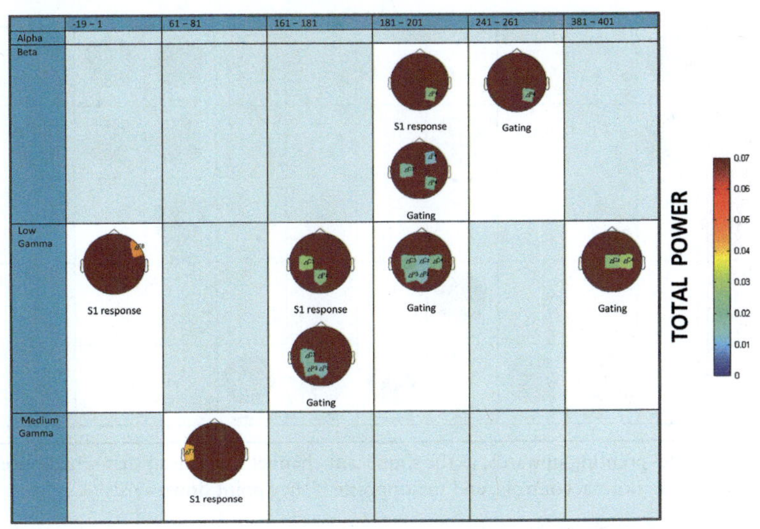

The triangles pointing upwards, at the significant channel sites, mean that schizophrenia patients showed higher values of total power than normal controls

Table 2
_p_-Values of statistically significant time-frequency bins of evoked power between schizophrenia patients and normal controls

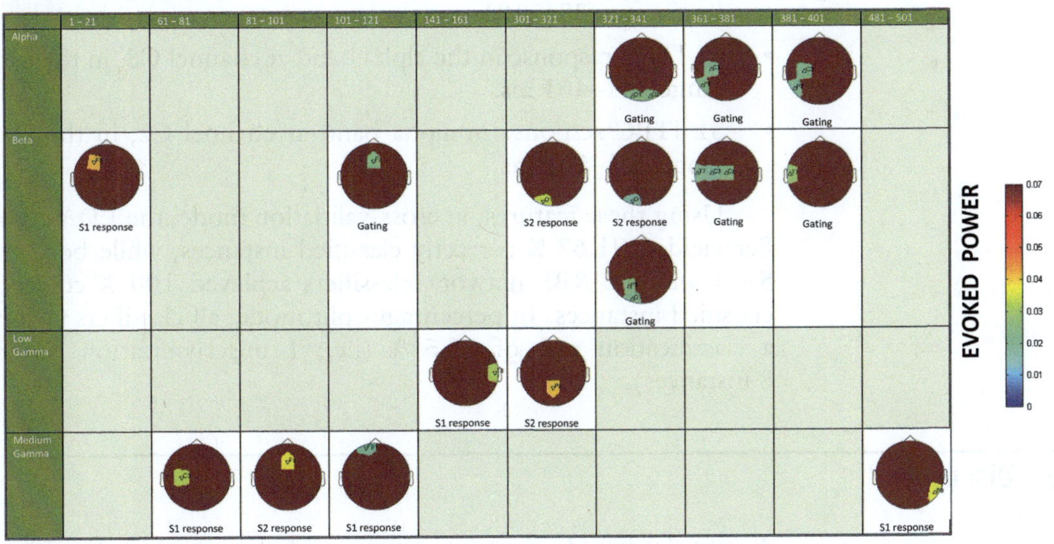

The triangles pointing upwards, at the significant channel sites, mean that schizophrenia patients showed higher values of evoked power than normal controls, and the opposite if they point downwards

Table 3

_p_-Values of statistically significant time-frequency bins of ITPC between schizophrenia patients and normal controls

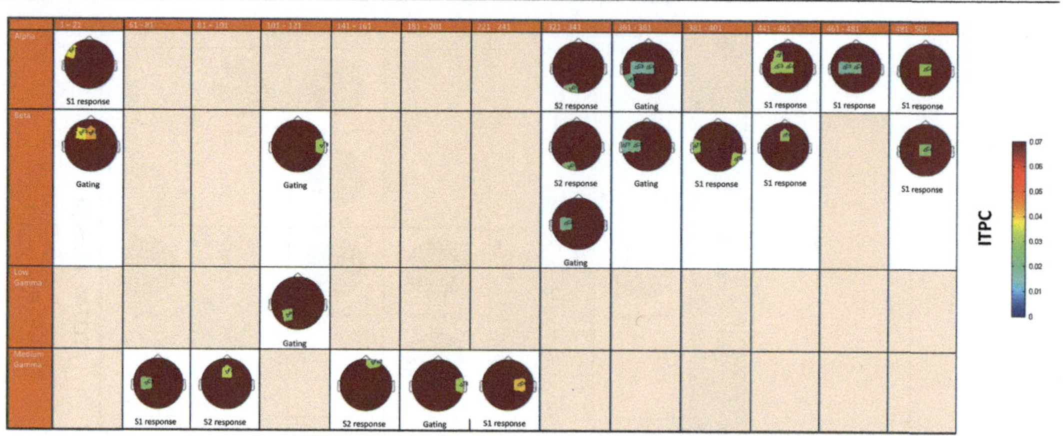

The triangles pointing upwards, at the significant channel sites, mean that schizophrenia patients showed higher values of ITPC than normal controls, and the opposite if they point downwards

- Gating of total response in low gamma band at channel C3, in the time range 161–181 ms.

- S2 evoked response in medium gamma band at channel Fz, in the time range 81–101 ms.

- Gating of evoked response in alpha band at channel P7, in the time range 381–401 ms.

- S1 ITPC response in the alpha band at channel C3, in the time range 441–461 ms.

- S1 ITPC response in alpha band at channel Cz, in the time range 481–501 ms.

Using these features, in cross-validation mode, the LDA classifier yielded 91.67 % correctly classified instances, while both the SVM and the RBF network classifiers achieved 100 % correctly classified instances. In percentage-split mode, all classifiers yielded a classification rate of 87.5 % (i.e., 1 misclassification out of 8 instances).

5　Discussion

A wavelet-based time-frequency analysis, combined with a nonparametric permutation statistical test, based on Monte Carlo randomization, was employed in this work, in order to study the alterations of the sensory gating process in schizophrenia patients, along different frequency bands. The methodology was applied on EEG data derived from a paired-click stimulus paradigm carried out on both healthy subjects and patients suffering from schizophrenia. The time-frequency measures used in the statistical analysis were the total oscillatory power, the evoked power, and the ITPC.

Compared to healthy controls, schizophrenia patients exhibited significantly lower gating of evoked power—in beta band—and ITPC—in beta and lower gamma bands—along the time bin 101–121 ms, which is an indication of a sensory gating deficit. However, this effect was localized in specific individual sensors and a generalization should be made with caution. In addition, schizophrenia patients showed reduced gamma band S1 responses of evoked power and ITPC, compared to normal controls; nevertheless it is not a generalized effect either.

Gamma event-related oscillations have been studied by several researchers in the late years, mostly focused on the examination of the so-called early auditory gamma band response (EABGR), which refers to the gamma response before 100 ms after the first stimulus. This response to simple stimuli occurs early in the information processing stream and like other sensory processes it appears to be enhanced by selective attention [63, 64] and possibly task difficulty [65]. EAGBR abnormalities in schizophrenia would be consistent with basic low level processing deficits in schizophrenia as suggested by recent studies [66, 67]. Consequently, it seems likely that the EAGBR would be reduced in schizophrenia; however results to date have been inconsistent. Johannesen et al. [33] found reduced spectral power of the EAGBR responses in SP, compared to healthy controls while Hall et al. [49] also reported reduced gamma band power to S1 stimulus in patients, compared to healthy subjects. On the contrary, Clementz and Blumenfeld [30] found no differences in gamma band responses between patients and healthy controls, but rather found differences in the low frequency components (theta and beta ranges). Likewise, Hong et al. [46] reported no difference in S1 gamma response between schizophrenia patients and controls whereas Brockhaus-Dumke et al. [39] reported similar phase locking (i.e., intertrial phase coherence in current study) after S1, in patients and controls.

The results in the current study are considered as indications of reduction of EABGR in schizophrenia patients. However, a surprising effect is reported after the 161 ms after the first stimulus: schizophrenia patients show consistently higher gating of total response, evoked response, and ITCP compared to healthy subjects. In the time range 161–181 ms SP show significantly higher gating of total response in low gamma band, at left central and parietal regions, probably due to the significantly higher S1 total response at the same regions. Higher gating of total power of SP is also noted during 181–201 ms in beta band at both left and right hemispheres, while the significantly higher S1 total response at P4 could also indicate the reason of such increased gating.

On the other hand, the increased gating of evoked power and ITPC of SP during the time range 321–381 ms, limited to alpha and beta bands, could be justified by the reduced S2 responses in the corresponding sensor areas.

It should be noted that no significant differences of either total, evoked power, or ITPC between the two groups are observed in alpha band before 320 ms, whereas differences in gamma band are evident mostly before that point.

Finally, ITCP S1 responses of schizophrenia patients in lower frequency bands (i.e., alpha and beta) are significantly increased compared to healthy subjects, during the three last time bins before the application of S2 stimulus (441–501 ms). This finding is partly consistent with Popov et al. [68], where the authors reported a pronounced decrease of the alpha band power in healthy controls, compared to patients, shortly before or at S2 onset, although this decrease was particularly pronounced at posterior sensors, as opposed to the current study where the same effect was localized around the central sensors. As the authors point out in [68], despite the fact that the paired-click experiment did not include any explicit form of memory task, S1 stimulus must have initiated memory trace formation and working memory processes [24], thereby modulating alpha desynchronization. Normally, any kind of stimulus recognition and comparison with the memory trace is expected to be manifested in the form of oscillatory desynchronization following (rather than before) S2 onset. Nevertheless, the decrease of alpha ITCP preceding S2 stimulus could reflect a preparatory state facilitating the comparison and memory retrieval of S2. Accordingly, less desynchronization in patients may reveal less efficient working memory processes.

Acknowledgments

This work was supported in part by an award to GZ from Ikerbasque—the Basque Foundation for Science in Spain—and by AI-CARE (Cooperation 2011—11SYN-6-2009) and predictES (Cooperation 2011—11SYN-10-998).

References

1. Voineskos AN et al (2013) Neuroimaging evidence for the deficit subtype of schizophrenia. JAMA Psychiatry 70(5):472–480

2. Seethalakshmi R et al (2006) Regional brain metabolism in schizophrenia: an FDG-PET study. Indian J Psychiatry 48(3):149–153

3. Case M et al (2011) The heterogeneity of antipsychotic response in the treatment of schizophrenia. Psychol Med 41(6):1291–1300

4. Javitt DC et al (2008) Neurophysiological biomarkers for drug development in schizophrenia. Nat Rev Drug Discov 7(1):68–83

5. Hallez H et al (2007) Review on solving the forward problem in EEG source analysis. J Neuroeng Rehabil 4:46

6. Khodayari-Rostamabad A et al (2010) Diagnosis of psychiatric disorders using EEG data and employing a statistical decision model. Conf Proc IEEE Eng Med Biol Soc 2010:4006–4009

7. Khodayari-Rostamabad A et al (2010) A pilot study to determine whether machine learning methodologies using pre-treatment electroencephalography can predict the symptomatic response to clozapine therapy. Clin Neurophysiol 121(12):1998–2006

8. Rotarska-Jagiela A et al (2010) Resting-state functional network correlates of psychotic symptoms in schizophrenia. Schizophr Res 117(1):21–30

9. Kindler J et al (2011) Resting-state EEG in schizophrenia: auditory verbal hallucinations are related to shortening of specific microstates. Clin Neurophysiol 122(6):1179–1182

10. Nikulin VV, Jonsson EG, Brismar T (2012) Attenuation of long-range temporal correlations in the amplitude dynamics of alpha and beta neuronal oscillations in patients with schizophrenia. Neuroimage 61(1):162–169

11. Hong LE et al (2012) A shared low-frequency oscillatory rhythm abnormality in resting and sensory gating in schizophrenia. Clin Neurophysiol 123(2):285–292

12. Koutsoukos E et al (2013) Indication of increased phase coupling between theta and gamma EEG rhythms associated with the experience of auditory verbal hallucinations. Neurosci Lett 534:242–245

13. Carlino E et al (2012) Nonlinear analysis of electroencephalogram at rest and during cognitive tasks in patients with schizophrenia. J Psychiatry Neurosci 37(4):259–266

14. McCormick LM et al (2012) Mirror neuron function, psychosis, and empathy in schizophrenia. Psychiatry Res 201(3):233–239

15. Leonard CJ et al (2013) Toward the neural mechanisms of reduced working memory capacity in schizophrenia. Cereb Cortex 23(7):1582–1592

16. Sakkalis V et al (2006) Time-significant wavelet coherence for the evaluation of schizophrenic brain activity using a graph theory approach. Conf Proc IEEE Eng Med Biol Soc 1:4265–4268

17. Luck SJ et al (2011) A roadmap for the development and validation of event-related potential biomarkers in schizophrenia research. Biol Psychiatry 70(1):28–34

18. Cho RY et al (2005) Functional neuroimaging and electrophysiology biomarkers for clinical trials for cognition in schizophrenia. Schizophr Bull 31(4):865–869

19. Braff DL, Geyer MA (1990) Sensorimotor gating and schizophrenia. Human and animal model studies. Arch Gen Psychiatry 47(2):181–188

20. Gjini K, Arfken C, Boutros NN (2010) Relationships between sensory "gating out" and sensory "gating in" of auditory evoked potentials in schizophrenia: a pilot study. Schizophr Res 121(1–3):139–145

21. Adler LE et al (1982) Neurophysiological evidence for a defect in neuronal mechanisms involved in sensory gating in schizophrenia. Biol Psychiatry 17(6):639–654

22. Boutros N et al (1993) The P50 component of the auditory evoked potential and subtypes of schizophrenia. Psychiatry Res 47(3):243–254

23. Zouridakis G, Boutros NN (1992) Stimulus parameter effects on the P50 evoked response. Biol Psychiatry 32(9):839–841

24. Lijffijt M et al (2009) P50, N100, P200 sensory gating: relationships with behavioral inhibition, attention and working memory. Psychophysiology 46(5):1059–1068

25. Patterson J et al (2008) P50 sensory gating ratios in schizophrenics and controls: a review and data analysis. Psychiatry Res 158(2):226–247

26. Light GA, Braff DL (1999) Human and animal studies of schizophrenia-related gating deficits. Curr Psychiatry Rep 1(1):31–40

27. Boutros NN et al (2009) Sensory-gating deficit of the N100 mid-latency auditory evoked potential in medicated schizophrenia patients. Schizophr Res 113(2–3):339–346

28. Clementz BA et al (2003) Ear of stimulation determines schizophrenia-normal brain activity differences in an auditory paired-stimuli paradigm. Eur J Neurosci 18(10):2853–2858

29. Blumenfeld LD, Clementz BA (2001) Response to the first stimulus determines reduced auditory evoked response suppression in schizophrenia: single trials analysis using MEG. Clin Neurophysiol 112(9):1650–1659

30. Clementz BA, Blumenfeld LD (2001) Multichannel electroencephalographic assessment of auditory evoked response suppression in schizophrenia. Exp Brain Res 139(4):377–390

31. Boutros NN et al (2004) Morphological and latency abnormalities of the mid-latency auditory evoked responses in schizophrenia: a preliminary report. Schizophr Res 70(2–3):303–313

32. Boutros NN et al (2004) Sensory gating deficits during the mid-latency phase of information processing in medicated schizophrenia patients. Psychiatry Res 126(3):203–215

33. Johannesen JK et al (2005) Contributions of subtype and spectral frequency analyses to the study of P50 ERP amplitude and suppression in schizophrenia. Schizophr Res 78 (2–3):269–284

34. Clementz BA, Geyer MA, Braff DL (1997) P50 suppression among schizophrenia and normal comparison subjects: a methodological analysis. Biol Psychiatry 41(10):1035–1044

35. Jin Y et al (1997) Effects of P50 temporal variability on sensory gating in schizophrenia. Psychiatry Res 70(2):71–81

36. Chang WP et al (2011) Probing the relative contribution of the first and second responses to sensory gating indices: a meta-analysis. Psychophysiology 48(7):980–992

37. de Wilde OM et al (2007) A meta-analysis of P50 studies in patients with schizophrenia and relatives: differences in methodology between research groups. Schizophr Res 97 (1–3):137–151

38. Mazhari S et al (2011) Evidence of abnormalities in mid-latency auditory evoked responses (MLAER) in cognitive subtypes of patients with schizophrenia. Psychiatry Res 187 (3):317–323

39. Brockhaus-Dumke A et al (2008) Sensory gating revisited: relation between brain oscillations and auditory evoked potentials in schizophrenia. Schizophr Res 99(1–3):238–249

40. Arnfred SM (2006) Exploration of auditory P50 gating in schizophrenia by way of difference waves. Behav Brain Funct 2:6

41. Iyer D, Zouridakis G (2008) Single-trial analysis of the auditory N100 improves separation of normal and schizophrenia subjects. Conf Proc IEEE Eng Med Biol Soc 2008:3840–3843

42. Patterson JV et al (2000) Effects of temporal variability on p50 and the gating ratio in schizophrenia: a frequency domain adaptive filter single-trial analysis. Arch Gen Psychiatry 57 (1):57–64

43. Makeig S et al (2004) Mining event-related brain dynamics. Trends Cogn Sci 8 (5):204–210

44. Roach B, Mathalon D (2008) Event-related EEG time-frequency analysis: an overview of measures and an analysis of early gamma band phase locking in schizophrenia. Schizophr Bull 34(5):907–926

45. Sakkalis V, Zervakis M, Micheloyannis S (2006) Significant EEG features involved in mathematical reasoning: evidence from wavelet analysis. Brain Topogr 19(1–2):53–60

46. Hong LE et al (2004) Gamma/beta oscillation and sensory gating deficit in schizophrenia. Neuroreport 15(1):155–159

47. Kopell N et al (2000) Gamma rhythms and beta rhythms have different synchronization properties. Proc Natl Acad Sci USA 97 (4):1867–1872

48. Brenner CA et al (2009) Event-related potential abnormalities in schizophrenia: a failure to "gate in" salient information? Schizophr Res 113(2–3):332–338

49. Hall MH et al (2011) Sensory gating event-related potentials and oscillations in schizophrenia patients and their unaffected relatives. Schizophr Bull 37(6):1187–1199

50. Grossman A, Morlet J (1984) Decomposition of Hardy functions into square integrable wavelets of constant shape. SIAM J Math Anal 15(4):723–736

51. Kronland-Martinet R, Morlet J, Grossmann A (1987) Analysis of sound patterns through wavelet transforms. Int J Patt Recogn Artif Intell 1:237–301

52. Grossmann A, Kronland-Martinet R, Morlet J (1987) Reading and understanding continuous wavelet transforms. In: Combes J-M, Grossmann A, Tchamitchian P (eds) Wavelets: time-frequency methods and phase space. Springer, Berlin, pp 2–20

53. Delorme A, Makeig S (2004) EEGLAB: an open source toolbox for analysis of single-trial EEG dynamics includingindependent component analysis. J Neurosci Methods 134 (1):9–21

54. Tallon-Baudry C, Bertrand O (1999) Oscillatory gamma activity in humans and its role in object representation. Trends Cogn Sci 3 (4):151–162

55. Truccolo WA et al (2002) Trial-to-trial variability of cortical evoked responses: implications for the analysis of functional connectivity. Clin Neurophysiol 113(2):206–226

56. Zervakis M et al (2011) Intertrial coherence and causal interaction among independent EEG components. J Neurosci Methods 197 (2):302–314

57. Makeig S (1993) Auditory event-related dynamics of the EEG spectrum and effects of exposure to tones. Electroencephalogr Clin Neurophysiol 86(4):283–293

58. Tallon-Baudry C et al (1996) Stimulus specificity of phase-locked and non-phase-locked 40 Hz visual responses in human. J Neurosci 16(13):4240–4249

59. Jung T et al (2000) Removing electroencephalographic artifacts by blind source separation. Psychophysiology 37(2):163–178

60. Oostenveld R et al (2011) Fieldtrip: open source software for advanced analysis of MEG, EEG, and invasive electrophysiological data. Comput Intell Neurosci 2011:9

61. Kohavi R, John GH (1997) Wrappers for feature subset selection. Artif Intell 97 (1–2):273–324

62. Bouckaert RR et al (2010) WEKA-experiences with a Java open-source project. J Mach Learn Res 11:2533–2541

63. Tiitinen H et al (1993) Selective attention enhances the auditory 40-Hz transient response in humans. Nature 364(6432):59–60

64. Debener S et al (2003) Top-down attentional processing enhances auditory evoked gamma band activity. Neuroreport 14(5):683–686

65. Mulert C et al (2007) Auditory cortex and anterior cingulate cortex sources of the early evoked gamma-band response: relationship to task difficulty and mental effort. Neuropsychologia 45(10):2294–2306

66. Leavitt VM et al (2007) Auditory processing in schizophrenia during the middle latency period (10-50 ms): high-density electrical mapping and source analysis reveal subcortical antecedents to early cortical deficits. J Psychiatry Neurosci 32(5):339–353

67. Javitt DC (2009) When doors of perception close: bottom-up models of disrupted cognition in schizophrenia. Annu Rev Clin Psychol 5:249–275

68. Popov T et al (2011) Evoked and induced oscillatory activity contributes to abnormal auditory sensory gating in schizophrenia. Neuroimage 56(1):307–314

Neuromethods (2015) 91: 231–248
DOI 10.1007/7657_2013_63
© Springer Science+Business Media New York 2013
Published online: 12 February 2014

Schizophrenia Assessment Using Single-Trial Analysis of Brain Activity

George Zouridakis, Darshan Iyer, and Javier Diaz

Abstract

In the quest for neurophysiological biomarkers that uniquely characterize schizophrenia subjects, auditory evoked potentials (EPs) have been extensively used during the past several decades. Typically, EPs are estimated using ensemble averaging to obtain robust components. Averaging, however, eliminates all temporal variability of the recorded signals and, therefore, hampers the study of the brain temporal dynamics underlying the generation of EP components. In this chapter, we present a methodology for analyzing EPs on a single-trial basis using an iterative independent component analysis procedure. The method is capable of identifying and measuring the amplitude, latency, and overall morphology of individual EP components in single trials and, as such, permits the study of phase characteristics among single trials while preserving known features of the average EPs. Recordings from schizophrenia patients and normal controls demonstrate that activity phase synchronization plays a crucial role in EP generation and explains the sensory gating deficits observed in schizophrenia subjects. Furthermore, the findings from this method are very robust across recordings from different labs and experimental protocols and can be used to separate schizophrenia patients from normal controls with 100 % classification accuracy.

Key words Evoked potentials, Single-trial EP, Independent component analysis, Schizophrenia, Gating deficit, P50, N100, Phase synchronization

---

1   Introduction

As a complex neuropsychiatric disorder, schizophrenia is characterized by essential impairments, including chronic neurocognitive deficits, which can be assessed using neurophysiological measures derived from electroencephalography (EEG) and evoked potentials (EPs). These measures can have significant diagnostic value, as they uniquely characterize schizophrenia patients by providing objective functional indices of the brain processes underlying symptoms, and in that respect, they can be used as neurophysiological *biomarkers*.

Following an auditory external stimulus, the information flow through the various stages of subcortical and cortical processing is reflected on the sequence of main EP components, such as the P50, N100, P200, and so on. Each stage is thought to reflect different aspects of information extraction that is passed on to the next level

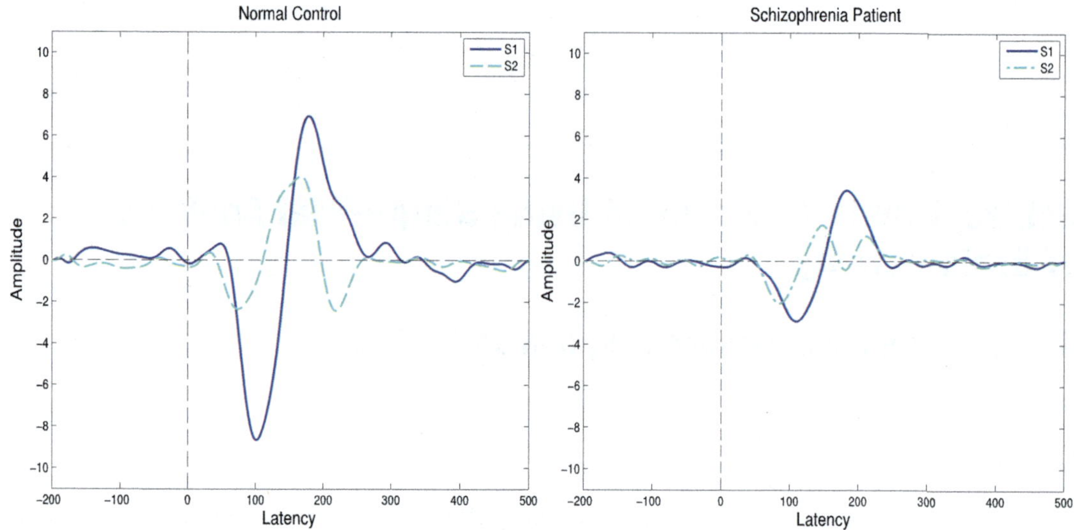

Fig. 1 Typical N100 components from normal control (*left*) and schizophrenia patient (*right*)

of processing. To avoid flooding with unnecessary or irrelevant information, high cortical centers are protected through an active filtering mechanism known as *sensory gating* [1], which shields higher order cognitive functions from low-level interference [2–6]. Sensory gating of the P50 and N100 components can be studied with a paired-stimulus paradigm [7, 8], in which repeated pairs of clicks or tones are presented at 0.5-s intervals. A minimum of 8-s intertrial interval is needed [9] so that stimulation effects are confined within a particular trial and do not carry over to the next one.

During the past several decades, a plethora of studies have used the paired-stimulus paradigm to show that the amplitude of the first (conditioning, or S1) response is significantly higher in normal controls than in schizophrenia subjects and that the amplitude of the second (test, or S2) response is significantly decreased compared to the first one in normal controls but not in schizophrenia patients [10–15]. These findings are summarized in Fig. 1, which shows examples of the N100 component recorded in a normal control (NC) and a schizophrenia (SZ) patient for the S1 and S2 responses. The higher S2/S1 ratios would indicate less effective sensory gating and an inability of schizophrenia patients to inhibit irrelevant sensory input, which is thought to result in sensory overload, as the amount of information reaching consciousness increases, possibly because of a defect in subcortical and cortical inhibitory pathways [16].

Both the P50 [2, 7, 8, 17–22] and the N100–P200 complex [18, 22–26] exhibit gating effects, even though it is now believed that the N100–P200 sensory gating could reflect different mechanisms than those reflected by the P50 gating [5, 6] and thus could account for different brain functions [10].

2 EP Estimation

The ongoing EEG activity is of considerably higher amplitude than EPs, which are typically buried in the background processes and require averaging of a large number of single-trial responses to become visible. Averaging, however, does not allow the study of changes in EP amplitude, latency, or phase characteristics from trial to trial, and thus single-trial analysis would be better suited for analyzing the dynamics of brain activation.

In addition to neural activity, scalp recordings often contain activity from eye blinks, slow eye movements, muscle and cardiac activity, interference from power lines, as well as background activity unrelated to the experimental task under study. Ideally, one would like to remove all these noise processes and record only the useful signal, i.e., the true EP activity of the cortical generators activated by the experimental task. Early attempts at estimating EPs included Fourier-based filtering techniques [27, 28], which, however, were not well suited to the nonstationary nature of the signals. Thus, time-adaptive filters were introduced [29, 30] that were better adjusted to the EP signal characteristics. Wavelets, which follow a similar concept, have been extensively used in the analysis of EEG and average EPs [31, 32] and, more recently, single-trial EPs [33]. Neural networks [34] and principal component analysis have also been employed mostly to remove artifactual activity [35, 36]. Another technique that is based on a combination of Woody filtering [37] for latency correction in the time domain and wavelet denoising has been proposed to enhance single-trial EPs [38]. This method, however, was shown to work mostly with high-amplitude EPs, while for low signal-to-noise ratios (SNRs) its performance deteriorated [38].

To explore possible phase relationships among single trials that would help improve EP estimation, we developed a selective averaging technique that involved fuzzy clustering [39, 40], in which each subject's single trials were clustered in two mutually exclusive groups and then two partial EPs were computed by averaging the trials in each group, separately for the S1 and S2 responses.

In normal controls, we found that individual responses had approximately the same amplitude for both the S1 and the S2 parts. However, the two partial EPs were *in phase* around the S1 P50 latency and *out of phase* around the S2 P50 latency. Phase synchronization produced a high-amplitude average S1 response, while phase desynchronization yielded a lower amplitude average S2, as seen previously with ensemble averaging. Figure 2 shows typical EPs from a control subject for the S1 (left) and S2 (right)

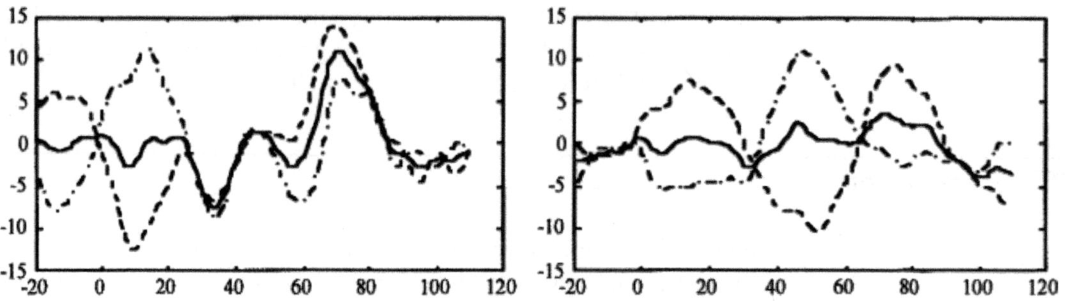

Fig. 2 Partial (*dashed lines*) and ensemble-average (*solid lines*) P50 components obtained from a normal subject, corresponding to the S1 (*left*) and S2 (*right*) responses, respectively

responses. Solid lines represent the ensemble average EPs, whereas dashed lines depict the two partial EPs computed from the same dataset.

In schizophrenia subjects, again partial EPs had approximately the same amplitude; however, partial EP synchronization was significantly reduced compared to normal controls, for both the S1 and the S2 responses. This reduced synchronization resulted in lower S1 amplitude and decreased S2 amplitude attenuation, as was typically seen in schizophrenia patients with ensemble averaging. Similar latency desynchronization effects were reported later [41] as increased latency jitter (i.e., latency variability) in S1 in schizophrenia subjects compared to controls, but not in S2, and could explain the decreased apparent amplitude of the average S1 response in patients. Overall, these studies show that *phase synchronization occurs primarily in normal controls and to a lesser degree in schizophrenia patients, and this difference can explain the gating discrepancies observed in the two groups.*

Most of the aforementioned EP estimation techniques attempt to improve the entire brain response waveform at once, i.e., all EP components simultaneously, without paying attention to difference in noise susceptibility of individual components. Contrary to all other methods, we have proposed an alternative methodology [42, 43] that can extract individual components out of the complete EP waveform. The method analyzes single-trial EPs based on independent component analysis (ICA) and relies on the hypothesis that brain activity resulting from an experimental stimulus is independent from neurophysiological artifacts and background activity [32, 44, 45]. The method provides clear estimates of individual EP components in single-trial responses, along with estimates of their amplitude, latency, and phase characteristics, and consequently allows studying the dynamics of the underlying cortical generators that give rise to specific EP components.

The method has been studied extensively with regard to its performance. In particular, we compared [46] the *iterative ICA* (*i*ICA)-based single-trial analysis against traditional ensemble averaging using synthetic and real data from normal subjects and found that the iICA method provides improved estimates of the N100 component. We also showed that these findings are independent of the model assumed for EP generation and remain true for both the phase-resetting and additive models. Moreover, when compared with time–frequency wavelet denoising, iICA performed equally well or better over a wide range of experimental conditions and SNRs [47]. The iICA method has been used to analyze EP activity with data obtained at different labs and under different experimental protocols, including a simple N100 auditory task and typical paired stimulus P50 and N100 paradigms [48, 49]. The following sections describe the various steps of the iICA method along with application examples on data recorded from normal control subjects and schizophrenia patients.

3 Methods

3.1 Fuzzy Clustering

Clustering is an algorithmic procedure used to partition a set of observations, or patterns, into clusters so that patterns within a class are more similar to each other than patterns in the rest of the classes (in our case, a pattern is a single-trial EP). In conventional clustering, a pattern can either belong or not to a particular class—in the former case, its degree of membership to that class is one; in the latter, it is zero. In fuzzy clustering, on the other hand, every pattern belongs to all possible classes, but membership values are allowed to vary continuously from zero to one. The higher the membership of a pattern to a particular class, the more "typical" the pattern for that class.

For each class of patterns, the clustering algorithm identifies a class representative or template. In conventional clustering a class template is computed as the average of the patterns assigned to that class. In contrast, in fuzzy clustering a class template is computed as a weighted average of all patterns, the weights being the membership values of the patterns to that class. In this way typical patterns contribute more to the template, whereas patterns that are less typical contribute less, and the resulting templates are more reliable representatives of their respective classes.

Mathematically the above procedure can be described assuming that a set of N patterns X_j, with $j = 1, 2, \ldots, L$, can be divided into K classes C_i with centroids V_i, where $i = 1, 2, \ldots, K$, respectively. The degree of membership μ_{ij} of vector X_j to class C_i, for $j = 1, 2, \ldots, L$, can be computed as

$$\mu_{ij} = \frac{\left[\dfrac{1}{d^2\left(X_j, V_i\right)}\right]^{1/q-1}}{\sum\limits_{i=1}^{K}\left[\dfrac{1}{d^2\left(X_j, V_i\right)}\right]^{1/q-1}},$$

where d^2 is a distance measure, often the common Euclidean distance, between the j-th pattern X_j and the i-th template V_i given by $d^2\left(X_j, V_i\right) = \sum\limits_{i=1}^{M}\left[X_j(t) - V_i(t)\right]^2$ and $q > 1$ is the fuzziness index whose value is to be determined empirically ($q = 1$ gives conventional "crisp" clustering). Typically, class templates V_i and pattern membership values μ_{ij} are identified using an iterative procedure. Bezdek's [50] fuzzy K-means algorithm computes membership values μ_{ij}, so that $\sum\limits_{i=1}^{K}\mu_{ij} = 1$ for $j = 1, 2 \ldots, N$, where $j = 1, 2, \ldots, L$. The fuzzy centroids V_i, with $i = 1, 2, \ldots, K$, are updated iteratively through the relation

$$V_i = \frac{\sum\limits_{j=1}^{N}\left(\mu_{ij}\right)^q X_j}{\sum\limits_{j=1}^{N}\left(\mu_{ij}\right)^q}$$

until the difference between two successive estimates of V_i is less than a predefined small value (e.g., $<10^{-3}$) or until a predefined maximum number of iterations has been reached.

3.2 Iterative Independent Component Analysis

Independent component analysis [51] is a method for solving the blind source separation problem [52], which tries to recover N independent source signals, $s = \{s_1, \ldots, s_N\}$, from N observations, $x = \{x_1, \ldots, x_N\}$ that represent linear mixtures of the independent source signals. The key assumption to separate sources from mixtures is that the sources are statistically independent, while the mixtures are not. Mathematically, the problem is described as a transformation $x = As$, where A is an unknown mixing matrix. The task then is to recover a version, u, of the original sources, similar to s, by estimating a matrix, W, which inverts the mixing process, i.e., $u = Wx$. The estimates u are called *independent components* (ICs). The extended infomax algorithm is the most efficient technique to solve this problem and relies on a neural network approach and information theory [53–56].

Briefly, for a discrete variable X the entropy H is defined as

$$H(X) = -\sum_i P(X = a_i) \log P(X = a_i),$$

where a_i are all possible values of X, whereas the mutual information between variables X and Y is a measure of the information that these random variables share and can be defined as

$$I(X, Y) = H(X) + H(Y) - H(X, Y),$$

where $H(X, Y)$ is the joint entropy of X and Y. For a random vector $x = \{x_1, x_2, \ldots, x_N\}$, the mutual information can also be seen as the distance between the joint distribution and the product of marginal distributions,

$$I(X) = \int p(x) \log \frac{p(x)}{\prod_{i=1}^{N} p_i(x_i)} \, dx.$$

Maximum information preservation suggests [57] that the transformation of a vector x observed in the input layer of a neural network to a vector y produced in the output layer jointly maximizes the information about the activities in the input layer. The parameter to be maximized is the average mutual information between x and y, in the presence of processing noise. Information maximization can solve the blind source separation problem [53]. Maximizing the joint entropy $H(y)$ of the output y can approximately minimize the mutual information among the output components $yi = gi(ui)$, where $gi(ui)$ is an invertible monotonic nonlinearity, $u = Wx$, and W are the synaptic weights of the neural processor. The joint entropy of the outputs is given by

$$H(y_1, , y_N) = H(y_1) + + H(y_N) - I(y_1, , y_N),$$

where $H(y_1), \ldots, H(y_N)$ are the marginal entropies of the output variables, and $I(y_1, \ldots, y_N)$ is their mutual information. Maximizing the joint entropy consists of maximizing the marginal entropies and minimizing the mutual information. The former will also decrease the latter, since the mutual information is always positive. In the limit, when $I(y_1, \ldots, y_N) = 0$ the joint entropy is the sum of the marginal entropies.

Two sets of parameters determine the maximum joint entropy, namely, the nonlinearity $yi = gi(ui)$ and the synaptic weights W in the neural network. The nonlinearity can be a fixed logistic function [53], while the synaptic weights W can be found by maximizing the joint entropy with respect to W. By computing the derivative of H with respect to W, an efficient way to maximize the joint entropy is to follow the "natural" gradient $W^T W$ [58] given by

$$\Delta W \propto \frac{\partial H(y)}{\partial W} W^T W.$$

The ideal form for the nonlinearity is the cumulative distribution function of the independent sources, but in practice, it is chosen to be a sigmoid function [53]. The extended infomax ICA

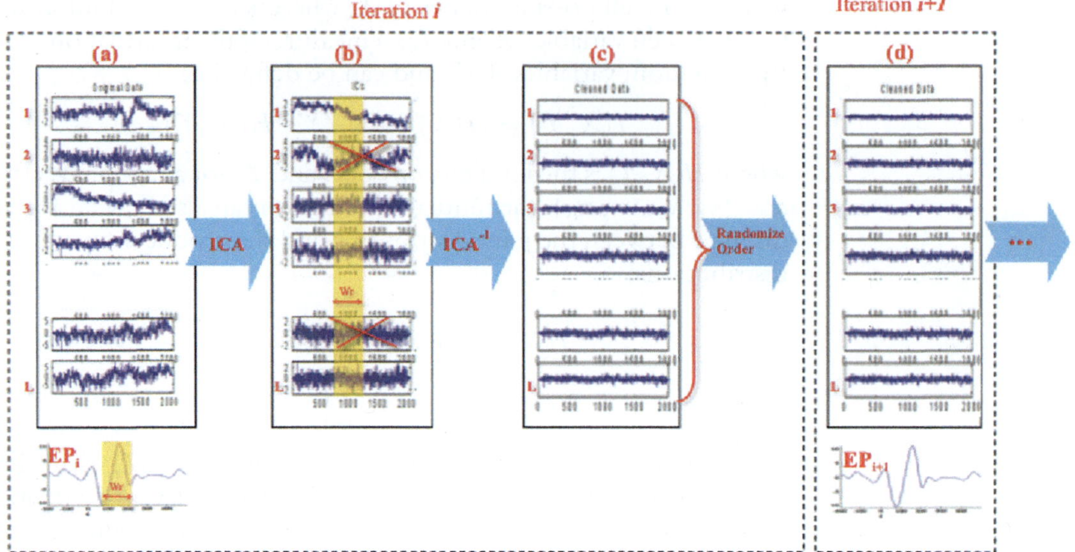

Fig. 3 *i*CA processing of a single-channel EEG at the *i*-th iteration of the algorithm: (**a**) *L* single trials are averaged to produce a template EPi and then (**b**) ICA transformed. The resulting ICs are compared with the template EP*i* within a window *Wr*, and those ICs with correlation less than a predefined threshold are dropped, while the rest are (**c**) inverse transformed, randomized in order, and (**d**) averaged to produce a new template EP*i* + 1, which is used in the next iteration *i* + 1

algorithm implemented in the software package EEGLab [59] used in our analyses allows for sources with sub-Gaussian or super-Gaussian distribution, and the maximization in these cases is given by [54]

$$\Delta W \propto \left[I - K \tanh(u)u^{\mathrm{T}} - uu^{\mathrm{T}}\right] W \begin{cases} k_i = 1 & \text{super-Gaussian} \\ k_i = -1 & \text{sub-Gaussian} \end{cases},$$

where k are the elements of an N-dimensional diagonal matrix K given by [55] $k_i = \mathrm{sign}[E\{\mathrm{sech}^2(u_i)\}E\{u_i^2\} - E\{\tanh(u_i)u_i\}]$, which ensures stability of the learning rule.

Our technique, termed *i*ICA, is an iterative implementation of this algorithm and is applied to a set of recordings consisting of L single trials obtained from N recording channels. All single trials obtained from a particular channel are processed in the following seven steps [42, 43]:

1. Compute an average EP from all trials in the entire set.

2. ICA-transform all single trials in blocks of 10.

3. Compute the absolute correlation values between the current average EP and the ICs in all blocks, within a predefined window *Wr*.

4. Set to zero those ICs with correlation less than a predefined threshold r_{th}.

5. Inverse-transform the updated ICs back to the time domain, separately in each block.

6. Randomize the order of the updated single trials in the entire set.

7. Repeat steps 1–6 until a convergence criterion is met.

The above steps involved with processing one channel of data containing L single trials are graphically illustrated in Fig. 3. At the i-th iteration, the procedure starts with the computation of an initial template EP_i, which is considered a rough estimate of the true EP component, and this estimate is gradually improved at subsequent iterations. The same procedure is then applied to the rest of the channels until all of them have been processed. Randomizing the order of single trials guarantees that each block will include different trials in the next iteration, and thus the resulting ICA system of equations will not be underdetermined. The window parameter values used in our studies were $Wr = 15$–90 ms poststimulus for the P50 waveform and $Wr = 50$–250 ms poststimulus for the N100–P200 complex, while the threshold was set to $r_{\mathrm{th}} = 0.15$. The convergence criterion required the root-mean-squared-amplitude difference of the average EP between the current and the previous iteration to be less than 0.001; thus, the algorithm converged when the average EP morphology was stabilized.

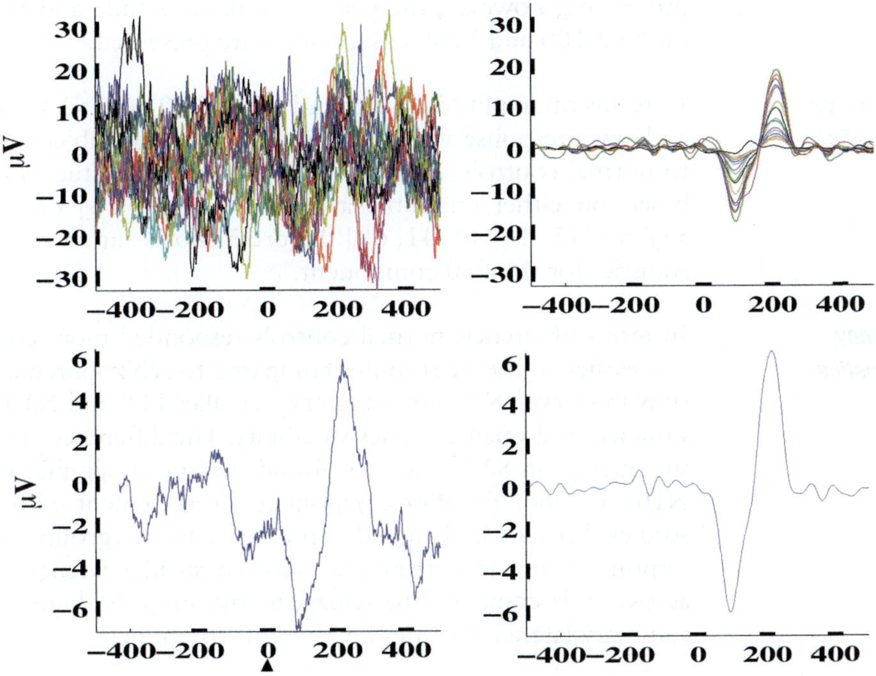

Fig. 4 Original (*left*) and ICA-processed (*right*) single-trial responses and the corresponding average EPs (*bottom*)

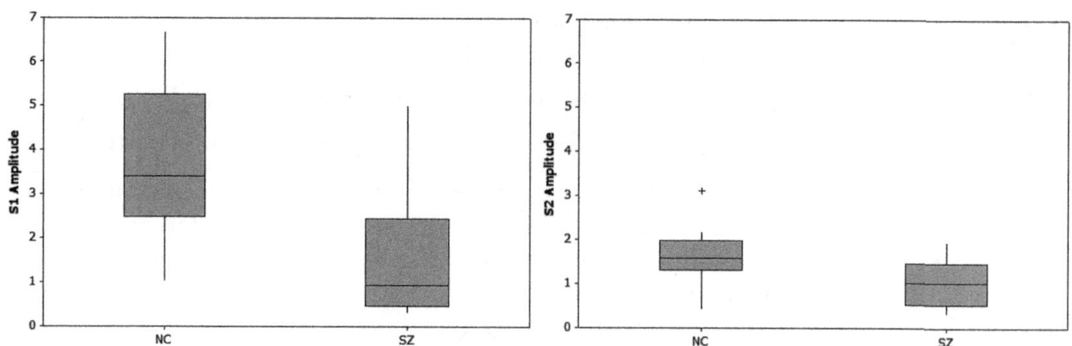

Fig. 5 P50 component S1 (*left*) and S2 (*right*) amplitude for normal controls (NC) and schizophrenia patients (SZ). *Lines* indicate the lower quartile, median, and upper quartile values; "+" indicates outliers

4   Application Examples

An example of the *i*ICA method performance in estimating components hidden in background noise is given in Fig. 4, where the original and *i*ICA-processed single-trial responses are shown, along with the corresponding average EPs. Clear N100 components are seen after processing in all single trials that were not discernable in the unprocessed data. The average response is less noisy after processing; however, the peak-to-peak amplitude and the latency of the N100 and P200 components are preserved.

4.1   Amplitude Characteristics

In terms of amplitude, we found lower P50 and N100 amplitude and lower response attenuation in schizophrenia subjects compared to normal controls, in agreement with previous studies that were based on either ensemble averaging [5, 19, 24] or single-trial analysis [15, 16, 40, 41, 60]. Figure 5 shows summary amplitude statistics for the P50 component.

4.2   Latency Characteristics

In terms of latency, normal controls responded more consistently and earlier to the S1 stimulus compared to schizophrenia subjects, showing lower N100 mean latency, smaller P50 and N100 latency variance, and smaller latency variability. The difference between the two groups in S2 latency was found statistically significant for the N100 but not for P50 component, in agreement with previous studies [16, 41]. Thus, the normal control group tended to respond earlier in general and showed smaller latency variability across trials compared to schizophrenia subjects. Figure 6 shows summary latency statistics for the P50 component.

4.3   Morphology Characteristics

One of the most striking findings of the *i*ICA analysis in normal controls and schizophrenia patients was the existence of aberrant (positive-going) N100 components that were intermixed with

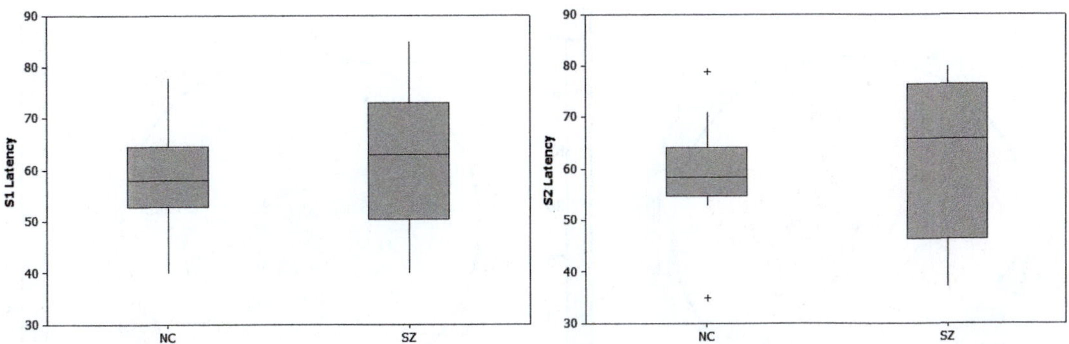

Fig. 6 P50 component S1 (*left*) and S2 (*right*) latency for normal controls (NC) and schizophrenia patients (SZ). *Lines* indicate the lower quartile, median, and upper quartile values; "+" indicates outliers

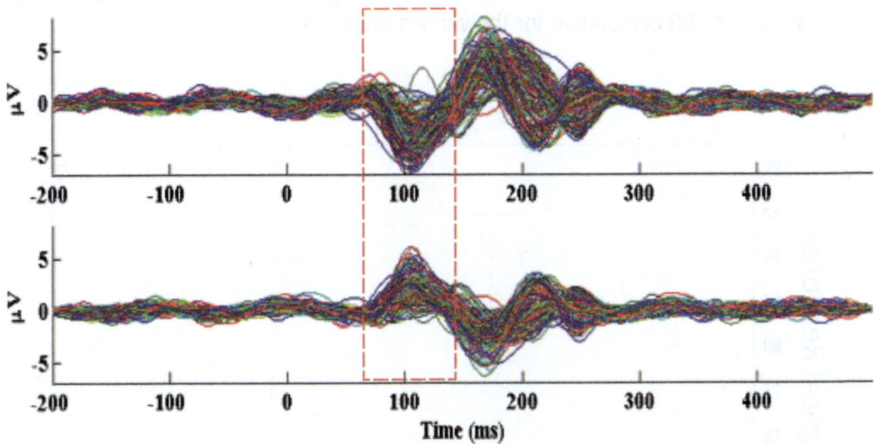

Fig. 7 Example of EP components within the N100 latency range obtained from 248 superimposed channels after separating the typical (*upper traces*) from the aberrant (*lower traces*) responses

typical (negative-going) ones [61]. Figure 7 shows an example from a normal control where the EP components within the N100 latency range obtained from 248 channels are superimposed, after separating in each channel the typical (upper traces) from the aberrant (lower traces) responses. Figure 8 shows the percentage of typical N100 components detected in single trials (left panel) and the corresponding amplitude of the average N100 component for the various channels (right panel). In general, central channels show higher percentages of typical (negative-going) components, which results in larger (more negative) average N100 amplitudes.

The box plots of Fig. 9 depict the average value and variation of the percentage of typical responses obtained from the Cz channel in normal controls and schizophrenia patients for the S1 and S2 responses. The S1 values were significantly higher than the S2 only within the group of normal controls but not in the patient group.

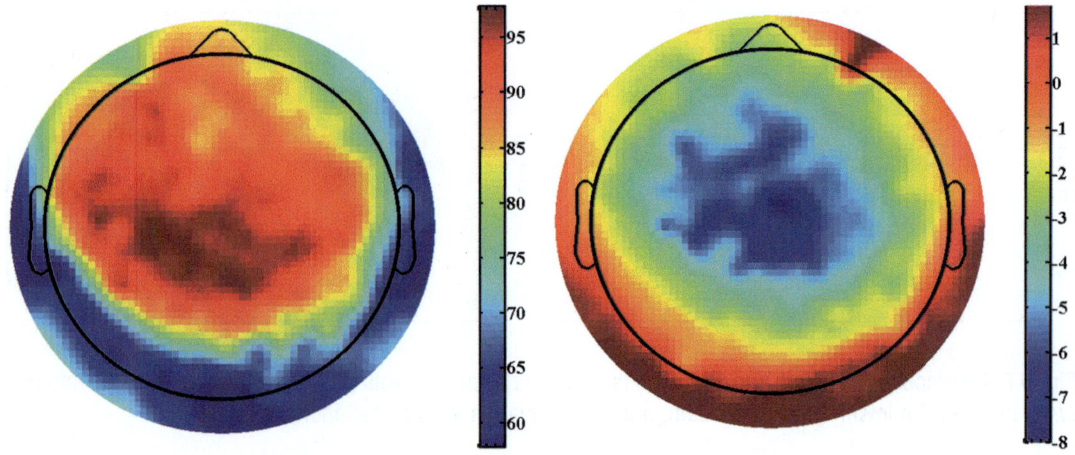

Fig. 8 Percentage of typical N100 components detected in single trials (*left panel*) and the corresponding amplitude of the average N100 component for the various channels

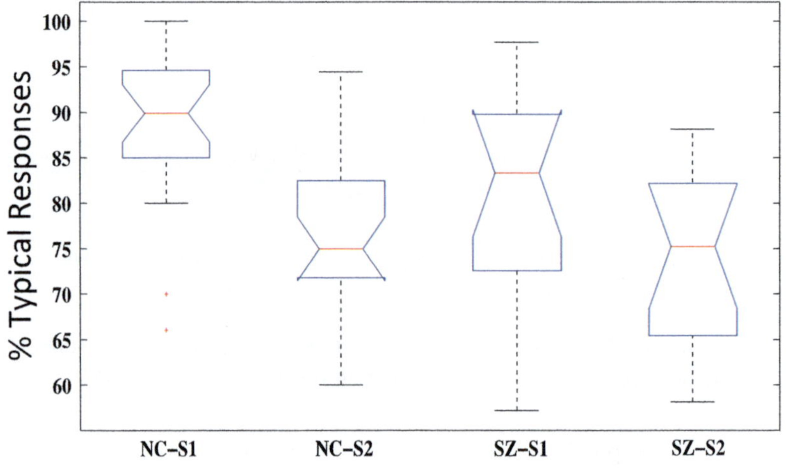

Fig. 9 Percentage of typical responses obtained from the Cz channel in normal controls (NC) and schizophrenia patients (SZ) for the S1 and S2 responses

Across groups, normal controls exhibited much higher percentage of typical responses for both the S1 and the S2 responses.

4.4 Phase Characteristics

In an effort to understand better the phenomenon of aberrant response generation, we considered the ongoing EEG activity as a sinusoidal-type signal and estimated its phase at the time of stimulus arrival in single trials, after iICA processing [62, 63]. We found that some phases of the EEG activity were more likely to result in typical responses than others. Specifically, the probability of obtaining a typical response for the N100 component was varying in a sinusoidal fashion and reached a maximum value of approximately

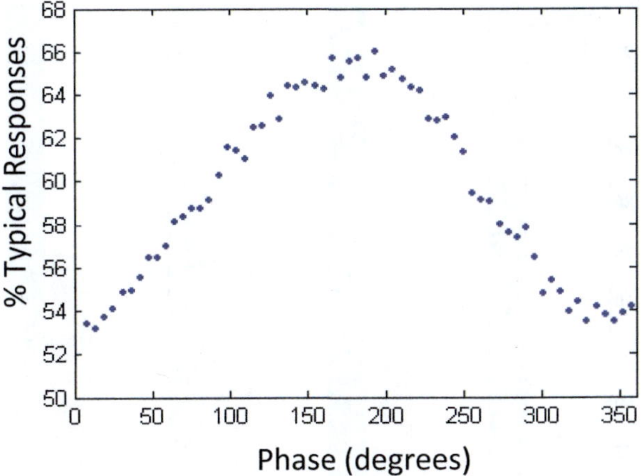

Fig. 10 Percentage of typical N100 components on single trials as a function of the phase of the ongoing EEG activity at the time of stimulus arrival

67 %, when the EEG phase was around 180°, and a minimum of only 53 %, when the phase was close to 0° or 360°, as shown in Fig. 10. Thus, the characteristics of the ongoing EEG activity do affect EP morphology.

4.5 Subject Classification

To explore whether the characteristics of the S1 and S2 responses were statistically distinct to allow automatic subject classification, we analyzed the amplitude and latency values of the P50 and N100 components seen in the S1 and S2 responses, when the latter were identified from ensemble averaging or from single-trial *i*ICA processing [48]. Feature ranking found S1 latency to be the best discriminatory feature, followed by S2 latency, while the S1 and S2 amplitudes had no significant contributions to classification accuracy. For a given subject, each single trial was labeled as normal or schizophrenia, and the final assignment as control or patient for that subject was determined by the majority (70 % or more) of single-trial labels.

Assessment of misclassification error rates was addressed using a tenfold stratified cross-validation scheme, in which the data were randomly divided into ten sets of approximately equal size, taking into account the percentage of subjects in each class. One of the ten sets was set aside as the test set. The classification rule was constructed on the remaining nine sets (training set), while the error rates were computed on the test set. This was repeated ten times, and the error rates were averaged to obtain the final estimate. The same procedure was used with several popular classifiers to avoid bias, including the random forest (RF) classifier [64] as implemented in the data mining software WEKA [65], *k*-nearest-neighbor (*k*-nnc) classifier [66], J48 classifier [67], and the pseudo-Fisher

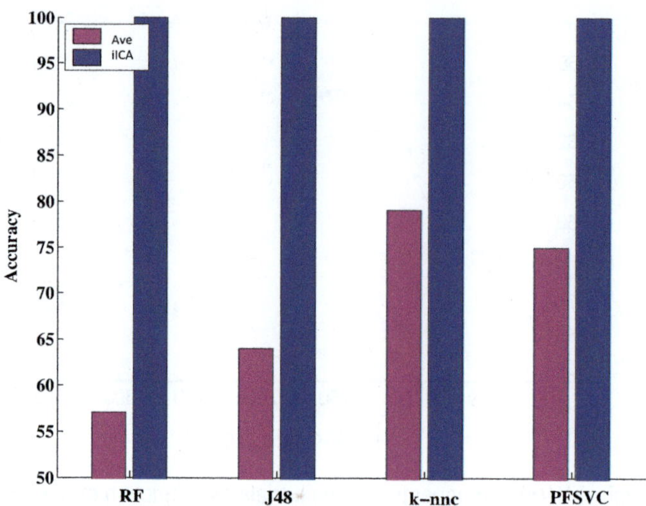

Fig. 11 Classification performance using N100 component measures from the original (Ave) and *i*ICA-processed (*i*ICA) single trials for four different classifiers

[68] support vector classifier (PFSVC, PR Toolbox, Matlab). With classical averaging, the best sensitivity, specificity, and overall accuracy values were 70, 85, and 79 %, respectively, and were obtained with the N100 component using the *k*-nnc classifier. With single-trial analysis all performance measures were 100 %, for both the P50 and the N100, with all classifiers, as shown in Fig. 11.

5   Discussion

Overall, our results using the *i*ICA method agree with previous findings that schizophrenia subjects show decreased S1 amplitude and decreased S2 attenuation compared to normal controls [12, 15, 16, 24, 40, 41, 61] and that normal controls tend to respond earlier with smaller latency variability across trials compared to schizophrenia subjects [16, 41, 61]. For subject classification, S1 latency was still the best and S2 amplitude the worst discriminatory feature in line with previous studies [15, 16, 25, 41, 69].

However, a unique contribution of the *i*ICA methodology is the complete separation of the schizophrenia and control groups based on single-trial measures. To our knowledge, none of the previous studies has been able to measure P50 amplitudes clearly in single trials or to separate the two groups with 100 % accuracy. Contrary to classical averaging which wipes out all trial-to-trial variability, the *i*ICA procedure can capture the dynamics of an EP component by estimating correctly its amplitude and latency in single trials. Another important factor to the success of our method is the two-step classification approach: by classifying a large number of single trials from each subject, instead of a single average response, a subject is allowed to have several responses that are atypical for its group and still be classified correctly, as long as the majority of the responses are typical for that group.

A second unique contribution of the iICA methodology is the notion that, apart from the typical negative polarity N100 components, there exist atypical positive polarity N100 responses, which we called aberrant. Previous studies have regarded these responses as noncontributing, nonresponsive, or inappropriate [70, 71].

Although fewer in number, aberrant responses are found in all subjects, both normal controls and schizophrenia patients, and play a crucial role in determining the amplitude of the ensemble average response. In general, schizophrenia subjects have a significantly higher number of aberrant responses compared to normal controls, and this can explain both the lower S1 amplitude and the decreased S2 attenuation seen in the patients. These findings are also consistent with other N100 studies that looked at activity phase synchronization to show population differences [15, 70, 72].

A third unique contribution of the iICA methodology is the confirmation of an earlier study [39] showing that EP amplitude is related to the phase of the ongoing EEG activity at the time of stimulus arrival. We now know that certain EEG phases can produce single-trial responses of diverse phases and even completely opposite polarity than expected, and this phase desynchronization among individual responses decreases the overall amplitude of the average EP.

Taken as a whole, the above findings suggest that single-trial analysis using iICA can detect and quantify brain activity synchronization, which plays a crucial role in EP generation, and can explain the sensory gating deficits observed in schizophrenia subjects. Thus, the iICA methodology may have a significant impact as a clinical tool in the quest for identifying physiological markers of schizophrenia.

Acknowledgment

This work has been partially supported by NSF-MRI grant no. 521527.

References

1. Venables PH (1964) Input dysfunction in schizophrenia. Prog Exp Pers Res 72:1–47

2. Freedman R, Waldo MC, Bickford-Wimer P, Nagamoto H (1991) Elementary neuronal dysfunctions in schizophrenia. Schizophr Res 4:233–243

3. Hsieh MH, Liu S, Chiu M, Hwu H, Chen ACN (2004) Memory impairment and auditory evoked potential gating deficit in schizophrenia. Psychiatry Res Neuroimaging 130:161–169

4. Kisley MA, Noecker TL, Guinther PM (2004) Comparison of sensory gating to mismatch negativity and self-reported perceptual phenomena in healthy adults. Psychophysiology 41:604–612

5. Boutros NN, Korzyukov O, Jansen BH, Feingold A, Bell M (2004) Sensory gating deficits during the mid-latency phase of information processing in medicated schizophrenia patients. Psychiatry Res 126:203–215

6. Wan L, Friedman BH, Boutros NN, Crawford HJ (2008) P50 sensory gating and attentional performance. Int J Psychophysiol 67:91–100

7. Adler LE, Pachtman E, Franks RD, Pecevich M, Waldo MC, Freedman R (1982) Neurophysiological evidence for a defect in neuronal mechanisms involved in sensory gating in schizophrenia. Biol Psychiatry 17:639–654

8. Boutros NN, Zouridakis G, Overall J (1991) Replication and extension of P50 findings in schizophrenia. Clin Electroencephalogr 22:40–45

9. Zouridakis G, Boutros NN (1992) Stimulus parameter effects on the P50 evoked response. Biol Psychiatry 32:839–841

10. Lijffijt M, Lane SD, Meier SL, Boutros NN, Burroughs S, Steinberg JL, Moeller FG, Swann AC (2009) P50, N100, and P200 sensory gating: relationships with behavioral inhibition, attention, and working memory. Psychophysiology 46(5):1059

11. Jutzeler CR, McMullen ME, Featherstone RF, Tatard-Leitman VM, Gandal MJ, Carlson GC, Siegel SJ (2011) Electrophysiological deficits in schizophrenia: models and mechanisms, psychiatric disorders—trends and developments. In: Toru Uehara (ed) ISBN: 978-953-307-745-1, InTech, DOI: 10.5772/27511. http://www.intechopen.com/books/psychiatric-disorders-trends-and-developments/electrophysiological-deficits-in-schizophrenia-models-and-mechanisms

12. Boutros NN, Korzyukov O, Oliwa G, Feingold A, Campbell D, McClain-Furmanski D et al (2004) Morphological and latency abnormalities of the mid-latency auditory evoked responses in schizophrenia: a preliminary report. Schizophr Res 70:303–313

13. Clementz BA, Geyer MA, Braff DL (1998) Poor P50 suppression among schizophrenia patients and their first-degree biological relatives. Am J Psychiatry 155:1691–1694

14. Freedman R, Adler LE, Myles-Worsley M, Nagamoto HT, Miller C, Kisley M et al (1996) Inhibitory gating of an evoked response to repeated auditory stimuli in schizophrenic and normal subjects. Human recordings, computer simulation, and an animal model. Arch Gen Psychiatry 53:1114–1121

15. Jansen BH, Hegde A, Boutros NN (2004) Contribution of different EEG frequencies to auditory evoked potential abnormalities in schizophrenia. Clin Neurophysiol 115:523–533

16. Patterson JV, Jin Y, Gierczak M, Hetrick WP, Potkin S, Bunney WE Jr, Sandman CA (2000) Effects of temporal variability on P50 and the gating ratio in schizophrenia: a frequency domain adaptive filter single-trial analysis. Arch Gen Psychiatry 57:57–64

17. Adler LE, Waldo MC, Freedman R (1985) Neurophysiologic studies of sensory gating in schizophrenia: comparison of auditory and visual responses. Biol Psychiatry 20:1284–1296

18. Boutros NN, Belger A, Campbell D, D'Souza C, Krystal J (1999) Comparison of four components of sensory gating in schizophrenia and normal subjects: a preliminary report. Psychiatry Res 88:119–130

19. Clementz BA, Geyer MA, Braff DL (1997) P50 suppression among schizophrenia and normal comparison subjects: a methodological analysis. Biol Psychiatry 41:1035–1044

20. Franks R, Adler L, Waldo M, Alpert J, Freedman R (1983) Neurophysiological studies of sensory gating in mania: comparison with schizophrenia. Biol Psychiatry 18:989–1005

21. Guterman Y, Josiassen RC, Bashore TR Jr (1992) Attentional influence on the P50 component of the auditory event-related brain potential. Int J Psychophysiol 12:197–209

22. Müller MM, Keil A, Kissler J, Gruber T (2001) Suppression of the auditory middle latency response and evoked gamma-band response in a paired-click paradigm. Exp Brain Res 136:474–479

23. Boutros NN, Brockhaus-Dumke A, Gjini K, Vedeniapin A, Elfakhani M, Burroughs S et al (2009) Sensory-gating deficit of the N100 mid-latency auditory evoked potential in medicated schizophrenia patients. Schizophr Res 113:339–346

24. Brockhaus-Dumke A, Mueller R, Faigle U, Klosterkoetter J (2008) Sensory gating revisited: relation between brain oscillations and auditory evoked potentials in schizophrenia. Schizophr Res 99:238–249

25. Clementz BA, Blumenfeld LD (2001) Multichannel electroencephalographic assessment of auditory evoked response suppression in schizophrenia. Exp Brain Res 139:377–390

26. Rosburg T, Boutros NN, Ford JM (2008) Reduced auditory evoked potential component N100 in schizophrenia—a critical review. Psychiatry Res 161:259–274

27. Walter DO (1969) A posteriori Wiener filtering of average evoked response. Electroencephalogr Clin Neurophysiol 27(1):61–70

28. Doyle DJ (1975) Some comments on the use of Wiener filtering in the estimation of evoked potentials. Electroencephalogr Clin Neurophysiol 38:533–534

29. de Weerd JP (1981) A posteriori time-varying filtering of averaged evoked potentials: I. Introduction and conceptual basis. Biol Cybern 41:211–222

30. de Weerd JP, Kap JI (1981) A posteriori time-varying filtering of averaged evoked potentials: II. Mathematical and computational aspects. Biol Cybern 41:223–234

31. Samar VJ, Bopardikar A, Rao R, Swartz K (1999) Wavelet analysis of neuroelectric waveforms: a conceptual tutorial. Brain Lang 66:7–60

32. Makeig S, Westerfield M, Jung T-P, Enghoff S, Townsend J, Courchesne E, Sejnowski TJ (2002) Dynamic brain sources of visual evoked responses. Science 295:690–694

33. Quiroga RQ, Garcia H (2003) Single-trial event-related potentials with wavelet denoising. Clin Neurophysiol 114:376–390

34. Nishida S, Nakamura M, Suwazono S, Honda M, Nagamine T, Shibasaki H (1994) Automatic detection method of P300 waveform in the single sweep records by using a neural network. Med Eng Phys 16(5):425–429

35. Lagerlund TD, Sharbrough FW, Busacker NE (1997) Spatial filtering of multichannel electroencephalographic recordings through principal component analysis by singular value decomposition. J Clin Neurophysiol 14(1):73–82

36. Casarotto S, Bianchi AM, Cerutti S, Chiarenza GA (2004) Principal component analysis for reduction of ocular artefacts in event-related potentials of normal and dyslexic children. Clin Neurophysiol 115:609–619

37. Woody CD (1967) Characterization of an adaptive filter for the analysis of variable latency neuroelectric signals. Med Biol Eng 5:539–553

38. Effern A, Lehnertz K, Grunwald T, Fernandez G, David P, Elger CE (2000) Time adaptive denoising of single trial even-related potentials in the wavelet domain. Psychophysiology 37:859–865

39. Zouridakis G, Jansen BH, Boutros NN (1997) A Fuzzy clustering approach to EP estimation. IEEE Trans Biomed Eng 44:673–680

40. Zouridakis G, Boutros NN, Jansen BH (1997) A fuzzy clustering approach to study the auditory P50 component in schizophrenia. Psychiatry Res 69:169–181

41. Jin Y, Potkin SG, Patterson JV, Sandman CA, Hetrick WP, Bunney WE Jr (1997) Effects of P50 temporal variability on sensory gating in schizophrenia. Psychiatry Res 70:71–81

42. Zouridakis G, Iyer D (2004) Improved estimation of evoked potentials using an iterative independent component analysis procedure. WSEAS Trans Signal Process Robot Automat 2(1):288–291

43. Iyer D (2005) Improved evoked potential estimation using iterative independent component analysis. PhD Dissertation, Department of Electrical and Computer Engineering, University of Houston

44. Vigario RN (1997) Extraction of ocular artifacts from EEG using independent component analysis. Electroencephalogr Clin Neurophysiol 103:395–404

45. Makeig S, Westerfield M, Jung T-P, Covington J, Townsend J, Sejnowski TJ et al (1999) Independent components of the late positive response complex in a visual spatial attention task. J Neurosci 19:2665–2680

46. Zouridakis G, Iyer D, Diaz J, Patidar U (2007) Estimation of individual evoked potential components using iterative independent component analysis. Phys Med Biol 52:5353–5368

47. Zouridakis I (2007) Clin Neurophysiol 118:495–504

48. Iyer D, Boutros N, Zouridakis G (2012) Single-trial P50 and N100 analysis improves separation of normal and schizophrenia subjects. Clin Neurophysiol 123:1810–1820

49. Iyer D, Boutros N, Zouridakis G (2011) Aberrant auditory evoked responses in schizophrenia: evidence from signal trial analysis. Conf Proc IEEE Eng Med Biol Soc 2011:4406–4409

50. Bezdek JC (1974) Numerical taxonomy with fuzzy sets. J Math Biol 1:57–71

51. Comon P (1994) Independent component analysis—a new concept? Signal Process 36:287–314

52. Jutten C, Herault J (1991) Blind separation of sources. I. An adaptive algorithm based on neuromimetic architecture. Signal Process 24:1–10

53. Bell A, Sejnowski T (1995) An information-maximization approach to blind source separation and blind deconvolution. Neural Comput 6:1129–1159

54. Girolami M (1998) An alternative perspective on adaptive independent component analysis algorithms. Neural Comput 10:2103–2114

55. Lee TW, Girolami M, Sejnowski TJ (1999) Independent component analysis using an extended infomax algorithm for mixed sub-Gaussian and super-Gaussian sources. Neural Comput 11:417–441

56. Jung T-P, Makeig S, Westerfield M, Townsend J, Courchesne E, Sejnowski TJ (2000) Removal of eye activity artifacts from visual

event-related potentials in normal and clinical subjects. Clin Neurophysiol 111:1745–1758

57. Linsker R (1989) An application of the principle of maximum information preservation to linear systems. Adv Neural Inform Process Syst 1:186–194

58. Amari S, Cichocki A, Yang H (1996) A new learning algorithm for blind signal separation. Adv Neural Inform Process Syst 8:757–763

59. Delorme A, Makeig S (2004) EEGLAB: an open source toolbox for analysis of single-trial EEG dynamics. J Neurosci Methods 134:9–21

60. Young KA, Smith M, Rawls T, Elliott DB, Russell IS, Hicks PB (2001) N100 evoked potential latency variation and startle in schizophrenia. Neuroreport 12:767–773

61. Zouridakis G, Iyer D (2005) Phase aspects and localization analysis of the auditory N100 component. In: The joint meeting of the 5th international conference on bioelectromagnetism and the 5th international symposium on noninvasive functional source imaging within the human brain and heart, Minneapolis, MN, 12–15 May

62. Diaz J and Zouridakis G (2006) Relationship between the phase of the ongoing activity at the stimulus time and the N100 detected response using EEG recordings. In: ISPRA'06 proceedings of the 5th WSEAS international conference on signal processing, robotics and automation, Madrid, Spain, p 317–322

63. Zouridakis G, Diaz J (2006) Latency characteristics of the auditory N1 component using an iterative independent component analysis procedure. In: IEEE EMBS annual international conference, New York, NY

64. Breiman L (2001) Random forests. Mach Learn 45:5–32

65. Witten IH, Frank E (2000) Data mining: practical machine learning tools with java implementations. Morgan Kaufmann, San Francisco

66. Rosenblatt M (1956) Remarks on some nonparametric estimates of a density function. Ann Math Stat 27:832–837

67. Quinlan JR (1993) C4.5: programs for machine learning. Morgan Kaufmann, San Francisco

68. Duin RPW, Juszcak P, Paclik P, Pekalska E, de Ridder D, Tax DMJ (2004) PR tools: a matlab toolbox for pattern recognition

69. Boutros NN, Zouridakis G, Rustin T, Peabody C, Warner D (1993) The P50 component of the auditory evoked potential and subtypes of schizophrenia. Psychiatry Res 47:243–254

70. Hu L, Boutros NN, Jansen BH (2009) Evoked potential variability. J Neurosci Methods 178:228–236

71. Rodionov V, Goodman C, Fisher L, Rosenstein GZ, Sohmer H (2002) A new technique for the analysis of background and evoked EEG activity: time and amplitude distributions of the EEG deflections. Clin Neurophysiol 113 (9):1412–1422

72. Jansen BH, Hu L, Boutros NN (2010) Auditory evoked potential variability in healthy and schizophrenia subjects. Clin Neurophysiol 121:1233–1239

Neuromethods (2015) 91: 249–265
DOI 10.1007/7657_2013_66
© Springer Science+Business Media New York 2014
Published online: 21 February 2014

Phase Variants of the Common Spatial Patterns Method

Kenneth P. Camilleri, Owen Falzon, Tracey Camilleri, and Simon G. Fabri

Abstract

Numerous studies on EEG data indicate that various brain processes are characterized by phase relationships between different regions of the brain. The development of analytic techniques that can provide a better detection of these phase relationships can lead to a better understanding of such brain processes. In this chapter two variants of the common spatial patterns (CSP) method which are designed to capture such phase relationships, namely, the "phase synchronization"-based CSP (P-CSP) algorithm and the analytic CSP (ACSP) algorithm, are presented. The P-CSP and ACSP methods are analyzed and tested on real EEG data. The nature of the results of the two methods is then discussed, highlighting the differences between them.

Key words Electroencephalogram (EEG), Brain–computer interfacing (BCI), Phase synchronization, Common spatial patterns (CSP), Phase-locking value (PLV), Spatial filtering, Steady-state visual evoked potential (SSVEP), Analytic common spatial patterns (ACSP), "Phase synchronization"-based common spatial patterns (P-CSP)

---

1 Introduction

Significant phase relationships within and between different regions of the brain have been observed in electroencephalographic (EEG) data recorded during various mental states [1–8].

In [1], Rodriguez et al. detected consistent patterns of phase synchronization during a visual perception task. Specifically, subjects were presented with ambiguous visual stimuli that could be perceived as either faces or meaningless shapes. During instances of face recognition, consistent phase synchronization links were detected between the occipital, parietal, and frontotemporal areas. In another visual processing study carried out by Busch et al. [2, 3], subjects were presented with brief flashes of light, the luminance of which was thresholded such that subjects could detect approximately half of the visual stimuli. In this case, a correlation between the detection of the visual stimuli and the phase of EEG signals in the theta and alpha bands was found. Most of the detection instances were found to occur at a particular phase value for each subject, and minimal detection was found to take place at the opposite phase, indicating that the hits and misses were correlated with opposite phase values.

A number of studies indicate that the disruption of normal phase synchronization patterns is linked to abnormal mental states. In [4], Bhattacharya analyzed the EEG signals of individuals suffering from seizures and mania and observed a significant reduction in the phase synchronization levels of pathological subjects when compared to a control group. Mormann [5] also detected a decrease in phase synchronization activity across EEG recording sites in epileptic patients prior to seizure onset. In another study by Gysels et al. [6], large variations in the phase synchronization level were observed in the pathological brain hemisphere of epileptic patients when the subjects transitioned from the awake state to the sleep state.

Ito et al. [7] also reported strong phase synchronization activity between EEG signals recorded from the frontal and occipital areas of the brain during an eyes closed resting activity. During these intervals, two spatial phase patterns emerged repeatedly across the subjects considered in the study. One of these patterns consisted of gradual phase variations, characteristic of travelling waves, moving from the posterior to the anterior region of the scalp or vice versa. The second typical spatial phase pattern that was observed exhibited sudden phase variations of approximately $180°$, characteristic of a standing wave, between electrode sites C3 and FC3. Similar standing and travelling wave phenomena, distinguished by sudden and gradual spatial phase changes, respectively, have also been observed in EEG recordings obtained from subjects who were exposed to repetitive flashing visual stimuli [8].

These studies indicate that several brain processes are characterized by consistent spatial phase relationships in EEG signals. Therefore, analytic techniques that can provide a better detection of phase relationships in EEG activity may lead to a better understanding of various brain processes. In this chapter, two variants of the widely used common spatial patterns (CSP) method are discussed. The proposed variants consist of the "phase synchronization"-based CSP (P-CSP) algorithm [9, 10] and the analytic CSP (ACSP) algorithm [11–13], each of which takes into account a different aspect of the phase relationships in the EEG data. In this chapter, these two algorithms are discussed, with a focus on the different information that the two methods can provide in relation to EEG data analysis.

1.1 Phase Estimation from EEG Signals

Phase components can be extracted from EEG signals using the discrete Hilbert transform [14], such that

$$x_h(k) = \frac{1}{\pi} \sum_{l=-\infty}^{\infty} \frac{x(l)\left[1 - e^{j\pi(k-l)}\right]}{k - l} \quad \text{for } l \neq k \tag{1}$$

where $x_h(k)$ is the discrete Hilbert transform of a signal $x(k)$ [15]. The above convolution effectively introduces a 90° phase shift at each frequency component of the signal $x(t)$.

For a discrete signal $x(k)$, the complex-valued representation of the signal can be obtained by

$$z(k) = x(k) + jx_h(k) = A(k)e^{j\phi(k)} \tag{2}$$

where the term on the right-hand side is the polar form representation of the analytic signal $z(k)$. The instantaneous amplitude $A(k)$ and instantaneous phase $\phi(k)$ components are computed as

$$A(k) = \sqrt{x(k)^2 + x_h(k)^2} \tag{3}$$

and

$$\phi(k) = \arctan\left(\frac{x_h(k)}{x(k)}\right), \tag{4}$$

respectively. The analytic signal representation allows the computation of the instantaneous phase values from frequency bands determined by the user according to the spectral filtering carried out on the original signal $x(k)$. In the case of EEG signals consisting of multiple frequency components, the phase values that result from the analytic signals are influenced by all the considered frequency components. In such a case, although the instantaneous amplitude and phase values are mathematically tractable, a physical interpretation of the underlying activity may be ambiguous [16].

1.2 Phase-Based Features for EEG Class Data Analysis

In recent years there has been a growing interest in the use of phase-based features for the analysis of EEG data as evidenced by the increasing number of publications on the subject matter [17–19]. However, with the exception of a few phase measures, the development of feature extraction methods that utilize explicit phase information in the EEG data remains largely unexplored [20, 21]. The phase-locking value (PLV), which quantifies the level of phase synchronization between two EEG channels, is one of the most widely used phase-based measures in EEG signal analysis [17, 22–25], and also forms the basis of our proposed P-CSP method [9, 10]. The details of the PLV measure are next described.

1.2.1 Single-Trial Phase-Locking Value

Developed by Lachaux et al. [26, 27], the PLV gives a measure of the phase synchronization between two signals. In contrast with classical measures of connectivity, such as the magnitude squared coherence (MSC), the PLV is not influenced by variations in the amplitudes of two signals but depends solely on the phase differences between the two considered signals.

The first step for determining the PLV between two signals, $x(k)$ and $y(k)$, involves the computation of their instantaneous phase values. This is given by

$$\phi_{xy}(k) = \phi_x(k) - \phi_y(k) \tag{5}$$

where $\phi_x(k)$ and $\phi_y(k)$ are the phase values for $x(k)$ and $y(k)$, respectively, obtained using the Hilbert transform.

Assuming a single trial, the phase-locking value is computed along a moving window to obtain a dynamic value capturing the phase-locking variation within that trial. In such a case, the PLV can be applied at each time sample, k, by using a moving window of size T to obtain the single-trial PLV (S-PLV) [27]

$$\text{S-PLV}_T(k) = \left| \frac{1}{T} \sum_{j=k-\frac{T}{2}}^{k+\frac{T}{2}} e^{j\phi_{xy}(j)} \right| \tag{6}$$

The size of the considered time window is typically based on the frequency of the signals under consideration. Specifically, the window is chosen to incorporate a selected number of oscillatory cycles. When a large window size (and a large number of cycles) is considered, a smoother S-PLV is obtained at the cost of greater computational requirements. On the other hand, shorter time windows incorporate fewer oscillations and are more likely to lead to spurious phase locking arising by chance alone [27].

2 CSP Incorporating Phase Information

The CSP method is a data analysis technique for two-class discrimination yielding spatial patterns of potential interest [28, 29]. The CSP method works by decomposing multichannel EEG data into a set of uncorrelated components that can be used to optimally discriminate between two classes of data in the least-squares sense. The CSP method will be briefly outlined next, and subsequently our proposed "phase-synchronization"-based CSP (P-CSP) and analytic CSP (ACSP) methods that may be used to analyze phase information in EEG signals will be presented and discussed.

2.1 The CSP Method

Let $\mathbf{X}_1, \mathbf{X}_2 \in \mathbb{R}^{N \times T}$ represent EEG trials for two classes of data, where N is the number of EEG channels and T is the number of samples per channel. The $N \times N$ averaged covariance matrices obtained by averaging the covariance matrices across the trials for each class of data can be denoted by $\mathbf{C}_1$ and $\mathbf{C}_2$, respectively. The CSP method can then be used to determine a set of spatial filters $\mathbf{w}$ that extremise the following generalized Rayleigh quotient [30, 31]

$$\frac{\mathbf{w}^{\mathrm{T}}\mathbf{C}_1\mathbf{w}}{\mathbf{w}^{\mathrm{T}}\mathbf{C}_2\mathbf{w}} \tag{7}$$

By simultaneously diagonalizing the averaged covariance matrices $\mathbf{C}_1$ and $\mathbf{C}_2$ this expression can be extremised yielding a matrix $\mathbf{W}^{\mathrm{T}} = [\mathbf{w}_1\,\mathbf{w}_2\,\ldots\,\mathbf{w}_N]^{\mathrm{T}}$ [29, 32]. Specifically, if $\mathbf{C}_c := \mathbf{C}_1 + \mathbf{C}_2$, the eigenvalue decomposition of $\mathbf{C}_c$ can then be represented by

$$\mathbf{C}_c = \mathbf{U}_c\Lambda_c\mathbf{U}_c^{\mathrm{T}} \tag{8}$$

where the columns of $\mathbf{U}_c$ are the eigenvectors corresponding to the eigenvalues in the diagonal matrix Λ_c. $\mathbf{C}_c$ is then whitened by $\mathbf{P} = \Lambda_c^{-\frac{1}{2}}\mathbf{U}_c^{\mathrm{T}}$ such that

$$\begin{aligned} \mathbf{I} &= \mathbf{P}\mathbf{C}_c\mathbf{P}^{\mathrm{T}} \\ &= \mathbf{P}\mathbf{C}_1\mathbf{P}^{\mathrm{T}} + \mathbf{P}\mathbf{C}_2\mathbf{P}^{\mathrm{T}} \\ &:= \mathbf{S}_1 + \mathbf{S}_2 \end{aligned} \tag{9}$$

An eigenvalue decomposition is subsequently carried out on $\mathbf{S}_1$ to obtain $\mathbf{U}_1\Lambda_1\mathbf{U}_1^{\mathrm{T}}$, where $\mathbf{U}_1$ and Λ_1 represent the eigenvector and eigenvalue matrices, respectively. Pre-multiplying and post-multiplying Eq. (9) by $\mathbf{U}_1^{\mathrm{T}}$ and $\mathbf{U}_1$, respectively, results in

$$\begin{aligned} \mathbf{I} &= \mathbf{U}_1^{\mathrm{T}}\mathbf{S}_1\mathbf{U}_1 + \mathbf{U}_1^{\mathrm{T}}\mathbf{S}_2\mathbf{U}_1 \\ \mathbf{I} &= \Lambda_1 + \mathbf{U}_1^{\mathrm{T}}\mathbf{S}_2\mathbf{U}_1 \\ \mathbf{I} - \Lambda_1 &= \mathbf{U}_1^{\mathrm{T}}\mathbf{S}_2\mathbf{U}_1 \\ \mathbf{I} - \Lambda_1 &= \Lambda_2 \end{aligned} \tag{10}$$

The corresponding real-valued diagonal elements in Λ_1 and Λ_2 must therefore add up to 1 and are ordered such that when the diagonal elements in Λ_1 decrease, those in Λ_2 increase, or vice versa. The projection matrix $\mathbf{W}^{\mathrm{T}}$ can then be obtained from

$$\mathbf{W}^{\mathrm{T}} = \mathbf{U}_1^{\mathrm{T}}\mathbf{P} \tag{11}$$

where the first and last rows of $\mathbf{W}^{\mathrm{T}}$ extremise Eq. (7), thereby providing a projection that maximizes the variance for one class of data while minimizing the variance for the other. In addition to the spatial filters, $\mathbf{W}$, a set of spatial patterns, $\mathbf{A} = \mathbf{W}^{-\mathrm{T}}$, can also be obtained from the CSP method, where the columns of $\mathbf{A} = [\mathbf{a}_1\,\mathbf{a}_2\,\ldots\,\mathbf{a}_N]$ represent a set of spatial patterns which consist of a reprojection of the most discriminative components onto the scalp electrode locations. The first and last columns of $\mathbf{A}$ can thus provide a spatial map of the discriminative EEG activity for the considered classes of data.

2.2 The P-CSP Method

In this section the details of the P-CSP proposed in [9, 10] are presented. The P-CSP method combines the discriminative feature extraction and dimensionality reduction properties of the

CSP framework with phase synchronization measures from the single-trial PLV approach. The P-CSP method utilizes the CSP framework on phase synchronization signals determined by computing the S-PLV between EEG signals. The P-CSP algorithm can thus provide a set of features which are related to the most discriminative phase synchronization links in the data, without requiring prior knowledge or additional processing steps for channel pair selection.

2.2.1 Method

Given two classes of data, $\mathbf{X_1}, \mathbf{X_2} \in \mathbb{R}^{N \times T}$, where N is the number of EEG channels and T is the number of samples per channel, the first step of the P-CSP method involves the computation of S-PLV signals for each possible pairwise combination of channels. For this purpose, the real-valued EEG signals are first converted into analytic signals by using the Hilbert transform, and the instantaneous phase component is then extracted from the computed analytic signals. Subsequently, the instantaneous phase difference for each possible combination of channel pairs is determined such that for EEG data from N channels, $\frac{N(N-1)}{2}$ phase difference signals are obtained. These signals can in turn be used to determine S-PLV signals representing the varying level of phase locking in the considered data as described in Sect. 1.2.

While the conventional CSP method uses the covariance matrices of the original EEG recordings to extremise a generalized Rayleigh quotient, the P-CSP method makes use of the computed S-PLV signals to extremise this quotient. However, in contrast with the CSP method, where the mean value of the EEG signals is not of relevance, the S-PLV signals used in the P-CSP method carry significant information in their mean value which would be lost during the computation of the covariance matrices. Therefore, in order to retain the mean component of the S-PLV signals, the "second moment" matrices $\mathbf{C}(i, j)$ are computed from the signals [10] as follows:

$$\mathbf{C}(i,j) = \frac{1}{T} \sum_{t=1}^{T} s_i(t) s_j(t) \tag{12}$$

Similar to the CSP method, the computed matrices are used as inputs for a simultaneous diagonalization procedure from which a transformation matrix $\mathbf{W}$ is obtained. The block diagram in Fig. 1 shows a representation of the steps involved in determining spatial filters using the P-CSP algorithm.

From the computed spatial filters for the P-CSP method, a set of spatial patterns $\mathbf{A} = \mathbf{W}^{-T}$ can also be derived. These constitute patterns of phase synchronization which can yield information on those synchronization links associated with the most discriminative components for the considered classes.

The P-CSP Algorithm

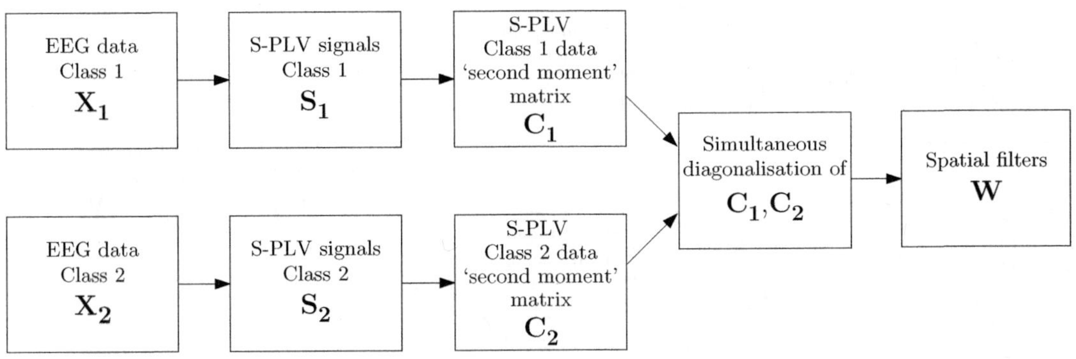

Fig. 1 The block diagram provides a representation of the P-CSP algorithm, where a set of spatial filters **W** that discriminate two classes of EEG data X_1 and X_2 based on the most discriminative phase synchronization information in the data are determined

2.3 The Analytic Common Spatial Patterns Method

Another variant of the CSP method that considers phase information in the EEG data is the analytic common spatial patterns (ACSP) method proposed in [11–13]. The ACSP method considers an analytic representation of the EEG signals in order to obtain an explicit representation of the amplitude and phase components of the data, thereby addressing the limitations of the conventional CSP technique when dealing with phase relationships between EEG signals from different spatial locations.

2.3.1 Method

The P-CSP method presented earlier was designed to represent the level of phase-locking between channels without taking into account signal amplitude and without providing information about the actual phase differences between the channels. The ACSP method [11] we have proposed follows the standard CSP method [29], with the distinct difference that the input signals are transformed into their analytic counterparts and the joint diagonalization process is then computed on the resulting complex-valued covariance matrices. The ACSP method thus involves carrying out the joint diagonalization computations with complex-valued matrices rather than the real-valued matrices that are typical for the standard CSP implementation. It is this complex representation and the resulting complex-valued spatial patterns and spatial filters that lead to a number of advantages of the proposed technique over the standard CSP method. In particular, this complex representation allows for the representation of class-specific phase difference spatial maps apart from the magnitude spatial maps typical for CSP.

Similar to the CSP method, the scope of the ACSP method is to discriminate between two classes of data by determining a set of spatial filters that maximize the variance for one class of data while minimizing the variance for the other. However, the first step in the

The ACSP Algorithm

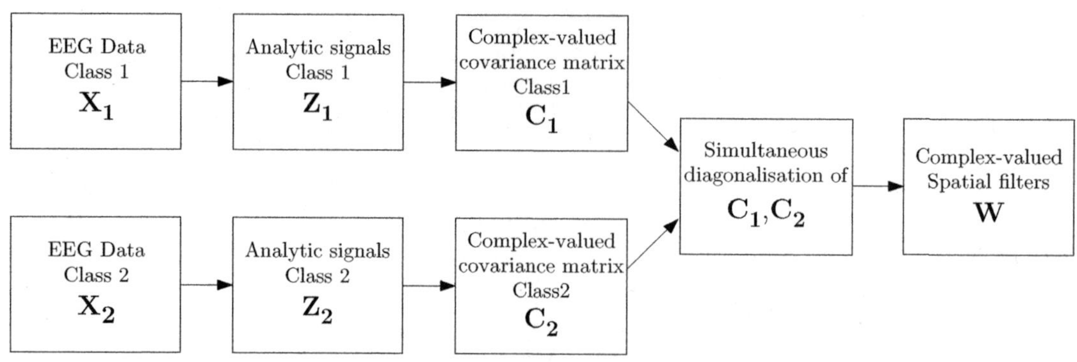

Fig. 2 The block diagram provides a representation of the ACSP algorithm where a set of complex-valued spatial filters **W** that discriminate two classes of EEG data $\mathbf{X}_1$ and $\mathbf{X}_2$ by considering an analytic representation of the data are determined

ACSP procedure involves a conversion of the real-valued EEG signals into complex-valued signals such that the EEG trials $\mathbf{X}_1$ and $\mathbf{X}_2$ are transformed into the complex-valued trials $\mathbf{Z}_1$ and $\mathbf{Z}_2$. This can be achieved by computing the Hilbert transform of the EEG signals filtered within the frequency band of interest. The covariance matrix of the complex-valued EEG data $\mathbf{X}_1$ is then given by

$$\mathbf{C}_1 = \mathrm{E}[(\mathbf{Z}_1 - \mathrm{E}[\mathbf{Z}_1])(\mathbf{Z}_1 - \mathrm{E}[\mathbf{Z}_1])^*] \tag{13}$$

where $\mathrm{E}[\cdot]$ represents the expectation operator. The same procedure can be applied for $\mathbf{Z}_2$ to obtain $\mathbf{C}_2$. Thus, in this case $\mathbf{C}_1$ and $\mathbf{C}_2$ constitute a pair of $N \times N$ Hermitian positive semi-definite matrices, with real-valued elements along the main diagonal and generally complex-valued off-diagonal elements.

For the ACSP method, the problem consists in finding a vector **w** that extremises

$$\frac{\mathbf{w}^*\mathbf{C}_1\mathbf{w}}{\mathbf{w}^*\mathbf{C}_2\mathbf{w}} \tag{14}$$

where the principal difference from the formulation in (7) for the CSP method lies in the complex-valued covariance matrices that are used and the complex-valued spatial filters $\mathbf{w}^*$ that are determined through a complex-valued implementation of the simultaneous diagonalization approach for the ACSP method [11, 12]. A projection matrix $\mathbf{W}^* = [\mathbf{w}_1\,\mathbf{w}_2\,\ldots\,\mathbf{w}_N]^*$ is obtained where the rows of $\mathbf{W}^*$ correspond to a set of complex-valued spatial filters that maximize the variance for one class of data and minimize the variance for the other class. The block diagram in Fig. 2 shows a representation of the steps involved in determining spatial filters using the ACSP algorithm.

In addition to these spatial filters $\mathbf{W}^*$, a set of complex-valued spatial patterns, $\mathbf{A} = (\mathbf{W}^*)^{-1}$ can also be determined from the ACSP method. Since for the ACSP method, the matrix $\mathbf{A}$ is a complex-valued array, it may be fashioned into a magnitude and a phase array. The spatial patterns associated with the most discriminative components, if cautiously considered, may thus give useful neurophysiological insight into the discriminative brain activity related to a given set of tasks [30] as described in the discussion in Sect. 3.2.

3   Results and Discussion

While the P-CSP and ACSP methods have been used for two-class EEG discrimination tasks, such as in brain–computer interfaces, this contribution focuses on the spatial patterns yielded by the two methods.

3.1   P-CSP Scalp Patterns for Real EEG Data

The P-CSP method was applied to motor imagery data to obtain a transformation that maximizes the difference between two motor imagery classes based on the PLV in order to obtain spatial patterns that distinctly characterize each class. Specifically, the P-CSP algorithm was tested on EEG data obtained from BCI Competition IV dataset IIa [33]. EEG recordings from 22 scalp locations from nine subjects, sampled at 250 Hz were considered. For each subject, data from 72 left and 72 right imagined hand movements were analyzed. The electrode placement and experimental protocol used for data acquisition are shown in Fig. 3. Additional details of the recording protocol can be found in [33].

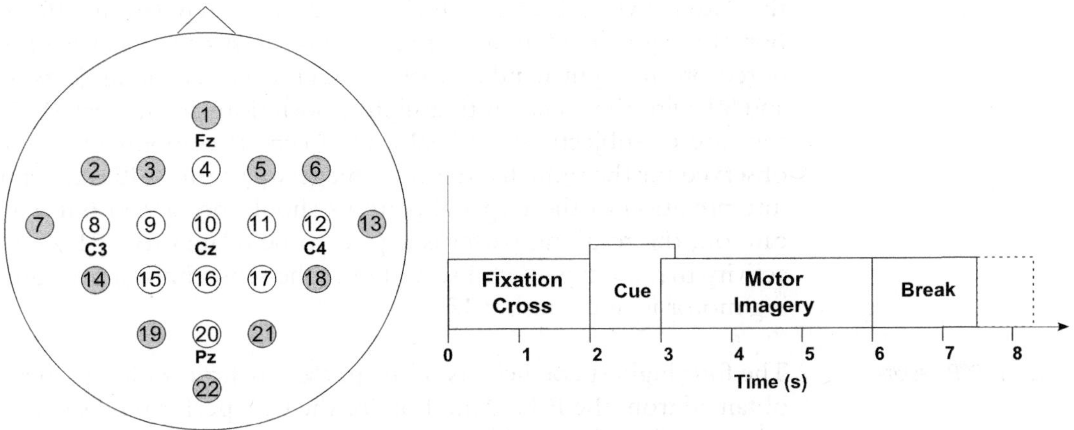

Fig. 3 The recording setup with 22 scalp electrodes. A Laplacian filter was applied to reduce volume conduction and reference effects. Boundary electrodes (*shaded*) were therefore excluded and the ten remaining electrodes were used for further processing. (Diagrams adapted from [33])

A spatial Laplacian filter was applied to the EEG data in order to reduce reference signal [34, 35] and volume conduction [36] effects, both of which can lead to spurious phase synchronization levels and misleading interpretations of results. Due to the Laplacian filtering, edge electrodes were subsequently ignored and signals from the ten remaining electrodes were considered for further processing.

An FIR least-squares filter of order 64, with forward and reverse filtering (to avoid phase distortion), was then used to band-pass filter the signals to the frequency range of 8–28 Hz. The methods tested in this study were applied to the broadband filtered EEG signals and subsequently repeated on data filtered at sub-bands of 4 Hz each. The broadband data gave the best discrimination performance in terms of classification accuracy and therefore only results from the broadband EEG data are considered here. From each trial a 2 s interval starting 0.5 s after cue onset (refer to Fig. 3) was considered for further analysis using the CSP and P-CSP methods. This interval was selected based on the best-performing algorithm in BCI Competition IV [33] for this dataset. For the P-CSP method, a 0.5 s moving window was used to compute the S-PLV signals.

3.1.1 CSP Results

Figure 4 shows the extreme spatial patterns, a_1 and a_N, obtained from the CSP method, corresponding to the left- and right-hand motor imagery tasks for the three best-performing subjects (S3, S8, and S9).

A higher weighting is observed over the C3 area for the left-hand motor imagery task for Subject S3. For the other two subjects a higher weighting is observed over Cz with a lower weighting exhibited on the posterior right-hand side of the scalp. The increasing amplitude over the right posterior areas of the scalp maps for the latter two subjects is mainly due to extrapolation effects. For the right-hand motor imagery task a strong coefficient is noted on the right-hand side of the scalp over C4 for subjects S3 and S9, with the focus shifting slightly posteriorly for subject S8. In the case of subjects S8 and S9 a mild central component is also observed for the right-hand motor imagery patterns. Although any interpretation of these spatial patterns should be carried out with caution, the resulting patterns appear to be related to ERD/ERS activity that are typically observed over the sensorimotor area during motor imagery tasks [37].

3.1.2 P-CSP Results

The four highest coefficients of the phase synchronization patterns obtained from the P-CSP method for the best-performing subjects (also S3, S8, and S9 in this case) are shown in Fig. 5. The computed phase synchronization patterns associated with the most discriminative components exhibit synchronization links that are markedly consistent across the subjects. For left-hand motor imagery data,

Spatial Patterns

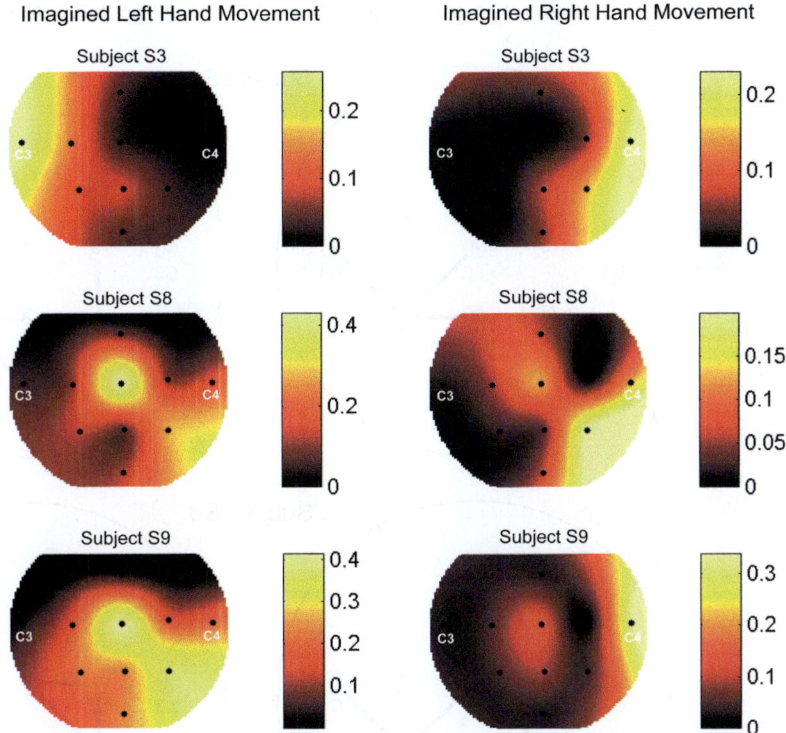

Fig. 4 Spatial patterns obtained from the CSP method for the three best-performing subjects, namely, S3, S8, and S9. The absolute values for the spatial patterns are displayed. The patterns on the left relate to the most significant activity related to imagined left-hand movement, while the ones on the right are those associated with imagined right-hand movements

strong interactions are obtained for channel pairs 4–9, 6–9, and 8–10 across most subjects. These prominent links are in fact present in the highest weighted phase synchronization links for the three best-performing subjects, shown in Fig. 5. Similarly, for right-hand motor imagery, consistent phase synchronization patterns were again determined across most subjects with strong links showing up repeatedly between channels 3–7 and 8–10 in the resulting patterns. The main differentiating characteristic between the two classes thus appears to consist of the contralateral synchronization links in the phase synchronization patterns corresponding to each motor imagery task. These links are consistent with a discriminative contralateral phase synchronization during the considered motor imagery tasks.

The P-CSP method provides an efficient way to analyze the PLV between all channels and obtain a maximally discriminant transformation allowing for the reduction of the dimensionality of the data. It is of particular interest as well that the P-CSP method

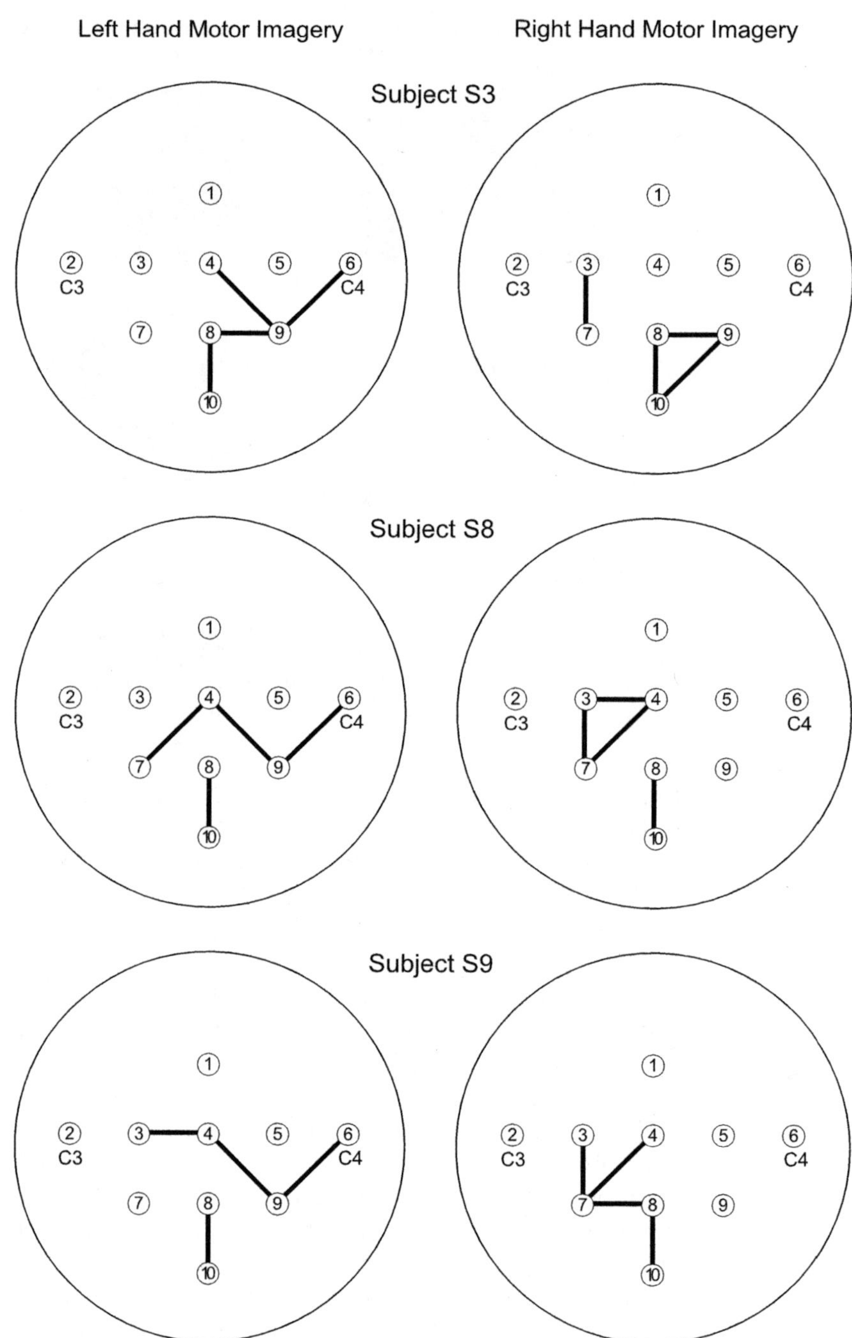

Fig. 5 The phase synchronization patterns obtained from the P-CSP method for the three best-performing subjects, namely, S3, S8, and S9. For clarity the figures above only show the links related to the channel pairs associated with the four highest phase synchronization pattern coefficients. A number of links related to each class of data show up consistently across subjects indicating the likeliness of common underlying synchronization processes taking place across the subjects

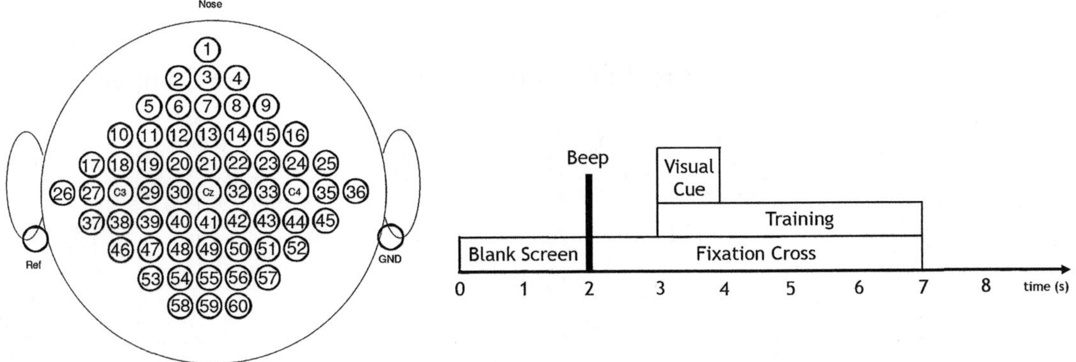

Fig. 6 The positions of the scalp EEG electrodes and the recording protocol for the motor imagery experiment. The diagrams are adapted from the description notes of Dataset IIIa from BCI Competition 2005 [38]

provides a graph of phase-lock-related measures that are distinct for each class of data. The method thus complements the information obtained from the CSP method by providing phase-lock-derived feature graphs. However, the P-CSP method does not provide any information regarding the actual phase differences between channels. Conversely, the ACSP method was designed to extend the CSP directly into the complex domain, thus obtaining separate magnitude and phase spatial maps.

3.2 ACSP Scalp Patterns for Real EEG Data

The ACSP method was applied to real EEG data in order to discriminate between left-hand and right-hand motor imagery data. The spatial patterns that result from the ACSP and CSP method are analyzed and compared.

3.2.1 Methodology

The ACSP algorithm was tested on motor imagery EEG data obtained from BCI Competition III, Dataset IIIa [38]. Data from three subjects, L1, K3, and K6, performing imagined left-hand and imagined right-hand movements were considered. The EEG signals were sampled at 250 Hz and obtained from 60 electrodes according to the recording protocol shown in Fig. 6 [38]. In our work the data were bandpass filtered from 8 to 30 Hz for analysis purposes. Further details on the recording protocol and preprocessing steps used in this study are provided in [11].

3.2.2 Results and Discussion

Considering the magnitude spatial patterns from the CSP method, and the separate magnitude and phase spatial patterns obtained using the ACSP method for Subject K3 shown in Fig. 7, it is evident that the spatial patterns from the CSP method and the amplitude component for the complex-valued ACSP spatial patterns exhibit similar activity. For both methods, the patterns associated with left-hand motor imagery indicate an increased activity around C3, while the patterns associated with right-hand motor imagery indicate

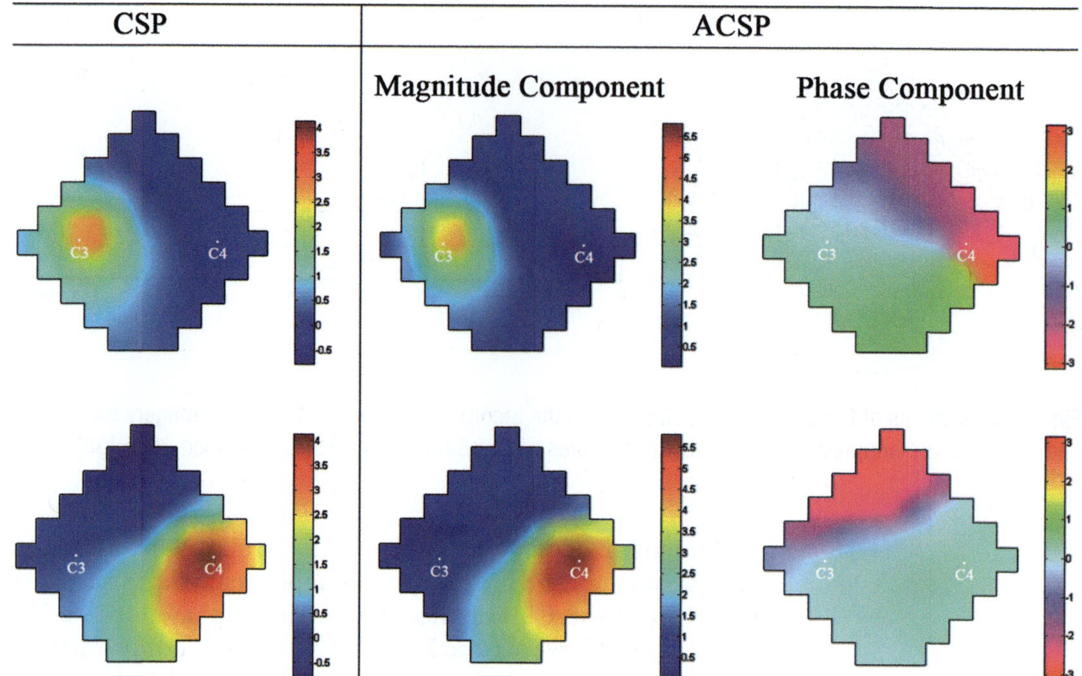

Fig. 7 The spatial patterns associated with the most discriminative components obtained from the CSP and ACSP methods for Subject K3. The top patterns associated with the left-hand imagery task, while the bottom patterns are associated with the right-hand motor imagery task. The angles for the phase components are in radians

increased activity around C4. These spatial patterns agree with the results obtained in similar motor imagery experiments reported in the literature involving the application of the CSP method on motor imagery data [28, 30]. The magnitude patterns obtained in fact indicate a relative increase in EEG variance over the ipsilateral hemisphere due to event-related desynchronization (ERD) on the contralateral hemisphere.

In addition to the magnitude patterns, phase patterns have also been extracted from the complex-valued spatial patterns obtained from the ACSP method. The phase components that result indicate abrupt phase variations of approximately 180° spanning from the contralateral sensorimotor region to the frontal region possibly related to the event-related desynchronization taking place in the contralateral area. In contrast, the ipsilateral sensorimotor areas exhibited much smoother phase variations possibly reflecting the lack of desynchronization in this region. Any interpretation of these phase maps should, however, be treated with caution considering that they were obtained from broadband data and may therefore be influenced by a number of different frequency components. To our knowledge no studies have taken into account such phase maps in the context of motor imagery; most studies considering phase in motor imagery were concerned with phase synchronization between channels [22, 24] rather than the phases of the signals themselves. Studies on signal phase have, however, been reported regarding steady state visual evoked potentials (SSVEPs) in [8, 36] where travelling and standing wave patterns were observed. When the ACSP method was applied to SSVEP data, phase patterns exhibiting posterior to anterior phase changes consistent with the phase changes reported in [8, 36] were obtained. Further details on this application of ACSP may be found in [12, 13].

The ACSP method therefore provides phase difference spatial maps which are complementary to the P-CSP in terms of its class phase discriminatory information. In cases where class phase discriminatory information exists, the CSP attempts to capture this implicitly in the magnitude spatial patterns, generally confounding the patterns. The complex-valued nature of the ACSP method keeps the magnitude and phase characteristics distinct, thus avoiding this problem and providing two separate, but complementary spatial maps to represent the distinct magnitude and phase class-discriminatory patterns.

4   Concluding Remarks

In this chapter we have reviewed two recent adaptations of the CSP method that, similar to the CSP method, yield spatial maps that may provide class discriminatory characteristics. The P-CSP method consists of a heuristic approach based on PLV features that provides class-specific phase-lock-derived spatial graphs, whereas the ACSP method consists of a natural complex extension of the CSP method that provides a class-discriminatory phase map in addition to the class-discriminatory magnitude map. In a sense, these two methods complement each other by providing spatial patterns which may give class discriminatory information that is related to the channel phase differences in the case of the ACSP method and to the degree of phase-lock between channels in the case of the P-CSP method.

References

1. Rodriguez E, George N, Lachaux JP, Martinerie J, Renault B, Varela FJ (1999) Perception's shadow: long-distance synchronization of human brain activity. Nature 397 (6718):430–433

2. Busch NA, Dubois J, VanRullen R (2009) The phase of ongoing EEG oscillations predicts visual perception. J Neuroscience 29 (24):7869–7876

3. Busch NA, VanRullen R (2010) Spontaneous EEG oscillations reveal periodic sampling of visual attention. Proc Natl Acad Sci 107 (37):16048–16053

4. Bhattacharya J (2001) Reduced degree of long-range phase synchrony in pathological human brain. Acta Neurobiol Exp 61(4): 309–318

5. Mormann F (2003) Synchronization phenomena in the human epileptic brain. Ph.D. dissertation, University of Bonn, March 2003

6. Gysels E, Le Van Quyen M, Martinerie J, Boon P, Vonck K, Lemahieu I, Van De Walle R

(2002) Long-term evaluation of synchronization between scalp EEG signals in partial epilepsy. In: Proceedings of the 9th international conference on neural information processing, 2002. ICONIP '02, vol 3, pp 1495–1498

7. Ito J, Nikolaev AR, Van Leeuwen C (2005) Spatial and temporal structure of phase synchronization of spontaneous alpha EEG activity. Biol Cybern 92(1):54–60

8. Burkitt G, Silberstein R, Cadusch P, Wood A (2000) Steady-state visual evoked potentials and travelling waves. Clin Neurophysiol 111 (2):246–258

9. Falzon O, Camilleri KP (2009) An algorithm for brain computer interfacing based on phase synchronization spatial patterns. In: Third international conference on advanced engineering computing and applications in sciences, ADVCOMP '09, October 2009, pp 142–147

10. Falzon O, Camilleri KP (2012) Phase synchronization features and common spatial patterns for the classification of motor imagery data. In: The Ninth IASTED international conference on biomedical engineering, BIOMED 2012, 2012

11. Falzon O, Camilleri KP, Muscat J (2010) Complex-valued spatial filters for task discrimination. In: Proceedings of the 2010 annual international conference of the IEEE engineering in medicine and biology society (EMBS), September 2010, pp 4707–4710

12. Falzon O, Camilleri KP, Muscat J (2012b) The analytic common spatial patterns method for EEG-based BCI data. J Neural Eng 9 (4):045009

13. Falzon O, Camilleri KP, Muscat J (2012c) Complex-valued spatial filters for SSVEP-based BCIs with phase coding. IEEE Trans Biomed Eng 59(9):2486–2495

14. Gold B, Oppenheim AV, Rader CM (1969) Theory and implementation of the discrete Hilbert transform. In: Symposium on computer processing in communications, Polytechnic Institute of Brooklyn, 1969, pp 235–250

15. Cohen L (1995) Time-frequency analysis, Chap 2. Prentice Hall, New Jersey, p 30–36

16. Pockett S, Bold GE, Freeman WJ (2009) EEG synchrony during a perceptual-cognitive task: Widespread phase synchrony at all frequencies. Clin Neurophysiol 120(4):695–708

17. Krusienski DJ, McFarland DJ, Wolpaw JR (2012) Value of amplitude, phase, and coherence features for a sensorimotor rhythm-based brain-computer interface. Brain Res Bull 87:130–134

18. Wei Q, Wang Y, Gao X, Gao S (2007) Amplitude and phase coupling measures for feature extraction in an EEG-based brain-computer interface. J Neural Eng 4(2):120–129

19. Townsend G, Graimann B, Pfurtscheller G (2006) A comparison of common spatial patterns with complex band power features in a four-class BCI experiment. IEEE Trans Biomed Eng 53(4):642–651

20. Krusienski DJ, Grosse-Wentrup M, Galán F, Coyle D, Miller KJ, Forney E, Anderson CW (2011) Critical issues in state-of-the-art brain-computer interface signal processing. J Neural Eng 8(2):025002

21. Bashashati A, Fatourechi M, Ward RK, Birch GE (2007) A survey of signal processing algorithms in brain-computer interfaces based on electrical brain signals. J Neural Eng 4(2): R32–R57

22. Wang Y, Hong B, Gao X, Gao S (2006) Phase synchrony measurement in motor cortex for classifying single-trial EEG during motor imagery. In: Proceedings of the 28th IEEE EMBS annual international conference, 2006, pp 75–78

23. Besserve M, Jerbi K, Laurent F, Baillet S, Martinerie J, Garnero L (2007) Classification methods for ongoing EEG and MEG signals. Biological Res 40:415–437

24. Gysels E, Celka P (2004) Phase synchronization for the recognition of mental tasks in a brain-computer interface. IEEE Trans Neural Syst Rehabil Eng 12(4):406–415

25. Brunner C, Scherer R, Graimann B, Supp G, Pfurtscheller G (2006) Online control of a brain-computer interface using phase synchronization. IEEE Trans Biomed Eng 53 (12):2501–2506

26. Lachaux JP, Rodriguez E, Martinerie J, Varela FJ (1999) Measuring phase synchrony in brain signals. Hum Brain Mapp 8(4):194–208

27. Lachaux J-P, Rodriguez E, Quyen MLV, Lutz A, Martinerie J, Varela FJ (1999) Studying single-trials of phase synchronous activity in the brain. Int J Bifurcat Chaos 10 (10):2429–2439

28. Ramoser H, Müller-Gerking J, Pfurtscheller G (2000) Optimal spatial filtering of single trial EEG during imagined hand movement. IEEE Trans Biomed Eng 8(4):441–446

29. Müller-Gerking J, Pfurtscheller G, Flyvbjerg H (1999) Designing optimal spatial filters for single-trial EEG classification in a movement task. Clin Neurophysiol 110(5):787–798

30. Blankertz B, Tomioka R, Lemm S, Kawanabe M, Müller K-R (2008) Optimizing spatial

filters for robust EEG single-trial analysis. IEEE Signal Process Mag 25(1):41–56

31. Lotte F, Guan C (2011) Regularizing common spatial patterns to improve BCI designs: Unified theory and new algorithms. IEEE Trans Biomed Eng 58(2):355–362

32. Koles ZJ, Lazar MS, Zhou SZ (1990) Spatial patterns underlying population differences in the background EEG. Brain Topogr 2:275–284

33. BCI Competition IV Data set IIa (2012) Data set provided by the institute for knowledge discovery (laboratory of brain-computer interfaces), Graz University of Technology, (Clemens Brunner, Robert Leeb, Gernot Müller-Putz, Alois Schlögl, Gert Pfurtscheller). http://www.bbci.de/competition/iv/desc_2a.pdf Last accessed: June 4, 2012

34. Guevara R, Velazquez JL, Nenadovic V, Wennberg R, Senjanovic G, Dominguez LG (2005) Phase synchronization measurements using electroencephalographic recordings: what can we really say about neuronal synchrony? Neuroinformatics 3(4):301–314

35. Schiff SJ (2005) Dangerous phase. Neuroinformatics 3(4):301–314

36. Nunez PL, Srinivasan R (2006) Electric fields of the brain. The neurophysics of EEG, 2nd ed. Oxford University Press, Oxford

37. Pfurtscheller G, Lopes Da Silva FH (2011) EEG event-related desynchronization (ERD) and event-related synchronization (ERS). In: Niedermeyer's electroencephalography: Basic principles, clinical applications, and related fields, 6th edn, Chap 45. Lippincott Williams & Wilkins, Philadelphia, USA

38. Pfurtscheller G, Schlögl A BCI Competition III, Data set IIIa (2012) Data set provided by the laboratory of brain-computer interfaces, Graz University of Technology.http://www.bbci.de/competition/iii/desc_IIIa.pdf Last accessed: June 4, 2012

Neuromethods (2015) 91: 267–289
DOI 10.1007/7657_2014_72
© Springer Science+Business Media New York 2014
Published online: 17 May 2014

Estimation of Regional Activation Maps and Interdependencies from Minimum Norm Estimates of Magnetoencephalography (MEG) Data

Abdou Mousas, Panagiotis G. Simos, Roozbeh Rezaie, and Andrew C. Papanicolaou

Abstract

Here, we describe the development of a minimally supervised pipeline for the analysis of event-related magnetoencephalography (MEG) recordings that preserves the temporal resolution of the data and enables estimation of patterns of regional interdependencies between activated regions. The method was applied to MEG recordings obtained from six right-handed young adults performing an overt naming task using a whole-head system with 248 axial magnetometers. Minimum norm estimates of distributed source currents were calculated in Brainstorm during the first 400 ms post-stimulus onset. A spatiotemporal source-clustering algorithm was applied to identify extended regions of significant activation as compared to the prestimulus baseline for each participant. Finally, regional interdependencies were estimated through cross-lag correlation analysis between time-series representing the time course of activity within each cluster. Consistently across participants activation loci were found in primary and association visual cortices, the fusiform and lingual gyri, the posterior portion of the superior and middle temporal gyri (MTG), the anterior, middle-inferior temporal lobe (ITG), and the inferior (IFG) and middle frontal gyri (MFG). Regions where dynamic activation patterns appeared to be "affected" by prior activation in primary/association visual cortices and the fusiform gyrus were found in the MTG/anterior ITG, and MFG. Immediately prior to articulation, activations were found in IFG, ITG, fusiform and supramarginal gyri.

Keywords Magnetoencephalography (MEG), Inverse problem, Spatiotemporal clustering, Naming, Language, Connectivity

---

1   Introduction

Object naming involves visual recognition as well as mnemonic and linguistic operations. Constituent operations include visual processing of the pictorial stimuli (or real three-dimensional objects), activation of stored modal (i.e., visual, auditory, somatesthetic, olfactory) or amodal ("conceptual" or verbal) representations associated with related episodic or semantic memories, activation of a phonological word form associated with the object's name, and construction and/or activation of an articulatory plan for vocalization of the name. Frequently, more than one such competing representations may be coactivated, requiring further evaluation and response selection. In the classic Broca–Wernicke model, as reiterated by Geschwind [1], these operations take place mainly in the following brain regions: Posterior portion of the superior temporal gyrus (STG) extending into the supramarginal gyrus (Wernicke's area), and the posterior portion of the inferior frontal gyrus (Broca's area), with contributions from the angular gyrus and the posterior portion of the inferior temporal gyrus (Brodmann's area 37). Large-scale studies of the effects of acute or progressive lesions and functional brain imaging studies in the intact brain have pointed to additional cortical regions that play an important role for object recognition and naming. These regions include the anterior temporal lobe, likely in both hemispheres, and prefrontal areas, such as the anterior/orbital portion of the inferior frontal gyrus and the middle frontal gyrus (e.g., [2]). Largely the outline of brain regions shown to be associated with naming deficits has been corroborated by hemodynamic studies [3, 4]. In view of the significant variability between studies in the overall profile of activation associated with simple naming tasks we aggregated data from 21 published studies (13 PET and 8 fMRI), describing a total of 71 experiments and reporting data from 302 subjects using BrainMap (www.BrainMap. org), a functional imaging database that currently includes 2,238 papers describing 10,646 experiments, carried out on 42,660 subjects. Using this method activation sites were identified in the following areas: Lateral occipital cortex (primarily middle occipital gyrus), left fusiform gyrus, left and right inferior temporal gyrus, right middle temporal gyrus, and left inferior and middle frontal gyri (BA 45 and 47).

To a large extent, delineation of the brain circuits underlying naming ability using hemodynamic imaging methods has relied on contrast-based statistical approaches, focusing on regions of activation and/or deactivation compared to a control task. However, adequate description of a given brain circuit (or network) requires identification of all key regions which are indispensable for the function in question, as well as of the manner in which these areas interact. Recent developments in white matter mapping techniques in the intact brain based on Diffusion Weighted Imaging (DTI) MRI sequences have been very helpful in delineating white matter tracts that support the purported functional network for basic language functions, including naming. Key white matter pathways include the arcuate, the fronto-occipital, middle longitudinal, and uncinate fasciculi. The arcuate fasciculus carries fibers directly connecting the posterior superior temporal gyrus with the inferior frontal region. An indirect route connecting the two key language regions through the inferior parietal lobule has also been established [5]. Direct connections between the two temporoparietal regions are also served by fibers running along the middle longitudinal fasciculus [6]. Further, signaling between occipitotemporal, anterior temporal, and inferior frontal regions is served primarily by the inferior longitudinal and uncinate fasciculi, respectively [7]. Occipitotemporal (lingual, middle and inferior occipital gyri) and inferior frontal regions are also directly connected through the ventral layer of the inferior fronto-occipital fasciculus (IFOF; [8, 9]). The crucial role of the inferior longitudinal fasciculus [10, 11] and the arcuate fasciculus [12] for object naming is supported by electrocortical stimulation data. Moreover, direct electrocortical stimulation of the ventral layer of the IFOF has been shown to induce difficulties in making nonverbal semantic judgments [13].

Methodological developments in the analysis of fMRI time-series have established various approaches for quantifying putative neural networks for naming, generally concurring with theoretical models emphasizing distinct posterior and anterior components. This can be accomplished by studying *functional connectivity*, in the form of patterns of associations between BOLD time series at two or more regions of interest (ROIs). In its simplest form cross-correlation indices are computed at zero lag, given the poor temporal resolution of the BOLD technique—as the signal evolves over several seconds during and after the target cognitive operations materialize. Principal Components Analysis can be applied to the variance–covariance matrix in order to extract patterns of regional interdependencies without making a priori assumptions regarding the outline of the brain network involved. For instance, Lu et al. [14] extracted two independent components summarizing regional BOLD interdependencies during object naming. The first component accounted for a pattern of covariation between bilateral superior temporal gyri, thalamus, and cerebellum, left inferior, middle, and superior frontal gyri, left motor cortex, superior parietal lobule, cingulate gyrus, and putamen. The second component accounted for correlated activity in bilateral angular, inferior and middle frontal, superior and middle temporal gyri, motor cortex, and cingulate.

Alternatively, hypothesis-driven path models can be developed and tested through structural equation modeling (SEM) algorithms on the group-average variance–covariance matrix of time points (BOLD time-series) and ROIs (e.g., [15, 16]). For instance, Zou et al. [17] reported significant positive correlations in bold signal intensity between the middle temporal, superior temporal, supramarginal and inferior occipital gyri, both within as well as between hemispheres, in healthy monolingual adults, during a standard naming task. While these approaches can, at least in principle, take into account temporal correlations in regional activity, in practice they are limited in that the directionality of connection paths is largely indeterminate.

Further insight into the dynamics of neural networks underlying object recognition and naming has been obtained using estimates of *effective connectivity*, which model networks based on statistical inferences made regarding the directional influence of one cortical region over another [18–20]. In their simpler forms, inferences regarding causal influences of one cortical region (typically a predetermined ROI) onto another employ autoregressive algorithms such as Directed Partial Correlation (e.g., [21]) within the context of Granger Causality (e.g., [22]). Effective connectivity algorithms, such as Dynamic Causal Modeling (DCM) aspire to model temporal, regional interdependencies in neural signaling by explicitly incorporating transfer functions linking these signals and the BOLD response [23]. For example, using fMRI and DCM, Abel et al. [24] identified two regions where the degree of activation was associated with naming performance and provided evidence of a predominantly feed-forward "connection" from fusiform to inferior frontal gyri (bilaterally).

Although group-averaged profiles of naming-related hemodynamic activation have been described by several studies, results of connectivity analyses have been quite variable. In part, inconsistencies across studies are expected due to reliance on radically different approaches (ranging from simple bivariate correlations to task-specific dynamic causal models). Such inconsistencies may also be attributed to the nature of the hemodynamic response which can, at best, provide an imperfect approximation to the true temporal dynamics of neurophysiological activity. On the other hand, the temporal resolution of magnetoencephalography (MEG) is practically perfect and the problem of estimating the true dynamics of the neural response (i.e., neuronal signaling) from the shape of the hemodynamic response (see [23]) is circumvented. The magnetic flux recorded in MEG varies continuously over time, coincident with the rise and evolution of the intracranial currents that produce the magnetic signals. In the past decades, a number of efforts have been made to solve the so-called "inverse problem" of MEG, substantially improving the spatial resolution of MEG-derived activation maps. These efforts utilized different computational algorithms, including the minimum norm estimation (MNE; [25]). These source-current estimation models assume a continuous distribution of current along the cortical surface which is modelled as a triangular mesh consisting of several thousand *vertices* (typically ~150,000; [26]). The source-space is subsequently reduced by down-sampling for computation purposes to approximately 3,000–10,000 vertices per hemisphere. Each vertex represents the location of a potential current dipole perpendicular to the cortical surface during the forward calculations.

Recent developments in the analysis of MEG time series have demonstrated the utility of this imaging method in studying brain networks, particularly those associated with reading [27, 28]. Specifically, Kujala and colleagues [29] applied a voxel-wise coherence analysis to MEG time series (Dynamic Imaging of Coherence Sources; DICS) reporting a densely connected network of regions, obtained in the context of a rapid serial visual presentation (RSVP) paradigm, to simulate natural reading. In accord with regions implicated in previous imaging studies of reading, these authors localized long-range neural connections, dominated by feed-forward connections from the left inferior occipitotemporal cortex to the cerebellum, superior temporal, anterior inferior temporal, precentral, insular, prefrontal, and orbitofrontal cortices, in the 8–13 Hz range. More recently, Campo et al. [30] estimated effective connectivity for object naming using Dynamic Causal Modeling [31]. Source current density data were obtained with the Statistical Parametric Mapping (SPM8) software based on a cortically constrained array of 512 potential sources. Three broad regions were selected in each hemisphere as showing the most prominent sources of magnetic activity during the first 400 ms after stimulus onset: lateral occipitotemporal cortex, encroaching into the middle temporal gyrus, anterior-inferior temporal lobe, and the inferior frontal gyrus. The best-fitting connectivity model for the group of healthy participants ($N = 10$) featured separate, duplicate networks in each hemisphere, each involving all three bidirectional paths between the three regions. Along the same lines, Clarke et al. [32] examined patterns of cortical connectivity during an object naming task in ten healthy participants, among three a priori-defined regions of interest in the left hemisphere (fusiform gyrus, anterior temporal lobe, and inferior frontal cortex). Functional regional interdependencies were assessed on a measure of phase locking computed on single-trial source-current estimates. Results revealed significant phase locking in the gamma band (25–100 Hz) between the left fusiform and anterior temporal sites, the degree of which was modulated by semantic integration demands posed by the stimuli.

One limitation of source reconstruction methods such as MNE and DICS is that they produce highly distributed activation maps that represent instantaneous source-current distributions over the cortical surface, i.e., several such maps are constructed independently starting before stimulus onset and continuing up to the participant's response (in the case of naming tasks, the response is vocal). To overcome this problem, the vast majority of studies have adopted a ROI approach, whereby the source-current distribution if further reduced to a predetermined set of ROIs determined automatically using published cortical maps (e.g., [33]). By relying on anatomical rather than functional criteria for selecting potential activation sites, however, one may introduce significant measurement error considering that only a subset of the vertices belonging to a particular ROI may actually be active, in a functionally meaningful manner, during the performance of the naming task.

Therefore a method is needed that will identify activation sites associated with the processing of the experimental stimuli and preparation of the participant's response in individual data (so that the activation maps could be used clinically), while being capable of providing group-level maps capturing elements of activation that are observed with reasonable consistency across participants. Specific desirable features of such a method include the following:

(a) Identification of hypothetical modelled sources (i.e., vertices) displaying estimated current that significantly exceeds a criterion level during the post-stimulus/pre-response period. In the method developed and presented here, this criterion level is computed in single-subject data based on the *distribution* of current values during the pre-stimulus baseline. To ensure that this routine selects only those vertices that display functionally significant activity (i.e., represent neurophysiological activity that actually relates to stimulus processing), an additional criterion may be introduced, requiring that elevated current levels be present for at least 40 ms over contiguous time points. In this manner, "noise" levels at the estimated source level are taken into account in determining what constitutes significantly elevated activity time-locked to the stimulus for a particular individual and recording session.

(b) Identification of spatially contiguous vertices (regardless of anatomical map boundaries) forming *source clusters* that show temporally overlapping elevations of estimated current. This requirement rests on the assumption that each vertex represents activity from a section of the neuronal population in a particular cortical patch which is engaged in stimulus processing/response preparation (i.e., rather than an entire, anatomically defined region). If the various sections of the neuronal population represented by each vertex are engaged in the same neurophysiological process then the activation peaks exhibited by each constituent vertex should be largely overlapping in time. In addition, the entire single-vertex time series should be highly intercorrelated (at zero time lag) reflecting activation of a relatively uniform population of neurons.

(c) Projection of the derived source clusters on the participant's MRI and identification of the anatomical region(s) where constituent vertices are located, using published cortical maps. This step would permit specification of the relative frequency with which source clusters appear in particular cortical regions across participants.

(d) Calculation of the temporal characteristics of activity for each source cluster. The procedure should be capable of determining the peak current-amplitude of activity for each cluster, as well as a time-window of potentially critical activity (i.e., section of the average time-series of amplitudes across vertices). This step is essential in order to establish the temporal progression of activity across clusters and compute measures of effective (i.e., directed) connectivity between source clusters.

In this work we describe the development of a data analysis pipeline that incorporates the aforementioned features and its application on a small series ($N = 6$) of MEG data obtained from a simple object-naming task and analyzed at the source-level using MNE. The study had two goals: First to examine the concurrent validity of the method as compared to aggregate data derived from hemodynamic imaging studies and, second, to explore the viability of activation data provided by this method in functional connectivity analyses (i.e., in estimating patterns of regional interdependencies between activated sites).

---

2 Materials

2.1 Subjects

Six right-handed adults (two men, mean age: 36.4 years, range: 27–42 years) with normal or corrected-to-normal vision and normal hearing participated in this study. All participants were right-handed according to the Edinburgh Inventory with a mean handedness quotient of 90.5 ± 10.1 % and were financially compensated for their participation providing written consent. In accordance with the Declaration of Helsinki, the protocol was approved by the Institutional Review Board of the University of Tennessee Health Science Center.

2.2 Experimental Design

Each participant performed an overt object naming task, consisting of 50 living objects (25 animals, 25 fruits-vegetables-plants) and 50 inanimate objects (25 objects, 25 tools), drawn from the CRL International Picture-Naming Project (http://crl.ucsd.edu/experiments/ipnp/index.html) [34]. The 100 stimuli were randomly arranged in four blocks of 25 items each, and were presented for 1 s, one at a time (with a randomly varied interstimulus interval of 3–4 s), through a Sony LCD projector (Model VPL-PX21) on a back-projection screen located approximately 60 cm in front of the participant.

2.3 Meta-analysis of Hemodynamic Studies

Regions that showed significantly elevated hemodynamic activity associated with object naming as compared to rest across studies were determined through BrainMap (www.BrainMap.org), a functional imaging database of 2,238 published studies. By searching this database, we identified 522 foci reported in 21 papers (13 PET and 8 fMRI), 71 experiments, and 302 subjects. Using these foci, an activation likelihood estimate (ALE) map of overt object naming was performed in Talairach space using GingerALE 2.1. In order to determine only the most strongly activated regions, the resultant

ALE map was thresholded to a cluster size >2,296 mm^3 (random effects analysis; [35]) and a false discovery rate (FDR, $q = 0.05$) corrected threshold of $p < 0.003$.

2.4 MEG Measurements

Brain activity was recorded using a whole-head magnetoencephalography system with 248 axial magnetometers (WH 3600, 4D Neuroimaging, San Diego, California, USA) housed in a magnetically shielded room. Signals were digitized at a sampling rate of 508 Hz and filtered online with a 0.1 Hz high-pass filter. The head position was computed prior to and at the end of acquisition (after both tasks) using coils placed at the nasion and ear canals bilaterally.

2.4.1 MEG Analysis Methods

MEG recordings were initially low pass filtered at 30 Hz with additional notch filters at 60 and 120 Hz. Channels that were either dead (zero amplitude) or presenting unexpectedly high amplitudes were excluded from subsequent analyses. Single trial epochs were then extracted from the continuous recordings each lasting 1 s (200 ms baseline and 800 ms after stimulus onset). Before averaging, epoch data were baseline corrected. By visual inspection for myogenic and ocular artifacts, a number of trials were rejected (Note 1). For every subject a set of digitized head points was available. Using these head points, the default Brainstorm anatomy (*Colin27*) was warped for best fit. The resulted, warped anatomy along with the averaged, time-locked, recordings, were used to compute the head model (lead field—solution of the forward problem). The pre-stimulus baseline was used to estimate the noise covariance matrix. In the final step, the wMNE algorithm, with loose source orientations, was used to solve the inverse problem. The final output of this procedure is a matrix with dimensions (number of sources) × (number of time points) describing the electrical activity of each current-source through time (Note 3). Only the first 400 ms after stimulus onset were analyzed (prior to the onset of articulation-related artifacts). This matrix served as the input to the first step of the analysis pipeline (agglomerative hierarchical clustering).

2.4.2 Agglomerative Hierarchical Clustering

Initially the program considers each vertex as a potential cluster and calculates time window(s) during which current significantly exceeds a baseline-defined threshold (Note 2). Then, the program identifies the nearest neighbor vertex (Note 4), using their respective source coordinates, and if the activation windows overlap sufficiently (Note 5), the two vertices are "merged" to a single cluster (Note 6). This cluster has as spatial coordinates and activation windows the mean value of coordinates and activation windows of its constituent vertices. The algorithm terminates when no other vertices can be linked to existing clusters. Because of the greedy vertex merging rules it can be the case that when the process is completed two clusters are found in close spatial proximity.

To compensate for this side-effect, the algorithm is complemented by a post-hoc cluster-merging test that takes into account dynamic similarity between cluster time series. Thus, two spatially adjacent clusters are merged if their respective time-series are highly cross-correlated at zero lag (Note 7). Each final cluster is represented by the set of its constituent vertices, the Euclidean center of their Cartesian coordinates in SCS/CTF space, and its mean source-current time-series.

2.4.3 Cluster Labelling	Using one of the available anatomical atlases in Brainstorm (Note 8), each cluster is identified anatomically after the cortical region where the majority of its vertices are located (Note 9). As the cortical parcellations given by the atlases are quite coarse, two or more clusters can be identified within the same brain region.

Next, the frequency of occurrence of *anatomical regions* where clusters were found across subjects is tabulated to determine the map of consistently active regions ("common map"; Note 10) in a particular data set (task and group of participants). In order, however, to preserve individual differences in the temporal progression of activity within individual clusters, this common map merely serves as a guide for selecting clusters to be included in the final step of the pipeline (computing measures of interdependence between activation sites). Here we have implemented the simplest method for estimating functional interdependencies, cross-lagged bivariate correlation, as described in more detail in the following section.

2.4.4 Cross-lagged Correlation Analysis	The rationale behind the final step of the current analysis pipeline rests on the assumption that each source cluster represents the *location* of a cortical patch where systematic neuronal signaling takes place during stimulus processing/response preparation. This regional neuronal activity takes place at one or more specific latencies (or time windows) that appear as corresponding distinct peaks in the time series of each cluster. In principle, each peak may represent a particular neurophysiological process performed by the same neuronal population—either the same process that takes place repeatedly (by the same population of neurons) or, alternatively, distinct processes performed by distinct neuronal populations located in the same anatomical region. Although the latter possibility cannot be precluded on the basis of neuroimaging data alone, the spatiotemporal clustering procedure implemented here was designed to distinguish between such functionally distinct activations. The goal of the final stage of the analysis pipeline is to establish, for each participant, a sequence of regional "bouts" of activity that follow a regular temporal progression and show significant cross-correlations at specific time lags.

The first step in this final stage of the analysis is to find, for every time-series, the peak(s) of activity in the post-stimulus period

(Note 11) using the MATLAB function *findpeaks* (Note 12). In order to minimize false positives that this function may produce, the 90th percentile of the distribution of post-stimulus current values was set as threshold. A genuine peak was defined by instantaneous current values exceeding this threshold for a minimum of 20 contiguous time points (40 ms). As it is common to obtain clusters with time series containing multiple peaks, the table of clusters (one for each subject) is updated to include multiple entries for clusters with multiple peaks and sorted in ascending temporal order.

Based on their respective peak latencies, clusters are then assigned to one of three broad time-windows, to which the post-stimulus epoch is divided, representing early (typically 0–150 ms), middle and late activity (Note 13). Next, cross-correlation values are calculated between all possible cluster pairs straddling the two time-window borders (early to middle and middle to late). In order to focus on activity around each peak, which presumably reflects neurophysiological activity characteristic of a particular cluster, correlations are computed on the section of the time-series around the peak (rather than the entire post-stimulus time series). Cross-correlations are computed at several time lags while taking into account the relative temporal order of cluster pairs (Note 14). The maximum cross-correlation value for each pair and the lag at which this value was achieved is stored. Subsequently, a statistical test is applied to the maximum cross-correlation values using a model-independent Monte-Carlo method (Random Subset Selection; RSS). For two time-series with N time-points each, from the set $\{1, 2,\ldots, N\}$ of the indices, a random sample of length N is taken with replacement using the MATLAB function *randsample*. In this sample some values are repeated and when only one is kept the resulting sample has size $<N$. These time tags are sorted to preserve time ordering and two new time-series are created from the original ones (obviously, with missing values). The cross-correlation at zero-lag is then recalculated and the process is repeated thousands of times, providing a distribution of cross-correlation values. For all possible pairs of cross-correlations, p-values are calculated and only statistically significant ($\alpha = 0.05$) values are kept. Additionally, a threshold of 0.7 in the cross-correlation values is applied, maintaining only substantial inter-cluster associations. The final table of cross-correlations represents functional, directed "pathways" between activations (i.e., "early" clusters potentially "affecting" "middle-latency" clusters and "middle-latency" clusters potentially influencing activity of "late" clusters. In this tabulation each cluster is represented by the anatomical area in which it is located, allowing the data to be collapsed across participants (Note 15).

At this point the frequency of occurrence of each pathway can be computed across participants and a simple statistical test (such as the binomial test) can be applied to identify potential directed,

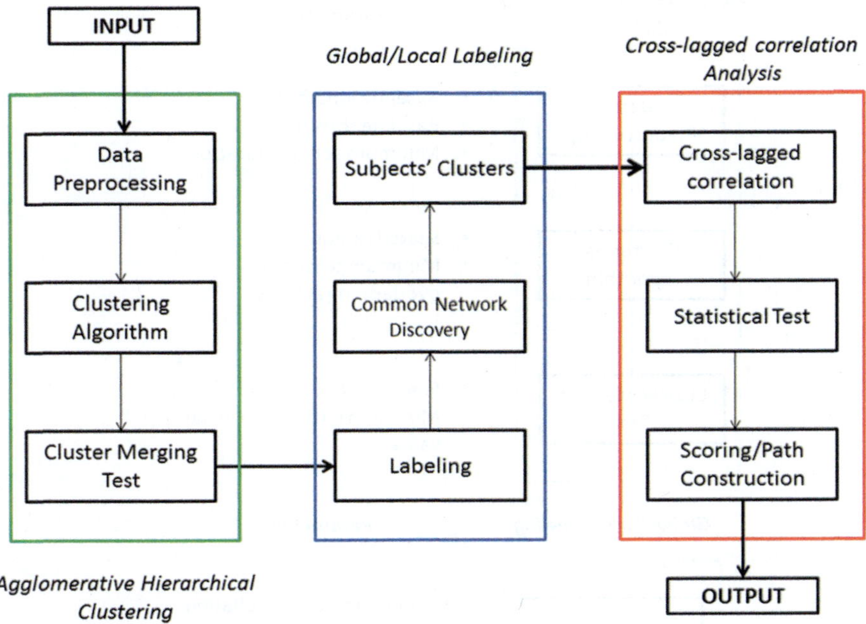

Fig. 1 Flowchart of the analysis pipeline

functional connections that appear with significant degree of consistency in the data set (Note 16). In the final stage of the pipeline, one can explore potential connectivity paths between three brain regions at a time, each containing source clusters with early, middle, and late activity peaks, respectively. The MEG analysis pipeline described above is schematically illustrated in the flowchart of Fig. 1.

2.4.5 The Influence of Parameters on Results

Each part of the analysis pipeline (clustering, labelling, connectivity inference) corresponds to a distinct and autonomous module of the method. In this respect, the influence of each parameter is confined inside the module it belongs to and the impact it has on the whole is exerted only through the output of that specific module. As illustrated in Fig. 2, the structure of the code affords significant flexibility to the user permitting relatively straightforward expansion and modification capacity. For instance, in the connectivity inference stage, instead of using the cross-correlation one could use a multitude of different measures (transfer entropy, Granger causality, DCM, coherence etc.). The replacement of the final module would not affect the previous stages of the pipeline.

In the data preprocessing stage all the parameters are involved in the definition of the activation windows of each source. Higher values will result in the exclusion of more sources and in lower number/duration of activation windows. The parameters of the clustering algorithm affect the size of the resulting clusters. Increasing the search radius and lowering the minimum temporal

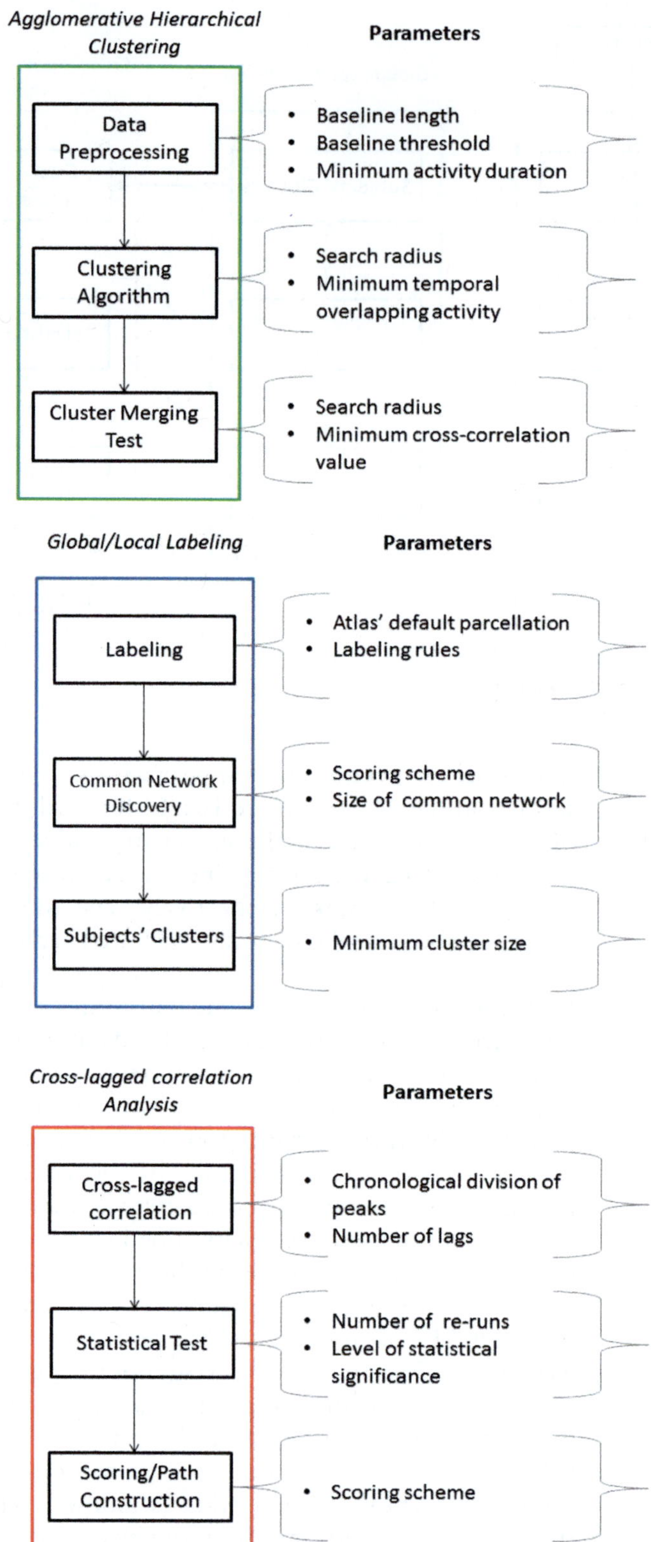

Fig. 2 List of the parameters that affect the clustering module (*upper panel*), labelling module (*middle panel*), and connectivity module (*lower panel*) of the analysis pipeline

overlap will result in bigger and more disperse clusters. The same holds for the parameters of the cluster merging test. Higher values of the search radius and lower of the minimum cross-correlation will output larger source clusters.

In the labelling phase major role plays, as expected, the cortical parcellation of the employed atlas. The naming rules affect merely the ease with which the label is given to each cluster. In the common network discovery, one can simply count the number of clusters in order to identify the most common brain regions. Alternatively, one could correct the count by dividing with the total number of clusters in every region or even introduce penalties that promote the importance of clusters in specific regions of the brain. At the end, different common networks may appear. The number of nodes in the network may also change and results in the exclusion or inclusion of more brain regions. Returning back at the subject level, the minimum cluster size parameter controls the minimum size of clusters one is willing to keep for the subsequent cross-lagged correlation analysis.

In the calculation of the cross-correlation values an obvious parameter is the number of lags on which the calculation takes place. The sections of the actual time-series selected depend on the rules that control the division of each cluster's activity-peaks into early, middle and late. In the statistical test (which is essentially a permutation method), the number of repetitions affects the precision of the results. Higher number of re-runs makes the results of test more trustworthy but also increases computation time. The statistical significance level reflects the strictness of the test. Finally, after the connectivity paths are constructed, in identifying the most commonly observable, one can just count or use different scoring schemes that take into account the relative importance of the paths inside each subject. In return, different paths will be considered more important and they will be ranked higher.

3 Results

3.1 Application of the Method and Preliminary Results

Figure 3 displays single-subject average, time-locked MEG recordings collapsed across the 248 magnetometer sensors. In Fig. 4 images in the four left hand columns display average source-current distributions in the form of Minimum Norm Estimates (MNE) at four representative time points, projected on the cortical surface of one participant. These maps, thresholded using a minimum current level that was determined for each participant on the basis of their respective baseline source-current estimates, were input to the agglomerative hierarchical clustering. This procedure identified groups of vertices that displayed spatial proximity *and* highly synchronous estimated activations. The time-series for each cluster was calculated as the mean of the time-series of all the constituent

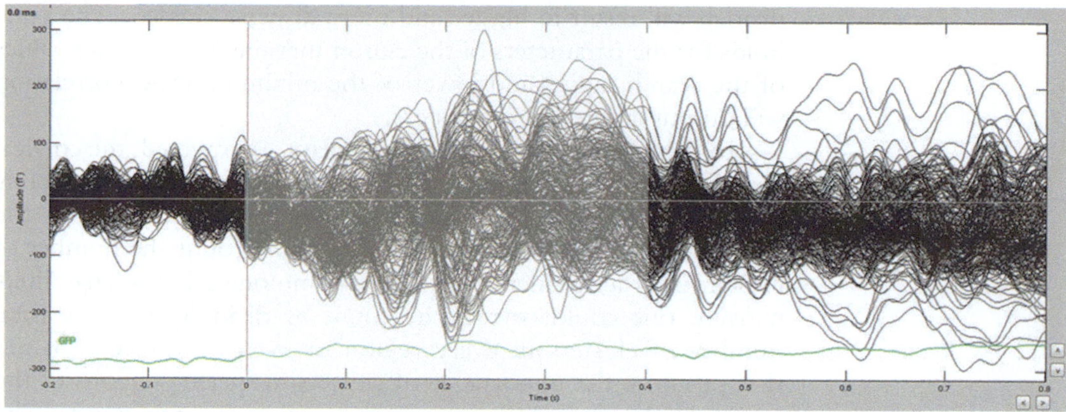

Fig. 3 Averaged MEG epochs recorded from one participant (#008) during performance of the naming task. Stimulus onset is at 0 ms. The *shaded region* indicates the time window submitted to further analyses (0–400 ms post-stimulus onset)

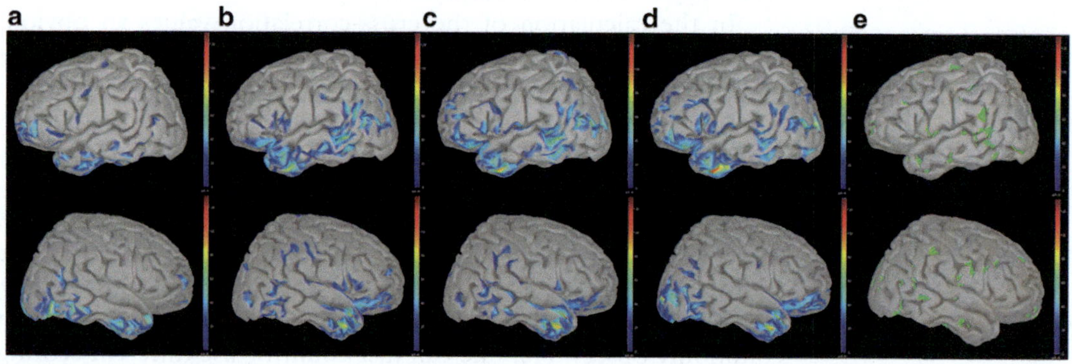

Fig. 4 Average cortical activations (across participants) derived through MNE at 100 (**a**), 150 (**b**), 230 (**c**) and 320 ms (**d**) after stimulus onset are shown on the *left* (*upper row*) and *right* hemisphere surface (*lower row*). (**e**): Anatomical locations of cortical source clusters (collapsed over time) which were identified through the hierarchical, spatiotemporal clustering procedure. Each cluster is composed of spatially contiguous current sources displaying highly inter-correlated time courses

vertices as illustrated in Fig. 5. It is important to note that clustering was performed on the individual-subject data and group-average maps are for viewing purposes only.

The locations of activity clusters that appeared consistently across subjects are shown in the composite image in the right-hand column of Fig. 4. The anatomical regions (as defined in the Tzourio-Mazoyer atlas available in Brainstorm [36]) where activity clusters were found consistently in at least 4/6 participants, regardless of the latency of significant activity characterizing each cluster, are listed in Table 1.

In the single-subject data, the program identified more than one clusters in several brain areas, while certain clusters showed

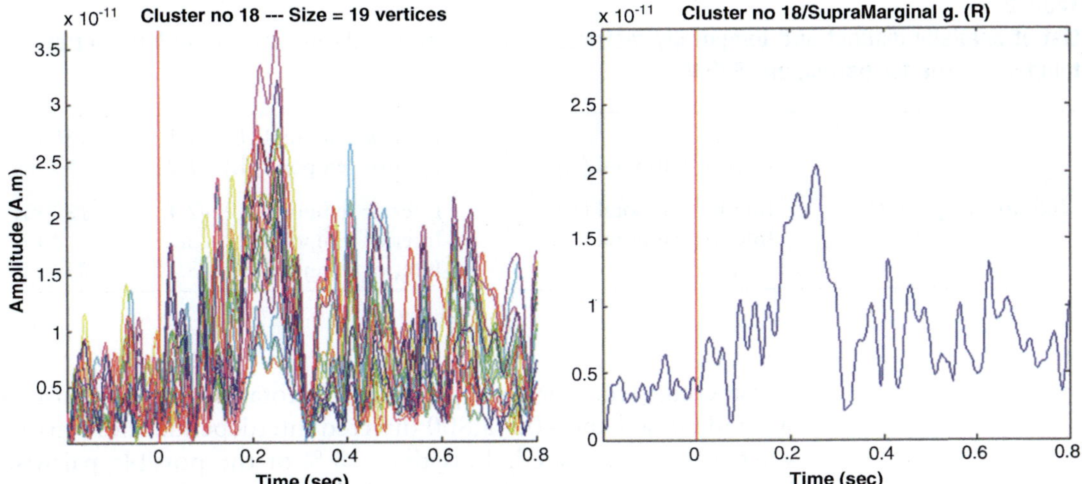

Fig. 5 Time series of estimated activity in each of the 19 vertices that formed cluster no.18 in participant S#008 located in the right supramarginal gyrus; *left hand panel*). The average time series for that cluster is shown in the *right-hand panel*

Table 1
Anatomical regions in the Tzourio-Mazoyer atlas where activity clusters were found consistently across subjects

Middle frontal (R)	Frontal inferior opercular (R)	Angular (R)	Frontal inferior triangular (L)	Middle occipital (R)
Inferior temporal (L)	Frontal inferior triangular (R)	Inferior temporal (R)	Calcarine (L)	Fusiform (R)
Middle occipital (L)	Lingual (R)	Precentral (R)	Calcarine (R)	Middle frontal (L)
Middle temporal (L)	Supra-marginal (R)	Frontal middle orbital (R)	Lingual (L)	Postcentral (R)

more than one discrete peaks of activity. In the sample participant displayed in Table 2 (S#008) three spatially distinct clusters were found within the anatomical borders of the left inferior temporal gyrus (middle column). The activity time series of two of these clusters (1 and 2) contained two peaks (at 188 and 259 ms for cluster 1 and at 130 and 314 ms for cluster 2). Accordingly, the final table of temporally restricted, spatially distinct clusters for this participant contained 57 entries (10 in the early time window, 24 in the middle time window, and 23 in the late time window).

In the current implementation of the pipeline, autocorrelations between successive time-series segments from the same cluster (indicative of self-loops) or between clusters located in the same anatomical region (indicative of feedback loops) were not

Table 2
List of spatially distinct and temporally restricted clusters of activity located in the left inferior temporal gyrus for participant S#008

		Inferior temporal (L)—1.1	188 ms
	Inferior temporal (L)—1	Inferior temporal (L)—1.2	259 ms
Inferior temporal (L)	Inferior temporal (L)—2	Inferior temporal (L)—2.1	130 ms
	Inferior temporal (L)—3	Inferior temporal (L)—2.2	314 ms
		Inferior temporal (L)—3.1	354 ms

considered. Thus for each participant a total of 47–106 clusters were identified and 400–2,800 directed interdependencies ("pathways") were calculated. Less than 15 % of the possible pairwise interdependencies met the statistical test (at $p < .05$) and exceeded the $r = .70$ threshold. The consistency with which these potential functional pathways between clusters appeared across subjects was subsequently assessed by representing each cluster by their respective anatomical area.

The final output of the algorithm returns the connectivity pathways in Fig. 6, which represents a hypothetical network of brain regions each containing at least one cortical patch that showed strong, directed, and reliable interdependence with at least one other brain region in the first 400 ms following presentation of a picture to be named. The actual time-series of clusters that participate in this hypothetical network in a given participant can then be visualized. For instance, Fig. 7 shows the entire time series of the three clusters of activity that contributed to the hypothetical pathway linking the left lateral occipital region, the right lingual gyrus, and the left inferior temporal gyrus in participant S#009.

4   Discussion

4.1 Discussion of Preliminary Results and Future Directions

By using a method that does not require a priori assumptions regarding the spatiotemporal profiles of activation in individual participants, activation loci associated with object naming were found in the following areas: medial and lateral occipital cortex, fusiform and lingual gyri, the posterior portion of the superior and middle temporal gyri (MTG), the anterior, middle-inferior temporal lobe (ITG), motor cortex, and the inferior (IFG) and middle frontal gyri (MFG). Meta-analytic data revealed hemodynamic activation foci in the majority of these regions. One notable exception was the anterior ITG where several fMRI studies have failed to find significant activation (e.g., [37–39]). This may have been due to increased susceptibility of fMRI signals to distortion by neighboring air-filled cavities. Conversely, activation sites in the anterior

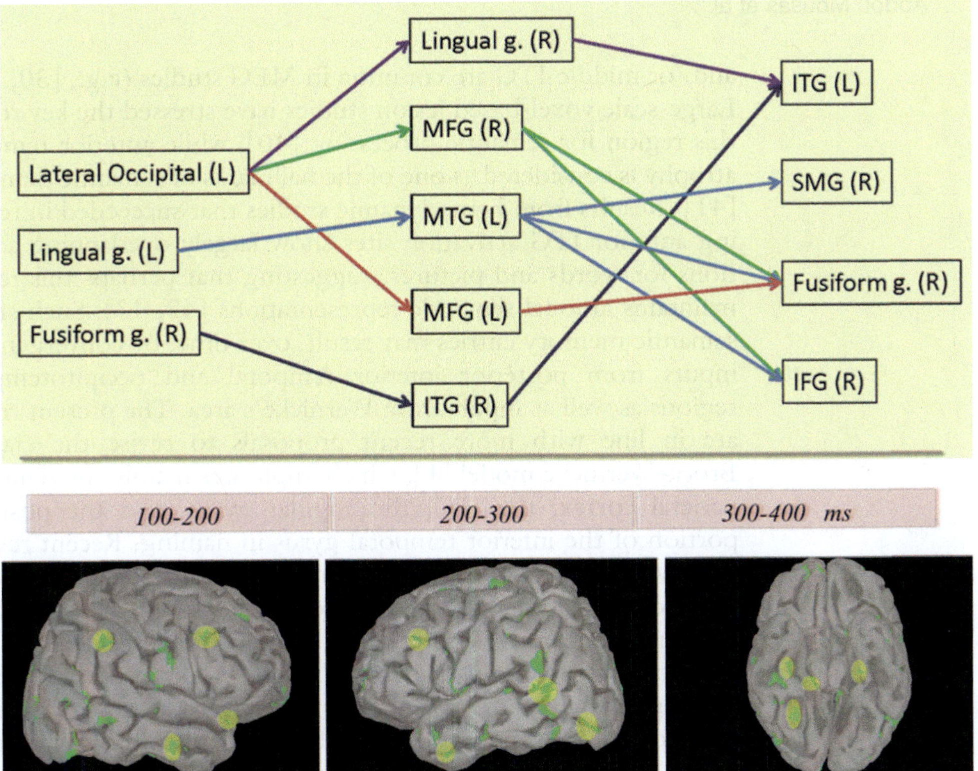

Fig. 6 *Upper panel.* Significant associations between regional activation time courses in single-subject analyses which were found consistently in the majority of participants (≥4/6 cases). Each box represents a single source-cluster peaking during the time window shown at the bottom of the panel. *Lower panel.* Clusters demonstrating significant interdependencies in pairwise cross-lag analyses are marked with oval yellow shapes. *g* gyrus, *L* left hemisphere, *R* right hemisphere, *MFG* middle frontal gyrus, *MTG* middle temporal gyrus, *ITG* inferior temporal gyrus, *SMG* supramarginal gyrus, *IFG* inferior frontal gyrus

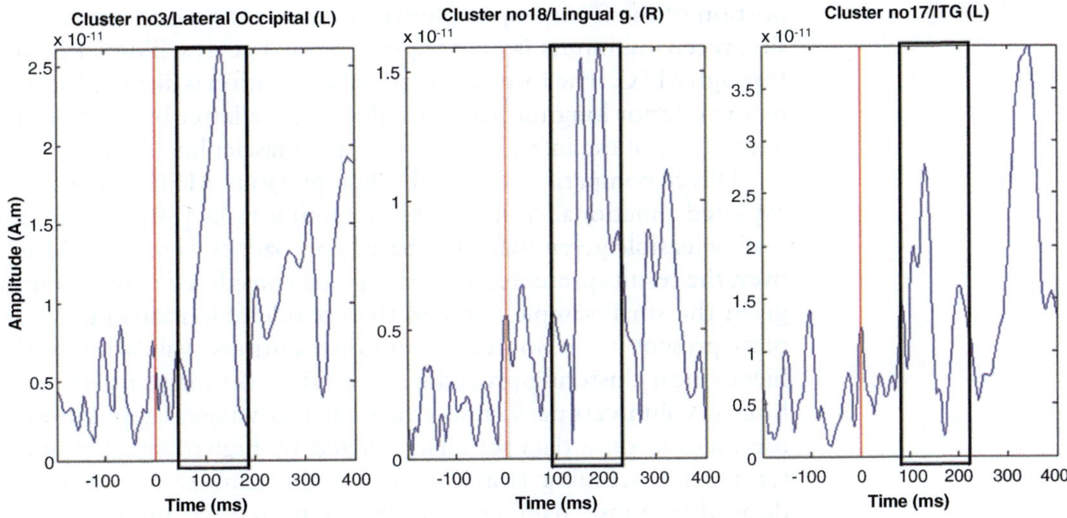

Fig. 7 Time-series for three clusters forming a hypothetical path (left Lateral Occipital region → right Lingual gyrus ($r = .77$ at 8 ms time lag) → left inferior temporal gyrus (ITG; $r = .72$ at 5 ms time lag) in participant S#009. *Boxes* indicate the sections of each time series on which cross-correlations were computed

and/or middle ITG are common in MEG studies (e.g., [30, 38]). Large-scale voxel-based lesion studies have stressed the key role of this region for semantic processing [40], while anterior temporal atrophy is considered as one of the hallmarks of semantic dementia [41]. Results from hemodynamic studies that succeeded in revealing anterior ITG activation sites show largely overlapping activations for words and pictures, suggesting that perhaps this region maintains amodal semantic representations [42, 43]. Such stored semantic memory entries may result, over time, by convergence of inputs from posterior inferior temporal and occipitotemporal regions as well as inputs from Wernicke's area. The present results are in line with more recent proposals to revise the classical Broca–Wernicke model [1], which implicates mainly the temporo-parietal cortex, the IFG, the angular gyrus, and the posterior portion of the inferior temporal gyrus in naming. Recent reviews and meta-analyses call for further revisions to this model by stressing additional roles of the temporoparietal and inferior frontal areas in the executive control over semantic processing [4].

Connectivity analyses identified a subset of these areas demonstrating reliable ($p < .05$) and substantial ($r > .70$) pair-wise interdependencies at time lags between 5 and 80 ms. Although both the extent (cortical surface) and peak source-current were greater in the left hemisphere, several right hemisphere regions were also identified in the connectivity analyses. Regions where dynamic activation patterns appeared to be "affected" by prior activation in primary and association visual cortices (Lateral Occipital, Lingual and Fusiform gyri in Fig. 6) were located in the MTG, anterior ITG, and MFG. The left anterior ITG emerged as one of the sites of converging late activation receiving input primarily from occipitotemporal (lingual gyrus) and right anterior ITG. The IFG (located in the orbital portion of the frontal operculum) emerged as a second converging site, receiving inputs from occipitotemporal cortex (lingual gyrus) through MTG. The former functional association is likely subserved by the inferior longitudinal fasciculus and the latter by fibers in the deep layers of the inferior fronto-occipital fasciculus [44].

Direct comparisons with the few previous MEG studies that explored functional connectivity in naming tasks [30, 32] are not readily feasible given that ROI-based approaches were used. Moreover, the results presented here should be considered as preliminary given the small sample size and the considerable individual variability present in the source-current distributions. Validation of the hierarchical clustering procedure is required on a larger data set of healthy volunteers performing the same task whereas application of the method on a data set characterized by higher signal-to-noise ratio is forthcoming (for instance, in the context of a modified delayed-response naming task design to reduce myogenic and motion artifacts associated with articulation). More importantly, validation of the functional importance of specific source clusters in individual patients requires direct comparison of cluster location with sites associated with naming deficits in electrocortical stimulation studies, much in the same way that consistently active sites during aural and visual word recognition tasks were validated by our group [45, 46].

5 Notes

1. Across participants the number of artifact-free epochs that were input to MNE varied between 43 and 67.

2. In the case of time-locked data, the baseline coincides with the pre-stimulus period. In our case, we used as a global threshold the 75th-percentile of the pre-stimulus, data distribution. In the case of continuous recordings, any threshold that represents baseline brain activity can be employed.

3. The matrix consisted of rows representing vertices (sources) and columns representing time points. In the present data, there were 15,028 covering the entire cortical surface and 510 time points including a 200 ms baseline and 400 ms of post-stimulus activity (at a 2 ms sampling interval).

4. A 2 cm radius was employed in this search.

5. Two vertices were merged if their respective activation windows overlapped by more than 50 %.

6. The minimum size of a cluster was set to seven vertices corresponding to approximately 1.5 cm^2. This resulted in 47–106 clusters, depending on the subject, entering the labelling phase. The largest clusters contained 30–50 vertices.

7. Two clusters were merged if their zero-lag, cross-correlation value was >0.7.

8. Cortical atlases such as the Destrieux [47], Desikan-Killiani [33] and Tzourio-Mazoyer [36] maps are available and provide different cortical parcellations and labels for each region.

9. A given cluster was considered to fall in a single anatomic area (e.g., right middle frontal gyrus), if more than 70 % of its vertices were located in that region. On several occasions, however, this criterion was not met and a particular cluster was formally identified to be located in two, or even three, contiguous anatomic areas (e.g., left inferior temporal gyrus and left fusiform gyrus).

10. Given that this procedure is entirely data-driven, the internal validity of the resulting map depends heavily upon data quality both in terms of noise levels and with respect to the task/ experimental conditions employed to ensure that, to a large extent, the same brain mechanism was engaged in every

participant. The number of participants providing useful data is also critical in this step to reduce sampling error and ensure that only the most consistently active anatomical areas will be part of the final activation map.

11. Every neuron or brain area can be thought of, from the systems neuroscience point of view, as a system that operates through time, receiving an input and producing an output at every given time point. To say that a neuron or a brain area responds to an input is to say that certain properties of the output are affected. The simplest such property is the amplitude of the output. Finding, thus, the peaks of the clusters' time-series corresponds to finding the time-point at which the brain area is activated, responding to a major input.

12. There are several ways to identify activity peaks. The function chosen here is a readily available, simple method that requires a single line of code.

13. The number of epochs as well as the actual time-limits that define them are arbitrary and depend on the specific experimental task and the length of the post-stimulus period. For shorter epochs such as the one adopted in the data example presented here, a division into early (0–150 ms), middle (150–250 ms) and late (250–400 ms) is probably appropriate given that the participants' response rarely occurred after 500 ms. For longer epochs, typically permitted in delayed-response tasks, a division into 0–150, 150–450 and 450–850 ms windows is suggested.

14. Around every peak 40 time-points were padded on each side. This number depends mostly on the length of the post-stimulus period. Higher values can be used. The resulting length of the time-series, $40 + 40 + 1 = 81$ time-points in our case, has to be adequate for a meaningful calculation of the cross-correlation. The maximum number of lags that was used was 25. The two peaks were not allowed to come closer than five time-points. This minimum time discrepancy represents the minimum time required by a signal from the first region to reach the second region.

15. The alternative method in order to collapse "connectivity" data across participants and perform group-level analyses would be to "match" clusters across participants on the basis of anatomical and functional criteria. However, individual variability in the location of clusters renders this procedure very difficult and seemingly futile.

16. This is accomplished by dividing the number of times each "pathway" appears by the total number of times the two brain regions appear in the table of clusters (taking into account multiple clusters within a particular anatomical regions and multiple peaks in the time series of a particular cluster).

Acknowledgments

This work was supported in part by the Shainberg Neuroscience Research Fund—Le Bonheur Children's Hospital Neuroscience Institute.

References

1. Geschwind N (1965) Disconnexion syndromes in animals and man. I. Brain 88:585–644

2. Damasio H, Tranel D, Grabowski T, Adolphs R, Damasio A (2004) Neural systems behind word and concept retrieval. Cognition 92:179–229

3. Price CJ, Devlin JT, Moore CJ, Morton C, Laird AR (2005) Metaanalyses of object naming: effect of baseline. Hum Brain Mapp 25:70–82

4. Visser M, Jefferies E, Lambon Ralph MA (2010) Semantic processing in the anterior temporal lobes: a meta-analysis of the functional neuroimaging literature. J Cogn Neurosci 22:1083–1094

5. Catani M, Jones DK, ffytche DH (2005) Perisylvian language networks of the human brain. Ann Neurol 57:8–16

6. Frey S, Campbell JS, Pike GB, Petrides M (2008) Dissociating the human language pathways with high angular resolution diffusion fiber tractography. J Neurosci 28: 11435–11444

7. Catani M, Thiebaut de Schotten M (2008) A diffusion tensor imaging tractography atlas for virtual in vivo dissections. Cortex 44: 1105–1132

8. Nigel I, Lawes C, Barrick TR, Murugam V, Spierings N, Evans DR, Song M, Clark CA (2008) Atlas-based segmentation of white matter tracts of the human brain using diffusion tensor tractography and comparison with classical dissection. Neuroimage 39:62–79

9. Sarubbo S, De Benedictis A, Maldonado IL, Basso G, Duffau H (2013) Frontal terminations for the inferior fronto-occipital fascicle: anatomical dissection, DTI study and functional considerations on a multi-component bundle. Brain Struct Funct 218:21–37

10. Mandonnet E, Gatignol P, Duffau H (2009) Evidence for an occipito-temporal tract underlying visual recognition in picture naming. Clin Neurol Neurosurg 111:601–605

11. Shinoura N, Suzuki K, Tsukada M, Yoshida M, Yamada R, Tabei Y, Saito K, Koizumi T, Yagi K (2010) Deficits in the left inferior longitudinal fasciculus results in impairments in object naming. Neurocase 16:135–139

12. Duffau H, Capelle L, Sichez N, Denvil D, Lopes M, Sichez JP, Bitar A, Fohanno D (2002) Interoperative mapping of the subcortical language pathways using direct electrical stimulations: an anatomo-functional study. Brain 125:199–214

13. Moritz-Gasser S, Herbet G, Duffau H (2013) Mapping the connectivity underlying multimodal (verbal and non-verbal) semantic processing: a brain electrostimulation study. Neuropsychologia 51:1814–1822

14. Lu C, Ning N, Peng D, Ding G, Li K, Yang Y, Lin C (2009) The role of large-scale neural interactions for developmental stuttering. Neuroscience 161:1008–1026

15. Bullmore E, Horwitz B, Honey G, Brammer M, Williams S, Sharma T (2000) How good is good enough in path analysis of fMRI data? Neuroimage 11:289–301

16. Mechelli A, Penny WD, Price CJ, Gitelman DR, Friston KJ (2002) Effective connectivity and intersubject variability: using a multisubject network to test differences and commonalities. Neuroimage 17:1459–1469

17. Zou L, Abutalebi J, Zinszer B, Yan X, Shu H, Peng D, Ding G (2012) Second language experience modulates functional brain network for the native language production in bimodal bilinguals. Neuroimage 62:1367–1375

18. Friston KJ, Harrison L, Penny W (2003) Dynamic causal modeling. Neuroimage 19: 1273–1302

19. Mechelli A, Price CJ, Noppeney U, Friston KJ (2003) A dynamic causal modeling study on category effects: bottom-up or top-down mediation? J Cogn Neurosci 15:925–934

20. Penny WD, Stephan KE, Mechelli A, Friston KJ (2004) Modeling functional integration: a comparison of structural equation and dynamic causal models. Neuroimage 23:S264–S274

21. Saur D, Schelter B, Schnell S, Kratochvil D, Küpper H, Kellmeyer P, Kümmerer D, Klöppel S, Glauche V, Lange R, Mader W, Feess D, Timmer J, Weiller C (2010) Combining functional and anatomical connectivity reveals brain

networks for auditory language comprehension. Neuroimage 49:3187–3197

22. Mader W, Feess D, Lange R, Saur D, Glauche V, Weiller C, Timmer J, Schelter B (2008) On the detection of direct directed information flow in fMRI. IEEE J Sel Top Signal Process 2:965–974

23. Stephan KE, Weiskopf N, Drysdale P, Robinson P, Friston KJ (2007) Comparing hemodynamic models with DCM. Neuroimage 38:387–401

24. Abel S, Huber W, Weiller C, Amunts K, Eickhoff SB, Heim S (2011) The influence of handedness on hemispheric interaction during word production: insights from effective connectivity analysis. Brain Connect 1:219–231

25. Hamalainen MS, Ilmoniemi RJ (1994) Interpreting magnetic fields of the brain: minimum norm estimates. Med Biol Eng Comput 32:35–42

26. Dale AM, Fischl B, Sereno MI (1999) Cortical surface-based analysis. I. Segmentation and surface reconstruction. Neuroimage 9:179–194

27. Gross J, Kujala J, Hamalainen M, Timmermann L, Schnitzler A, Salmelin R (2001) Dynamic imaging of coherent sources: studying neural interactions in the human brain. Proc Natl Acad Sci U S A 98:694–699

28. Salmelin R, Kujala J (2006) Neural representation of language: activation vs. long-range connectivity. Trends Cogn Sci 10:519–525

29. Kujala J, Pammer K, Cornelissen PL, Roebroeck A, Formisano E, Salmelin R (2007) Phase coupling in a cerebro-cerebellar network at 8–13 Hz during reading. Cereb Cortex 17:1476–1485

30. Campo P, Poch C, Toledano R, Iqoa GM, Belinchon M, Garcia-Morales I, Gil-Nagel A (2013) Anterobasal temporal lobe lesions alter recurrent functional connectivity within the ventral pathway during naming. J Neurosci 33:12679–12688

31. Stephan KE, Friston KJ (2007) Models of effective connectivity in neural systems. In: Jirsa VK, McIntosh AR (eds) Handbook of brain connectivity: understanding complex systems. Springer, Berlin, pp 303–326

32. Clarke A, Taylor KI, Tyler LK (2011) The evolution of meaning: spatio-temporal dynamics of visual object recognition. J Cogn Neurosci 23:1887–1899

33. Desikan RS, Ségonne F, Fischl B, Quinn BT, Dickerson BC, Blacker D, Buckner RL, Dale AM, Maguire RP, Hyman BT, Albert MS, Killiany RJ (2006) An automated labeling system for subdividing the human cerebral cortex on MRI scans into gyral based regions of interest. Neuroimage 31:968–980

34. Szekely A, D'Amico S, Devescovi A, Federmeier K, Herron D, Iyer G, Jacobsen T, Arévalo A, Vargha A, Bates E (2005) Timed action and object naming. Cortex 41:7–25

35. Eickhoff SB, Laird AR, Grefkes C, Wang LE, Zilles K, Fox P (2009) Coordinate-based activation likelihood estimation meta-analysis of neuroimaging data: a random-effects approach based on empirical estimates of spatial uncertainty. Hum Brain Mapp 30:2907–2926

36. Tzourio-Mazoyer N, Landeau B, Papathanassiou D, Crivello F, Etard O, Delcroix N, Mazoyer B, Joliot M (2002) Automated anatomical labeling of activations in SPM using a macroscopic anatomical parcellation of the MNI MRI single-subject brain. Neuroimage 15:273–289

37. Basso G, Magon S, Reggiani F, Capasso R, Monittola G, Yang FJ, Miceli G (2013) Distinguishable neurofunctional effects of task practice and item practice in picture naming: a BOLD fMRI study in healthy subjects. Brain Lang 126:302–313

38. Liljeström M, Hultén A, Parkkonen L, Salmelin R (2009) Comparing MEG and fMRI views to naming actions and objects. Hum Brain Mapp 30:1845–1856

39. Garn CL, Allen MD, Larsen JD (2009) An fMRI study of sex differences in brain activation during object naming. Cortex 45:610–618

40. Dronkers NF, Wilkins DP, Van Valin RD Jr, Redfern BB, Jaeger JJ (2004) Lesion analysis of the brain areas involved in language comprehension. Cognition 92:145–177

41. Hoffman P, Evans GA, Lambon Ralph MA (2014) The anterior temporal lobes are critically involved in acquiring new conceptual knowledge: evidence for impaired feature integration in semantic dementia. Cortex 50:19–31

42. Bright P, Moss H, Tyler LK (2004) Unitary vs. multiple semantics: PET studies of word and picture processing. Brain Lang 89:417–432

43. Tranel D, Grabowski TJ, Lyon J, Damasio H (2005) Naming the same entities from visual or from auditory stimulation engages similar regions of left inferotemporal cortices. J Cogn Neurosci 17:1293–1305

44. Duffau H, Herbet G, Moritz-Gasser S (2013) Toward a pluri-component, multimodal, and dynamic organization of the ventral semantic stream in humans: lessons from stimulation

mapping in awake patients. Front Syst Neurosci 7:44

45. Papanicolaou AC, Simos PG, Breier JI, Zouridakis G, Willmore LJ, Wheless JW, Constantinou JE, Maggio WW, Gormley WB (1999) Magnetoencephalographic mapping of the language-specific cortex. J Neurosurg 90:85–93

46. Simos PG, Papanicolaou AC, Breier JI, Wheless JW, Constantinou JE, Gormley WB, Maggio WW (1999) Localization of language-specific cortex using MEG and intraoperative stimulation mapping. J Neurosurg 91:787–796

47. Destrieux C, Fischl B, Dale A, Halgren E (2010) Automatic parcellation of human cortical gyri and sulci using standard anatomical nomenclature. Neuroimage 53:1–15

Neuromethods (2015) 91: 291–322
DOI 10.1007/7657_2013_64
© Springer Science+Business Media New York 2013
Published online: 17 January 2014

Blind Signal Separation Methods in Computational Neuroscience

Mujahid N. Syed, Pando G. Georgiev, and Panos M. Pardalos

Abstract

In this paper we present a survey of Blind Signal Separation (BSS) methods based on independence (Independent Component Analysis) and sparsity (Sparse Component Analysis). The presentation covers the most important methods described in the literature and gives a mathematical justification of the most used algorithms. We provide an experimental justification for the linear mixing in neurological data. Furthermore, we show the applicability of nonnegative source decomposition approaches in demixing neural images.

Keywords: Blind source separation, Independent component analysis, Sparse component analysis

1 Introduction

Signal separation is a specific case of signal processing, which aims at identifying unknown source signals $s_i(t)(i = 1, \ldots, n)$ from their observable mixtures $x_j(t)(j = 1, \ldots, m)$. In this problem, a mixture is assumed to be a linear transformation of sources, i.e., $\mathbf{x}(t) = \mathbf{A}\mathbf{s}(t)$, where $\mathbf{A} \in \mathbb{R}^{m \times n}$ is mixing matrix (or sometimes called as dictionary). Typically, t is any acquisition variable, over which a sample of mixture (a column for discrete acquisition variable) is collected. The most common types of acquisition variables are time and frequency. However, position, wave number, and other indices can be used depending on the nature of the physical process under investigation. Apart from identification of the sources, knowledge about mixing is assumed to be unknown. The generative model of the problem in its standard form can be written as:

$$\mathbf{X} = \mathbf{A}\mathbf{S} + \mathbf{N} \tag{1}$$

where $\mathbf{X} \in \mathbb{R}^{m \times N}$ denotes the mixture matrix, $\mathbf{A} \in \mathbb{R}^{m \times n}$ is the mixing matrix, $\mathbf{S} \in \mathbb{R}^{n \times N}$ denotes the source matrix, and $\mathbf{N} \in \mathbb{R}^{m \times N}$ denotes uncorrelated noise. Since both $\mathbf{A}$ and $\mathbf{S}$ are unknown, the signal separation problem is called "Blind" Signal Separation (BSS) problem. The BSS problem first appeared in (1), where the authors have proposed the seminal idea of BSS via an

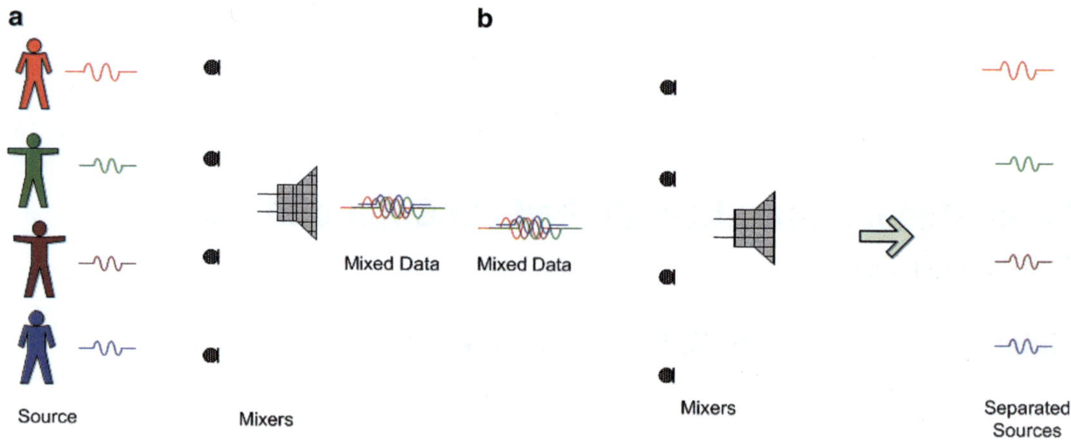

Fig. 1 Cocktail party problem (2). (**a**) Setup. (**b**) Problem

example of two source signals ($n = 2$) and two mixture signals ($m = 2$). Their objective was to recover source signals from the mixture signals, without any further information.

A classical illustrative example for the BSS model is the cocktail party problem where a mixture of sound signals from simultaneously speaking individuals is available (see Fig. 1 for a simple illustration). In a nutshell, the goal in BSS is to identify and extract the sources (Fig. 1b) from the available mixture signals (Fig. 1a). This problem caught the attention of many researchers, due to its wide applicability in different scientific research areas. A general setup of the BSS problem in computational neuroscience is depicted in Fig. 2. Any surface (or scalp) noninvasive cognitive activity recording can be used as a specific example. Depending upon the scenario, the mixture can be EEG, MEG, or fMRI data. Typically, physical substances like skull, brain matter, muscles, and electrode–skull interface act as mixers. The goal is to identify the internal source signals, which hopefully reduce the mixing effect during further analysis.

Currently, most of the approaches of BSS in computational neuroscience are based on the statistical independence assumptions. There are very few approaches that exploit the sparsity in the signals. Sparsity assumptions can be considered as flexible approaches for BSS compared to the independence assumption, since independence requires the sources to be at least uncorrelated. In addition to that, if the number of sources is larger than the number of mixtures (underdetermined case), then the statistical independence assumption cannot reveal the sources, but can reveal the mixing matrix. For sparsity-based approaches, there are very few papers in the literature (compared to independence-based approaches) that have been devoted to develop identifiability

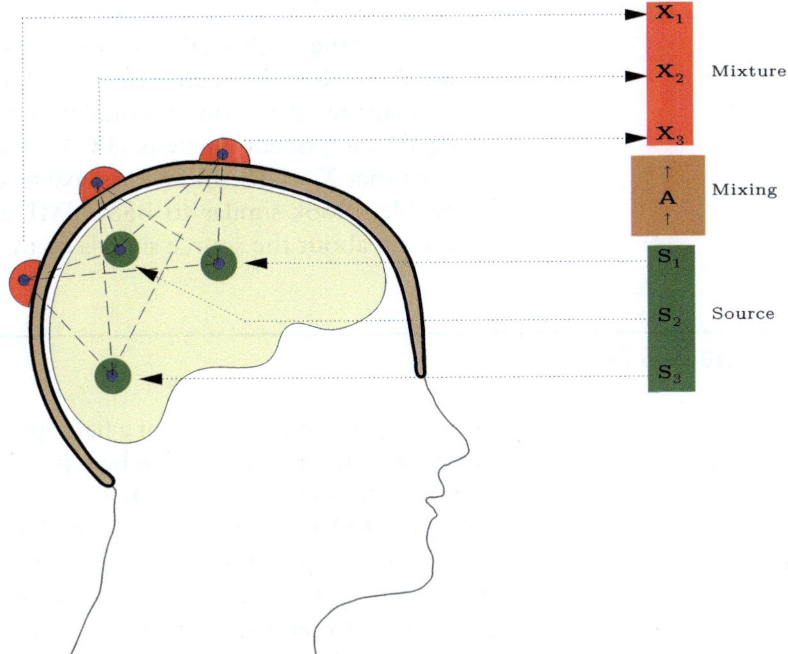

Fig. 2 BSS Setup for human brain

conditions, and to develop the methods of uniquely identifying (or learning) the mixing matrix (3–6).

In this paper, an overview of the BSS problem will be presented. Sufficient identifiability conditions will be revised, and their implication on the solution methodology will be discussed. Different well-known approaches that are used to find the solution of BSS problem will also be briefly presented. Application of sparsity-based BSS methods in computational neuroscience will be highlighted by presenting various experiments on real-world data. Due to the large number of approaches for Independent Component Analysis (ICA), only the most important from our point of view will be considered.

1.1 Other Look-Alike Problems

BSS is a special type of Linear Matrix Factorization (LMF) problem. There are many other methods that can be described in the form of LMF. For instance, Nonnegative Matrix Factorization (NMF), Morphological Component Analysis (MCA), Sparse Dictionary Identification (SDI), etc. The three properties that differentiate BSS from other LMF problems are:

- The model is assumed to be generative
  In BSS, the data matrix $\mathbf{X}$ is assumed to be a linear mixture of $\mathbf{S}$.

- Completely unknown source and mixing matrices.
  Some of the LMF methods (like MCA) assume partial knowledge about mixing.

- Identifiable source and mixing matrices
  Some of the LMF methods (like NMF, SDI) focus on estimating $\mathbf{A}$ and $\mathbf{S}$ without any condition for identifiability. NMF can be considered as a dimensionality reduction method like Principal Component Analysis (PCA). Similarly, SDI estimates $\mathbf{A}$ such that $\mathbf{X} = \mathbf{AS}$, and $\mathbf{S}$ is as sparse as possible. Although the problem look similar to BSS, NMF and SDI have no precise notion about the source signals or their identifiability.

2 The BSS Problem

From this point we assume that a flat representation of mixture data is given, i.e. mixture signals can be represented by a matrix containing finite number of columns. Before presenting the formal definition of the BSS problem, consider the following notations that will be used throughout the paper: A scalar is denoted by a lowercase letter, such as y. A column vector is denoted by a bold lowercase letter, such as $\mathbf{y}$, and a matrix is denoted by a bold uppercase letter, such as $\mathbf{Y}$. For example, in this chapter, the mixtures are represented by matrix $\mathbf{X}$. An ith column of matrix $\mathbf{X}$ is represented as $\mathbf{x}_i$. An ith row of matrix $\mathbf{X}$ is represented as $\mathbf{x}_{i\cdot}$. An ith row jth column element of matrix $\mathbf{X}$ is represented as $x_{i,j}$.

Now, the BSS problem can be mathematically stated as: Let $\mathbf{X} \in \mathbb{R}^{m \times N}$ be generated by a linear mixing of sources $\mathbf{S} \in \mathbb{R}^{n \times N}$. Given $\mathbf{X}$, the objective of BSS problem is to find two matrices $\mathbf{A} \in \mathbb{R}^{m \times n}$ and $\mathbf{S}$, such that the three matrices are related as $\mathbf{X} = \mathbf{A}\,\mathbf{S}$. In the rest of the paper, we ignore the noise factor. Although, without noise, the problem may appear easy. However, from the very definition of the problem, it can be seen that the solution of the BSS problem suffers from uniqueness and identifiability. Thus the notion of "good" solution to the BSS problem must be precisely defined. In the following part of this section, we explain in brief the uniqueness and identifiability issues.

- *Uniqueness:*
  Let $\Lambda, \Pi \in \mathbb{R}^{n \times n}$ be a diagonal matrix and permutation matrix, respectively. Let $\mathbf{A}$ and $\mathbf{S}$ be such that, $\mathbf{X} = \mathbf{AS}$. Consider the following:

$$\mathbf{X} = \mathbf{A}\,\mathbf{S}$$

$$= (\mathbf{A}\,\Pi\,\Lambda)\left(\Lambda^{-1}\Pi^{-1}\mathbf{S}\right)$$

$$= \mathbf{A}_{\circlearrowright}\,\mathbf{S}_{\circlearrowright}$$

Thus, even if $\mathbf{A}$ and $\mathbf{S}$ are known, there can be infinite equivalent solutions of the form $\mathbf{A}_{\circlearrowright}$ and $\mathbf{S}_{\circlearrowright}$. The goal of good BSS solution algorithm should be to find at least one of the equivalent solutions. Due to the inability of finding the unique solution, we not only lose the information regarding the order of sources but also lose the information of energy contained in the sources. Generally, normalization of rows of $\mathbf{S}$ may be used to tackle scalability. Also, relative or normalized form of energy can be used in the further analysis. Theoretically any information pertaining to order of source is impossible to recover. However, practically, problem-specific knowledge will be helpful in identifying correct order for the further analysis.

- *Identifiability:*
  Let $\Theta \in \mathbb{R}^{n \times n}$ be any nonsingular matrix. Let $\mathbf{A}$ and $\mathbf{S}$ be such that, $\mathbf{X} = \mathbf{AS}$. Consider the following:

$$\mathbf{X} = \mathbf{A}\,\mathbf{S}$$
$$= (\mathbf{A}\,\Theta)\,(\Theta^{-1}\mathbf{S})$$
$$= \mathbf{A}_\delta\,\mathbf{S}_\delta$$

Thus, even if $\mathbf{A}$ and $\mathbf{S}$ are known, there can be infinite non-identifiable solutions of the form $\mathbf{A}_\delta$ and $\mathbf{S}_\delta$. The goal of BSS solution algorithm is to avoid the non-identifiable solutions. Typically, the issue of identifiability arises from the dimension and structure of $\mathbf{A}$ and $\mathbf{S}$. The key idea to rightly identify both the matrices (of course with unavoidable scaling and permutation ambiguity) is to impose structural properties on $\mathbf{S}$ while solving the BSS problem (see Fig. 3). Some widely known BSS solution approaches from the literature are summarized below:

- Statistical Independence Assumptions:
  One of the earliest approaches to solve the BSS problem is to assume statistical independence among the source signals. These approaches are termed the ICA approaches. The fundamental assumption in ICA is that the rows of matrix $\mathbf{S}$ are statistically independent and non-Gaussian (7, 8).

- Sparse Assumptions:
  Apart from ICA, the other types of approaches, which provide sufficient identifiability conditions are based on the notion of sparsity in the $\mathbf{S}$ matrix. These approaches can be named as Sparse Component Analysis (SCA) approaches. There are two distinct categories in the sparse assumptions:

  1. Partially Sparse Nonnegative Sources (PSNS):In this category, along with certain level of sparsity, the elements of $\mathbf{S}$ are assumed to be nonnegative. Ideas of this type of approach can be traced back to the Nonnegative Matrix Factorization (NMF) method. The basic assumption in NMF is that the elements of $\mathbf{S}$ (and $\mathbf{A}$) are assumed to be nonnegative (9). However, in the case of BSS problem the nonnegativity assumptions on the elements of matrix $\mathbf{A}$ can be relaxed (10) without damaging the identifiability of $\mathbf{A}$ and $\mathbf{S}$.

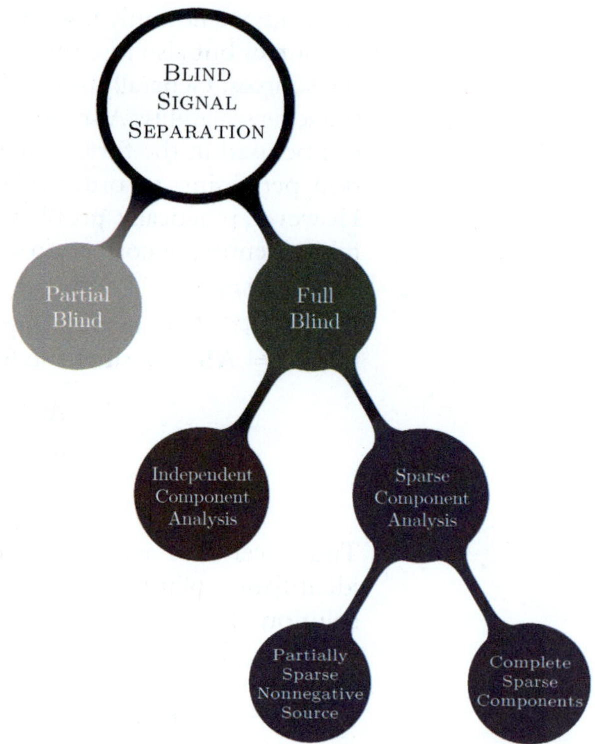

Fig. 3 Overview of different approaches to solve the BSS problem

2. Completely Sparse Components (CSC):
    In this category, no sign restrictions are placed on the elements of **S**, i.e. $s_{i,j} \in \mathbb{R}$. The only assumption used to define the identifiability conditions is the existence of certain level of sparsity in every column of **S** (11).

At present, these are the only known BSS approaches that can provide sufficient identifiability conditions (uniqueness up to permutation and scalability). In fact, the sparsity-based approaches (see (4, 10)) are relatively new in the area of BSS when compared to the traditional statistical independence approaches (see (8)). For neurological data, recent studies have shown relevance of sparsity assumption when compared to statistical independence assumption (12). In the following sections, brief discussion on the important issues of the above listed methods will be presented.

2.1 Statistically Independent Component Analysis

ICA is one of the oldest and widely known solution approaches to the BSS problem. Herault and Jutten (1) proposed the idea of statistical independent components in sound signal processing. Since then, the method has been updated and modified in number of ways (13) and applied to different areas of signal processing (8, 14, 15). The basic approach is to transform a given data matrix

X into matrix **S**, whose rows are as statistically independent as possible. It would be out of scope of this chapter to enumerate all the possible methods to obtain such transformation. Interested readers may refer the works (13, 16, 17) for various approaches in ICA, and some introductory level texts and tutorials (18, 19) for understanding mathematical concepts. Nevertheless, in this chapter, widely known ICA methods will be presented. Next, the basic identifiability conditions for successful implementation of any ICA approach are enlisted.

2.1.1 Sufficient Identifiability Conditions on A and S for ICA

ICA1. Not more than one row of **S** may follow Gaussian distribution.

ICA assumes no knowledge of distributions on the rows of **S**. However, typically it is assumed that the rows do not follow Gaussian distribution. Practically, a single row with Gaussian distribution is permissible for successful identification.

ICA2. $m \geq n$.

Typically, most of the ICA algorithms are designed for $m = n$. There have been many algorithms that are designed for $m > n$, the overdetermined case (20). Moreover, there are very few algorithms that are developed for the case of $m < n$, the underdetermined case (21). However, for the underdetermined case, the mixing matrix is only identifiable, whereas the source matrix cannot be fully recovered.

ICA3. **A** has full rank.

This is another critical criterion that is absolutely necessary for the successful identification of the sources.

2.1.2 Implication of the Identifiability Conditions of ICA

If one of the source signals (a single row in **S** matrix) follows Gaussian distribution, then it is still possible to identify sources. If more than one source follow Gaussian distribution, then identification is not possible with the high order statistics. However, the second order statistics methods can be used, which guarantee independence when all the sources are Gaussian. Therefore it is assumed that the sources do not follow Gaussian distribution. This fundamental assumption is used in order to exploit the center limit theorem in source identification. Thus, many classes of algorithms are designed to find a matrix W that minimizes the Gaussianity in $W^T X$, or maximizes the nongaussianity in $W^T X$.

A practical approach to incorporate independence is to consider uncorrelatedness. Although uncorrelation does not imply independence, but the vice versa is true. Thus, enforcing uncorrelatedness is one of the practical methods to incorporate independence. Another key implication of uncorrelatedness is the orthogonal nature of the mixing matrix **A**. Even if **A** is not orthonormal, prewhitening method can be used to create orthonormal columns of $\mathbf{A}$. This key factor, along with the deflation (22, 23) is one of the exploited criteria in the sequential extraction of the sources.

2.1.3 Centering, Prewhitening, and Deflation

Typically, in any ICA-based approaches, the basic preprocessing is to convert data in each row of $\mathbf{X}$ to zero mean, i.e., use the transformation $\mathbf{x}_{i\bullet} = \mathbf{x}_{i\bullet} - \text{mean}(\mathbf{x}_{i\bullet})$. This process is called as centering of $\mathbf{X}$, and this centering will in turn result in centering of $\mathbf{S}$. Once $\mathbf{A}$ and $\mathbf{S}$ are recovered from the centered data, $\mathbf{s}_{i\bullet}$ can be updated as $\mathbf{s}_{i\bullet} = \mathbf{s}_{i\bullet} + \mathbf{A}^{-1}\text{mean}(\mathbf{x}_{i\bullet})$, to recover the original uncentered sources.

In prewhitening, the idea is to convert $\mathbf{A}$ into an orthonormal matrix. This is done by eigenvalue decomposition:

$$\Theta = \mathbf{X}\mathbf{X}^T = \mathbf{Q}\Lambda\mathbf{Q}^T \qquad (2)$$

where Λ is a square diagonal matrix whose elements are eigenvalues of Θ and $\mathbf{Q}$ is a square orthonormal matrix of eigenvectors of Θ. Suppose Θ is positive definite, i.e., all the elements of Λ are positive. A transformation matrix Φ can be defined as:

$$\Phi = \Lambda^{-\frac{1}{2}}\mathbf{Q}^T \qquad (3)$$

Now the BSS problem can be transformed as:

$$\Phi\mathbf{X} = \Phi\mathbf{A}\mathbf{S} \qquad (4)$$

$$\hat{\mathbf{X}} = \hat{\mathbf{A}}\,\mathbf{S} \qquad (5)$$

The above transformation reduces the ill-conditioning effect due to the mixing matrix. Since ICA assumes independence, the original sources are uncorrelated, i.e., $\mathbf{S}\,\mathbf{S}^T = \mathbf{I}$. Therefore, $\hat{\mathbf{A}}\hat{\mathbf{A}}^T = \mathbf{I}$ as shown below:

$$\begin{aligned}\hat{\mathbf{A}}\hat{\mathbf{A}}^T &= \Phi\mathbf{A}\mathbf{A}^T\Phi^T \\ &= \Phi\mathbf{A}\mathbf{S}\mathbf{S}^T\mathbf{A}^T\Phi^T \\ &= \Phi\mathbf{X}\mathbf{X}^T\Phi^T \\ &= \Lambda^{-\frac{1}{2}}\mathbf{Q}^T\mathbf{Q}\Lambda\mathbf{Q}^T\mathbf{Q}\Lambda^{-\frac{1}{2}} \qquad (6) \\ &= \mathbf{I} \qquad\qquad (7)\end{aligned}$$

Typically, prewhitening is not uniquely defined and can be seen by the following example: Let $\mathbf{B}$ be any orthornormal matrix. The $\mathbf{B}\hat{\mathbf{X}}$ is another pre-whited data. Furthermore, eigenvalue transformation is one of the methods of prewhitening, and many other methods without the usage of eigenvalue decomposition can be designed for prewhitening.

Deflation, based on Gram–Schmidt-like decorrelation scheme, is another technique often used in ICA approaches based on one unit algorithms. The key idea is that the source signals can be extracted one by one. The idea can be described in brief as follows:

Let $\mathbf{w}_1 \in \mathbb{R}^m$ be a vector such that $\mathbf{s}_1{}_\bullet = \mathbf{w}_1{}^T\mathbf{X}$. It means that $\mathbf{w}_1$ is proportional to one of the columns of $\mathbf{A}$, say $\mathbf{a}_{i_1}$. Let $\mathbf{T} \in \mathbb{R}^{m \times (m-1)}$ be a transformation matrix, formed by orthonormal columns spanning the orthogonal complement of $\mathbf{w}_1$.

Consider, the following transformation of the data matrix:

$$\mathbf{X}_1 = \mathbf{T}^T \mathbf{X}, \tag{8}$$

The matrix $\mathbf{X}_1 \in \mathbb{R}^{(m-1)\times N}$ is a new mixture data matrix, which does not contain any traces of $\mathbf{s}_{i_1 \bullet}$, i.e.,

$$\mathbf{X}_1 = \mathbf{T}^T \mathbf{A} \mathbf{S} = \mathbf{A}_{i_1} \mathbf{S}, \tag{9}$$

where $\mathbf{A}_{i_1} \in \mathbb{R}^{(m-1)\times m}$ is a new mixing matrix, whose i_1th column consists of zeros due to the orthonormal construction of $\mathbf{T}$. This all zeros column removes the traces of $\mathbf{s}_{i_1 \bullet}$ in $\mathbf{X}_1$.

A particular transformation matrix, in which the columns span the orthogonal complement of $\mathbf{w}_1$, can be generated as:

$$\mathbf{t}_{i\bullet} = \begin{cases} \mathbf{e}_i^T & \forall i < q \\[2ex] -\dfrac{[\mathbf{w}_1^{\{\backslash q\}}]^T}{w_{1,q}} & \forall i = q \\[2ex] \mathbf{e}_{i-1}^T & \forall i > q. \end{cases} \tag{10}$$

where $\mathbf{w}^{\{\backslash q\}} \in \mathbb{R}^{(m-1)}$ represents $\mathbf{w}$ without the qth element, and $\mathbf{e}_i \in \mathbb{R}^{m-1}$ is a vector containing all zeros, except 1 at the ith position. The qth element is selected such that $w_{1,\,q} \neq 0$. The above process can be repeated until the last source is extracted.

2.1.4 Typical Objective Functions

- Kurtosis Minimization (or Maximization)
  Kurtosis is the classical measure of Gaussianity, i.e. for a Gaussian random variable kurtosis is zero. Usually, absolute or square value of kurtosis is used to measure nongaussianity. However, it should be noted that there exists some nongaussian random variables with zero kurtosis. These nongaussian variables are assumed to be very rare for practical scenarios. The basic structure of kurtosis minimization(or maximization) problem can be written as:

$$\min / \max :$$

$$E\left[(\mathbf{w}^T X)^4\right] \tag{11a}$$

$$\text{subject to} :$$

$$E\left[(\mathbf{w}^T X)^2\right] = \tag{11b}$$

$$\mathbf{w} \in \mathbb{R}^m \tag{11c}$$

where $\mathbf{w} \in \mathbb{R}^m$ is the only variable. The above minimization (or maximization) problem will result in obtaining one of the sources, $\mathbf{s}_{i\bullet} = \left(\mathbf{w}^T X\right)^T$. Kurtosis minimization(or maximization) is theoretically simple and computationally inexpensive. However, the kurtosis measure itself is non-robust (24), i.e., kurtosis is very sensitive to the outliers.

- Negentropy MinimizationFrom the information theory, it is affirmed that among all the random variables of equal variances, a Gaussian random variable is known to have the largest entropy (25). Using the idea of entropy, a related measure called Negentropy, J, is defined as:

$$J(r) \;=\; H(z) - H(r) \tag{12}$$

where H is the entropy function of a random variable r, and z is the Gaussian random variable of the same covariance matrix as r. From Eq. (12), it can be stated that Negentropy is zero for Gaussian variables and negative for all other variables. Negentropy (or entropy) is one of the optimal measures of nongaussianity with respect to statistical properties. However, the measure is computationally expensive due to the estimation of probability distribution function. Therefore, a practical approach is to use some approximate measures of negentropy. For example, in (8) a general structure of approximate negentropy is defined as:

$$J(r) \;\approx\; \sum_i k_i (E[G_i(r)] - E[G_i(z)])^2 \tag{13}$$

where k_i is any positive weight, G_i are some nonquadratic functions, and random variables r and z have zero mean with unit variance. The generic function defined in Eq. (13) has a zero value for Gaussian variable and positive for any other random variable. In addition to that, suitable selection of G_i will result in a robust measure, which is an added advantage apart from computational simplicity.

- Likelihood MaximizationPham et al. (26, 27) defined the likelihood function as:

$$L \;=\; \sum_{j=1}^{N} \sum_{i=1}^{n} \log\!\left(f_i(\mathbf{w}_i^T \mathbf{x}_j)\right) \;+\; N \, \log(|\det W|) \tag{14}$$

where f_i is the probability density function of component $\mathbf{s}_{i\bullet}$. Practically, the definition given in Eq. (14) cannot be used since f_i's are unknown. Thus, they are estimated as from the data.

- Mutual information Minimization
  A general measure of dependence between n random variables $\mathbf{r} = (r_1, r_2, \ldots, r_n)^T$ is the mutual information I, defined as:

$$I(\mathbf{r}) \;=\; \sum_{i=1}^{n} H(r_i) - H(\mathbf{r}) \tag{15}$$

The key property of mutual information measure is it is zero for statistically independent variables and always positive otherwise. Typically, while minimizing the mutual information, it is preferable to add a constraint that emphasizes the uncorrelation between rows the estimated sources.

2.1.5 Typical ICA Approaches

The typical objective functions (described above) are minimized by different optimization algorithms. Here we discuss some well-known algorithms. Let us remind one of the possible definitions of the *cumulant* for given k random variables $x_1, \ldots, x_k$:

$$\mathbf{cum}\{x_1, \ldots, x_k\} = \sum_{(P_1, \ldots, P_m)} (-1)^{m-1}(m-1)! E\left[\prod_{i \in P_1} x_i\right] \cdots E\left[\prod_{i \in P_m} x_i\right],$$

where the summation is taken over all possible partitions $\{P_1, \ldots, P_m\}$, $m = 1, \ldots, k$ of the set of the natural numbers $\{1, \ldots, k\}$; $\{P_i\}_{i=1}^{m}$ are disjoint subsets, whose union is $\{1, \ldots, k\}$, E is the expectation operator.

Of particular interest for us is the fourth-order cumulant,

$$\begin{aligned}
\mathrm{cum}&\{x_1, x_2, x_3, x_4\} \\
&= E\{x_1 x_2 x_3 x_4\} - E\{x_1 x_2\} E\{x_3 x_4\} \\
&\quad - E\{x_1 x_3\} E\{x_3 x_4\} - E\{x_1 x_4\} E\{x_2 x_3\}
\end{aligned}$$

When $x_i = x, i = 1, \ldots, 4$, it is called *kurtosis* of x:

$$kurt(x) = E\{x^4\} - 3(E\{x^2\})^2$$

Define a fourth order cumulant matrix $\mathbf{C}^{2,2}_{\mathbf{x}, \mathbf{x}_p}(\mathbf{B})$ of the sensor signals as follows (see (28)):

$$\begin{aligned}
\mathbf{C}^{2,2}_{\mathbf{x}, \mathbf{x}_p}(\mathbf{B}) &= E\{\mathbf{x}\mathbf{x}^T \mathbf{x}_p^T \mathbf{B} \mathbf{x}_p\} - E\{\mathbf{x}\mathbf{x}^T\}\mathrm{tr}(\mathbf{B} E\{\mathbf{x}_p \mathbf{x}_p^T\}) \\
&\quad - E\{\mathbf{x}\mathbf{x}_p^T\}\mathbf{B} E\{\mathbf{x}_p \mathbf{x}^T\} - E\{\mathbf{x}\mathbf{x}_p^T\}\mathbf{B}^T E\{\mathbf{x}_p \mathbf{x}^T\}
\end{aligned}$$

where $\mathbf{B} \in \mathbb{R}^{n^2}$ is a parameter matrix, $\mathbf{x}_p$ is the vector of delays signals (with delay p), $\mathbf{x}_p = \mathbf{x}(. - p)$, and E is the mathematical expectation.

The (i, j)th element of $\mathbf{C}^{2,2}_{\mathbf{x}, \mathbf{x}_p}(\mathbf{B})$ is

$$\mathbf{C}^{2,2}_{\mathbf{x}, \mathbf{x}_p}(\mathbf{B})_{i,j} = \sum_{k,l=1}^{n} \mathbf{cum}\{x_i(t), x_j(t), x_k(. - p), x_l(. - p)\} B_{k,l},$$

where $\mathbf{cum}\{x_i, x_j, x_k(. - p), x_l(. - p)\}$ denotes the fourth-order cumulant.

Main property: if

$$\mathbf{x} = \mathbf{H}\mathbf{s} \tag{16}$$

and $\mathbf{s}$ has independent components, then

$$\mathbf{C}^{2,2}_{\mathbf{x},\mathbf{x}_p}(\mathbf{B}) = \mathbf{H}\Delta(\mathbf{B})\mathbf{H}^T,$$

where $\Delta(\mathbf{B}) = diag\{\mathbf{cum}_{s_1}(p)\mathbf{h}^T_{*1}\mathbf{Bh}_{*1},\ldots,\mathbf{cum}_{s_n}(p)\mathbf{h}^T_{*n}\mathbf{Bh}_{*n}\}$,
$\mathbf{cum}_{s_i}(p) = \mathbf{cum}\{s_i, s_i, s_i(.-p), s_i(.-p)\}$ and $\mathbf{h}_{*i}$ denotes the ith column of $\mathbf{H}$.

Therefore, if the mixing matrix $\mathbf{H}$ is orthogonal, we can separate the sources by:

- eigenvalue decomposition of $\mathbf{C}^{2,2}_{\mathbf{x},\mathbf{x}_p}(\mathbf{B})$ (if its eigenvalues are distinct), which estimates $\mathbf{H}$ up to multiplication with permutation and diagonal matrices—this method works if the initial sources are independent enough,

- joint diagonalization of several cumulant matrices: find orthogonal matrix $\mathbf{H}$ such that the matrices

$$\mathbf{H}^T\mathbf{CH}: \qquad \mathbf{C} \in \mathcal{C}$$

are diagonal as much as possible.

We will consider two classes $\mathcal{B}$ of parameter matrices, which give two types of algorithms.

1. JADE algorithm, see (28, 29): $\mathcal{C} = \{\mathbf{C}^{\in,\in}_{\mathbf{x},\mathbf{x}}(\mathbf{B}), \mathbf{B} \in \mathcal{B}\}$ when $\mathcal{B}$ consists of eigen-matrices $\mathbf{B}$ of the cumulant tensor defined by the linear operator $\mathbf{F}$ on all matrices by $\mathbf{F}(\mathbf{M})_{i\,j}: = \sum_k \iota c\, u\, m\, (x_i, x_j, x_k, x_l)M_{k\,l}$, i.e. $\mathbf{F}(\mathbf{B}) = \lambda\mathbf{B}$.

   The joint digitalization procedure is performed by an elegant method (30), which itself is used in other joint digitalization tasks, for instance in Second Order Blind Identification (SOBI) algorithm (31).

2. JADETD algorithm, see (32): when $\mathbf{B} = \mathbf{I}$ (identity matrix) and we jointly diagonalize the class of matrices $\mathcal{C} = \{\mathbf{C}^{2,2}_{\mathbf{x},\mathbf{x}_p}(\mathbf{I}), p = 1,\ldots,L\}$. It can separate colored sources of order 4, which are white of order 2.

One-Unit Contrast Functions: One Source Extraction via Maximization of the Absolute Value of the Cumulants

Consider again the model: $\mathbf{x} = \mathbf{A}\,\mathbf{s}$ of linear mixture of unknown signals $\mathbf{s}$.

Consider the maximization problems

OP(p)   maximize $|\varphi_p(\mathbf{w})| = |\mathbf{cum}_p(\mathbf{w}^T\mathbf{x})|$
        under constraint $\|\mathbf{w}\| = 1$.

and

DP(p)   maximize $|\psi_p(\mathbf{c})| = |\mathbf{cum}_p(\mathbf{c}^T\mathbf{s})|$
        under constraint $\|\mathbf{c}\| = 1$,

where $\mathbf{cum}_p$ means the self-cumulant of order p:

$$\mathbf{cum}_p(s) = \mathrm{cumulant}(\underbrace{s,\ldots,s}_{p})$$

They are related to the change of variables: $\mathbf{w} = \mathbf{A}\,\mathbf{c}$.

We emphasize the following property of the cumulants, which is used essentially: if s_i, $i = 1,\ldots,n$ are statistically independent, then

$$\mathbf{cum}_p\left(\sum_{i=1}^{n} c_i s_i\right) = \sum_{i=1}^{n} c_i^p \mathbf{cum}_p(s_i),$$

Without loss of generality we may assume that the matrix $\mathbf{A}$ is orthogonal. We have

$$\varphi_p(\mathbf{w}) = \psi_p(\mathbf{A}^T \mathbf{w}).$$

The problems DP(p) and OP(p) are equivalent in sense that $\mathbf{w}_0$ is a solution of OP(p) if and only if $\mathbf{c}_0 = \mathbf{A}^T \mathbf{w}_0$ is a solution of DP(p).

A very useful observation is the following: if a vector $\mathbf{c}$ contains only one nonzero component, say $c_{i_0} = \pm 1$, then the vector $\mathbf{w} = \mathbf{A}\mathbf{c}$ gives extraction (say $y(k)$) of the source with index i_0, since

$$\begin{aligned} y(k) &:= \mathbf{w}^T \mathbf{x}(k) \\ &= \mathbf{c}^T \mathbf{A}^T \mathbf{x}(k) \\ &= \mathbf{c}^T \mathbf{s}(k) = \pm s_{i_0}(k) \quad \forall k = 1, 2, \ldots. \end{aligned}$$

Fact: any solution $\mathbf{c}$ of DP(p) have exactly one nonzero element, so taking a vector $\mathbf{w} = \mathbf{A}\mathbf{c}$ as a solution of the original problem OP(p), by $y(k) = \mathbf{w}^T \mathbf{x}(k)$ we achieve extraction of one source signal.

A mathematical justification of one source extraction via maximization of the absolute value of the cumulants is the following lemma (33).

Lemma 1: Consider the optimization problem:

minimize (maximize) $\sum_{i=1}^{n} k_i v_i^p$
subject to $\|\mathbf{v}\| = c > 0,$
where $p > 2$ *and* $\mathbf{v} = (v_1, \ldots, v_n).$
Denote $I^+ = \{i \in \{1, \ldots, n\} : k_i > 0\}$, $I^{-1} = \{i \in \{1, \ldots, n\} : k^i < 0\}$
and $e_i = (0, \ldots 0, 1, 0, \ldots, 0), (1$ *is the ith place).*

If p is even, then the points of local minimum are exactly the vectors $\pm c\, \mathbf{e}_i, i \in I^-$ and the points of local maximum are exactly the vectors $\pm c\, \mathbf{e}_j, j \in I^+$.

A Generalization of the Fixed-Point Algorithm (33)

Consider the following algorithm:

$$\mathbf{w}(l) = \frac{\varphi_p'(\mathbf{w}(l-1))}{\|\varphi_p'(\mathbf{w}(l-1))\|}, \quad l = 1, 2, \ldots,$$

which is a generalization of the fixed-point algorithm of Hyvärinen and Oja (34). The name is derived by the Lagrange equation for the optimization problem OP(p), since the algorithm tries to find a solution of it iteratively, and this solution is a fixed point of the operator defined by the right-hand side.

The next theorem gives precise conditions for convergence of the fixed-point algorithm of Hyvärinen and Oja and its generalization.

Theorem 1 ((33)): Assume that s_i are statistically independent, zero mean signals and the mixing matrix $\mathbf{A}$ is orthogonal. Let $p \geq 3$ be a given even integer number, $\mathbf{cum}_p(s_i) \neq 0, i = 1,\ldots,n$ and let

$$I(\mathbf{c}) = \arg \max_{1 \leq i \leq n} c_i \left| \mathbf{cum}_p(s_i) \right|^{\frac{1}{p-2}}.$$

Denote by W_0 the set of all elements $\mathbf{w} \in \mathrm{IR}^n$ such that $\|\mathbf{w}\| = 1$, the set $I(\mathbf{A}^\mathrm{T}\mathbf{w})$ contains only one element, say $i(\mathbf{w})$, and $c_{i(\mathbf{w})} \neq 0$. Then

(a)   The complement of W_0 has a measure zero.

(b)   If $\mathbf{w}(0) \in W_0$, then

$$\lim_{l \to \infty} y_l(k) = \pm s_{i_0}(k) \quad \forall k = 1, 2, \ldots,$$

where $y_l(k) = \mathbf{w}(l)^\mathrm{T}\mathbf{x}(k)$ and $i_0 = i(\mathbf{w}(0))$.

(c)   The rate of convergence in (b) is of order $p - 1$.

Examples

1) $p = 4$. Then

$$\varphi_4(\mathbf{w}) = \mathbf{cum}_4(\mathbf{w}^T\mathbf{x}) = E\{(\mathbf{w}^T\mathbf{x})^4\} - 3\left(E\{(\mathbf{w}^T\mathbf{x})^2\}\right)^2$$

and

$$\varphi_4'(\mathbf{w}) = 4E\{(\mathbf{w}^T\mathbf{x})^3\mathbf{x}\} - 12E\{(\mathbf{w}^T\mathbf{x})^2\}E\{(\mathbf{x}\mathbf{x}^T)\}\mathbf{w}.$$

We note that if the standard prewhitening is made (i.e., $E\{\mathbf{x}\mathbf{x}^T\} = \mathbf{I}_n$, $\mathbf{A}$ is orthogonal), the algorithm recovers the fixed-point algorithm of Hyvarinen and Oja, i.e.

$$\mathbf{w}(l+1) = \frac{E\{(\mathbf{w}(l)^T\mathbf{x})^3\mathbf{x}\} - 3\mathbf{w}(l)}{\|E\{(\mathbf{w}(l)^T\mathbf{x})^3\mathbf{x}\} - 3\mathbf{w}(l)\|}$$

2) $p = 6$. Then

$$\varphi_6(\mathbf{w}) = E\{(\mathbf{w}^T\mathbf{x})^6\} - 15E\{(\mathbf{w}^T\mathbf{x})^2\}E\{(\mathbf{w}^T\mathbf{x})^4\}$$
$$- 10\left(E\{(\mathbf{w}^T\mathbf{x})^3\}\right)^2 + 30\left(E\{(\mathbf{w}^T\mathbf{x})^2\}\right)^3,$$

$$\varphi_6'(\mathbf{w}) = 6E\{(\mathbf{w}^T\mathbf{x})^5\mathbf{x}\} - 30E\{\mathbf{x}\mathbf{x}^T\}\mathbf{w}E\{(\mathbf{w}^T\mathbf{x})^4\}$$
$$- 60E\{(\mathbf{w}^T\mathbf{x})^2\}E\{(\mathbf{w}^T\mathbf{x})^3\mathbf{x}\}$$
$$- 60E\{(\mathbf{w}^T\mathbf{x})^3\}E\{(\mathbf{w}^T\mathbf{x})^2\mathbf{x}\}$$
$$+ 180\left(E\{(\mathbf{w}^T\mathbf{x})^2\}\right)^2 E\{\mathbf{x}\mathbf{x}^T\}\mathbf{w}.$$

If, in addition, $E\{\mathbf{x}\mathbf{x}^T\} = \mathbf{I}$, then

$$\varphi_6'(\mathbf{w}) = 6E\{(\mathbf{w}^T\mathbf{x})^5\mathbf{x}\} - 30\mathbf{w}E\{(\mathbf{w}^T\mathbf{x})^4\}$$
$$- 60E\{(\mathbf{w}^T\mathbf{x})^3\mathbf{x}\}$$
$$- 60E\{(\mathbf{w}^T\mathbf{x})^3\}E\{(\mathbf{w}^T\mathbf{x})^2\mathbf{x}\}$$
$$+ 180\mathbf{w}.$$

The crucial issue while using ICA-based approaches is the fundamental assumption of statistical independence of the finite data. Practically, it boils down to the notion that all the rows of $\mathbf{S}$ are uncorrelated. Furthermore, for biological and physiological signals measured from a single individual, independence is hard to justify. Typically, in such cases ICA tries to find the source containing least dependent rows. Although, if part of the rows of $\mathbf{S}$ are taken as noise and/or artifact, then ICA can be applied to filter the signal. On the other hand, when it is known from the theory that the sources are correlated, then substituting a minimal dependency source as true source is not appropriate for analysis. Thus, a fundamentally different assumption which considers structure of the sources based on their origin is key requirement for BSS methods to operate on physiological data. Sparsity is one of the promising criteria used to exploit structure of the physiological data. In the following subsection, the prominent methods that are based on sparsity assumptions will be summarized.

2.2 Sparse Component Analysis

The earliest methods that proposed the notion of sparsity and the identifiability conditions for BSS problems can be found in (4, 10, 35). From the literature, different approaches to solve SCA problem can be grouped into two distinct classes. The main difference between the two classes is based on the nonnegativity assumption of the elements in the $\mathbf{S}$ matrix. The reason for such division is due to the structure of the resulting SCA problem. Typically, when the sources are nonnegative the SCA problem can be boiled down to a convex programming problem. Thus, the algorithms for the class with nonnegativity assumptions are computationally inexpensive whereas, for the other class, the SCA problem generally results in nonconvex optimization problem. Therefore, finding a global optimal solution when the source elements are real is a computationally expensive task.

SCA can be considered as a flexible method for BSS than ICA. ICA requires the source to be statistically independent, whereas SCA requires sparsity of sources (a weaker assumption). In addition to that, ICA is not suitable if the number of sources is larger than number of mixtures (underdetermined case). Typical ideas of SCA can be found in (4–6). Furthermore, the identifiability conditions on $\mathbf{X}$ that improve the separability of sources are studied by few researchers (3, 4).

2.2.1 Partially Sparse Nonnegative Sources

In many physiological data scenarios, the notion that the source signal is nonnegative seems to be valid; for example, medical imaging, NMR, ICP, HR, etc. Using this ideology, and the fact that ICA at least requires complete uncorrelated source signals, a partial correlated BSS method can be developed. A source matrix $\mathbf{S}$ is defined to be partially correlated, when rows of certain set of columns of $\mathbf{S}$ are uncorrelated. However, the rows of full $\mathbf{S}$ matrix are correlated. For the sources on which nonnegativity assumption holds, partially correlated assumption is less restrictive than ICA. The primary idea on which this class of SCA methods works can be summarized as: any vector $\mathbf{x}_i \forall i = 1, \ldots, N$ is nothing but a nonnegative linear combination of vectors $\mathbf{a}_j \forall j = 1, \ldots, n$. Thus, sparse assumptions on $\mathbf{S}$ that may lead to proper identification of $\mathbf{A}$ can be exploited in order to identify $\mathbf{A}$ and $\mathbf{S}$. One of the earliest approaches towards this method, presented by Naanaa and Nuzillard (10), is called Positive and Partially Correlated (PPC) method. Next, the sufficient identifiability condition for PPC will be discussed.

Sufficient Identifiability Conditions on A and S for PPC (10)

PPC1. There exists a diagonal submatrix in $\mathbf{S}$. For each row $\mathbf{S}_i$ there exists a $j \in \{1, \ldots, N\}$ such that $s_{i,\,j} = 0$ and $s_{k,\,j} > 0$ for $k = 1, \ldots, i-1, i+1, \ldots, N$.

PPC2. Columns of $\mathbf{A}$ are linearly independent.

Implication of the Identifiability Conditions for PPC

Due to the restriction given in PPC1, the PPC BSS problem boils down to the following: All the columns of matrix $\mathbf{X}$ span a cone in $\mathbb{R}^m$, where the edges of the cone are nothing but the columns of matrix $\mathbf{A}$. Using this simplification, suitable linear or convex programming problems can be solved to identify the edges of cone spanned by the columns of $\mathbf{X}$. Finding these edges results in identification of $\mathbf{A}$. Matrix $\mathbf{S}$ can be obtained by using Moore–Penrose pseudoinverse of $\mathbf{A}$.

PPC Approaches

In (10) a least square minimization problem is proposed to solve the PPC problem. The formulation is given as:

$$\text{minimize} : \left\| \sum_{\substack{i = 1 \\ i \neq j}}^{N} \alpha_i \mathbf{x}_i - \mathbf{x}_j \right\|_2 \tag{17a}$$

$$\text{subject to} : \quad \alpha_i \geq 0 \quad \forall\, i \tag{17b}$$

In addition to the above formulation, based on the same edge extraction idea, many recent works are directed towards efficient edge extraction from $\mathbf{X}$ (36, 37).

Another recent modification of PPC approach called Positive everywhere Partially orthogonal Dominant intervals (PePoDi) can be seen in (38). In PePoDi, the PPC1 condition is modified by stating that the last row of $\mathbf{S}$ is positive dominant and does not satisfy PPC1.

However this modification comes with the price of restricting $\mathbf{A}$ to be nonnegative. Thus, PePoDi method can be seen as a special case of the NMF problem.

2.2.2 Complete Sparse Component Sources (39)

When the sources are not nonnegative, then the BSS problem transforms into a nonconvex optimization problem. In fact, the only identifiable condition that is known for real sources is to have sparsity in each column of $\mathbf{X}$. Before defining the complete sparse component (CSC) criteria, consider the following definitions:

CSC-conditioned: A matrix M is said to be *CSC-conditioned* if every square submatrix of M is nonsingular.

CSC-sparse: A matrix M is said to be *CSC-sparse* if every column of M has at most $m - 1$ nonzero elements.

CSC-representable: A matrix M is said to be *CSC-representable* if for any $n - m + 1$ selected rows of M, there exists m columns such that:

- All the m columns contain zeros in the selected rows, and

- Any $m - 1$ subset of the m columns is linearly independent.

Sufficient Identifiability Conditions on **A** and **S** for CSC

CSC1. $\mathbf{A}$ is *BSS-conditioned*,

CSC2. $\mathbf{S}$ is *BSS-sparse*,

CSC3. $\mathbf{S}$ is *BSS-representable*

Implication of the Identifiability Conditions for CSC

Due to the restriction given in CSC2 and CSC3, the CSC BSS problem boils down to the following: All the columns of matrix $\mathbf{X}$ lie on m-hyperplanes passing through origin, where the normal vectors of the hyperplanes are nothing but the orthonormal complement of the matrix $\mathbf{A}$. Using this transformation, suitable hyperplane clustering methods can be used to identify the hyperplanes defined by $\mathbf{X}$. Since the hyperplane clustering is non-convex, the CSC BSS problem is relatively difficult to solve than compared to PPC BSS problem.

CSC Approaches

The initial approaches to CSC BSS problem is based on bilinear hyperplane clustering approach (11). The method is a sequential approach that typically converges to a local minimum. In fact, most of the hyperplane clustering methods in the literature are confined to seven to eight dimensions. Another recent approach, presented in (2), is based on $0 - 1$ integer programming transformation of the bilinear formulation. Currently, the $0 - 1$ integer programming approach solves problems of dimensions 14–16.

In the following part of this chapter, we present a brief literature highlighting the usage of BSS methods in neuroscience.

3   Applications in Neuroscience

It is a common observation in computational biomedicine, specifically in neuroscience, that invasive data recordings enhance the analysis and predictability than compared to noninvasive data recordings. On the other hand, invasive data recordings are not easily available and are recorded only in certain specific cases. Thus a transformation of noninvasive data that results in a similar information as that of invasive data will be a boon to computational biomedicine community. Consider the case of EEG, where the source signals are smeared by volume conduction. The recordings collected on scalp involve some sort of mixtures of the source signal. Moreover, the volume conduction from the origin of the signal to its destination on skull is assumed to have no time delays. Thus, if the mixing of source signal is assumed to be "linear," then the BSS problem aptly fits the above scenario. Of course, the linear mixing is a hypothetical assumption, and justification of the assumption is usually case specific. On the other hand, linear mixing is the only known way to efficiently separate the sources. Even for the case of quadratic mixing, there have been no successful results in the literature that depicts successful unmixing of the sources. Moreover, the linear mixing assumption is generally acceptable for the physiological signals, since most of the mixing can be attributed to superimposition of the source signals along with artifacts.

Makeig et al. (40) were pioneers in explaining the usage of BSS for biological signals. They identified four critical properties that should be satisfied by source signals so as to successfully implement BSS approach. However, their idealogy was limited to ICA view of BSS. In the following we extend the properties in terms of general BSS, which is more broader area than ICA.

1. The source signals should have *at least one* of the following properties:

   (a) The source signals are statistically independent, and not more than one source signal follows Gaussian distribution.

   (b) The source signals are nonnegative, and partial spatial sparsity exists among the sources.

   (c) The source signals are always spatially sparse.

2. The mixing mechanism should have *all* of the following properties:

   (a) Mixing is linear.

   (b) Mixing is not ill-conditioned.

   (c) The propagation delays occurring in the mixing medium are negligible.

3. The number of source signals and observed signals is nearly (not necessarily exact) equal in number.

These are the typical assumptions on data, which should be satisfied for successful extraction of the sources. Apart from source extraction, another major area where BSS finds its application in computational neuroscience is data filtering. Typically, BSS methods are used to filter out artifacts from the EEG, MEG, and other cortical activity recordings. The fundamental differences between actual source signals and artifacts in terms of frequency, amplitude, distribution, localization, sparsity, etc., are exploited to identify and winnow source signals. In the following parts of this section, important results from the literature that depicts the success of BSS methods in computational neuroscience will be presented.

3.1 Application in EEG and MEG Data Analysis

MagnetoEncephaloGraphy (MEG) is another noninvasive method (similar to EEG) which is used to measure cortical neuronal magnetic activity. So, similar to EEG, MEG suffers from the presence of artifacts. One of the earlier approaches of application of ICA to MEG can be found in (41, 42), where the authors were able to separate brain activity from eye blinks, or sensor malfunctions. The basic assumption of the authors was that the artifacts are anatomically and physiologically separate processes.

MEG signals are recorded in a magnetically shielded room using neuromagnetometer. Since MEG data is expensive to collect and requires a special facility, results on MEG with BSS approaches are comparatively less in number than compared to EEG with BSS approaches. Typically, the number of channel in MEG is very high than compared to EEG. Thus, dimensionality reduction methods like PCA are first used to reduce the dimensions. ICA is implemented on the resulting reduced data, in order to separate the existing independent components.

In order to validate the separation of MEG data with ICA, specialized experiment has to be designed. In a typical experimentation process (see (41)), the test person may be asked to blink, to turn head, to produce muscle movements, to bite teeth, etc. Each of the above activity is performed sequentially, for a specific period of time. The goal of the research was to separate the induced artifact from the pure source signals.

After Bell and Senjowski's infomax approach to ICA (43), EEG is identified as a possible application area for ICA (40). Most of the research experimentally provided validation for the usage of ICA (44). At present, it has been experimentally shown that the distribution of artifacts like ocular activity (eye movements, eye blinks), myographic activity (muscle, jaw tightening), cardiac cycle activity, electrical activity (50–60 Hz noise) from the devices is statistically, anatomically, and functionally independent when compared to the actual brain sources. Thus, artifact removal has been shown successful using the ICA method (45, 46).

**3.2 Application
in fMRI Data Analysis**

ICA has been applied to fMRI data under the assumption that the fMRI data is composed of sources, which are linearly independent. Furthermore, the mixing is assumed to be linear. Several papers have shown the results of applying ICA to fMRI data without validating the assumptions. McKeown and Sejnowski in (47) presented an experimental study towards the validation of the fMRI data assumptions. Spatial statistical independence between the consistently task-related components have been reported to be a reasonable assumption (48). However, in (47, 49) the dependence between consistently task-related and transient task-related components was also inferred. Nonlinear mixing between spatially independent components has also been identified as one of the reasons for poor performance of ICA.

Currently, ICA can be applied to fMRI data in two basic directions, i.e. in identifying spatially independent components or identifying temporally independent components. Methods that attempt to find simultaneously (or sequentially) spatial and temporal independent components have also been proposed (50, 51). However, the sparse spatial structure of fMRI, which is the typical structure for cognitive activities, indicates the suitability of spatially independent components over the other approaches. The BSS mapping for identifying spatial independent components is given as: $\mathbf{X} \in \mathbb{R}^{m \times N}$, m is the number of time points, and N is the number of voxels. The mixing matrix $\mathbf{A} \in \mathbb{R}^{m \times n}$ can be seen as time courses, where n represents the number of components. The goal of ICA is to find $\mathbf{A}$ and $\mathbf{S}$ matrices, where $\mathbf{S}$ represents spatial independent patterns. Typically, number of independent components are estimated prior to ICA. Some of the information theoretic methods of estimating number of independent components are presented in (52).

Another class of techniques that have been specifically developed for fMRI is group ICA. Group ICA is nothing but multisubject ICA approaches (53). Calhoun et al. provided a recent review of some of the well-known methods in the group ICA(54).

4 Simulations and Results

In this section, we will present some experimental results on real-world neural data. The objective of this section is to show usage of sparsity-based methods discussed in the previous sections. We specifically highlight the usage of SCA-based methods, since they are not very well known in the neuroscience community.

**4.1 Linear Mixing
EEG–ECoG Data**

The aim of this experiment is to understand the nature of mixing across the skull. In particular, our objective is to assert the validity of the linear mixing assumption in BSS problem. Since linear mixing is assumed in almost all the BSS methods, it will be of primary interest to show the validity of the assumption with respect to neural data. The idea of this experiment is to consider a neural data set which contains the information regarding the source as well as the mixture signals from brain. Based on the available information from the data set, the goal is to extract the mixing matrix. However, the mixing matrix itself may not provide a significant information, when compared to the total error from the linear mixing assumption. Therefore, in the following experiment, we consider a suitable publicly available data set (which contains both source and mixture data) and examine the linear mixing assumption across the skull by minimizing the total error.

In this experiment, simultaneous electrical activity over the scalp (EEG) and over the exposed surface of the cortex (ECoG) form a monkey is recorded. The information regarding experimental setup and position of electrodes is available here.[1] Since the data from this experiment is simultaneously collected from above the scalp and under the scalp, it opens the door to understand the mixing mechanism across the brain. Typically, we assume that the mixing over the skull is linear. Mathematical advantages in formulating the problem, developing the algorithms, and identifying the unknown source and mixing matrices are obtained through the linear mixing assumption. In fact, the only known successful results in BSS problem is obtained from linear mixing assumption. By analyzing the data of this experiment, our goal is to experimentally verify the validity of the linear mixing assumption.

The data consists of ECoG and EEG signals, which were simultaneously recorded from the same monkey. 128 channels ECoG array that covered entire lateral cortical surface of left hemisphere with every 5 mm spacing was implanted in the monkey. The EEG signal was recorded from 19 channels. The location of the EEG electrodes was determined by 10–20 systems without the Cz electrode (because the location of the Cz electrode interfered with a connector of ECoG). In the present simulation, we will present results on a particular data set, where the monkey is blindfolded, seated in a primate chair, and hands are tied to the chair. Figure 4a shows the 8 EEG channels of the left hemisphere, and Fig. 4b shows the 128 ECoG channels from the left hemisphere.

During the recording, the monkey is in resting condition. In such scenario, it is assumed that the theta and alpha bands should be dominant in a normal healthy primate. Thus, our goal is to see how particular frequency bands mix over the skull. Basically, our formulation is of the following form:

$$\text{minimize} : \quad |\mathbf{X}_{\text{EEG}} - \mathbf{A} \times \mathbf{X}_{\text{ECoG}}| \tag{18}$$

---

[1] http://wiki.neurotycho.org/EEG-ECoG_recording.

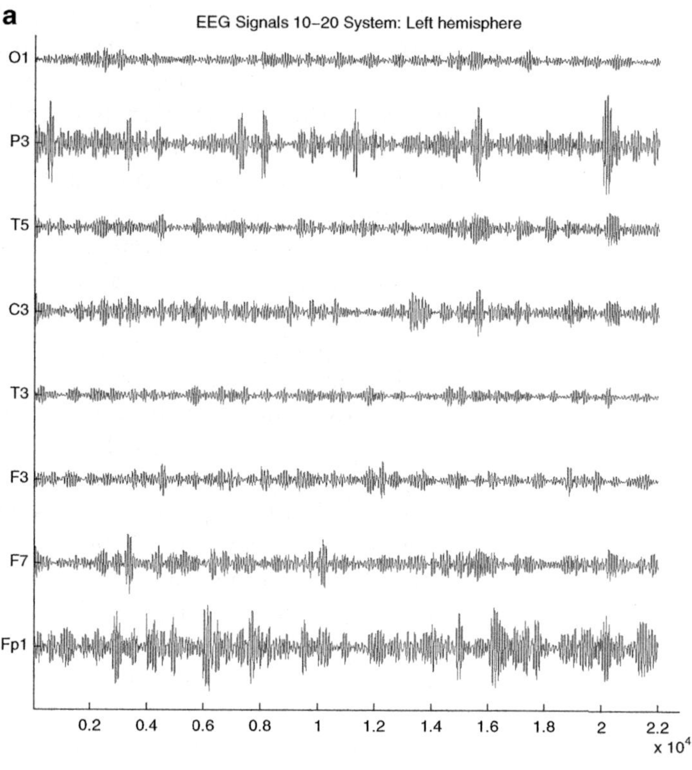

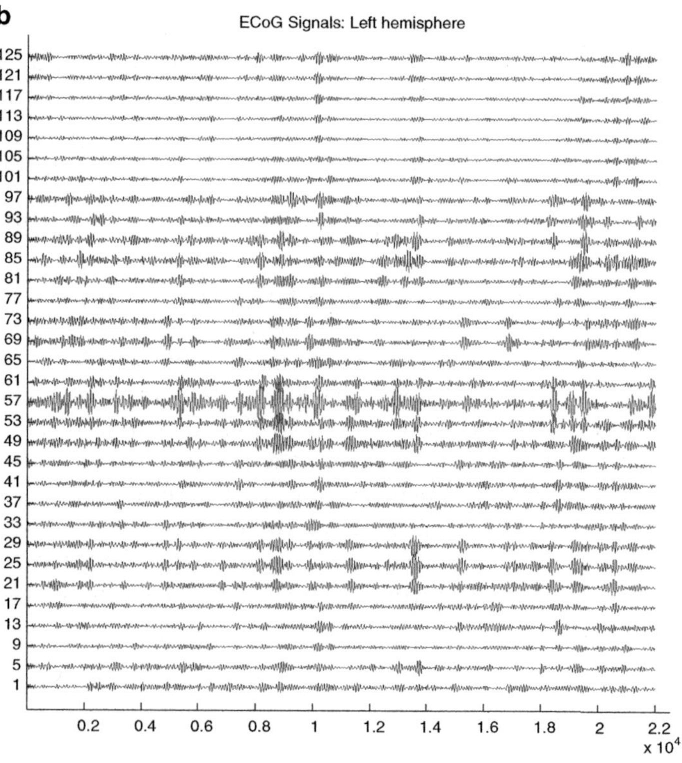

Fig. 4 Simultaneous EEG and ECoG. (**a**) EEG. (**b**) ECoG

Table 1
Error mean and variances for linear mixing assumption

	THETA	ALPHA	BETA
Frequency	3.5–7.5 Hz	8–13 Hz	14–30 Hz
Activity	Falling asleep	Closed eyes	Concentration
Error (mean)	7.32E−04	0.001	3.3539
Error (variance)	1.04E−06	2.06E−06	97.8081

where $\mathbf{X}_{\mathrm{EEG}} \in \mathbb{R}^{18 \times N}$ represents the EEG data from 18 channels (each column represents a channel), $\mathbf{X}_{\mathrm{ECoG}} \in \mathbb{R}^{128 \times N}$ represents the ECoG data from 128 channels (each column represents a channel), and $\mathbf{A} \in \mathbb{R}^{18 \times 128}$ is the unknown mixing matrix. Before solving optimization problem (18), the data has been filtered to remove high(≥ 45 Hz) and low(≤ 0.5 Hz) frequencies. In addition to that, we have used 50 and 60 Hz notch filters to remove the noise induced by the electric current. Furthermore, we have used average reference for all the channels before conducting the analysis, i.e., for example, we referenced EEG data value from a particular channel at a given time instance with average EEG from all the EEG channels at the same time instance. Similarly, we referenced the ECoG data.

Instead of solving optimization problem (18) with respect to the whole data, we have solved the optimization problem multiple times, with reduced data sets. Typically, the reduced data sets are nothing but smaller chunks of data with the window size of $N = 2,000$ points for a particular frequency band, taken from the original data. The objective of optimization problem (18) is to calculate the total absolute error due to linear mixing assumption in different frequency bands. Thus, this experiment provides a mechanism to understand mixing around the skull. A low error shows that linear mixing assumption is valid whereas a high error indicates that the linear mixing assumption is invalid. Moreover, the ultimate goal is to show if the mixing is constant over the time. However, to develop such results, complete understanding regarding total number of sources should be available. At this point, we will present a simple experiment, where we assume that all the ECoG electrodes are sources, and all the EEG electrodes are mixtures. Thus, the model is highly under-determined, but due to the availability of both source and mixture information, optimization problem (18) transforms to a convex programming problem.

The results of the experiment are shown in Table 1. While calculating the error, we have considered only those channels that are placed on the left hemisphere, i.e., 8 EEG channels, and 128 ECoG channels. Since ECoG data is available for the left hemisphere, we have neglected the right hemispherical channels in EEG. In Table 1, the third row presents the mean value of the total absolute error over all the multiple runs on the reduced data set. The fourth row provides the corresponding variance of the total absolute errors for the multiple runs. The low average error and negligible variance in alpha and theta bands suggest the existence of linear mixing across the skull. At this stage, from the experiment, the linear mixing assumption is validated in the neural data. However, it is far from theoretical validation and generalization to other neural data sets. Furthermore, the other critical question, which directs towards the constancy in mixing is open for further investigation.

4.2 fMRI Data Analysis

In this, and in the following experiments, we focus on nonnegative sources. Most of the time, images fall under the nonnegative sources category. The aim of the current experiment is to examine the validity of PPC sparsity assumption in fMRI data. Generally, sparsity in fMRI images is a more plausible assumption than independence (12). However, the PPC sparsity may not be applicable to fMRI data. Through this experiment, we hope to conclude the applicability of PPC method on fMRI data.

For this experiment, an fMRI data set examined previously in the literature is considered. The description of experimental setup and data collection of the fMRI data is available in (55), where the authors compare ICA and SCA methods. Here, we use the same data to analyze the convex hull of the fMRI data. The basic idea is if PPC assumptions are valid, then the convex hull should be a simplex. Furthermore, if the convex hull is simplex in n dimensions, then an affine transformation to lower dimensions, like PCA, should result in a simplex in lower dimensions. Furthermore, the extreme points (or vertices) of simplex (or convex hull) is nothing but the columns of the mixing matrix. Thus, finding convex hull leads to the identification of mixing matrix (Fig. 5).

The fMRI data from a single subject consists of 98 images taken every 50 ms. We vectorize the images by scanning the image vertically from top left to bottom right. Next, the dimensionality of data is reduced to three principal components using PCA for ease of identifying the convex hull. Since the images are vectorized, the relation between fMRI data and PCA components is intangible. However, the usage of PCA has a crucial advantage in visualization of the data, which in turn leads to easy identification of the convex hull in the lower dimensions. Figure 6a shows the scatter plot of three principal components. Now, taking the three principal components, we project the data on a two-dimensional plane. This projection of the three principal components into two dimensions is shown in Fig. 7. First thing to notice is the projection in two dimensions is different from Fig. 6b, which shows the scatter plot of two principal components. Next, a simplex that fits all the points in Fig. 7 gives the information pertaining to the columns of the

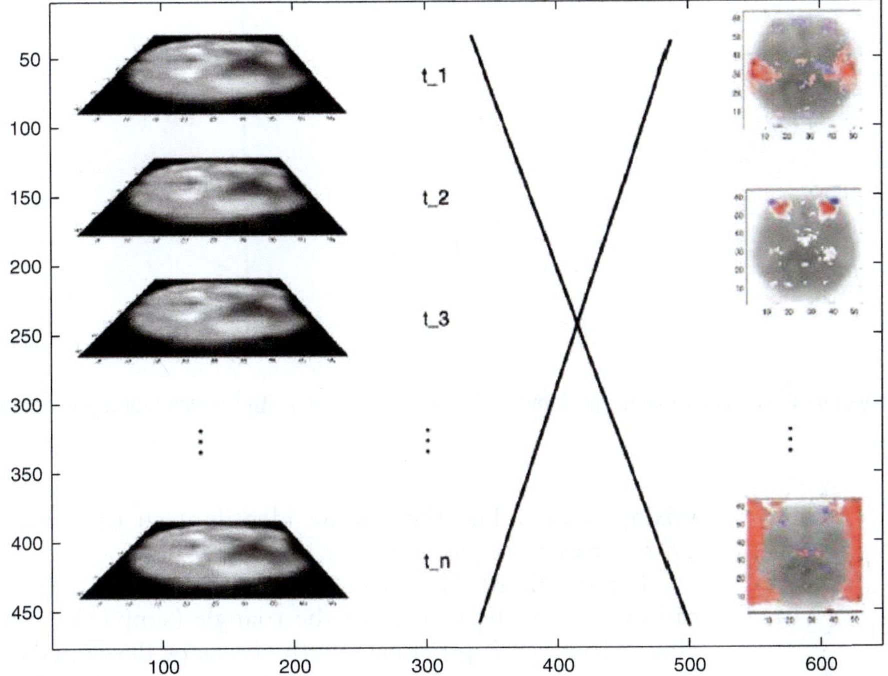

Fig. 5 fMRI components (55)

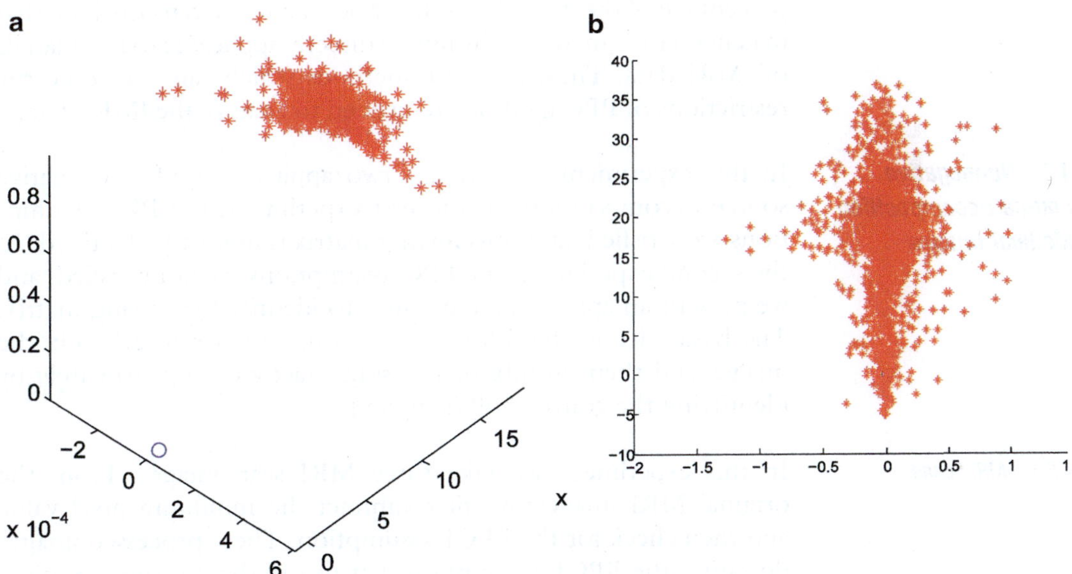

Fig. 6 fMRI data visualization. (**a**) PCA reduction to three dimensions. (**b**) PCA reduction to two dimensions

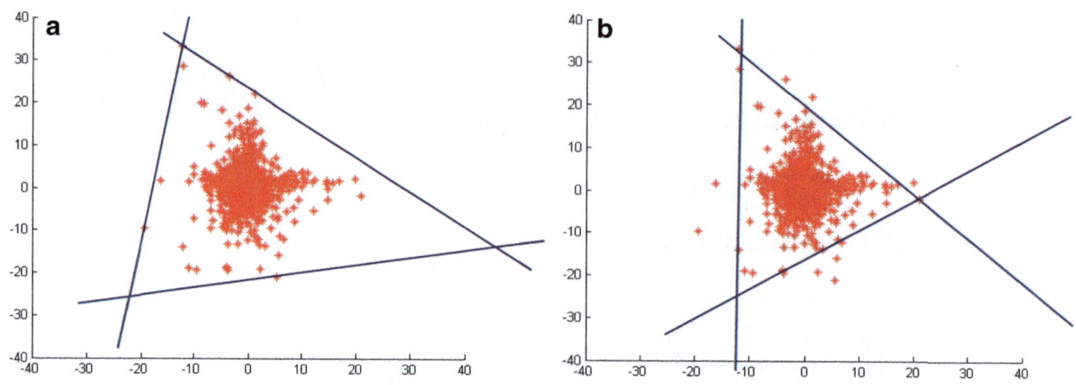

Fig. 7 Convex hull PPC1 assumption. (**a**) Convex hull representation 1. (**b**) Convex hull representation 2

mixing matrix. For the unique identification of mixing matrix, existence of a unique simplex is necessary.

For the fMRI data, the PPC1 conditions are not completely satisfied [since the vertices of the triangle (simplex) are not available]. However, approximate methods can be developed to identify the extreme points of the triangle. For example, Fig. 7a, b shows different ways of extrapolating the data, to obtain the vertices. Obviously, this idea can be extended in high dimensions, by defining objectives like finding a simplex of minimum volume containing all the data, or finding a simplex of minimum volume containing high percentage of the data. From this experiment, we can conclude that in general PPC method may not be directly applicable to the analysis of fMRI data. Thus, alternate methods which can overcome the restrictions of PPC method are needed to analyze the fMRI data.

4.3 Nonnegative Source Decomposition Medical Images

In this experiment, we present two applications of nonnegative source decomposition. In the first experiment, the PPC assumptions are satisfied, and thus mixing matrix is uniquely identified. In the second experiment, the PPC assumptions are not satisfied, and we present an approximate method to identify the mixing matrix. The basic idea in the following experiment is to linearly mix the images and then un-mix them using exactly or approximately by identifying the convex hull (simplex).

4.3.1 MRI Scans

In this experiment we take three MRI scan images. From the original MRI images, we first subtract the minimum pixel value and then check for the PPC1 assumption. These processed images do satisfy the PPC1 assumption. Let us call these images as initial images. Now the initial images are linearly mixed to obtain three mixture images. The goal is to extract the pure source images from the mixture images. Figure 8a displays the initial sources, and Fig. 8b represents a mixtures.

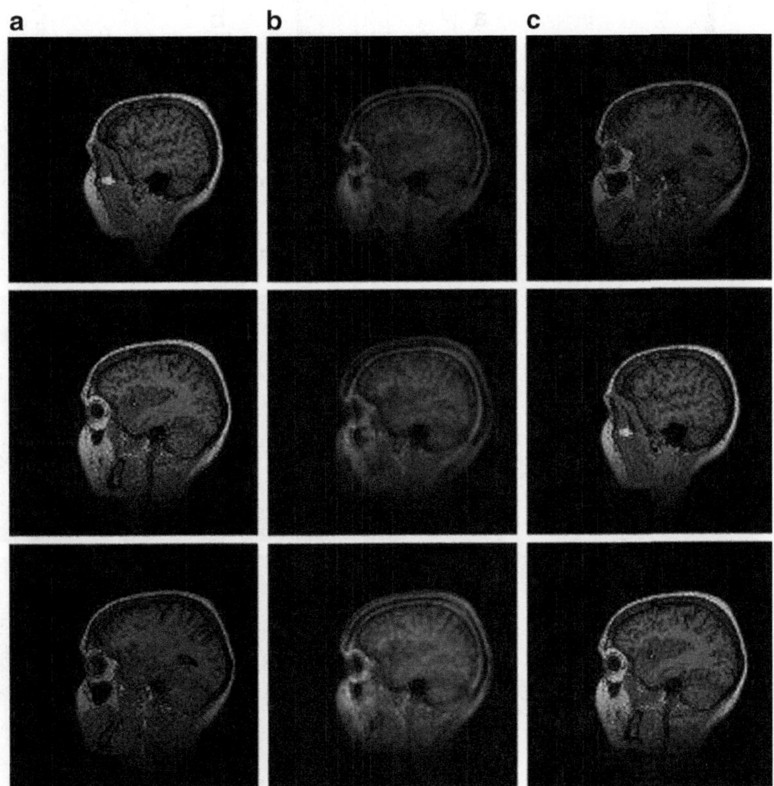

Fig. 8 Mixing and unmixing of MRI scans. (**a**) Original. (**b**) Mixture. (**c**) Recovered

The mixture images are vectorized and reduced to three dimensions using PCA. Now, the three principal components are projected on the two-dimensional space. From the projected data, we identify the three unique vertices of the simplex (triangle). Since the initial images satisfy the PPC1 assumption, we have unique vertices, i.e., no extrapolation is needed. From the vertices of the simplex, we construct the mixing matrix. Using the information of the mixing matrix, the source images are recovered. Figure 8c shows the recovered source images. Since the PPC1 assumptions were satisfied initially, except the ordering and intensity (ambiguity of permutation and scalability) of the images, all the other information is recovered from the mixture images.

4.3.2 Finger Prints

Similar to the MRI scans, in the finger print scans, the minimum pixel in each image is first subtracted from the images, and then checked for the PPC1 assumption. These processed images do not satisfy the PPC1 assumption. Let us call these images as initial images. Now we repeat the linear mixing operation as in the MRI scan images experiment, to obtain three mixture images. Since the PPC1 assumption is not satisfied, the goal is to approximately

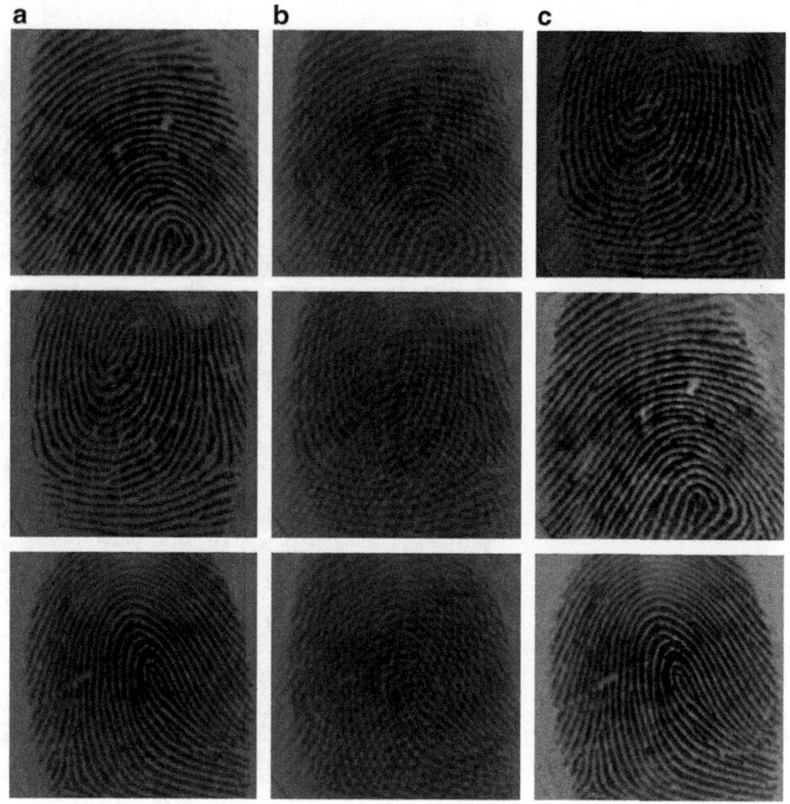

Fig. 9 Mixing and unmixing of finger prints. (**a**) Original. (**b**) Mixture. (**c**) Recovered

extract the pure sources from the mixture images. Figure 9a displays the initial sources, and Fig. 9b represents a mixture.

The mixture images are vectorized and reduced to three dimensions using PCA. Now, the three principal components are projected on the two-dimensional space. We use simple k-means algorithm to find the three best extreme points, from among all the extreme points, of the convex hull on the projected data. Taking the three points as the vertices of simplex, we construct the mixing matrix. Sources are recovered using the information of mixing matrix. Figure 9c shows the extracted sources. It can be seen from the recovered sources that, apart from intensity and ordering, the recovery is not perfect.

The purpose of presenting the two experiments (MRI scans and finger prints) was to highlight the applicability of the novel BSS methods that are based on nonnegative sources. MRI scan experiment shows that when the PPC1 assumption satisfies, perfect recovery is obtained whereas from the finger print experiment, we can see that the approximate recovery with little modification to the PPC method performs well on unmixing the data.

5   Criticism

There are two basic types of criticism that has been commented on the usage of BSS approaches in computational neuroscience. The primary comment is on the loss of order in the sources. As discussed in the Sect. 2, the scalability issue of BSS method can be overcome by using suitable normalization approaches. However, identifying appropriate in general is not possible. Furthermore, BSS for EEG or MEG is underdetermined, thus knowing of order is practically important. Makegi et al. in (44) discussed the above issue and stated the importance of knowing "what the sources are?" instead of knowing "where the sources are?" in understanding cortical activity. Furthermore, the underdetermined case is usually resolved by experimental design, where artifacts are induced into the data while recording to reduce the underdeterminacy.

The other type of criticism that is received on the usage of BSS approaches is on the validity of the assumptions imposed on mixing and source matrices. The smearing of signal by volume conduction is instantaneous, thus no delay assumption is not much of a concern. The linear mixing assumption is the critical one and is hard to validate experimentally. However, superposition of signals (a typical natural phenomenon) can be used to support the notion of linear mixing. In addition to that, the assumptions imposed on source signals are often objected. Statistical independence among the neuronal signals is hard to justify. Therefore, researchers working with ICA directed the research justifying statistical independence among artifacts and neuronal signals. On the other hand, the assumptions of the novel SCA approaches are yet to be experimentally validated on neurological data. Furthermore, sparsification methods transforming the given problem into a sparse source problem are yet to be explored.

6   Conclusion

BSS approaches based on ICA are well known in the computational neuroscience. However, sparsity-based BSS methods are relatively new, and their potential is yet to be explored in the area of computational neuroscience. Through the systematical overview presented in this paper, we hope to increase awareness of the novel sparsity-based BSS methods and highlight the differences between ICA and SCA methods. The primary difference between ICA-based methods, compared to SCA-based methods is that the ICA-based methods are mostly suitable for artifact filtering. However, the striking difference is that the SCA-based methods may be suitable for separating pure sources, which are not necessarily statistically independent. Similar to EEG/MEG analysis with ICA, where artifacts are induced into the signal via strategic experiments, efficient experiments for SCA can be designed, where sparsity can be induced into the source signals. Furthermore, sparsification methods (like wavelet transforms) that can efficiently sparsify source signals can also be used to analyze non-sparse source signals. To sum up, SCA-based methods may open a new door for understanding the mysteries of the brain.

Acknowledgements

Research has been partially supported by DTRA, Air Force and NSF grants. The last author is also partially supported by LATNA Laboratory, NRU HSE, RF government grant, ag. 11. G34.31.0057.

References

1. Hérault J, Jutten C, Ans B (1985) Détection de grandeurs primitives dans un message composite par une architecture de calcul neuromimétique en apprentissage non supervisé. In: 10 Colloque sur le traitement du signal et des images, FRA, 1985, GRETSI, Groupe dEtudes du Traitement du Signal et des Images

2. Syed M, Georgiev P, Pardalos P (2013) A hierarchical approach for sparse source blind signal separation problem. Comput Oper Res 41:386–398

3. Aharon M, Elad M, Bruckstein A (2006) On the uniqueness of overcomplete dictionaries, and a practical way to retrieve them. Linear Algebra Appl 416(1):48–67

4. Georgiev P, Theis F, Cichocki A (2005) Sparse component analysis and blind source separation of underdetermined mixtures. IEEE Trans Neural Network 16(4):992–996

5. Gribonval R, Schnass K (2010) Dictionary identification - sparse matrix-factorization via l_1-minimization. IEEE Trans Inform Theory 56(7):3523–3539

6. Kreutz-Delgado K, Murray J, Rao B, Engan K, Lee T, Sejnowski T (2003) Dictionary learning algorithms for sparse representation. Neural Comput 15(2):349–396

7. Te-Won L (1998) Independent component analysis, theory and applications. Kluwer Academic, Boston

8. Hyvarinen A, Oja E (2000) Independent component analysis: algorithms and applications. Neural Network 13(4):411–430

9. Cichocki A, Zdunek R, Amari S (2006) New algorithms for non-negative matrix factorization in applications to blind source separation. In: IEEE international conference on acoustics, speech and signal processing, 2006. ICASSP 2006 Proceedings, vol 5. IEEE, pp V–V

10. Naanaa W, Nuzillard J (2005) Blind source separation of positive and partially correlated data. Signal Process 85(9):1711–1722

11. Georgiev P, Pardalos P, Theis F (2007) A bilinear algorithm for sparse representations. Comput Optim Appl 38(2):249–259

12. Daubechies I, Roussos E, Takerkart S, Benharrosh M, Golden C, D'Ardenne K, Richter W, Cohen J, Haxby J (2009) Independent component analysis for brain fMRI does not select for independence. Proc Natl Acad Sci USA 106 (26):10415–10422

13. Hyvarinen A (1999) Fast and robust fixed-point algorithms for independent component analysis. IEEE Trans Neural Network 10 (3):626–634

14. Sameni R, Jutten C, Shamsollahi M (2006) What ICA provides for ECG processing: application to noninvasive fetal ECG extraction. In: 2006 I.E. international symposium on signal processing and information technology, pp 656–661. IEEE

15. Bartlett M, Movellan J, Sejnowski T (2002) Face recognition by independent component analysis. IEEE Trans Neural Network 13 (6):1450–1464

16. Comon P (1994) Independent component analysis, a new concept? Signal Process 36 (3):287–314

17. Lee T, Girolami M, Sejnowski T (1999) Independent component analysis using an extended infomax algorithm for mixed subgaussian and supergaussian sources. Neural Comput 11 (2):417–441

18. Hyvärinen A, Karhunen J, Oja E (2002) Independent component analysis. Wiley, New York

19. Stone J (2004) Independent component analysis: A tutorial introduction. Bradford Books, Cambridge

20. Joho M, Mathis H, Lambert R (2000) Overdetermined blind source separation: Using more sensors than source signals in a noisy mixture. In: Proceedings of the international conference on independent component analysis and blind signal separation, pp 19–22

21. De Lathauwer L, De Moor B, Vandewalle J, Cardoso J (2003) Independent component analysis of largely underdetermined mixtures. In: Proceedings of the 4th international symposium on independent component analysis and blind signal separation (ICA 2003), pp 29–34

22. Zhang K, Chan L-W (2003) Dimension reduction based on orthogonality - a decorrelation method in ICA. In: Artificial neural networks and neural information processing - ICANN/ ICONIP 2003, pp 173–173

23. Zhang K, Chan L-W (2006) Dimension reduction as a deflation method in ICA. IEEE Signal Process Lett 13(1):45–48

24. Huber P (2012) Data analysis: what can be learned from the past 50 years, vol 874. Wiley, New York

25. Cover T, Thomas J (2006) Elements of information theory. Wiley-Interscience, New York

26. Pham D, Garat P (1997) Blind separation of mixture of independent sources through a quasi-maximum likelihood approach. IEEE Trans Signal Process 45(7):1712–1725

27. Pham D, Garat P (1997) Blind separation of mixture of independent sources through a maximum likelihood approach. In: Proceedings of EUSIPCO, Citeseer

28. Cardoso J-F (1999) High-order contrasts for independent component analysis. Neural Comput 11(1):157–192

29. Cardoso J-F, Souloumiac A (1993) Blind beamforming for non-Gaussian signals. IEE Proc F Radar Signal Process (IET) 140:362–370

30. Cardoso J-F, Souloumiac A (1996) Jacobi angles for simultaneous diagonalization. SIAM J Matrix Anal Appl 17(1):161–164

31. Belouchrani A, Abed-Meraim K, Cardoso J-F, Moulines E (1997) A blind source separation technique using second-order statistics. IEEE Trans Signal Process 45(2):434–444

32. Georgiev P, Cichocki A (2003) Robust independent component analysis via time-delayed cumulant functions. IEICE Trans Fund Electron Comm Comput Sci 86(3):573–579

33. Georgiev P, Pardalos P, Cichocki A (2007) Algorithms with high order convergence speed for blind source extraction. Comput Optim Appl 38(1):123–131

34. Hyvärinen A, Oja E (1997) A fast fixed-point algorithm for independent component analysis. Neural Comput 9(7):1483–1492

35. Georgiev P, Theis F (2004) Blind source separation of linear mixtures with singular matrices. Indepen Comp Anal Blind Signal Separ 121–128

36. Chan T-H, Ma W-K, Chi C-Y, Wang Y (2008) A convex analysis framework for blind separation of non-negative sources. IEEE Trans Signal Process 56(10):5120–5134

37. Yang Z, Xiang Y, Rong Y, Xie S (2013) Projection-pursuit-based method for blind separation of nonnegative sources. IEEE Trans Neural Netw Learn Syst 24(1):47–57

38. Sun Y, Xin J (2012) Nonnegative sparse blind source separation for NMR spectroscopy by data clustering, model reduction, and ℓ_1 minimization. SIAM J Imag Sci 5(3):886–911

39. Georgiev P, Theis F, Ralescu A (2007) Identifiability conditions and subspace clustering in sparse BSS. Indepen Comp Anal Signal Separ 357–364

40. Makeig S, Bell A, Jung T, Sejnowski T et al (1996) Independent component analysis of electroencephalographic data. Adv Neural Inform Process Syst 145–151

41. Vigário R, Jousmäki V, Haemaelaeninen M, Haft R, Oja E (1998) Independent component analysis for identification of artifacts in magnetoencephalographic recordings. Adv Neural Inform Process Syst 229–235

42. Vigário R, Sarela J, Jousmiki V, Hamalainen M, Oja E (2000) Independent component approach to the analysis of EEG and MEG recordings. IEEE Trans Biomed Eng 47 (5):589–593

43. Bell A, Sejnowski T (1995) An information-maximization approach to blind separation and blind deconvolution. Neural Comput 7 (6):1129–1159

44. Makeig S, Jung T-P, Ghahremani D, Bell A, Sejnowski T (1996) What (not where) are the sources of the EEG? In: The 18th annual meeting of the cognitive science society

45. Jung T, Makeig S, Westerfield M, Townsend J, Courchesne E, Sejnowski T, et al (2000) Removal of eye activity artifacts from visual event-related potentials in normal and clinical subjects. Clin Neurophysiol 111 (10):1745–1758

46. Croft R, Barry R (2000) Removal of ocular artifact from the EEG: a review. Neurophysiol Clin/Clin Neurophysiol 30(1):5–19

47. McKeown M, Sejnowski T, et al. (1998) Independent component analysis of fMRI data: examining the assumptions. Hum Brain Mapp 6(5–6):368–372

48. Formisano E, Esposito F, Kriegeskorte N, Tedeschi G, Di Salle F, Goebel R (2002) Spatial independent component analysis of functional magnetic resonance imaging time-series: characterization of the cortical components. Neurocomputing 49 (1):241–254

49. McKeown M, Jung T-P, Makeig S, Brown G, Kindermann S, Lee T-W, Sejnowski T (1998) Spatially independent activity patterns in functional MRI data during the stroop color-naming task. Proc Natl Acad Sci USA 95 (3):803–810

50. Stone J, Porrill J, Buchel C, Friston K (1999) Spatial, temporal, and spatiotemporal independent component analysis of fMRI data. In: Proceedings of the Leeds statistical research workshop, Citeseer

51. Seifritz E, Esposito F, Hennel F, Mustovic H, Neuhoff J, Bilecen D, Tedeschi G, Scheffler K, Di Salle F (2002) Spatiotemporal pattern of neural processing in the human auditory cortex. Science 297(5587):1706–1708

52. Li Y, Adalı T, Calhoun V (2007) Estimating the number of independent components for functional magnetic resonance imaging data. Hum Brain Mapp 28(11):1251–1266

53. Calhoun V, Adali T, Pearlson G, Pekar J (2001) Group ICA of functional MRI data: separability, stationarity, and inference. In: Proceedings of the international conference on ICA and BSS, vol 155, San Diego

54. Calhoun V, Liu J, Adalı T (2009) A review of group ICA for fMRI data and ICA for joint inference of imaging, genetic, and ERP data. Neuroimage 45(Suppl 1):S163

55. Georgiev P, Theis F, Cichocki A, Bakardjian H (2007) Sparse component analysis: a new tool for data mining. Data Min Biomed 91–116

Neuromethods (2015) 91: 323–350
DOI 10.1007/7657_2013_67
© Springer Science+Business Media New York 2013
Published online: 15 December 2014

Current Trends in ERP Analysis Using EEG and EEG/fMRI Synergistic Methods

K. Michalopoulos, M. Zervakis, and N. Bourbakis

Abstract

Analysis of electroencephalography (EEG) and functional magnetic resonance imaging (fMRI) data provides useful insight on how the brain works and provides important information for the diagnosis of different brain pathologies. EEG provides excellent temporal resolution for the study of brain activity while, on the other hand, fMRI provides good spatial localization of different cognitive functions. Although the combination of the two modalities looks very promising, there are still problems and challenges that need to be addressed in order to take advantage of their supplementary nature. In this chapter, we review the various methods and techniques used for the joint analysis of EEG and fMRI.

Key words ERP, EEG, fMRI, EEG–fMRI integration

---

1 Introduction

The human brain is the most complex organ known. It consists of millions of neurons interconnected with each other, forming a large network capable to store and process information from the environment. Towards understanding how the brain works different techniques have been developed which allow us to capture different manifestations of the operating brain.

Electroencephalography (EEG) is one of the first techniques used to study the living brain by measuring the electric current reaching the scalp as a result of the brain activity [1]. This electric current is recorded using electrodes attached in the human scalp. Electrical oscillations have been associated with different brain functions or cognitive states and provided insight on the way the brain works. On top of that, EEG allows the characterization of different brain pathologies and a lot of effort has been devoted to the research of markers that can characterize a brain disorder and help the diagnosis [2]. Different studies have been conducted in search of discriminant features for diseases like epilepsy, Alzheimer disease, and schizophrenia [2]. We will present in more detail the pros and cons of EEG in the analysis of brain activity in Section 2 along with measures and techniques that allow us to analyze and detect specific brain activations.

Magnetic resonance imaging (MRI) is an imaging technique that takes advantage of the properties of hydrogen and its interaction with a large external magnetic field and radio waves to produce detailed images of the tissue under consideration. Functional MRI (fMRI) is an imaging technique that measures the change in blood flow that occurs alongside with neuronal activations. There exist different techniques that enable the measurement of changes in the brain flow like blood oxygen-level dependent (BOLD) measurements [3], arterial spin labeling [4], and injected contrast agents techniques [5]. Regardless of the method used, fMRI provides an indirect measure of neuronal activity and lacks the temporal resolution of EEG. BOLD is widely used in the different multi-modal studies, so we are going to focus mainly on BOLD studies for the remaining of the article. We will present the properties of BOLD fMRI along with its strengths and limitations in more detail in Section 3.

Combining information obtained from different modalities seems really promising, since fMRI and EEG (or MEG) seem to be complementary in nature and are ideal candidate modalities for such integration. EEG provides excellent temporal resolution of neural activations and MRI/fMRI provides structural and spatial accurate information about metabolic changes in different brain regions—that can be attributed to neural activation.

We will focus on a specific category of experiments, known as event related, that involve repetition of a specific input or test multiple times in order to infer information about the EEG/MRI response under certain conditions. These conditions can be correlated with specific mental states or cognitive functions and have been proven extremely useful in clinical and physiological research and are well known as event-related potentials (ERPs). There exists a rich literature about the functional meaning of the different peaks of ERP in EEG (such as the P1, N1, and P3), which are thought to reflect different aspects of information processing in the brain.

In this chapter we are going to discuss methods and techniques used for the analysis of EEG and fMRI. We will focus on methods for analysis and characterization of event-related experiments in EEG as well in fMRI. Then, we will discuss the different approaches to EEG–fMRI integration and fusion along with their strengths and limitations. The methods presented have been used in order to isolate, identify, and characterize neuronal activations in terms of their functional role. The last section is devoted to methods used for the integration of the individual brain activations in terms of a distributed network and attempts to integrate EEG and fMRI in the network space.

2 Event-Related Experiments Using EEG

The unmatched feature of EEG measures the electrical activity of the brain that reaches the scalp and as a result presents excellent temporal resolution. EEG measures the electrical activity of the brain that reaches the scalp and as a result presents excellent temporal resolution, of the millisecond scale. However, the electrical signature of single neurons is too weak to be recorded in the scalp. The electrical signal we capture with EEG is the combined, synchronous firing of large group of neurons, mainly large pyramidal neurons oriented perpendicular to the scalp [6].

Being a direct measure of neuronal activity with great temporal resolution, EEG is used for studying the temporal dynamics of neuronal activity. The main drawback of EEG is that it presents poor spatial resolution and we cannot directly attribute EEG features to a certain brain area, without prior knowledge. More specifically, as the generated electrical potentials from a certain brain region have to travel through brain tissue to scalp, the end result is spatial dispersion and mixing with concurrent electrical activations of other nearby regions. The mixing effect due to the transmission process of the generated electromagnetic field is known as volume conduction effect. As a result of this, the final recorded signal captures the sum of multiple activations propagated to the scalp from nearby regions. The problem of recovering the unknown sources from the observed EEG signal in the electrodes is an ill-posed problem known as the EEG/MEG inverse problem [7].

2.1 Analysis and Characterization of Event-Related Potentials in EEG

The ERP represents the brain response under a specific input. It has been proven extremely useful in clinical and physiological research. There are a lot of studies about the functional meaning of the different characteristics of the ERP such as peaks and valleys of the time signal or increases/decreases of a specific band with stimulus. Such characteristics are considered as manifestations of specific aspects of information processing in the brain [1]. Based on this modular view of the processing that takes place in the brain ERPs were considered to be generated by fixed latency, phase-locked responses [1]. The underlying assumption is that a certain task will evoke a specific brain response. Repeating the same experiment multiple times we will be able to detect this response by averaging the recorded signals over trials in an effort to increase the signal-to-noise ratio (SNR) in the average signal [8].

Induced activities are expressed through the increase or decrease of energy in a specific band post-stimulus, denoted as event-related synchronization (ERS) or desynchronization (ERD), respectively [9]. Induced activities are oscillations non-phase-locked to the stimulus and have been associated with a variety of different functions related to perception and different types of cognitive processes [10]. Studies on induced oscillations put the classical, modular ERP paradigm into question. In [11], authors correlate alpha energy and alpha phase on stimulus onset with the ERP amplitude, indicating that the ERP and EEG oscillations interact and relate to each other. Evoked and induced oscillations may be considered as coupled processes progressing in time, with different spatial localization of origin and partially overlapping frequency content [12].

Nevertheless, there is an ongoing debate regarding the generative model of ERP activations and there is no definite answer on the generating mechanism of the ERP. The models that have been proposed can be grouped in two categories. The additive model considers the evoked response as completely independent from the ongoing background EEG, whereas the phase reset model suggests a phase reorganization of ongoing EEG oscillations as the generative mechanism of the ERP. Although this debate has attracted a lot of attention, there is not enough evidence which model is more valid. The evoked model has enabled the extraction of useful information about the neurophysiologic origins of ERP, mainly through averaging over trials. Underlying the evoked model approach is the assumption that the event itself gives rise to certain brain processes at fixed latency and with similar phase independently from ongoing EEG. This activation serves the response to the event and then vanishes. Averaging of the single trials and inspection of the characteristics of the resulting ERP waveform results in a significant increase of the SNR, so that useful information can be easily extracted about the brain process and its characteristic markers.

On the other hand, the phase-resetting model assumes that the phase-locked phenomenon is not different from the induced activations that appear to be time-locked but not phase-locked. This model states that the different ongoing oscillations either synchronize in phase or time modulated by the stimulus. This model reflects a more dynamic approach on how the different brain regions are synchronizing in order to respond to the given input.

Until now there is not a single answer to this ongoing debate. Indications that show the validity of one or the other model have been reported in many cases and each is strong enough to rule the other model out. On top of that there exist indications that maybe both of these models are valid under different circumstances, representing different types of activations. In the next section, we will present measures that try to quantify such events.

2.2 ERP Analysis Measures and Methods Based on EEG Analysis

A variety of methods and measures to characterize the nature of EEG activity in terms of their major time/frequency activity and topographic origin have been employed [13–16]. EEG is a highly non-stationary signal and time-frequency transforms have been employed in order to examine the time evolution of its power spectrum. Short time Fourier transform and the wavelet transform are two of the most widely used time-frequency transforms for EEG analysis [16–19].

The number of brain activations that are recorded simultaneously at the electrodes is generally unknown. The spatial mixing of these activations by volume conduction and the need to distinguish brain activity from signals originating from other parts of the human brain brought into the foreground techniques for the decomposition or de-mixing of the multichannel EEG. The most prominent techniques include principal component analysis (PCA) and independent component analysis (ICA) [20].

PCA operates in the multichannel EEG signal and transforms the data into uncorrelated mutually orthogonal components. PCA is a well-known technique for dimensionality reduction and has many applications in the field of pattern recognition. It has been used in this context as a preprocessing step before the application of ICA.

ICA is increasing in popularity in the field of EEG signal processing. ICA is a statistical signal processing technique which tries to solve the so-called blind source separation (BSS) problem. In the BSS problem the goal is to recover/estimate the original sources by using only the information available from the observed mixed signals without having any information about the mixing process or the sources. In general, the classic BSS model where we observe k signals which are generated by the linear mixture of m source signals can be described as

$$x(t) = A \times s(t). \tag{1}$$

Any approach used to solve this kind of problem has to make some assumptions regarding the source signals, the mixing process, or both. The main assumption of ICA is that the source signals are statistically independent. The success of ICA in the decomposition of many natural signals is that the assumption of independence seems to be more realistic than other mathematical constraints, like orthogonality.

One of the first applications of PCA and ICA was the denoising of the multi-channel EEG signal. PCA was the first technique employed for artifact correction and removal from EEG recordings [21, 22] but soon it was found that ICA was more efficient for this purpose [23]. PCA is commonly used for reducing the dimensions of the problem before applying ICA or for data reduction and summarization of time-frequency transformed data [13, 24].

ICA has been proven very successful in the analysis of EEG data and especially in the exploration of the dynamics of ERP data. It is being used for removal of artifacts from the EEG recordings with success without losing relevant information. Also, it has been successfully applied on continuous or event-related EEG to decompose it into a sum of spatially fixed and temporally independent components that can lead in different spatial distribution patterns, which in turn may be directly attributed to underlying cortical activity. The classical procedure for quantifying evoked responses is through averaging over trials. This procedure enhances the task-related response and filters out the irrelevant background EEG. As we noted earlier, this approach assumes that background EEG behaves as random noise and that the task-related response is approximately identical from trial to trial. Phase relevant measures as intertrial coherence (ITC) have been used to characterize the phase consistency of the detailed time frequency content throughout trials [19]. ITC can be expressed as following:

$$\text{ITC}(k, n) = \left| \frac{1}{T} \sum_i \frac{X_i(n, k)}{|X_i(n, k)|} \right| \tag{2}$$

In [16] a similar measure for quantifying phase-locked activity in ERP trials called phase intertrial coherence (PIC).

$$\text{PIC}(k, n) = \frac{\sum_i X_i(n, k)}{\sum_i |X_i(n, k)|} \tag{3}$$

Though the two measures are similar, PIC measure takes under consideration the amplitude of the measured signal and not only the phase [16]. The first measure for quantifying event-related oscillations is the ERD and ERS which measures increase or decrease in the power of specific bands relative to some baseline pre-stimulus power [9]. Event-related spectral perturbations (ERSP) is an extension of these measures in the time-frequency domain, enhanced with tests for significance [18]. In [16] a measure for the evaluation of consistent oscillations across trials is introduced under the term phase-shift intertrial coherence (PsIC) which is defined as following:

$$\text{PsIC}(k, n) = \sum_i \frac{X_i(n, k)|^2}{\max \sum_i |X_i(n, k)|^2} \tag{4}$$

This measure cannot be directly compared with the other two since it does not take into account variations in the power of a certain band but rather examines whether a narrow band oscillation is present in single trials, in a consistent manner. In Fig. 1 we tried to formulate the representation in the measures of the different assumptions about the generation of the ERP covering the spectrum between the two models. We can see that even though we can distinguish evoked from induced activations through the PIC measure the underlying generative model cannot be distinguished using only these measures. Further research is needed in order to shed light on this complex debate.

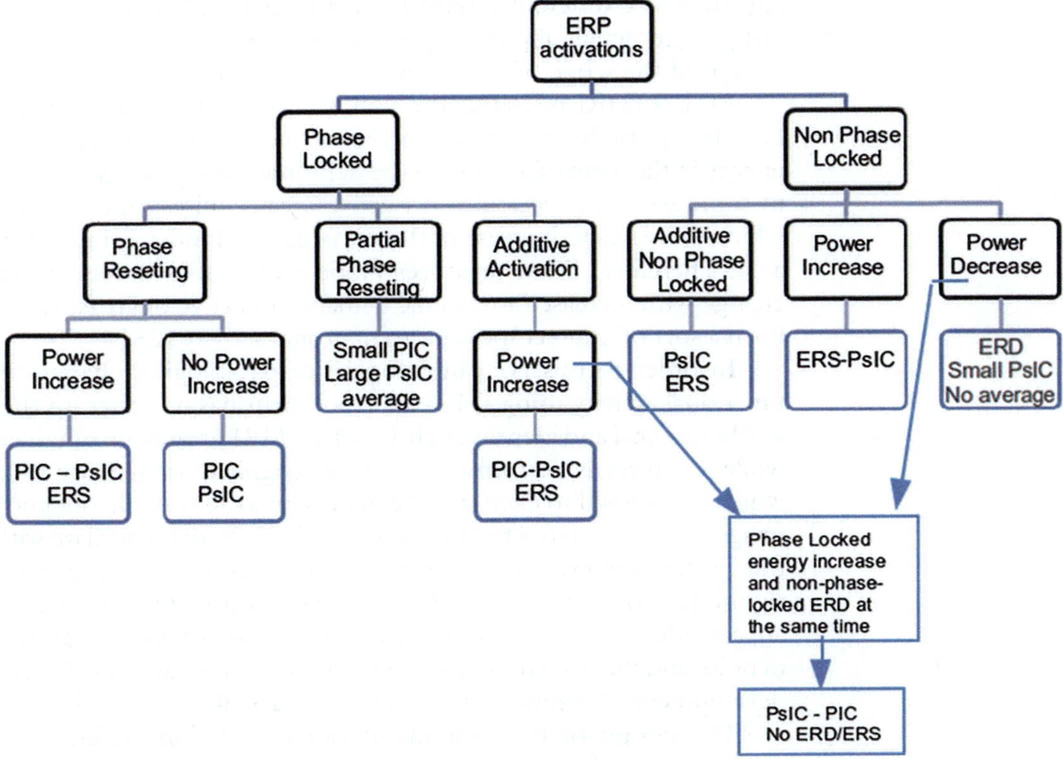

Fig. 1 Summary of the different ERP activations and how they are reflected in the measures. We can see that the measures discussed above cannot sufficiently characterize the nature of the activations.

3   Methods for fMRI Analysis

The main goal of fMRI is to identify and map brain regions to certain brain functionalities. We will focus on fMRI using BOLD, as the majority of fMRI-related studies are based on this approach.

It is important to describe the basic principles behind fMRI since it will allow us to better understand the nature of the measured signal and will allow us to identify the level at which the two modalities can be combined. BOLD fMRI takes advantage of the magnetic properties of the hemoglobin which are used as a natural contrast agent. Hemoglobin is the protein responsible for the transport of oxygen from the respiratory organs (lungs) to the rest of the body. The oxygenated hemoglobin presents different magnetic properties than the deoxygenated one. Deoxygenated hemoglobin is paramagnetic while the oxygenated presents dia-magnetic properties. In practice, this means that the magnetic field is distorted around blood vessels containing deoxygenated hemoglobin and this distortion results in reduced relaxation time [3]. The tissues around blood vessels with different concentrations of (de)oxygenated hemoglobin will have different relaxation times and therefore different intensities in the final MR image. We will not go into details about the process of image formation since this is beyond the scope of our study.

BOLD fMRI measures the flow or the change in flow of the blood in the brain. Increased blood flow occurs as the brain requires energy in the form of glucose, to be delivered in brain areas involved in some sort of processing. An active area requires more energy, which translates to increase of the oxygenated blood arriving in this area. Therefore, fMRI measures blood oxygenation levels, which change with increased metabolic demands of active brain areas and form a sort of indirect measure to neuronal activity [25].

In order to observe physiological or pathological changes in functional activity using fMRI an experiment has to be set up that will help reveal and identify such activities. MRI images are obtained while a subject is performing a motor, cognitive or sensory task, which is designed to elicit specific brain activity. Using the obtained images, the next step is to find patterns of activity that correlate with the performed task. Voxels whose intensities over time present significant correlation with the time evolution of the experiment are considered as related to the task and marked as active. The effect of noise and the need to perform complex experiments has led to the development of complex preprocessing and analysis methods.

The design of the experiment can be categorized into two distinct categories depending on the way that the stimuli are presented to the subject: block and event-related designs. In block designs the stimuli of a certain class are presented continuously for a certain period/block of time [26]. Usually, periods of rest are alternating with periods of task. The idea behind this type of design is to increase the signal-to-noise ratio of the hemodynamic response by requesting continuously the same response and therefore a steady level of BOLD signal. This comes in the expense of completely disregarding the temporal evolution of the hemodynamic response.

On the other hand, event-related designs alternate the order of presented stimuli, which are presented in random order separated by a short period of rest. The response to each stimulus is measured and the hemodynamic response function (HRF) can be estimated. Event-related designs allow the execution of more complex experiments in the expense that it is not always possible to detect these activations due to low signal-to-noise ratio [27].

3.1 fMRI Data Preprocessing

Analysis of fMRI data involves a series of preprocessing steps necessary for the actual statistical processing. The series and order of such steps are also known as preprocessing pipeline [28]. We will not deal with preprocessing steps taken in k-space, only with methods used in the image space.

These preprocessing steps involve temporal and spatial registration of the acquired MR signals in order to compensate for noise and variations due to the measurement process [28]. Noise in fMRI can be introduced due to MR signal strength fluctuations throughout the session known as thermal noise [25]. This kind of noise is due to thermal motion of the electrodes in the scanner or the tissue under examination. Thermal noise increases as the temperature and the strength of the magnetic field increases. Another source of noise originates from the system itself and is related to inhomogeneities of the magnetic field and variations of the gradient fields used to spatially target the measurement [25]. Thermal and system noise are unavoidable under a particular setup and do not relate to the experimental task and their effects can be easily mitigated.

A different source of noise is due to subject movements. Since we are dealing with a living organism we cannot expect that the subject can remain still throughout the whole session. Head movement is a major source of noise in fMRI studies and excessive head motion during an experiment may render the data unusable [25]. Since the whole procedure relies on radio frequency pulses in order to localize the recorded activity, even small movements will have the effect of transferring activation from one location to nearby ones resulting in blurring of the obtained signal.

An additional problem with head movement is that sometimes is related to the task and its effects can influence the final result. This effect is mostly apparent in visual tasks where the subject has to guide its gaze to various presented targets. Even though it is required only to move its eyeballs most of the time this movement is accompanied by a small involuntary displacement of the head. Also, even small movements due to breathing or heart beating introduce noise. The effect of breathing and heart beating on the results is rather complex since they make the brain inside the skull to move and expand [29].

In order to compensate for noise and cancel any non-task related influences the preprocessing of the data includes noise correction, slice-timing correction, motion correction, and registration [28]. The final pipeline sequence depends largely on the experiment design, the strength of the magnetic field, and the pulse sequence used for acquiring the k-space data.

Slice-timing correction adjusts the data of each slice as each slice was sampled at the same time. This procedure tries to correct the fact that each slice is sampled separately at a slight different time interval than the previous one. Slices can be collected either sequentially, one adjacent slice after the other or interleaved, odd slices first or then the even ones. If the time needed to collect the whole brain volume is TR, then depending on the manner that the slices are collected the last slice is collected TR or $TR/2$ from the first one. Interpolation between voxels of the same slice acquired in adjacent time frames is used to estimate the signal for the specific acquisition time [30]. Different interpolation techniques have been proposed for slice-timing correction like linear, cubic, and spline [31]. Usually slice-timing correction is used as the first step since it simplifies next steps in the preprocessing procedure [28].

As we mentioned, both due to system inaccuracies or subject movement, voxels are displaced and therefore we have to align them so that the time series of each one represents the BOLD signal from the same brain region throughout the session. Usually, rigid-body registration is used in order to correct head motion and a large number of techniques have been proposed [28]. We can classify the algorithms into those that use intensity-based measures and those that use landmarks in order to register the slices together. The algorithms used in order to calculate the rigid body parameters and the interpolation method are other factors that differentiate the different algorithms. Many algorithms have been proposed and almost all fMRI analysis packages include their own implementation of algorithms like mutual information-based algorithms, normalized correlation and AIR automated image registration just to name a few [32–35]. All of these algorithms assume that all the slices of a single stack have been collected at the same time and they consider a rigid movement of the whole brain volume. This is a reason why slice-timing correction usually precedes head motion correction.

Most of the fMRI studies use multiple subjects to infer information about activating regions. Since the size and shape of brains of different subjects vary along with the relative location of several anatomical brain structures a sort of normalization and registration has to be performed before inferring conclusions from the results. This preprocessing step is achieved by bringing the different volumes in the same coordinate system and then use some linear or nonlinear registration to a common brain atlas, so that their shape, size, and direction are the same [27, 36]. Figure 2 displays the results of these preprocessing steps. A frequently used coordinate system is the Talairach stereotactic coordinate system.

3.2 fMRI Analysis Approaches

After a series of preprocessing steps have been carried out, the data are ready for statistical analysis. The data that we have to work with include brain captions in different time points. Alternatively, we can consider our data as the time evolution of the intensity of each pixel in the captured brain slices. A typical fMRI dataset consists of thousands time-series; if we consider that we have 32 slices with each slice being a 128×128 matrix.

Typical fMRI analysis involves the detection of changes in the mean signal intensity of the MR data during different behavioral conditions. Different statistical methods have been employed to accomplish this task, ranging from simple correlation analysis to more complex models that take under consideration the temporal and spatial correlations of the data [30]. The final output is an

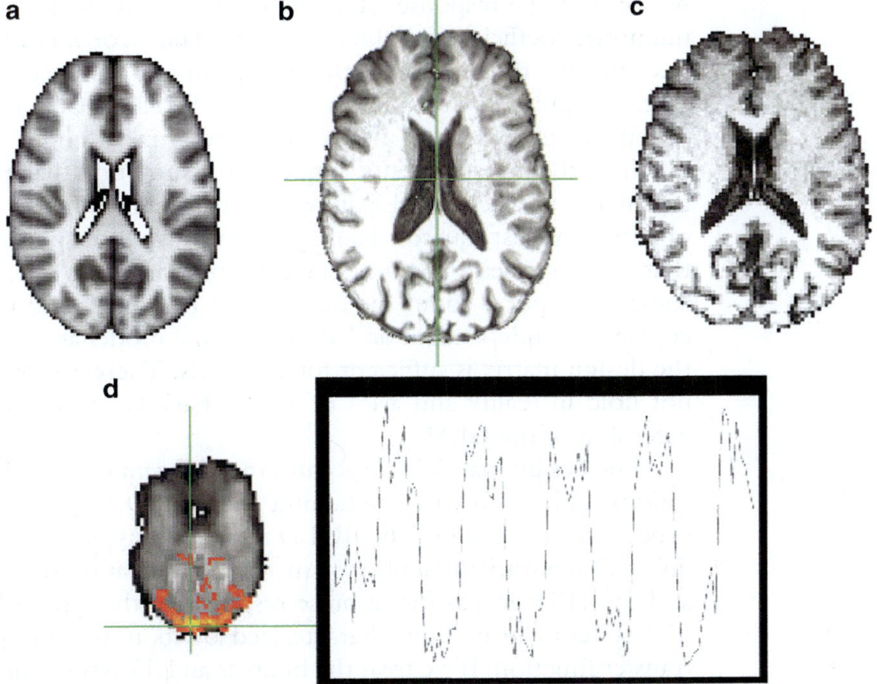

Fig. 2 *First row:* Example of affine registration using the FSL toolbox [43]. Figure (**a**) is the template brain we want to register to. Figure (**b**) is the brain we want to register and in figure (**c**) we can see the registration result. Affine registration is used as a first step before applying nonlinear techniques. *Second row* (**d**) shows the time course of a specific voxel with high correlation with a block experimental design

activation map generated for each condition or experimental task. Voxels, which present increased activity during a certain condition are marked as active. Voxels, which present decreased activity during the task condition are marked as deactivated; these are of less interest and are rarely exploited.

An approach for detecting significant changes is to use Student's t-test. t-Test is commonly used in fMRI analysis due to its simplicity and easy interpretation of the results. Since we have multiple recordings in time an alternative approach is to use the Kolmogorov–Smirnov (K–S) test to compare the distribution of the MR signal intensity between the two conditions [37]. This test can be used to detect changes in the variance along with changes in the mean [38]. Split-half test is another option that has been used to determine significantly different voxels [39]. Many studies have extended in these techniques by adding spatial or other constraints. Finally, an approach was proposed that used multiple linear regressions in order to take under consideration the time course of the fMRI signal [40, 41].

The formula of the general linear model is

$$Y = X \times b + n \tag{5}$$

where Υ is the response, X is the matrix of predictors, and b the unknown coefficients of the predictors. The error n is assumed to have normal distribution with zero mean and variance s^2. In our case, Υ represents the time course of the fMRI signal in the voxels; X represents the design matrix of the experiment under which we obtained the measurements. Under the assumption that each measurement and each voxel is independent, the parameters b can be obtained by least squares. The GLM model is a more general model that encompasses t-test and correlation approaches. GLM makes some assumption about the data as that the voxels and their time courses are independent and that the same model as described by the design matrix is sufficient for all voxels. These assumptions do not hold in reality and a lot of efforts have been devoted to the extension of the GLM.

The design matrix incorporates the different experimental conditions and is a matrix containing the BOLD response for each experimental stimulus. The BOLD response is typically modeled as linear time invariant (LTI) system, where the stimulus is the input and the HRF acts as the impulse response of the system [37]. An LTI system can be fully characterized by its impulse response or transfer function. If we treat the brain as an LTI system, finding the transfer function of this system would allow us to predict the response of the brain in different conditions and under complex inputs. Of course no actual physical system can be considered as a LTI one, but it has been shown that an LTI approximation can characterize quite well the behavior of many systems.

The GLM formulation described is the most popular technique for detecting active regions in an fMRI experiment. The main drawback is that it is a very strict model and any mismodeling will result in an increase of the false positive rate. Also the way that the HRF is calculated plays a significant role to the final result. Nonlinear models like the Balloon model [42] that models changes in the flow and volume and how these changes affect the BOLD response are more realistic but they require the estimation of a lot of parameters and they are sensitive to noise. The linear model is very popular mainly due to its simplicity, robustness and interpretability.

A different approach is to use the information of the data without imposing any strict, specific model. This kind of techniques are very popular since they allow the execution of experiments with complex stimuli that makes the difficult to estimate the time of activation and therefore makes the use of models not practical. Data-driven approaches have been used in psychological studies involving emotion, motivation, or memory. On top of that they allowed the study of the fMRI of the resting state.

Popular component decomposition techniques include PCA and ICA. We have already described the basic principles behind PCA and ICA in the EEG section. In the fMRI context, there exist two different approaches, mathematically equally. PCA or ICA can be either temporal or spatial [44]. In the first case we are looking at the temporal structure of the data in order to find voxels/regions that present the same time evolution while, on the other hand, we are looking in the spatial structure of the data and we are looking for similar spatial patterns through time. PCA was one of the first techniques to be employed in the voxel time series in order to extract spatial regions with similar temporal evolution, allowing exploration of the functional connectivity of brain regions, as we will discuss later.

ICA is more popular and has numerous applications in all kinds of studies, from event-related to resting state studies. Temporal and spatial ICA have been used extensively, although spatial ICA is more often encountered in studies. For the spatial case the ICA model is $X = As$ where X is a $t \times n$ matrix where t represents the time points and n the voxels, A is $t \times t$ temporal mixing matrix for the $t \times n$ spatial independent components/images of the matrix S. Temporal ICA presents the symmetrical configuration where X is an $n \times t$ matrix and A is an $n \times n$ spatially mixing matrix for the $n \times t$ independent time courses. Since the number of voxels is much larger than the number of time points, temporal ICA is much more computationally intensive than the spatial one.

The problem with ICA is that since it is a stochastic method, different runs would produce different results [45]. Before using the calculated independent components in our analysis we have to evaluate the reliability of those components. Different methods have been proposed for evaluating the consistency of the results. The most popular technique runs ICA multiple times in bootstrapped data and then clusters the independent components. ICs in clusters with small inter-cluster and high intra-cluster distance are considered reliable for further evaluation [45].

In contrast to the PCA where the significance, as variance explained, of each principal component is already known, in ICA each component has to be evaluated separately in order to distinguish task-related ICs from noise. In [46] spatial ICA was directly compared to the GLM and is used as a way to solve the GLM problem without using a fixed design matrix; it is directly computed by the ICA.

Analysis and evaluation of Independent components can be distinguished into two approaches. The one is inspired from the extensive work in the EEG Event-related experiments where independent components are separated into task-related and noise [14] and thus ICA is treated as a filtering procedure. In a different context, the filtering aspects of ICA were used in order to remove task-related activity in order to study the rest state of the brain [47].

Other methods that have been applied to fMRI data and not presented here include canonical correlation analysis (CCA) [48] with extensions to accommodate group analysis like in [49]. These approaches led to new algorithms that were able to incorporate data not only from multiple subjects but also from multiple modalities [50].

4 Electroencephalography–Functional Magnetic Resonance Imaging

Combining information obtained from different modalities seems really promising especially in the study of the brain. fMRI and EEG (or MEG) seem to be complementary in nature and form an ideal candidate pair of modalities for such integration. EEG provides excellent temporal resolution of neural activations and MRI/fMRI provides structural and spatial accurate information about metabolic changes in different brain regions—that can be attributed to neural activation.

It is apparent that the two modalities describe and represent different phenomena that there is no assurance that they are directly and uniquely associated/correlated. The most promising results supporting EEG–fMRI integration come from studies that combine fMRI with invasive electrode data [51]. These studies show significant correlations between the time course of activations of fMRI and electrophysiological signals.

On the other hand, there exist several studies that suggest that such integration is not as straightforward as it seems. A one-to-one correspondence between ERP peaks with fMRI activations cannot be assumed as underlined in [52, 53]. In [54] simultaneous recordings from a single subject are used in order to demonstrate that EEG significant features, as peak amplitude, are not likely to be correlated with BOLD signals.

Nevertheless, there are different views regarding the relationship between the local neuronal activity captured by the EEG and related changes in the cerebral blood flow. Another point that needs attention is how to treat the absence of any relation or correlation between the two modalities. This could be attributed in algorithmic limitations, meaning that have a false negative situation. The most difficult question is, though, what if a failure to associate the two modalities is by itself a significant finding or an indication of pathology. There are no straightforward answers in this problem and there is an ongoing research regarding these questions. In general, we can assume that EEG activity is not necessarily co-localized with fMRI activations and also certain fMRI activations do not correlate with EEG. An illustrative model to describe the overlap of EEG explained activations and fMRI ones is described in Fig. 3. It is obvious that parallel analysis of the two modalities will help us to understand better the activations reflected by these modalities. In this section we are going to discuss methods for identifying and characterizing brain activations using information from both modalities.

Towards this goal, several methods have been employed in order to take advantage of the extra information that each modality provides to the other. On the one hand we have the methodologies that use information from one modality in order to constrain or

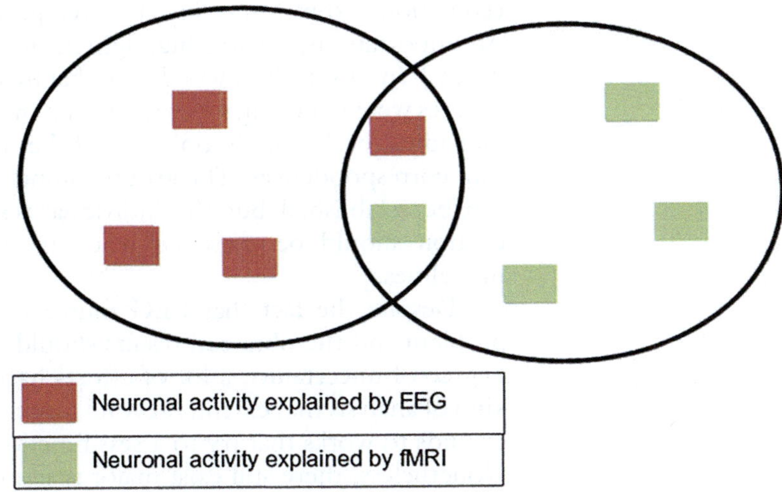

Fig. 3 Illustration of activities explained by each modality. Activities in the cross-section are reflected in both modalities and are the ones that we can use in fusion. Illustration is based on [55]

explain results derived from the other. This approach is known as information integration [55] and includes the methods that we will discuss in the next two sections. The other approach tries to find common patterns of activation in the two modalities in parallel. This approach is characterized as information fusion [55] and includes data-driven and model-based methods. We are interested mainly in the data-driven techniques that have been employed towards this direction.

4.1 EEG Localization Through fMRI Constraints

One of the first attempts for EEG–fMRI integration was to use the fMRI spatial information in order to constrain the problem of EEG source localization. Early attempts used dipole modeling to solve the source localization problem [56] and then regions extracted from the fMRI were used to constrain the dipole location inside the head [57, 58]. This methodology assumes that the possible EEG dipoles express hemodynamic changes reflected in the fMRI. As we mentioned earlier this is an assumption that does not hold in general. On the other hand, dipole modeling localization assumes that the observed EEG is created by a couple of dipoles which seems to be an unrealistic assumption.

In order to overcome the limitation of dipole modeling, current density modeling approaches were employed for the source localization of the EEG [59]. LORETA [60] is the most popular technique for localization through current density modeling. A major disadvantage of the proposed methodologies is that in order to compare the sources calculated from the EEG we have to collapse the EEG in time, either by calculating the sources using the average over time or by using the average LORETA source estimations, thus canceling the temporal advantages of EEG. An important aspect was highlighted by the findings in the study of [59]. In this study, it was shown that fMRI regions and LORETA sources were matching when the group mean data was used. On the individual level though, only half of the subjects presented significant correspondences. The group finding shows that such relations can be established but the individual results stress the fact that caution should be exercised when trying to combine the two modalities.

Despite the fact that EEG source localization is an ill-posed problem and the obtained results should be treated with a certain degree of uncertainty, a lot of efforts have been dedicated to this kind of analysis that extends beyond functional characterization and extends to works that try to assess functional integration [61, 62]. Nonetheless, there still exist major issues in the application of this approach that seem difficult to be transcended soon.

4.2 fMRI Prediction Using EEG Information

A different approach to the problem of EEG–fMRI integration is gaining grounds lately, primarily due to technology advances that allow simultaneous recording of EEG and fMRI. This approach uses EEG features in order to infer fMRI activity.

The work in [63] was the first to display that using fMRI activations we can localize EEG bands without the use of complex and ambiguous methods of source localization. In this work, authors used the alpha band power as a predictor in order to identify regions that changed with alpha band power modulation [63]. Following works extended the study in other bands [64, 65]. This technique has been extremely useful in the analysis of the brain rest state or in complex experiments without a specific stimulus or task. An important application is the presurgical evaluation of pharmacoresistant focal epilepsy where the accurate localization of the epileptic region is needed[66–68]. Actually this clinical application was the driving force behind the development of the needed hardware that allowed simultaneous recordings [69].

A different application is based on the examination of EEG–fMRI event-related single trial covariation. The goal is to identify brain regions that the BOLD response shows the same modulation as a specific single trial ERP component (peak). The basic idea is to use features of the single trial ERPs and use them as predictors for producing fMRI maps related to each single trial feature [70, 71].

4.3 Data-Driven Fusion Approaches

The approaches described above use one modality in order to constrain or predict the other. Models that use a common generative model, explaining the data of both modalities would be the ideal solution to the fusion problem. There have been made efforts towards this direction, without any model reaching a sufficient maturity level [72–74]. On top of that, the complexity of these models renders their application difficult and therefore many models exist only in the theoretical domain or have found limited applications.

Recently, inspired from the advances in the application of multivariate methods in EEG and fMRI, data-driven fusion is gaining a lot of attention. A lot of effort has been attributed to methods that extend application of ICA to multi-modal data. In this category we have a series of ICA-based methods developed for this end [50, 75]. Other methods include multi-set CCA [76, 77], which has been applied to single trial ERP and fMRI. The application of these methods has been focused primarily on the analysis of ERP data, in an effort to explore and exploit trial to trial variations.

Based on the success of ICA in analyzing EEG and fMRI separately, there have been an effort to extend ICA for the fusion of EEG and fMRI. There are studies that use ICA to decompose EEG data and extract useful features that can be used to predict the fMRI activation [70, 78]. Other studies used ICA to extract distributed regional networks from fMRI and the BOLD signals were correlated with power fluctuations in different EEG bands [79]. The aforementioned approaches use ICA in one or the other modality and then use an asymmetrical approach as the ones described earlier. We will focus on algorithms that operate in both modalities and offer a symmetric approach to the problem.

In this context joint ICA was proposed in [80]. The joint ICA algorithm assumes that EEG and fMRI features share the same mixing matrix. In joint ICA we assume that the modalities are jointly temporal or spatial independent and therefore increased BOLD activity will be reflected in increased amplitude of a certain ERP peak. Joint ICA operates in the space defined by the features of both modalities. In order to avoid bias we need to transform EEG and fMRI data and bring them in the joint space so that we can recover the common mixing matrix and independent sources that explain the features. Usually the ERP data from selected channels are used as features while the fMRI activation maps extracted in a previous step (using GLM for example) are the corresponding features for fMRI. The ERP data used are up sampled in order to match the number of fMRI voxels and the data from both modalities are normalized and concatenated into a single matrix, in which joint ICA is going to be applied.

The problem with joint ICA is that the assumptions regarding the generation of the observations are too strict and possibly are not physiologically plausible. A method proposed for relaxing these assumptions and to provide a more flexible estimation is parallel ICA [75]. This method identifies components in both modalities simultaneously and constrains the solution so that maximum correlation is achieved between the mixing matrices of the two decompositions. The correlation constraint is defined by the maximally correlated components in each iteration. The indices of constrained components as well as the number of them are allowed to vary from one iteration to the other. The correlation threshold is chosen manually and prior knowledge about the experiment is required for choosing the appropriate threshold. Parallel ICA has been reported to provide stable results and has been used extensively for the fusion of other modalities as well [81].

One of the first attempts of EEG–fMRI fusion was proposed in [82], using N-partial least squares (N-PLS) [83]. N-PLS is a general multi-way extension of PLS regression. N-PLS describes the covariance of both the independent and dependent variables. This is achieved by fitting the multi-linear models for the independent and dependent variables simultaneously, constrained by a regression model relating the two models. A three-way model was used for the EEG data, having spectral, temporal, and spatial atoms and a linear model was used for fMRI with spatial and temporal atoms. N-PLS was used to find correlations between fMRI time courses and spectral components of the EEG. The problem with this approach is that each modality is decomposed separately and then each decomposition is correlated with the other, a procedure that does not guarantee that the optimal relations will be discovered [84].

Multimodal CCA (mCCA) [77] assumes a different mixing matrix for each modality and transforms the data so that the correlation in the trial-to-trial variation between the two modalities is maximum. It has been proposed for group analysis of fMRI data and it has been used for feature-based multimodal fusion [77]. An extension of mCCA is multi-set CCA, which operates in the raw fMRI data and has been extended to incorporate more than two modalities. In [76] fusion between single trial ERP and raw fMRI has been explored. In order to work with the two modalities a series of preprocessing steps are needed in order to transform the data for common analysis. fMRI data are preprocessed (motion corrected, slice-timing correction, etc.) and regions of interest are extracted based on the anatomical automatic labeling atlas. The normalized mean of each ROI is used as input for the mCCA algorithm. It has been observed that the rate of stimulus presentation and subsequent learning of the stimulus appearance pattern modulate the amplitude of certain components of the single trial ERP. The amplitude of the single trial ERP from selected channels is used to calculate the modulation function that will be used in conjunction with the fMRI. The resulting signals over trials are convolved with a standard HRF function and then averaged over the channels and down sampled to reduce the dimensionality of the data and also match the spatial dimension of the fMRI data.

Multi-set CCA provides a flexible framework for information fusion. It can be used for multimodal and group analysis using raw or feature-based inputs. Unlike the ICA methods that promote sparsity of the results M-CCA produces component maps that may not be as sparse, thus making the interpretation of the results difficult.

An in-depth review of these methods can be found in [84] where their performance is evaluated in simulated scenarios, although there is not any absolute conclusion about the adequacy of each method. This is somewhat expected if we consider that the success of each method depends on the plausibility of the underlying assumptions.

The main drawback with the majority of these approaches is that all the methods described above consider a generic HRF function with its parameters like undershoot and peak latency to be fixed across regions and subjects. This is in contrast to the studies that show that there exists variability in the HRF response not only between subjects but also between regions of the brain [85, 86]. In order to reduce bias and increase sensitivity, it is important to incorporate subject and region specific hemodynamic functions in the methods. A recent work that is moving in that direction, using a model of neurovascular coupling for the constraint of the Parallel ICA can be found in [87]. Nevertheless, data-driven methods provide a flexible framework for exploring EEG–fMRI fusion with many directions that require further investigation.

5   Connectivity, EEG, and fMRI

The methods presented earlier try to identify relationships and common patterns in the data or features extracted from them. So far, the methods presented were considered for the identification of brain activations for EEG or active brain areas for fMRI and the subsequent functional characterization of such regions or activities. This approach is known as functional segregation and its goal is to identify brain activations and assign a specific functionality to them [88]. The complementary task is to describe the functional interactions of the different brain activities that together form large functional networks [88]. It is known as functional integration and it is the subject that we are going to discuss in this section and how fusion of EEG and fMRI can be achieved in this context.

Functional integration involves the study of networks formed between brain areas in the case of fMRI [89] or between EEG bands if we are dealing with EEG [90]. Such networks are thought to play significant role in information processing of the brain [16, 90].

Brain networks are studied using the functional connectivity (FC) and effective connectivity (EC) of brain activities [88, 91]. Functional connectivity is defined as the temporal correlations between remote brain activations. On the other hand, effective connectivity is defined as the causal influence of one system to another [91, 92]. Functional connectivity captures the correlation between remote brain activities; in essence identifies the nodes of the network while effective connectivity reveals the integration and hierarchy inside the network.

5.1 Functional Connectivity

Functional connectivity in EEG has been applied to identify interactions between single electrodes, predefined groups of electrodes or between independent components of the data [93–95]. In order to assess Functional connectivity in EEG different signal analysis techniques have been proposed. The earliest is through the calculation of the autocorrelation between signals captured in selected electrodes [96]. Magnitude squared coherence (MSC) is another measure that is used to measure linear connectivity in the frequency domain [9]. Coherence captures the variations in the relation of power and phase between two signals and takes its maximum value of one when the power and phase of the two signals are stable. A direct extension in the time-frequency domain is the wavelet coherence, which can quantify time-varying coherence [97]. Nonlinear techniques include phase synchronization measures as the phase locking value (PLV) [91, 98] and generalized synchronization (GS) [91, 99]. Finally, information-based techniques like cross mutual information and minimum description length have been applied [91].

Functional connectivity in fMRI allows the characterization of interactions between remote brain regions during a certain task. The earliest methods used to identify interacting regions relied on cross-correlation of initial seed voxels and voxels in other brain regions [100, 101]. In [102], authors indicated that this procedure is biased since the results depend on the initial choice of seed and completely ignore any relations between the voxels.

In order to avoid such biases, PCA was used in order to decompose the BOLD time-series into orthogonal components which represent spatial patterns with the greatest amounts of functional connectivity in descending order [89]. Spatial ICA as we saw earlier can decompose the data into independent maps which can be used to evaluate functional connectivity between brain regions [44, 46]. ICA has also been used to find and characterize functional networks (FN) in the data.

A natural extension of this approach was to examine the functional relation between FN networks instead of voxels or regions in the brain, a technique known as functional network connectivity (FNC) [103]. This technique could provide insight on complex relationships between remote cerebral sites.

5.2 Effective Connectivity

The methods described do not provide any information about the temporal precedence of the different brain activations and therefore cannot efficiently describe the dynamic interactions between brain activations in order to accomplish a certain mental task. This synchronization can be expressed as phase alignment between two remote neural assemblies or as the influence of one system to another [89]. Methods for assessing effective connective include dynamic causal modeling (DCM) [104] and Granger causality (GC) [105].

DCM employs a generative forward model explaining how the data were caused. DCM treats the brain as an input-state-output system where the deterministic inputs correspond to experimental stimulus that evokes brain responses [104]. It was first introduced for analysis of effective connectivity in fMRI and later was extended to EEG [106].

Granger causality is a data-driven technique and is based on the concept that cause precedes effects. Quantification of this concept is achieved by using the notion of cross-prediction. If incorporating past values of a time series s1 improves future predictions of time series s2, then we can saythat s_1 has a causalinfluence in s_2[92]. Since this relation is not reciprocal the relations depicted from this measure can be efficientlyrepresented as a directed graph. The Granger causality concept has been generalized to multivariate signalsby techniques as the directed transfer function (DTF) [107] and partial directed coherence (PDC)[108].

For the estimation of effective connectivity in fMRI, DCM and Granger causality are the most widely used techniques. DCM was originally developed for analysis of effective connectivity in fMRI. The main problem with the application of DCM is that it requires an a priori definition of the anatomical network and its computational complexity restricts its inference capability to a limited number of networks [109].

Granger causality has been applied in fMRI data using more or less the same methodologies used for functional connectivity. For example, it has been applied to calculate effective connectivity of a target region to the other voxels in the brain [110, 111]. A promising approach, inspired from FNC, uses Granger causality in order to quantify effective connectivity between derived Functional Networks in patients with schizophrenia and provides a different point of view in the fusion of EEG and fMRI[112].

5.3 Fusion in the Network Space

Measures of functional and effective connectivity, as described earlier, can be used to construct directed or undirected graphs between brain regions or electrode sites. Individually in each modality, there exist a number of studies that use graph theoretical approaches in order to describe the underlying networks [90]. The extracted networks from each modality cannot be directly compared, though. Regardless of the differences in their temporal scale, the nodes of the extracted network in each case represent different things. For the fMRI the nodes of the network are brain regions while in the EEG case the nodes are mapped to electrode sites. Although a direct comparison is not possible, the general topologies can be used to combine information from the micro to the macro level.

Few studies exist that attempt to fuse EEG and fMRI in the context of functional integration. The majority of studies presented are calculating functional or effective connectivity in only one modality. An early attempt to study EEG and fMRI under the light of

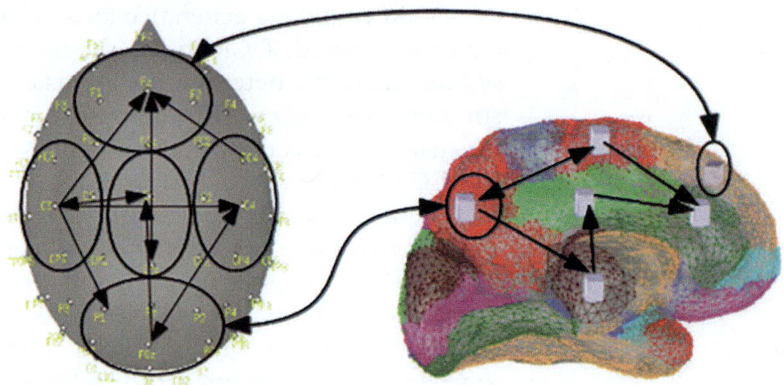

Fig. 4 Illustration of the explained activations and their connectivity. Notice that the activations related to only one modality can now be indirectly linked

functional integration is presented in [61]. In this work, fMRI was used to inform and constrain the inverse source localization of the EEG. The time courses of the EEG were used to calculate causality using the Direct Transfer Function. Nevertheless causality was computed based only on the EEG data and information from fMRI is used only to constrain the source localization solution.

Recently, the work presented in [113] proposes the fusion of EEG and fMRI in network space. They use the term multimodal FNC (mFNC) to describe their methodology. They use spatial ICA to decompose both modalities into independent components and extract FNs from each modality. Then, Granger causality is calculated between independent components of each modality. Source localization is used in order to localize the EEG independent components and test whether can be associated with an fMRI FN.

Matched EEG–fMRI FNs are assumed to represent the same neuronal populations. This approach is consistent with the model that explains the neuronal activity as expressed by fMRI and EEG, presented in Fig. 3. Figure 4 presents an illustrative description of the concept of FNs in EEG–fMRI integration. The problem of how to correspond network nodes between modalities is solved using source localization in order to map the scalp surface EEG activity to brain regions. This is an active field and the use of networks and graph theory for fusing information between modalities looks promising.

6   Discussion

We described methods and measures used for the analysis of ERPs and discussed the complex activations that take place and the debate about their generating mechanism. The excellent temporal resolution of the EEG allows the study of the different brain activations in the millisecond scale and new techniques and methodologies enhance our knowledge and understanding of how brain works. Moving beyond isolation and connection of specific EEG features to specific cognitive processes, functional integration as expressed by functional connectivity and effective connectivity will help to solve or clarify how the individual brain components work together and synchronize in time.

Analysis of fMRI has made great advances toward identifying brain regions activating in response to a task and provides supplementary spatial information that EEG cannot provide. The advances in fMRI depend not only on the accuracy of the system or the magnetic strength but also on the tools for analysis. ICA allowed the performance of complex experiments which would be difficult to model under the GLM and allowed the exploration of complex mental states and perhaps more significantly, the exploration of the so-called rest-state.

Fusion between EEG and fMRI looks very promising since the strong feature of the one is the weak of the other. Reality though, proved more complex and EEG and fMRI fusion is still an open area for research. The main problem is that there is not a one-to-one correspondence between active fMRI regions and EEG activations. Data-driven techniques that look for common pattern between the two modalities look very promising and provide a different view in the single trial analysis which could help solve open problems in both fields.

Of course, there is room for improvement especially in the part of the EEG where a lot of information has been left unexploited. In contrast to the wealth of information revealed in the single trial analysis of EEG data the majority of studies of the EEG–fMRI fusion seem to use only a portion of the available features, disregarding a lot of information and in a sense, canceling the advantages gained by the analysis of EEG.

Finally, the emerging trend of fusion of the two modalities in the network space seems promising and will allow for a more in-depth understanding of the underlying mechanisms.

References

1. Heinze HJ, Münte TF, Mangun GR (1994) Cognitive electrophysiology. Birkhäuser, Boston
2. Sakkalis V (2011) Applied strategies towards EEG/MEG biomarker identification in clinical and cognitive research. Biomarkers 5 (1):93–105
3. Ogawa S, Lee T, Kay A, Tank D (1990) Brain magnetic resonance imaging with contrast dependent on blood oxygenation. Proc Natl Acad Sci USA 87(24):9868–9872
4. Detre J, Leigh J, Williams D, Koretsky A (2005) Perfusion imaging. Magn Reson Med 23(1):37–45
5. Blockley N, Francis S, Gowland P (2009) Perturbation of the BOLD response by a contrast agent and interpretation through a modified balloon model. Neuroimage 48(1):84–93

6. Nunez P, Srinivasan R (2005) Electric fields of the brain: the neurophysics of EEG. Oxford University Press, New York

7. Baillet S, Mosher J, Leahy R (2001) Electromagnetic brain mapping. IEEE Signal Process Mag 18:14–30

8. Tallon-Baudry C, Bertrand O (1999) Oscillatory gamma activity in humans and its role in object representation. Trends Cogn Sci 3 (4):151–162

9. Pfurtscheller G, Lopes da Silva FH (1999) Event-related EEG/MEG synchronization and desynchronization: basic principles. Clin Neurophysiol 110(11):1842–1857

10. Klimesch W (1999) EEG alpha and theta oscillations reflect cognitive and memory performance: a review and analysis. Brain Res Rev 29(2–3):169–195

11. Klimesch W, Sauseng P, Hanslmayr S, Gruber W, Freunberger R (2007) Event-related phase reorganization may explain evoked neural dynamics. Neurosci Biobehav Rev 31 (7):1003–1016

12. Yordanova J, Kolev V, Polich J (2001) P300 and alpha event-related desynchronization (ERD). Psychophysiology 38(1):143–152

13. Bernat EM, Malone SM, Williams WJ, Patrick CJ, Iacono WG (2007) Decomposing delta, theta, and alpha time-frequency ERP activity from a visual oddball task using PCA. Int J Psychophysiol 64(1):62–74

14. Jung T-P, Makeig S, Westerfield M, Townsend J, Courchesne E, Sejnowski TJ (2001) Analysis and visualization of single-trial event-related potentials. Hum Brain Mapp 14 (3):166–185

15. Miwakeichi F, Marínez-Montes E, Valdés-Sosa PA, Nishiyama N, Mizuhara H, Yamaguchi Y (2004) Decomposing EEG data into space-time-frequency components using parallel factor analysis. Neuroimage 22:1035–1045

16. Zervakis M, Michalopoulos K, Iordanidou V, Sakkalis V (2011) Intertrial coherence and causal interaction among independent EEG components. J Neurosci Methods 197:302–314

17. Demiralp T, Ademoglu A, Comerchero M, Polich J (2011) Wavelet analysis of p3a and p3b. Brain Topogr 13(4):251–267

18. Makeig S, Debener S, Onton J, Delorme A (2004) Mining event-related brain dynamics. Trends Cogn Sci 8(5):204–210

19. Tallon-Baudry C, Bertrand O, Delpuech C, Pernier J (1996) Stimulus specificity of phase-locked and non-phase-locked 40 Hz visual responses in human. J Neurosci 16 (13):4240–4249

20. Bell A, Sejnowski T (1997) The "independent components" of natural scenes are edge filters. Vis Res 37(23)3327

21. Lagerlund T, Sharbrough F, Busacker N (1997) Spatial filtering of multichannel electroencephalographic recordings through principal component analysis by singular value decomposition. J Clin Neurophysiol 14 (1):73–82

22. Lins O, Picton T, Berg P, Scherg M (1993) Ocular artifacts in EEG and event-related potentials. I: Scalp topography. Brain Topogr 6(1):51–63

23. Jung T, Humphries C, Lee T, Makeig S, McKeown M, Iragui V, Sejnowski T (1998) Removing electroencephalographic artifacts: comparison between ICA and PCA. In: Neural networks for signal processing VIII, 1998. Proceedings of the 1998 I.E. signal processing society workshop, pp 63–72

24. Michalopoulos K, Iordanidou V, Zervakis M, Pardalos PM, Xanthopoulos P (2012) Application of decomposition methods in the filtering of event-related potentials. In: Pardalos PM, Xanthopoulos P, Zervakis M (eds) Data mining for biomarker discovery. 65 of Springer optimization and its applications. Springer, Boston

25. Lazar N (2008) The statistical analysis of functional MRI data. Statistics for biology and health. Springer, New York

26. Skup M (2010) Longitudinal fMRI analysis: a review of methods. Stat Interface 3 (2):235–252. PMID: 21691445; PMCID: PMC3117465

27. Lindquist MA (2008) The statistical analysis of fMRI data. Stat Sci 23(4):439–464. arXiv:0906.3662 (June 2009)

28. Strother S (2006) Evaluating fMRI preprocessing pipelines. Eng Med Biol Mag IEEE 25:27–41

29. Le TH, Hu X (1996) Retrospective estimation and correction of physiological artifacts in fMRI by direct extraction of physiological activity from MR data. Magn Reson Med 35 (3)290–298

30. Smith S (2001) Preparing fMRI data for statistical analysis. In: Functional MRI: an introduction to methods. Oxford University Press, Oxford, pp 229–241

31. Sladky R, Friston KJ, Tröstl J, Cunnington R, Moser E, Windischberger C (2011) Slice-timing effects and their correction in functional MRI. Neuroimage 58:588–594

32. Ashburner J, Friston K (1997) Multimodal image coregistration and partitioning—a unified framework. Neuroimage 6(3):209–217

33. Cox R, Jesmanowicz A et al (1999) Real-time 3D image registration for functional MRI. Magn Reson Med 42(6):1014–1018

34. Jenkinson M, Bannister P, Brady M, Smith S et al (2002) Improved optimization for the robust and accurate linear registration and motion correction of brain images. Neuroimage 17(2):825–841

35. Woods R, Dapretto M, Sicotte N, Toga A, Mazziotta J (1999) Creation and use of a talairach-compatible atlas for accurate, automated, nonlinear intersubject registration, and analysis of functional imaging data. Hum Brain Mapp 8(2–3):73–79

36. Lazar N, Luna B, Sweeney J, Eddy W (2002) Combining brains: a survey of methods for statistical pooling of information. Neuroimage 16(2):538–550

37. Cohen MS (1997) Parametric analysis of fMRI data using linear systems methods. Neuroimage 6:93–103

38. Cohen M, Kosslyn S, Breiter H, DiGirolamo G, Thompson W, Anderson A, Bookheimer S, Rosen B, Belliveau J (1996) Changes in cortical activity during mental rotation a mapping study using functional MRI. Brain 119(1):89–100

39. Constable RT, Skudlarski P, Gore JC (1995) An roc approach for evaluating functional brain mr imaging and postprocessing protocols. Magn Reson Med 34(1):57–64

40. Friston K, Holmes A, Poline J-B, Grasby P, Williams S, Frackowiak R, Turner R (1995) Analysis of fMRI time-series revisited. Neuroimage 2:45–53

41. Worsley K, Friston K (1995) Analysis of fMRI time-series revisited—again. Neuroimage 2:173–181

42. Buxton RB, Wong EC, Frank LR (1998) Dynamics of blood flow and oxygenation changes during brain activation: the balloon model. Magn Reson Med 39(6):855–864

43. Smith SM, Jenkinson M, Woolrich MW, Beckmann CF, Behrens TE, Johansen-Berg H, Bannister PR, Luca MD, Drobnjak I, Flitney DE, Niazy RK, Saunders J, Vickers J, Zhang Y, Stefano ND, Brady JM, Matthews PM (2004) Advances in functional and structural MR image analysis and implementation as FSL. Neuroimage 23(Suppl 1):S208–S219

44. Calhoun VD, Adali T, Pearlson GD, Pekar JJ (2001) Spatial and temporal independent component analysis of functional MRI data containing a pair of task-related waveforms. Hum Brain Mapp 13:43–53. PMID: 11284046

45. Himberg J, Hyvärinen A, Esposito F (2004), Validating the independent components of neuroimaging time series via clustering and visualization, NeuroImage, 22(3):1214–1222, ISSN 1053-8119, http://dx.doi.org/10.1016/j.neuroimage.2004.03.027

46. McKeown MJ, Sejnowski TJ (1998) Independent component analysis of fMRI data: examining the assumptions. Hum Brain Mapp 6:368–372

47. Arfanakis K, Cordes D, Haughton VM, Moritz CH, Quigley MA, Meyerand ME (2000) Combining independent component analysis and correlation analysis to probe interregional connectivity in fMRI task activation datasets. Magn Reson Imaging 18:921–930

48. Friman O, Cedefamn J, Lundberg P, Borga M, Knutsson H (2001) Detection of neural activity in functional MRI using canonical correlation analysis. Magn Reson Med 45(2):323–330

49. Calhoun V, Adali T, Pearlson G, Pekar J (2001) A method for making group inferences from functional MRI data using independent component analysis. Hum Brain Mapp 14(3):140–151

50. Calhoun V, Silva R, Liu J (2007) Identification of multimodal MRI and EEG biomarkers using joint-ICA and divergence criteria. In: 2007 I.E. workshop on machine learning for signal processing, pp 151–156

51. Logothetis N, Pauls J, Augath M, Trinath T, Oeltermann A et al (2001) Neurophysiological investigation of the basis of the fMRI signal. Nature 412(6843):150–157

52. McCarthy G (1999) Event-related potentials and functional MRI: a comparison of localization in sensory, perceptual and cognitive tasks. Electroencephalogr Clin Neurophysiol Suppl 49:3–12. PMID: 10533078

53. Woldorff M, Matzke M, Zamarripa F, Fox P (1999) Hemodynamic and electrophysiological study of the role of the anterior cingulate in target-related processing and selection for action. Hum Brain Mapp 8(2–3):121–127

54. Makeig S, Jung T-P, Sejnowski TJ (2003) Having your voxels and timing them too? In: Sommer FT, Wichert A (eds) Exploratory analysis and data modeling in functional neuroimaging. MIT, Cambridge, pp 195–207

55. Rosa MJ, Daunizeau J, Friston KJ (2010) EEG-fMRI integration: a critical review of biophysical modeling and data analysis approaches. J Integr Neurosci 9(4):453

56. Liu AK, Belliveau JW, Dale AM (1998) Spatiotemporal imaging of human brain activity using functional MRI constrained

magnetoencephalography data: Monte Carlo simulations. Proc Natl Acad Sci USA 95:8945–8950

57. Babiloni F, Babiloni C, Carducci F, Romani G, Rossini P, Angelone L, Cincotti F (2003) Multimodal integration of high-resolution EEG and functional magnetic resonance imaging data: a simulation study. Neuroimage 19:1–15

58. Opitz B, Mecklinger A, von Cramon D, Kruggel F (1999) Combining electrophysiological and hemodynamic measures of the auditory oddball. Psychophysiology 36(1):142–147

59. Vitacco D, Brandeis D, Pascual-Marqui R, Martin E (2002) Correspondence of event-related potential tomography and functional magnetic resonance imaging during language processing. Hum Brain Mapp 17:4–12. PMID: 12203683

60. Pascual-Marqui R, Esslen M, Kochi K, Lehmann D et al (2002) Functional imaging with low-resolution brain electromagnetic tomography (LORETA): a review. Methods Find Exp Clin Pharmacol 24(Suppl C):91–95

61. Babiloni F, Cincotti F, Babiloni C, Carducci F, Mattia D, Astolfi L, Basilisco A, Rossini P, Ding L, Ni Y, Cheng J, Christine K, Sweeney J, He B (2005) Estimation of the cortical functional connectivity with the multimodal integration of high-resolution EEG and fMRI data by directed transfer function. Neuroimage 24:118–131

62. Lin F-H, Witzel T, Hämäläinen MS, Dale AM, Belliveau JW, Stufflebeam SM (2004) Spectral spatiotemporal imaging of cortical oscillations and interactions in the human brain. Neuroimage 23:582–595

63. Goldman RI, Stern JM, Engel J, Cohen MS (2002) Simultaneous EEG and fMRI of the alpha rhythm. Neuroreport 13:2487–2492. PMID: 12499854; PMCID: PMC3351136

64. Koch S, Steinbrink J, Villringer A, Obrig H (2006) Synchronization between background activity and visually evoked potential is not mirrored by focal hyperoxygenation: implications for the interpretation of vascular brain imaging. J Neurosci 26(18):4940–4948

65. Parkes L, Bastiaansen M, Norris D et al (2006) Combining EEG and fMRI to investigate the post-movement beta rebound. Neuroimage 29(3):685

66. Elshoff L, Groening K, Grouiller F, Wiegand G, Wolff S, Michel C, Stephani U, Siniatchkin M (2012) The value of EEG-fMRI and EEG source analysis in the presurgical setup of children with refractory focal epilepsy. Epilepsia 53(9):1597–1606

67. Kobayashi E, Hawco C, Grova C, Dubeau F, Gotman J (2006) Widespread and intense BOLD changes during brief focal electrographic seizures. Neurology 66(7):1049–1055

68. Salek-Haddadi A, Merschhemke M, Lemieux L, Fish D (2002) Simultaneous EEG-correlated ictal fMRI. Neuroimage 16(1):32–40

69. Blinowska K, Müller-Putz G, Kaiser V, Astolfi L, Vanderperren K, Van Huffel S, Lemieux L (2009) Multimodal imaging of human brain activity: rational, biophysical aspects and modes of integration. Comput Intell Neurosci 2009:1–10

70. Debener S, Ullsperger M, Siegel M, Engel AK (2006) Single-trial EEG-fMRI reveals the dynamics of cognitive function. Trends Cogn Sci 10:558–563

71. Eichele T, Specht K, Moosmann M, Jongsma MLA, Quiroga RQ, Nordby H, Hugdahl K (2005) Assessing the spatiotemporal evolution of neuronal activation with single-trial event-related potentials and functional MRI. Proc Natl Acad Sci USA 102:17798–17803

72. Ostwald D, Porcaro C, Bagshaw AP (201) An information theoretic approach to EEG-fMRI integration of visually evoked responses. Neuroimage 49:498–516

73. Trujillo-Barreto N, Martinez-Montes E, Melie-Garcia L, Valdés-Sosa P (2001) A symmetrical Bayesian model for fMRI and EEG/MEG neuroimage fusion. Int J Bioelectromagn 3(1)

74. Valdes-Sosa PA, Sanchez-Bornot JM, Sotero RC, Iturria-Medina Y, Aleman-Gomez Y, Bosch-Bayard J, Carbonell F, Ozaki T (2009) Model driven EEG/fMRI fusion of brain oscillations. Hum Brain Mapp 30(9):2701–2721

75. Liu J, Calhoun V (2007) Parallel independent component analysis for multimodal analysis: application to fMRI and EEG data. In: 4th IEEE international symposium on biomedical imaging: from nano to macro, 2007 (ISBI 2007), April 2007, pp 1028–1031

76. Correa NM, Eichele T, Adalí T, Li Y-O, Calhoun VD (2010) Multi-set canonical correlation analysis for the fusion of concurrent single trial ERP and functional MRI. Neuroimage 50:1438–1445

77. Correa N, Li Y-O, Adali T, Calhoun V (2008) Canonical correlation analysis for feature-based fusion of biomedical imaging modalities and its application to detection of associative networks in schizophrenia. IEEE J Sel Topics Signal Process 2:998–1007

78. Benar C-G, Gunn R, Grova C, Champagne B, Gotman J (2005) Statistical maps for EEG dipolar source localization. IEEE Trans Biomed Eng 52:401–413

79. Mantini D, Marzetti L, Corbetta M, Romani G, Del Gratta C (2010) Multimodal integration of fMRI and EEG data for high spatial and temporal resolution analysis of brain networks. Brain Topogr 23(2):150–158

80. Calhoun V, Adali T, Giuliani N, Pekar J, Kiehl K, Pearlson G (2006) Method for multimodal analysis of independent source differences in schizophrenia: combining gray matter structural and auditory oddball functional data. Hum Brain Mapp 27(1):47–62

81. Meda SA, Narayanan B, Liu J, Perrone-Bizzozero NI, Stevens MC, Calhoun VD, Glahn DC, Shen L, Risacher SL, Saykin AJ, Pearlson GD (2012) A large scale multivariate parallel ICA method reveals novel imaging-genetic relationships for Alzheimer's disease in the ADNI cohort. Neuroimage 60:1608–1621

82. Martínez-Montes E, Valdés-Sosa PA, Miwakeichi F, Goldman RI, Cohen MS (2004) Concurrent EEG/fMRI analysis by multiway partial least squares. Neuroimage 22:1023–1034

83. Bro R (1996) Multiway calibration. Multilinear PLS. J Chemom 10(1):47–61

84. Sui J, Adali T, Yu Q, Chen J, Calhoun VD (2012) A review of multivariate methods for multimodal fusion of brain imaging data. J Neurosci Methods 204:68–81

85. Aguirre G, Zarahn E, D'Esposito M (1998) The variability of human, BOLD hemodynamic responses. Neuroimage 8:360–369

86. Handwerker DA, Ollinger JM, D'Esposito M (2004) Variation of BOLD hemodynamic responses across subjects and brain regions and their effects on statistical analyses. Neuroimage 21:1639–1651

87. Wu L, Eichele T, Calhoun V (2011) Parallel independent component analysis using an optimized neurovascular coupling for concurrent EEG-fMRI sources. In: 2011 annual international conference of the IEEE engineering in medicine and biology society (EMBS), September 2011, pp 2542–2545

88. Friston KJ (2009) Modalities, modes, and models in functional neuroimaging. Science 326:399–403

89. Friston KJ (1994) Functional and effective connectivity in neuroimaging: a synthesis. Hum Brain Mapp 2(1–2):56–78

90. Bullmore E, Sporns O (2009) Complex brain networks: graph theoretical analysis of structural and functional systems. Nat Rev Neurosci 10:186–198

91. Sakkalis V (2011) Review of advanced techniques for the estimation of brain connectivity measured with EEG/MEG. Comput Biol Med 41:1110–1117

92. Deshpande G, LaConte S, James GA, Peltier S, Hu X (2009) Multivariate granger causality analysis of fMRI data. Hum Brain Mapp 30 (4):1361–1373

93. Micheloyannis S, Pachou E, Stam CJ, Breakspear M, Bitsios P, Vourkas M, Erimaki S, Zervakis M (2006) Small-world networks and disturbed functional connectivity in schizophrenia. Schizophrenia Res 87:60–66

94. Sakkalis V, Oikonomou T, Pachou E, Tollis I, Micheloyannis S, Zervakis M (2006) Time-significant wavelet coherence for the evaluation of schizophrenic brain activity using a graph theory approach. In 28th annual international conference of the IEEE engineering in medicine and biology society, 2006 (EMBS'06), September 2006, pp 4265–4268

95. Stam CJ, Nolte G, Daffertshofer A (2007) Phase lag index: assessment of functional connectivity from multi channel EEG and MEG with diminished bias from common sources. Hum Brain Mapp 28(11):1178–1193

96. Adey W, Walter D, Hendrix C (1961) Computer techniques in correlation and spectral analyses of cerebral slow waves during discriminative behavior. Exp Neurol 3 (6):501–524

97. Ding M, Bressler S, Yang W, Liang H (2000) Short-window spectral analysis of cortical event-related potentials by adaptive multivariate autoregressive modeling: data preprocessing, model validation, and variability assessment. Biol Cybern 83(1):35–45

98. Pereda E, Quiroga RQ, Bhattacharya J (2005) Nonlinear multivariate analysis of neurophysiological signals. Prog Neurobiol 77(1–2):1–37

99. Arnhold J, Grassberger P, Lehnertz K, Elger C (1999) A robust method for detecting interdependences: application to intracranially recorded EEG. Phys D Nonlinear Phenom 134(4):419–430

100. Biswal BB, Kylen JV, Hyde JS (1977) Simultaneous assessment of flow and BOLD signals in resting-state functional connectivity maps. NMR Biomed 10(4–5):165–170

101. Cordes D, Haughton VM, Arfanakis K, Carew JD, Turski PA, Moritz CH, Quigley MA, Meyerand ME (2001) Frequencies contributing to functional connectivity in the

cerebral cortex in "Resting-state" data. Am J Neuroradiol 22:1326–1333

102. van de Ven VG, Formisano E, Prvulovic D, Roeder CH, Linden DE (2004) Functional connectivity as revealed by spatial independent component analysis of fMRI measurements during rest. Hum Brain Mapp 22 (3):165–178

103. Jafri MJ, Pearlson GD, Stevens M, Calhoun VD (2008) A method for functional network connectivity among spatially independent resting-state components in schizophrenia. Neuroimage 39:1666–1681. PMID: 18082428; PMCID: PMC3164840

104. Friston K, Harrison L, Penny W (2003) Dynamic causal modelling. Neuroimage 19:1273–1302

105. Granger CWJ (1969) Investigating causal relations by econometric models and cross-spectral methods. Econometrica 37 (3):424–438

106. David O, Kiebel SJ, Harrison LM, Mattout J, Kilner JM, Friston KJ (2006) Dynamic causal modeling of evoked responses in EEG and MEG. Neuroimage 30:1255–1272

107. Kaminski M, Blinowska K (1991) A new method of the description of the information flow in the brain structures. Biol Cybern 65 (3):203–210

108. Sameshima K, Baccalá L (1999) Using partial directed coherence to describe neuronal ensemble interactions. J Neurosci Methods 94(1):93–103

109. Lohmann G, Erfurth K, Müller K, Turner R (2012) Critical comments on dynamic causal modelling. Neuroimage 59:2322–2329

110. Goebel R, Roebroeck A, Kim D-S, Formisano E (2003) Investigating directed cortical interactions in time-resolved fMRI data using vector autoregressive modeling and granger causality mapping. Magn Reson Imaging 21:1251–1261

111. Roebroeck A, Formisano E, Goebel R et al (2005) Mapping directed influence over the brain using granger causality and fMRI. Neuroimage 25(1):230–242

112. Demirci O, Stevens MC, Andreasen NC, Michael A, Liu J, White T, Pearlson GD, Clark VP, Calhoun VD (2009) Investigation of relationships between fMRI brain networks in the spectral domain using ICA and granger causality reveals distinct differences between schizophrenia patients and healthy controls. Neuroimage 46:419–431. PMID: 19245841; PMCID: PMC2713821

113. Lei X, Ostwald D, Hu J, Qiu C, Porcaro C, Bagshaw AP, Yao D (2011) Multimodal functional network connectivity: an EEG-fMRI fusion in network space. PLoS One 6(9):e24642

Neuromethods (2015) 91: 351–383
DOI 10.1007/7657_2014_70
© Springer Science+Business Media New York 2014
Published online: 17 May 2014

Computer-Based Assessment of Alzheimer's Disease Employing fMRI and/or EEG: A Comprehensive Review

Evanthia E. Tripoliti, Michalis Zervakis, and Dimitrios I. Fotiadis

Abstract

This chapter presents the state-of-the-art methodologies applied for the assessment of Alzheimer's disease. First, the problem of Alzheimer's disease and its fundamental complications are presented. The assessment methodologies based on the analysis of functional Magnetic Resonance Imaging (fMRI) and on the analysis of Electroencephalography (EEG) are then reviewed. Finally the methodologies relying on the data fusion produced by the two above-mentioned techniques (fMRI and EEG) are presented.

Keywords Alzheimer's disease, Early assessment of dementia, Functional magnetic resonance imaging (fMRI), Electroencephalography (EEG), Multimodal data fusion, Neurogenerative modeling, Asymmetric data integration

1 Introduction

As the population of our society is aging, the number of people with chronic diseases is growing. Dementia [1] is one of the major public health problems affecting mainly the developed world. Statistics indicate that 35.6 million people all over the world are living with dementia. This number will be double by 2030 and more than triple by 2050 [2]. Dementia is not a specific disease but a term that is used medically to describe a syndrome (set of symptoms) caused by various diseases of the brain and conditions—usually of a chronic or progression nature—that result in damaged brain cells and thus significant impairment of personal, social, or occupational functions [3–5]. It is characterized by the loss of or decline in multiple higher cortical functions, including memory, thinking, orientation, comprehension, calculation, learning capacity, language, judgement, and information processing. The above-mentioned impairments of cognitive functions are commonly accompanied by deterioration in emotional control social behavior, or motivation [3, 6].

Different types of dementia have been associated with distinct symptom patterns. The symptoms for different types of dementia may overlap and can be further complicated by coexisting medical conditions [6]. The types of dementia can be split into two broad

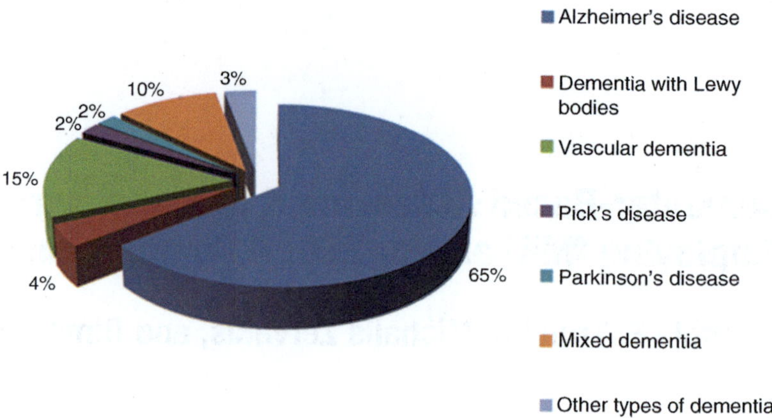

Fig. 1 Incidence of known types of dementias

categories: (a) those that affecting the outer layers of the brain that play a critical role in thinking abilities like memory and language (cortical dementias) and (b) those which result from dysfunction in parts of the brain which are beneath the cortex (subcortical dementias). The Alzheimer's disease (AD), the Creutzfeldt–Jakob disease, and the Pick's disease are three forms of cortical dementias. Typical examples of the second category are the Huntington's disease and the Parkinson's disease [3]. Figure 1 presents the incidence of known type of dementias.

The AD is the most common type of dementia accounting for an estimated 60–80 % of all cases. In the AD, information transfer at the synapses begins to fail, the number of synapses gradually declines and eventually brain cells die [4–6]. According to International Classification of Diseases—ICD-10 [7], the AD is classified as mental and behavioral disorder. The disease took its name by the German neurologist Alois Alzheimer, who first described the symptoms and the neuropathology of the disease in 1906. He identified the two major abnormalities in the brain that characterize the disease—senile plaques and neurofibrillary tangles—localized to specific regions of the brain which control memory and learning [3–5].

The AD is a gradually progressive disorder with a course of 6–12 years. It is diagnosed in people over 65 years of age, although the early-onset of the disease can occur much earlier. There is evidence that brain abnormalities are present for at least 30 years before symptoms become apparent [3, 8]. Although AD develops differently for every individual, memory loss, difficulty performing familiar tasks, problems with language, disorientation to time and place, poor or decreased judgment, problems with abstract thinking, misplacing things, changes in mood and behavior, changes in personality, and loss of initiative are the ten most common warning signs of the disease [3–6].

The development of AD seems to correlate well with increasing age; 10 % of people over the age of 65 and 50 % over 85 suffer from AD. While scientists know that AD involves failure of nerve cells, they do not know why this happens. They agree that the disease is not only a result of a single cause but a combination of multiple factors, with race, occupation, level of education, sex, geographic location, lifestyle, socioeconomic status contributing to the onset of the disease but are not considered decisive risk factors. Other important factors for the disease include family history, brain trauma, metabolic disorders, regulation of amyloid precursor protein and tau protein, as well as, the genetic factors [3–6, 9–13].

Current consensus statements have emphasized the need for early diagnosis of AD [13]. However, there is no single, comprehensive method to diagnose AD. Based on medical history, physical examination, cognitive assessment tests, laboratory and brain imaging results, a properly trained physician with expertise in dementia may diagnose AD. The diagnosis is made using the National Institute of Neurological and Communicative Disorders and Stroke—Alzheimer's Disease and Related Disorders Association criteria (NINCDS-ADRDA) [14–16] and the criteria of Diagnostic and Statistical Manual of Mental Disorders (DSM-IV) of the American Psychiatric Association [17]. Typically, it takes a few weeks to complete a diagnostic evaluation of the disease and most often diagnosis is achieved at late stages of the disease. Thus, it is important to support the clinician with tools and methods that facilitate more effective diagnosis at earlier stages of the disease. Some of these tools relate to anatomical/structural and functional characteristics of the brain that have been associated with and degraded by the disease.

Imaging assessment plays an important role in the diagnostic workup of patients with AD. A variety of neuroimaging techniques, including positron emission tomography (PET), functional magnetic resonance imaging (fMRI), single-photon emission computed tomography (SPECT) and structural magnetic resonance imaging (MRI) have made significant advances over the last years towards the early diagnosis of the disease and the monitoring of its progression [13, 18–21].

Imaging modalities allow clinicians and researchers to better examine, measure, and understand how different cognitive functions are affected by the presence of AD. Such methods allow for the assessment of "where" activity is localized in the brain. In an attempt to understand "when" such activity happens—how brain response changes over time—typical modalities measure the electrical fields generated by the active neural populations [22]. For this purpose, electroencephalogram (EEG) has rapidly gained growing interest and it is used more and more in clinical practice as a method for the assessment of AD. Abnormalities of cortical activity, due to

AD (cortical dementia), are depicted as abnormalities in EEG, since the electrical signals recorded in EEG are scalp potentials reflecting underlying cortical activity.

The aim of this chapter is to broadly examine and critically review research utilizing fMRI and EEG techniques separately to better understand AD and improve the effectiveness of medical diagnosis of the disease. Attention is given to studies that are based on the EEG-fMRI information fusion, which exploit the complementary nature of the two modalities in order to support diagnosis of AD. We should notice here that in this chapter we focus at the functional level of brain operation as affected by the disease. In a holistic consideration of the disease and its diagnosis, this functionality should be combined with information at cellular and genomic level, as well as with anatomic effects (e.g., size of white matter) and with regular cognitive tests, but this is out of the scope of this chapter.

2 Assessment of Alzheimer's Disease Based on Functional Magnetic Resonance Imaging

Functional Magnetic Resonance Imaging (fMRI) is a noninvasive neuroimaging technique which uses the physical phenomenon of Nuclear Magnetic Resonance (NMR) and the associated technology of MRI to measure and localize specific human brain functions. It works by detecting regional changes in cerebral metabolism, in blood flow, volume, or oxygenation in response to task activation. It relies on the fact that cerebral blood flow and neuronal activation are coupled. When a brain area is more active, it consumes more oxygen and to meet this increase demand blood flow increases to the active area [23–25].

The fMRI signal depends on the vascular response to functional brain activation and it is typically implemented by imaging of the Blood Oxygen Level Dependent (BOLD) contrast, which is based on the differentiation of the magnetic properties of oxygenated (diamagnetic) and deoxygenated (paramagnetic) blood. These magnetic susceptibility differences lead to small but detectable changes in weighed MR image intensity. More specifically, when the human brain receives a stimulus an increase in neuronal activation takes place. Since neurons do not have internal reserves of energy their activation causes a need for more energy. The energy is supplied in the form of glucose and oxygen, which is carried in hemoglobin. The resulting increased need for oxygen is over-compensated by a large increase in perfusion. As a result, the venous oxyhemoglobin concentration increases and the deoxyhemoglobin concentration decreases. The presence of deoxyhemoglobin causes local field inhomogeneities, which are responsible for a dephasing of the local transversal magnetization, leading to a reduction in the transverse relaxation time T2. As a diamagnetic molecule oxyhemoglobin does not produce the same dephasing. Thus, changes in deoxyhemoglobin can be observed as the BOLD contrast in T2-weighted magnetic resonance images, serving as an indirect measure of neuronal activity [23–26].

Among the several distinct advantages of the fMRI image modality (it is considerable safer since no contrast agent is needed in order the signal to be administrated, the total scan time is very short, it has high spatial and temporal resolution) is that its application in clinical practice allows neuroscientists to investigate the relationship between brain and behavior. It acts as a tool for identifying brain regions that are associated with certain perceptual, cognitive, emotional, and behavioral functions. It is suitable for accessing many aspects of human cognition and plays an important role in assisting pre-surgical planning in neurosurgery and in providing additional diagnostic information in the clinical environments of patients who have functional disorders due to neurological disease and mental illness. Some other fundamental application of fMRI is the management of pain, the improved assessment of risk, and the improved seizure localization [23, 25].

In this section we focus on how fMRI can contribute and improve the assessment of AD. A large number of studies have been reported in the literature regarding the analysis of fMRI in patient with AD. The studies can be grouped into two broad categories. The first category includes studies concentrated on identifying what is differentiated in brain function between healthy subjects and subjects with AD. The second category includes studies that are trying to express the above-mentioned differences through an index that can serve as a biomarker for the diagnosis of AD.

2.1 Functional Magnetic Resonance Imaging Analysis Studies Revealing Differentiations Between Healthy and Alzheimer's Disease Subjects

Clinical studies have shown that the posterior cingulate cortex (PCC) presents reduced activity in patient with AD, as well as, hypo metabolism in cognitively intact subjects with genetic susceptibility to the disease [27–32]. The combination of clinical findings led Raichle et al. [33] to propose that the PCC forms part of a "default mode" network, a network of brain regions that are active when the individual does not perform any task and the brain is at wakeful rest. Greicius et al. [34] using functional connectivity analysis of fMRI data observed that there is significant coactivation of several regions within this network.

Rombtous et al. [35] studied the activity and reactivity of the default mode network in the brain in patients with mild cognitive impairment (MCI), with mild AD and in healthy controls (HC). The three groups are differentiated in the early phase of deactivation. More specifically, the response of the default mode network in anterior frontal cortex distinguished MCI from HC and AD, while the response in the precuneus could only distinguish between patients and HC. These differences reflect the reactivity and adaptation of the network. For the conduction of the study they utilized fMRI analysis methods included in software packages like FEAT and FSL.[1]

The study of Sorg et al. [36] revealed the reduction or absence of functional connectivity and the atrophy of the regions consisting resting state networks in patients with AD. More specifically, Independent Component Analysis (ICA) of resting-state fMRI data was utilized in order eight spatially consistent resting state networks to be identified. Only selected areas of the networks demonstrated reduced activity in the patient group. Atrophy was presented in both medial temporal lobes of the patients through voxel-based morphometry. Regarding functional connectivity between medial temporal lobes and the posterior cingulate of the network it was present in healthy controls but not in patients.

Oghabian et al. [37] focused their study on the evaluation of the fMRI in differentiating between Alzheimer's, MCI and healthy aging. They applied ICA to compare the resting-state brain activation patterns between groups of subjects examining in this way the ability of fMRI to differentiate the three conditions mentioned above. The Minimum Description Length algorithm was employed to determine the number of independent components [38]. The components were transformed to Z space and then a Gaussian Mixture Model was applied to define the value of the threshold determining the creation of the activation maps [39]. Their studies revealed that healthy aging brain presents activation areas with larger area and greater intensity of activation compared with the MCI and Alzheimer's group in PCC region of the brain. It must be mentioned that the observation is valid when the subjects are under a certain resting-state session.

Small et al. [40, 41] studied how the hippocampal regions are interconnected to form a circuit. Using fMRI they evaluated the hippocampal regions in vivo, and they used their method to study three different groups of subjects, elderly with normal memory, elderly with isolated memory decline, and elderly with probable AD. Two distinct patterns of regional dysfunction were revealed among the last two groups. The first pattern involved all hippocampal regions, while the second one present dysfunction restricted to the subiculum. The processing of fMRI data was achieved using appropriate software packages (MEDx Sensor Systems, Boulder, CO; IDL Research Systems, Sterling, VA).

It has been observed that the development of AD causes neuropathological changes which are followed by the decline of basic cognitive processes. The two questions that arise are the

---

[1] FEAT (fMRI Expert Analysis Tool) v. 5.1, part of FSL (FMRIB's Software Library)—http://www.fmrib.ox.ac.uk/fsl

following: (a) Does the functional alterations in neural circuitry accompany these neuropathological changes? and if so (b) Can they be detected before onset of symptoms?. Based on these observations Smith et al. [42] used fMRI to examine if and how cortical activation is differentiated between cognitively normal subjects and subjects at risk for developing AD. The analysis of the data revealed that the two groups present similar patterns of brain activation, while the high risk group showed areas of significantly reduced activation in the mid- and posterior inferotemporal regions. AFNI[2] software and more specifically, cubic interpolation with splines, was utilized for the resampling of the activation maps and appropriate MATLAB 5 (MathWorks, Natick, MA) functions for the smoothing of the maps. The results of the previous processes were compared voxel by voxel using a two sample t-test.

Machulda et al. [43] based on the fact that the main feature of patients with AD is memory impairment studied whether an fMRI memory encoding task can distinguish cognitively healthy elderly individuals, patients with MCI, and patients with early AD. The following conclusions have been obtained: (a) MCIsubjects and subjects with AD had less medial temporal lobe activation on the memory task than the healthy subjects, but similar activation as healthy subjects on the sensory task, (b) decreased medial temporal activation may be a specific marker of limbic dysfunction due to the neurodegenerative changes ofAD, (c) fMRI can be used to detect changes in the prodromal, MCI, phase of the disease. The fMRI data were processed using AFNI software.

Petrella et al. [44] exploited the potentiality of the fMRI to examine how memory networks break down as AD progresses. They tried to detect brain regions in which changes in activation correlate with degree of memory impairment across Alzheimer disease, MCI, and elderly control subjects. They utilized a face-name associative encoding paradigm and the results revealed preservation of some areas of activation with an overall decrease in activation of medial temporal lobe structures from control subject to patient with MCI to patient with mild AD. Increase of activation magnitude was noted in the posteromedial cortex (PMC) region. PMC activation changes were larger in both magnitude and extent than those in the MTL. In addition, changes in PMC regions are significantly correlated with neuropsychological test performance.

Gron et al. [45] assessed episodic memory in older subjects that were diagnosed with probable Alzheimer's or major depressive disorder. In the fMRI paradigm repetitive learning and free recall of abstract geometric patterns were used. The analysis of fMRI data

---

[2] Cox RW. AFNI: software for analysis and visualization of functional magnetic resonance neuroimages. Comput Biomed Res, 29, 162–173, 1996.

was performed using Statistical Parametric Mapping (SPM99, Welcome Department of Cognitive Neurology, London, UK) and revealed that: (a) healthy seniors or depressive patients present superior hippocampal activation than patients with AD, (b) patients with AD showed bilateral prefrontal activity while healthy seniors did not, (c) patients who had major depressive disorder, in contrast to healthy seniors or patients who had AD, showed activation of the orbitofrontal cortex and the anterior cingulate in the hemispheres of the brain. In similar conclusions, regarding the differentiation of activation patterns of memory correlated regions, were presented by Sperling et al. [46, 47].

Differences between patients with probable AD (pAD) and cognitively able elderly volunteers, based on a visuospatial cognition task, were studied by Thulborn et al. [48]. They analyzed fMRI data received during an eye movement paradigm. The study concluded that: (a) statistically significant differences existed between the activation patterns of the patients with pAD and those of the volunteers, and (b) a left-dominant parietal activation pattern and an enhanced prefrontal cortical activation were observed in most patients with pAD but not in the control group. A three-phase analysis was followed. First the artefcats were removed and then the activation maps were created using a voxel-wise t-test. Finally, the AFNI software was used to detect the regions with higher activation. Talairach coordinates, activation volumes, and laterality ratios were used to characterize the activation patterns.

Buckner et al. [49] examined if the properties (amplitude and variance)) of hemodynamic response are differentiated between non demented and demented older adults. They conducted an fMRI event-related design paradigm involving repeated presentation of sensory-motor response trials. The results showed that visual cortex present significant reduction in the amplitude of the hemodynamic response function in contrast to the variance of the hemodynamic response that was significantly increased in demented than in non demented older adults. In the motor cortex the characteristics of the hemodynamic response were not affected in both groups of subjects. For the extraction of the results the fMRI data were preprocessed using SPM99 and ANOVA software.

Prulovic et al. [50] examined functional activation patterns in patients with AD during active visuospatial processing. They examined also how the local cerebral atrophy affects the strength of the activation. The fMRI data were recorded while subjects performed an angle discrimination task. Patients with AD and healthy subjects presented overlapping networks (superior parietal lobule, frontal and occipito-temporal cortical regions, primary visual cortex, basal ganglia, and thalamus). Superior parietal lobule and occipito-temporal cortical were the regions of the network that presented the most pronounced differences between the two groups of subjects. The differences can be attributed to the differences in individual atrophy of superior parietal lobule region. They suggested that the local cerebral atrophy should be considered as an expected result of all functional imaging studies of neurodegenerative disorders. The fMRI data were analyzed using the Brain Voyager Software.[3]

The conclusions presented in the study of Hao et al. [51] were based on the processing of fMRI data produced during two types of visual search tasks. The detection of the anatomical areas of activation that are associated with visual attention processing and the detection of changes that may occur in a group of AD patients compared with a control group was the purpose of the study. The findings of the study suggested that patients with AD present reduced activation in both parietal lobes and the left frontal regions. On the other hand, the healthy subjects presented increased activation in the right frontal lobes and the right occipito-temporal cortical regions with the conjunction task.

The relationship between fMRI activation and measures of global and regionally specific atrophy in normal aging and in Alzheimer disease was studied in [52]. The study was conducted by Jonshon et al. [52] and revealed that during a semantic process the left inferior frontal and the left superior temporal gyri were activated. Thus, the evaluation of the correlations between those regions and measures of local atrophy followed. No significant correlation exist in healthy subjects, in contrast to AD patients that present positive and negative correlation between atrophy and activation in left inferior frontal and left superior temporal gyri, respectively.

Rombouts et al. [53] analyzed fMRI data from healthy controls and patients with AD in order to study brain activation during a learning task. The aim of the study was to test the hypothesis that brain activation is decreased in the medial temporal lobe memory system in patients with AD compared to controls. The study revealed that the fusiform, parietal and occipital parts, the hippocampal formation and the frontal cortex present activation. Regarding the medial temporal lobe the results were differentiated between the two encoding tasks. In the first task a significant bilateral decrease of the activation in the left hippocampus and parahippocampal gyrus was observed in patients with AD, compared to control volunteers. Nothing of the above was observed the second encoding task. The analysis was conducted with the SPM99 software. Next, the data regarding each participant were modeled using a box car design, convolved with the hemodynamic response function. Then for each participant a "contrast image," was

---

[3] BrainVoyager 2000- http://www.brainvoyager.com.

calculated. Each participant's contrast-enhanced image was then fed into a statistical test. A one-sample *t*-test was used for the assessment of the average brain activation and a two-sample *t*-test for the assessment of the differences in brain activation between patients and control volunteers.

The effect of AD on functional neuroanatomical processing of semantic and phonological information was studied by Saykin et al. [54]. Healthy controls and AD patients present predominant activation foci in the inferior and middle frontal gyrus. However, patients present additional activation in the left prefrontal cortex.

The relationship between brain responses to memory tasks and the genetic risk factor (epsilon4 allele of the apolipoprotein E gene—APOE) is examined in the study of Bookheimer et al. [55]. AD affects both the magnitude and the extent of brain activation. More specifically, subjects with the APOE epsilon4 allele present larger brain activation, in terms of magnitude and extent, than subjects with the APOE epsilon3 allele. The signal intensity presented a larger, on average, increase during periods of recall in the carriers of the APOE epsilon4 allele. In addition, they presented a higher mean number of activated regions throughout the brain than the carriers of the APOE epsilon3 allele.

Important differences between healthy and demented subjects, as far as it concerns the anatomic distribution of the activation especially in the regions of posterolateral temporal and inferior frontal cortex, were presented in the study of Grossman et al. [56]. The patterns of neural activation were created by the analysis of fMRI data, using SPM99, which is formed during a verb processing task.

The studies reported until now examined the differentiation in brain regions which show positive activations, defined as increases in signal during an active task compared with a more passive baseline. Lustig et al. [57] focused on examining how negative activations, or deactivations, defined as decreases in signal during the task as compared with baseline, are differentiated between healthy (young and older adults) and demented subjects. Their study showed that deactivation in medial frontal regions was reduced due to aging but was not affected by the AD. On the other hand, the medial parietal/posterior cingulated region presents significant differences between healthy and demented subjects. Furthermore, the temporal profile of the region response was initially activated by all groups, but the response in young adults quickly reversed sign, whereas AD individuals maintained activation throughout the task block.

Celone et al. [58] studied the positive and negative activations that take place in memory related regions. The independent component analyses, performed on fMRI data, revealed specific memory-related networks that activated or deactivated during an associative memory paradigm. During the course of MCI and AD there is a direct relation between the loss of functional integrity of the hippocampal systems, responsible for memory functions, and the alterations of neural activity in parietal regions. These data may also provide functional evidence of the interaction the between neocortical and the medial temporal lobe pathology in early AD.

The studies reported above reveal significant differences between healthy and AD patients. The differences are related to:

(a) the activation patterns of specifically selected brain regions such as memory related regions, regions of the visual and motor cortex as well as regions responsible for the learning processes,

(b) the functional connectivity of the default mode network,

(c) the task-induced deactivation of brain networks, and

(d) the regional homogeneity of the fMRI signal.

A detailed review can be found in [59, 60]. For the detection of the differences fMRI data, that was received during resting-state or during task requiring the stimulation of cognitive functions, were preprocessed and analyzed through the utilization of statistical and connectivity analysis methods [26]. The studies revealed the potentiality of fMRI as a tool for the assessment of AD in contrast to the methods which are described in the next section that quantify differences detected using fMRI data and propose an index for the diagnosis of AD.

2.2 Functional Magnetic Resonance Imaging Analysis Studies Proposing Markers for the Diagnosis of Alzheimer's Disease

As already mentioned above a large number of studies is concentrated in the alternations of functional connectivity and activation patterns of brain networks that are activated even though the subjects are in a resting state and they do not perform a cognitive demanding task. Greicius et al. [61] examined the functionality of such a network and revealed that all group of subjects (healthy adults and AD patients) presented coactivation in the region of hippocampus indicating that the default-mode network is closely involved with episodic memory processing, the disrupted connectivity between posterior cingulate and hippocampus is responsible for the posterior cingulate hypometabolism and the activity in the default-mode network may be a biomarker of AD with 85 % sensitivity and 77 % specificity. The results of their study were based on the analysis of fMRI data. More specifically, they employed independent component analysis using FSL MELODICA ICA,[4] and the four best components were selected.

---

[4] FSL MELODIC ICA software—http://www.fmrib.ox.ac.uk/fsl/melodic2/index.html.

The selected components were combined using random effect analysis methods to create the activations maps for each group of subjects. A two sample *t*-test was applied for the comparison of the maps of each group. For further comparison between groups a goodness-of-fit analysis was applied.

The study of Supekar et al. [62] focused on the same network. The network was created for 18 healthy and 12 demented subjects and it was studied as an undirected graph. Wavelet analysis was applied for the computation of frequency-dependent correlation matrices that were thresholded in order the undirected graph of the functional brain network to be created. The characteristic path length and the clustering coefficient were computed using graph analytical methods. The analysis of those coefficients revealed significant differences between the two groups, differences capable to distinguish AD participants from the controls with a sensitivity 72 % and specificity 78 %.

The study of Wang et al. [63] was based in the regions of one task-positive and one task-negative network. The networks consisted of five bilateral homologous regions and 6 bilateral homologous regions, respectively. The mean time series of each of the regions were extracted by averaging the fMRI time series over all voxels in the region. Correlation coefficients were computed between each pair of the regions and a Fisher's r-to-z transformation was applied to improve the normality of these coefficients. The correlation coefficients between each pair of regions were entered into an one-sample two-tailed *t*-test to determine the existence or not statistical significant differences between the coefficients. Then Pseudo-Fisher Linear Discriminative Analysis (pFLDA) was performed to generate a linear classifier. The application of the above procedure to 14 health and 14 demented subjects showed a correct classification rate 72 % for the healthy subjects and 93 % for the demented.

Zhang et al. [64] studied alterations in functional connectivity in the resting brain networks in healthy elderly volunteers and patients with mild, moderate, or severe Alzheimer Disease. They preprocessed the fMRI data to remove artifacts and they segmented the data in gray matter and cerebrospinal fluid. The segmented gray matter data were filtered using a phase-insensitive band-pass filter in order the effect of low-frequency drift and high-frequency physiologic noise to be reduced. Posterior cingulated cortex was selected as the region of interest and Pearson Linear correlation coefficients between the time series of each scaled voxel and the time series of the average signal of the posterior cingulated cortex were computed. A Fisher z transform was applied to improve the normality of these correlation coefficients. The whole analysis revealed that the functional connectivity of the default mode network is affected by the presence and progression of AD.

Li et al. [65] tried to express the differentiations presented between nine healthy subjects, five subjects with mild cognitive impairment and ten subjects with AD using cross-correlation coefficients of spontaneous low frequency—COSLOF index, which is defined as the mean of the cross-correlation coefficients of spontaneous low frequency components between possible pairs of voxel time courses in a brain region. The two-tailed Student t-test was used to determine differences in the COSLOF index between the three groups. The results showed that the COSLOF index can differentiate the three groups with 80 % sensitivity and 90 % specificity.

Based on the previous work, Xu et al. [66] studied the effect of signal to noise ratio and the phase shift of spontaneous low-frequency (SLF) components on the index. They propose an new index called phase shift index—PHI. The application of the PHI on three groups of subjects (the control, mild cognitive impairment and AD) demonstrated that PHI is high in patients with AD (72.6 ± 11.3) in contrast to patients with mild cognitive impairment (58.6 ± 5.7) and healthy subjects (40.6 ± 8.4). They claimed that the PHI is more reliable index than the COSLOF index, however, they do not provide evaluation measures to support this fact.

Chen et al. [67] and Burge et al. [68] employed the properties of a Bayesian network classifier to diagnose AD. Chen et al. [67] utilized a Bayesian network classifier with inverse-tree structure to detect the brain regions with activation maps that can lead to the diagnosis of the disease. Burge et al. [68] constructed a dynamic discrete Bayesian network classifier which recognizes the functional correlations between neuroanatomical regions of interest and examines if those correlations can distinguish healthy from demented subjects. The above-mentioned studies diagnosed AD with 81 and 70 % accuracy, respectively.

Tripoliti et al. [69] proposed a six-step method (Fig. 2). First, spatial and temporal preprocessing of fMRI data were conducted to remove artifacts and the subsequent analysis to lead to reliable results (Step 1). A generalized linear model was applied to the voxel time courses extracted from the preprocessed images (Step 2). At the end of the second step an activation map was created for each subject. The analysis of each map allowed the extraction of features describing the activation patterns and the hemodynamic response of the brain. An index (path length), expressing the head movement during the conduction of the fMRI experiment, was calculated from the functional images. Additionally, the anatomic MRI images were exploited to extract features describing the atrophy of gray matter (Step 3). The extracted features, as well as, features recorded during the experiment (demographic details—age and sex and behavioral data—median and

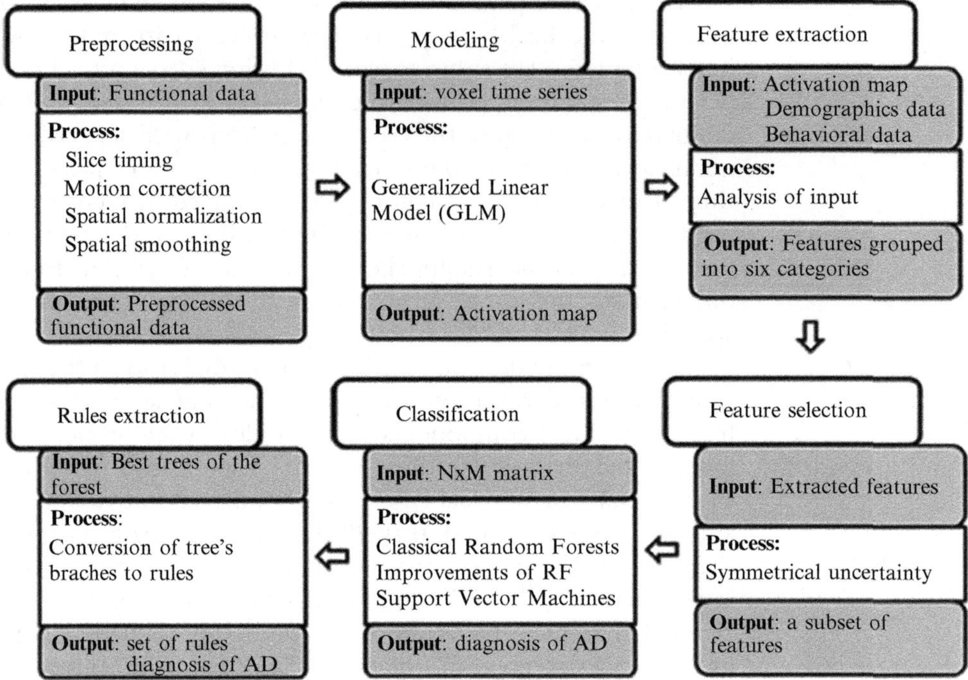

Fig. 2 A flowchart of the method proposed by Tripoliti et al. [69]

average reaction time) were evaluated for their potential to diagnose AD and monitor its progression. The subset of features, selected by the application of an information theory based selection approach (Step 4) was supplied as input to the classification step (Step 5). The Random Forests algorithm and Support Vector Machines (SVMs) were evaluated for their classification accuracy, sensitivity and specificity. Finally, the trees that consist of the forest were converted to rules (Step 6). The rules combine the information, which is encompassed to the features, and they provide the decision if the subject is suffering from AD. In the case of positive diagnosis the rules determine the stage of progression (very mild or mild). The method is evaluated using a dataset of 41 subjects. The results of the proposed method indicate the validity of the method in the diagnosis (accuracy 94 %) and the monitoring of the AD (accuracy 97 and 99 % for the three and four class problem, respectively). By modifying the Random Forests algorithm the disease was diagnosed with 98 % accuracy [70].

A short review of the above-mentioned studies can be found in Table 1. In the next section we focus on the potential of EEG to provide biomarkers for the assessment of AD.

Table 1
A short review of studies regarding assessment of Alzheimer disease based on fMRI

	Authors	Subjects	Design	fMRI analysis methods	Main conclusions
1.	Small et al. [40]	4 healthy 12 with MCI 4 with AD	Memory related task	Pixel by pixel t-test	Memory impairment is associated with the dysfunction of the hippocampus
2.	Saykin et al. [54]	6 healthy 9 with AD	Auditory stimulation tasks	Statistical analysis—SPM96	Patients presented additional activation in left prefrontal cortex
3.	Smith et al. [42]	12 low risk 14 high risk	Visual naming and letter fluency tasks	Cubic interpolation with splines, appropriate MATALAB 5 functions, two sample t-test	The high risk group showed areas of significantly reduced activation in the mid- and posterior inferotemporal regions
4.	Bookheimer et al. [55]	14 no carriers of ApoE4 16 carriers of ApoE4	Memory related paradigm	SPM96—ROI analysis	The carriers of the APOE epsilon 4 allele presented greater activation in terms of magnitude and extent than the carriers of the APOE epsilon3 allele. The signal intensity presented a greater, on average, increase during periods of recall in the carriers of the APOE epsilon4 allele. They presented a greater mean number of activated regions throughout the brain than the carriers of the APOE epsilon3 allele
5.	Buckner et al. [49]	14 healthy young 14 healthy old 13 with AD	Sensory-motor task	Hemodynamic Response Function Analysis	Motor cortex: no differences Visual cortex: significant decrease was presented in the amplitude of the hemodynamic response function in AD patients
6.	Johnson et al. [52]	16 healthy 8 with AD	Visual search tasks	SPM96	AD patients presented positive and negative correlation between atrophy and activation in left inferior frontal and left superior temporal gyri, respectively

(continued)

Table 1
(continued)

	Authors	Subjects	Design	fMRI analysis methods	Main conclusions
7.	Small et al. [41]	4 healthy 4 with AD	Memory related task	Independent t-test ROI analysis	The patients with AD presented reduced activation in the hippocampus and the subiculum
8.	Thulborn et al. [48]	10 healthy 18 with AD	Visually guided saccade paradigm	Voxel wise t-test	The patients with pAD present significant differences in the activation patterns (a left-dominant parietal activation pattern and enhanced prefrontal cortical activation) compared to the volunteers
9.	Rombouts et al. [53]	10 healthy 12 with AD	Two learning tasks	SPM99, one sample t-test and two sample t-test	The fusiform, parietal and occipital parts, the hippocampal formation and the frontal cortex presented activation AD patients vs. healthy controls: decrease activation in the left hippocampus and parahippocampal gyrus during the first encoding task but not during the second
10.	Prvulovic et al. [50]	14 healthy 14 with AD	Angle discrimination task	Brain Voyager software	Superior parietal lobule and occipito-temporal cortical were the regions of the network that presented significant differences between the two groups of subjects
11.	Gron et al. [45]	12 healthy 12 with MCI 12 with AD	Repetitive learning and free recall of abstract geometric patterns	Statistical analysis—SPM99	Memory presents significant dysfunction in patients with Alzheimer's disease in contrast to healthy subjects and to patients with memory decline due to major depressive disorders
12.	Li et al. [65]	9 healthy 5 with MCI 10 with AD	Resting-state	ROI analysis COSLOF index	COSLOF index is significant lower in patient with Alzheimer's disease compared with health subjects and subjects with MCI

13. Grossman et al. [56]	16 healthy 11 with AD	Subjects judged the pleasantness of verbs, including MOTION verbs and COGNITION verbs	SPM99	The anatomic distribution of the activation especially in the regions of posterolateral temporal and inferior frontal cortex, presented significant differences
14. Sperling et al. [46]	10 healthy young 10 healthy old 7 with AD	Face–name association encoding task	Statistical analysis—SPM99	Healthy old adults vs. younger controls: (a) reduced activation in both superior and inferior prefrontal cortices, (b) greater activation in parietal regions Alzheimer patients: (a) reduce activation in the hippocampal formation, (b) greater activation in the medial parietal and posterior cingulate regions
15. Lustig et al. [57]	32 healthy young 27 healthy old 23 with AD	Active semantic classification task and a passive fixation baseline	Statistical analysis—Generalized linear models	Deactivation in medial frontal regions was reduced due to aging but was not affected by AD The medial parietal/posterior cingulated region presents significant differences between healthy and demented subjects The temporal profile of the region response was initially activated by all groups. In young adults the response present reversed sign after a very short time. AD individuals maintained activation throughout the task block
16. Machulda et al. [43]	11 healthy 9 with MCI 9 with AD	Memory encoding task and passive sensory task	AFNI software package	Decreased medial temporal activation can act as a marker of limbic dysfunction due to the neurodegenerative changes of AD
17. Greicius et al. [61]	14 healthy young 14 healthy old 13 with AD	Simple sensory-motor processing task	ICA, Goodness-of-fit analysis	The default-mode network is related with episodic memory processing. The disrupted connectivity between posterior cingulate and hippocampus is responsible for the posterior cingulate hypometabolism

(continued)

Table 1
(continued)

Authors	Subjects	Design	fMRI analysis methods	Main conclusions
18. Hao et al. [51]	13 healthy 13 with AD	Visual search tasks	SPM99	Patients with AD present reduced activation in both parietal lobes and the left frontal regions. Healthy subjects present increased activation in the right frontal lobes and the right occipito-temporal cortical regions with the conjunction task
19. Rombouts et al. [35]	91 healthy 28 with MCI 18 with AD	Visual encoding task and a non spatial working memory task	FEAT v5.1 and FSL software packages	The most pronounced differences between Mild Cognitive Impairment, healthy controls, and AD occurred in the very early phase of deactivation, reflecting the reactivity and adaptation of the network
20. Celone et al. [58]	15 healthy 15 with low MCI 12 with high MCI 10 with AD	Memory related task	Independent component analysis	Differences between AD patient and healthy controls: (a) the activation patterns of specifically selected brain regions (memory related regions, regions of the visual and motor cortex, regions responsible for the learning processes), (b) the functional connectivity of the default mode network, (c) the task-induced deactivation of brain networks, and (d) the regional homogeneity of the fMRI signal
21. Wang et al. [63]	14 healthy 14 with AD	Resting-state	Pseudo-Fisher Linear Discriminative Analysis	The two intrinsically anti-correlated networks are the part of the brain network that is affected first in the early stage of AD

22. Petrella et al. 44]	28 healthy 34 with MCI 13 with AD	Novel encoding and familiar encoding of face-name pairs	Statistical analysis—Generalized Linear Models	Reduced activation in the medial temporal lobe and increased activation in the posteromedial cortex The changes in activation in the posteromedial cortex regions were larger in both magnitude and extent than those in the medial temporal lobe. They are significantly correlated with the California Verbal Learning Test II—CVTL II[a]
23. Sorg et al. [36]	16 healthy 25 with MCI	Resting-state	Independent component analysis and voxel-based morphometry	Selected areas of eight spatially consistent resting state networks demonstrated reduced activity in the patient group. Atrophy presented in both medial temporal lobes of the patients. Functional connectivity between medial temporal lobes and the posterior cingulate of the network absent in patients
24. Sperling et al. [47]	n/a	Memory related tasks	n/a	An unexpected phase of increased activation is presented early in the course of pAD
25. Chen et al. [67]	14 healthy young 14 healthy old 13 with AD	Sensory motor process	Bayesian network classifier with inverse tree structure	There are regions of the brain that their activation can act as an index for the diagnosis of AD with accuracy 81 %
26. Supekar et al. [62]	18 healthy 21 with AD	Resting-state	Characteristic path length and clustering coefficient were extracted from an undirected graph	Healthy controls presented high clustering coefficient and a low characteristic path length AD patients presented a significantly lower clustering coefficient
27. Xu et al. [66]	9 healthy 8 with MCI 10 with AD	Resting-state	ROI analysis Phase shift index	Phase shift index is highest in patient with AD (72.6 ± 11.3) in contrast to patient with MCI (58.6 ± 5.7) and healthy subjects (40.6 ± 8.4)

(continued)

Table 1
(continued)

	Authors	Subjects	Design	fMRI analysis methods	Main conclusions
28.	Burge et al. [68]	14 healthy young 14 healthy old 13 with AD	Sensory-motor task	Discrete dynamic Bayesian network	The detection of functional correlations between neuroanatomical regions of the brain can lead to the diagnosis of AD with 70 %
29.	Oghabian et al. [37]	15 healthy 11 with MCI 14 with AD	Resting-state	Independent Component Analysis	Healthy aging brain presented higher activation, in terms of intensity and extent, compared with MCI and Alzheimer's group in PCC region of the brain
30.	Zhang et al. [64]	16 healthy 16 with mild AD 11 with moderate NA 12 with severe AD	Resting-state	Time series correlation analysis	The functional connectivity of the default mode network is affected by the presence and progression of AD. The effect of AD in functional connectivity is most apparent in PCC
31	Tripoliti et al. [69]	14 healthy young 14 healthy old 13 with AD	Sensory-motor task	Classification using random forests and SVM Rule extraction	Diagnosis of Alzheimer's disease with accuracy 94 % Monitoring of the progression with accuracy 97 and 99 %
32	Tripoliti et al. [70]	14 healthy young 14 healthy old 13 with AD	Sensory-motor task	Modifications of the random forests algorithm	Diagnosis of Alzheimer's disease with accuracy 98 %

[a]D.C. Delis, J.H. Kramer, E. Kaplan, B.A. Ober, "The California Verbal Learning Test: research edition", New York, NY: Psychological Corporation, 1987

3   Assessment of Alzheimer's Disease Based on Electroencephalogram (EEG)

The realization of any cognitive, motor or sensory process is based on cerebral dynamics and turns the interest in the study of bioelectric signals measured with EEG [22]. EEG is the recording of cortical activity along the scalp. The activity is measured by sensors or electrodes placed on the scalp and pick up very small fluctuations resulting from ionic current flows within the neurons of the brain [71]. In clinical terms it refers to the recording of brain's spontaneous electrical activity over a short period of time, usually 20–40 min [22, 71].

EEG signals are derived from a summation of excitatory and inhibitory post synaptic potentials, in optical dendrites of pyramidal neurons in the more superficial layers of cortex. The activity reflects the summation of synchronous activity of thousands or millions of neurons that have similar spatial orientation. If the cells do not have similar spatial orientation, their ions do no line up and create waves to be detected [22]. EEG scalp recordings present oscillations at different frequencies. These oscillations can be described by the following features: (a) they have characteristic frequency ranges, (b) they have characteristic spatial distributions, and (c) they are related with different states of brain function [22]. The spontaneous and event-related oscillatory activity ranging exist for the following frequency ranges: 2–4 Hz (delta activity), 8–13 Hz (theta activity), 8–13 Hz (alpha activity), 13–30 Hz (beta activity), and >30 Hz (gamma activity). Beta and gamma activities represent a state of wake period and indicate increased brain activity needed for a cognitive process. The first three activities (delta, theta and alpha activities) represent state of sleep, drowsiness and relaxation, respectively [72].

EEG is usually measured in two domains, frequency and time. Changes in the power of the frequency spectra are analyzed in the frequency domain, while measurements in time domain are generally performed be evoked potentials with a specific stimulus (event related potentials—ERPs). The power spectrum reflects the general condition of the brain, while absolute power and/or phase measure the stability of EEG channels overtime, in a series of repeated trials [22, 73]. In addition, coherence of EEG channels is also measured to reflect the functional connectivity between brain regions, while the nature of ERP provides information about the source of the stimulus signal.

EEG is the most direct measure of cortical activity. The low spatial resolution, the poor signal to noise ration and the fact that it cannot efficiently be used for the study of neuronal activity that occurs below the cortex is some factors that limit the application of EEG. Despite these limitations EEG possesses multiple advantages over other techniques studying brain function (fMRI, PET, SPECT, MEG, etc.). More specifically, its high temporal resolution, low cost, tolerance to subject movement, portability of the equipment, as well as the fact that no contrast agent and no complex paradigms are needed, render EEG measures widely used in clinical practice.

EEG finds extensive application in the detection of epileptic seizures and localization of the seizure foci. Another well-known application is the study of sleep and sleep disorders. It is also used in the diagnosis of coma, encephalopathies, and brain death. The use of EEG-based neuro-modulation for stroke rehabilitation, in conjunction with brain stimulation, has also been examined. More recently several research groups have been investigating the potential of EEG as a biomarker for neurodegenerative diseases [74].

There are several reasons for this. The AD is a type of cortical dementia, thus EEG as a direct measure of cortical activity presents abnormalities reflecting anatomical and functional deficits of the cerebral cortex damaged by the disease. Coherence analysis of the EEG signal in AD patients permits the assessment of synaptic dysfunction in a noninvasive manner. The disturbances of synaptic connections come as consequences of several neurological disorders including AD. Finally, the study of nonlinear dynamics of EEG allows the understanding the nonlinearity in brain functions. The decrease of complexity of EEG patterns as well as the reduced functional connectivity in AD can also provide valuable information about the progress of the disease [72, 76].

Clinical studies have examined: (a) the slowing down effect of EEG signal, (b) the reduced complexity, and (c) the reduced synchrony and found them to be helpful in supporting medical diagnosis of AD. In 2004 Jeong [76] provided a detailed review of the main findings concerning EEG abnormalities due to AD. The review was enhanced in 2009 by Deursen [73] and later by Lisioz et al. [72] and Waser et al. [75]. The findings, reported above, were assessed by several studies in order to determine the potential and limitation of EEG in the differentiation between healthy and AD patients, and in the monitoring the progress of the disease and in the discrimination between different dementia diagnoses [72–76].

3.1 The Slowing Down Effect of the EEG Signal

The first observations were made by Hans Berger through visual analysis of EEG in patient with AD and later were confirmed and enriched through linear and nonlinear computerized analysis methods of EEG signals [76]. According to the observations, patients with AD, compared to healthy old adults, presented a slowing of the dominant posterior rhythm. Slowing down of the EEG signal is due to a decline in cortical brain activity, which mainly results in a decline of high-frequency signals. This leads to a relative increase of the slow-wave activity which is expressed in terms of relative power changes in different frequency bands [73]. More specifically, the slowing effect is presented as a decrease in power in alpha and beta bands and as an increase in power in delta and theta bands.

The observations of these changes in spectral power could allow for the determination of the severity of the AD. Studies have shown that the earliest changes are as follows: (a) an increase in theta activity and (b) a decrease in beta activity followed by a decrease in alpha activity. Later during the course of the disease an increase in delta activity is presented. Patients with mild dementia have shown a decrease in beta and an increase in theta activity, whereas in the later stages of the disease (severe dementia) a decrease in alpha and an increase in delta activity are exhibited. This suggests that the slowing of EEG signal is a marker for the subsequent rate of a cognitive and functional decline in early AD patients. The correlation between EEG spectral measures and the severity of the disease, which is expressed through cognitive deterioration scores, has been examined in many studies. The studies revealed that the existence of the above-mentioned correlation confirms that a disruption of information processing in cortical networks significantly contributes to the cognitive dysfunction seen in AD [72, 73, 76].

The effectiveness of EEG, in the differential diagnosis of AD from other causes of dementia, has been extensively studied. More specifically, mild AD subjects can be differentiated from cerebrovascular dementia, fronto-temporal dementia, and normal elderly subjects with similar cognitive impairment based on the peculiar features that are presented in mild AD subjects due to the considerable decline of posterior slow-frequency alpha power. Furthermore, patients with cerebrovascular dementia presented a marked increased amplitude of the theta sources [72, 73, 76]. However, it must be mentioned that the differentiation of AD from vascular dementia is not a straightforward task [73].

Different procedures can be applied in order the spectral density to be estimated. Three common procedures are: (a) the periodogram, (b) the AR-fitting using Yule-Walker equations, and (c) the AIC criterion. However, the interest is in estimating the power in the respective frequency bands. The cumulative periodogram is a consistent estimator for this purpose [75].

3.2 The Reduced Complexity of the EEG Signal

The high nonlinear nature of neuronal interactions lead a lot of researchers to apply nonlinear dynamic analysis methods to EEG signal aiming to the better comprehension of complex cortical dynamics of the underlying processes. Through their studies revealed that nonlinear dynamic analysis methods of the EEG in patients with AD might provide valuable information on the progress of the disease that cannot be assessed by conventional analyses. According to those studies AD patients present globally decreased complexity of brain electrical activity compared to age-approximated non-demented controls as well as, in information processing flexibility. Since the complexity of the EEG is a measure of signal dynamic coordination its reduction may be associated with deficient information processing in the brain damaged by AD [72, 76].

Through nonlinear dynamic analysis methods (a detailed description of the theoretical concepts and algorithms are presented in Abarbanel et al. [78] and in Faure et al. [79]) the correlation measure D_2 and the first positive Lyapunov exponent L_1 were measured. The AD patients presented reduced values of correlation measures D_2 in the occipital EEG. This difference was observed not only between healthy and patients with AD but also between patients with AD and patients with probable AD. Since the correlation measure D_2, in EEG analysis expresses the complexity of the underlying cortical dynamics its reduction lead to the conclusion that AD patients exhibit a decrease in the complexity of brain electrical activity. This reduction can be attributed to many reasons such as neuronal death, deficiency of neurotransmitters, loss of connectivity of local neuronal networks, inactivation of previously active networks, and loss of dynamical brain response to stimuli. The last two are considered the most probable causes responsible for the decrease in EEG complexity in AD patients. The combination of these reasons suggests that the ability of information processing of the cortex in AD patients is deficient [72, 76].

Besides the correlation measure D_2, which is a static geometric measure, attention was given to the first positive Lyapunov exponent L_1, which is a relatively dynamic measure of the flexibility of information processing of the brain. It describes the divergence of trajectories starting at nearby initial states. Decrease of L_1, in AD patients indicates a drop of the possibility of the central nervous system to reach different states of information processing form similar initial states [72, 76].

The potentiality of those measures for differential and early diagnosis of AD attracted the interest of researchers. After the completion of their studies conclude that D_2 can correctly classify the AD and normal subjects with 70 % accuracy while good correlations are found between nonlinear measures and the severity of the disease, a slowing of EEG rhythms and the neuropsychological performance. Regarding L_1 values, they are lower in the AD than in patients with Parkinson, while L_1 and D_2 are lower in patients in AD compared to those with vascular dementia. Despite the fact that nonlinear measures can provide important information for differential and early diagnosis of the disease its combinations with linear measures can improve the classification accuracy between AD and control patients [76].

3.3 The Reduced Synchrony of EEG Signal

The EEG coherence analysis is designed to examine whether brain waves from different brain areas are synchronized. This synchrony reflects the degree of functional connectivity between different brain regions and it constitutes a fundamental mechanism for implementing coordinated communication between spatially distributed brain networks. It is investigated separately for each of the frequency banks, or for specific pairs of electrodes, or averages over all electrode pairs for each frequency band [73, 76, 79]. The functional connectivity depends on the integrity of anatomical connections in the neuronal network and on the neurotransmission. Decreased coherence depicts reduced functional connections between cortical areas or reduced common modulation of two areas by one-third. This may be indicative of a loss of functional integrity of the neuronal connections between different brain areas [73, 76].

The AD is characterized by neuronal loss that leads to disconnection and loss of synchronization between brain regions. Coherence decrease in the alpha and theta bands is another characteristic feature of AD [72, 73, 76]. This decrease coherence is significantly correlated with cognitive impairment and can be a result from a loss of cortico-cortical association fibers [73]. Stam et al. [80] examined the hypothesis that the decreased coherence is not only a result of cortical neurons loss since all frequencies are not equally affected. They suggested that neurotransmitters are also involved. The AD patients presented changes in synaptic couplings among cortical neurons. This leads to the reduction of long distance functional connection even when the anatomical connections are intact. The abnormal pattern of EEG functional coupling can be attributed to the loss of acetylcholine, an excitatory neurotransmitter of the cerebral cortex and a distinct feature of AD [72, 73, 76].

In order linear and nonlinear dependencies in electrical activity across different brain regions to be taken in to account, mutual information [81] and synchronization methods [80, 82] were applied. Mutual information analysis showed that the independencies between different electrodes are reduced in patients with AD compared with normal subjects. Stam et al. [80, 82] measured synchronization likelihood and they observed that it was significantly decreased at specific bands (alpha and low beta) when patients with AD with patients with mild cognitive impairment and normal subjects are compared.

An open issue regarding the utilization of EEG in the assessment of AD is the improvement of the accuracy of the differential diagnosis of AD and the early detection in preclinical stage. Towards this direction several studies have examined the possibility of combining the EEG signal together with biological/neuropsychological markers and structure/functional imaging. The aim of the next section is to review integration strategies of EEG-fMRI to achieve an assessment of AD.

4   Assessment of Alzheimer's Disease Based on the Combination of fMRI with EEG

Numerous techniques have been developed to combine EEG-fMRI data [83, 84]. A major motivation for this combination is the desire to overcome the limitation of poor spatial and temporal resolution of EEG and fMRI, respectively. EEG-fMRI integration techniques can be divided into two categories: (a) techniques using separate recordings of EEG and fMRI, and (b) techniques using concurrent recordings. Each category has its own advantages associated with the area of application and the particular aims of a study.

Separate recordings of EEG and fMRI provide several advantages regarding mainly the signal to noise ratio and the spatial resolution of EEG. Separately recorded data usually have higher signal to noise ration since it is not degraded by residual or overcompensates scanner-related artifacts. High signal to noise ratio is extremely important in paradigms where the utilization of double differences in the analysis of data is necessary and in cases where weak signals must be detected. Separate recordings permit the utilization of a large number of electrodes something that is not possible with concurrent recordings since some electrodes are not compatible with a use inside the scanner. Furthermore, they allow researchers to optimize the paradigm independently for each recording while simultaneous recordings a priori require the same paradigm for both measurement modalities. The optimization of the paradigm leads to the acquisition of data with high signal to noise ratio. Finally, data obtained with the specific technique can be more easily analyzed with independent component analysis (a common approach for the analysis of EEG-fMRI data) because no artifact-related components, introduced by the scanner artifacts or their reminders after artifact removal, are presented in the data. This has as consequence the results of the analysis to be more reliable and easily interpreted [83].

Concurrent recordings of EEG and fMRI provide the ability the data to be acquired, noninvasively, from both modalities simultaneously allowing the researchers to achieve high 3D spatial resolution via fMRI and high temporal resolution via EEG. In addition, they are more sensitive to deep neural sources. The concurrent recording of EEG and fMRI should be preferred when the information obtained from electrophysiological measurements is necessary to drive BOLD fMRI analysis, because no other predictors for statistical analysis can be derived. Also, they dominate, in comparison to separate recordings, in paradigms aiming to investigate rapid learning processes that occur within a limited amount of time or trials. Finally, they are very useful in examining the correlation between fluctuations of EEG and fMRI signals that are presented from trial to trial, even if the simulation conditions and behavioral results are identical [83].

The analysis methods of concurrent recordings of EEG and fMRI allow for direct integration of information across modalities, in contrast to classical techniques that focus on the quantitative and qualitative comparison of the results produced by the separate analysis of EEG and fMRI data, which were acquired simultaneously [85]. The techniques are divided to: (a) asymmetric data integration, (b) neurogenerative modeling and (c) multimodal data fusion techniques depending on their necessity for prior knowledge and on biophysical and mathematical assumptions they rely on. Asymmetric data integration techniques, depending on how one modality is used to guide the analysis of the other are divided to (a) fMRI-informed EEG and (b) to EEG-informed fMRI approach [84, 85]. An introductory review of the above-mentioned approaches, as well as, a comparison of those approaches is provided in [85].

A common approach for the fusion of EEG and fMRI is independent component analysis (ICA) and some variations of it, joint ICA and parallel ICA [86–89]. Another data driven approach for the multimodal fusion is canonical correlation analysis (CCA) and its extension to multiple datasets mutliset CCA (M-CCA) [90]. Both Joint ICA and CCA are second-level analysis approaches. They are based on lower-dimensional features of the data, while the associations across the two modalities are based on inter subject covariations. However, they present a number of differences. For a detailed comparison of the two models refer to [91]. All the above-mentioned methodologies belong to the multimodal data fusion category.

Independent Component Analysis has been widely utilized as a tool for evaluating the hidden spatiotemporal structure contained within brain imaging data. It can discover hidden sources or features from a set of measurements such that the sources are maximally independent. In order this to be achieved, it works with higher order statistics [87]. It assumes that the observations are linear mixtures of independent sources [87]. Joint ICA [86, 87] enables the users to jointly analyze multiple modalities. It is based on the following assumptions: (a) data from different modalities have been collected in the same set of subjects, and (b) the sources associated with the two data types have the same linear covariation. This assumption can be relaxed by assuming that the modulation across samples for the sources from the two data types is correlated but it is not necessarily the same. Parallel ICA [87–89] provides this flexibility in the estimation. Parallel ICA aims to find hidden factors from both modalities and connections between them by computing either the correlation measure between mixing coefficients of the two modalities or the correlation measure between the mixing coefficient the one modality with the source of the other modality.

CCA is used to measure the relationships between two multi-dimensional variables. It works by finding the optimal basis for each variable under the constraints the correlation matrix between the variables to be diagonal and the correlations in the diagonal to be maximized. In contrast, M-CCA seeks for linear transformations that simplify the correlation structure among multiple random vectors. In order this to be achieved multiple stages are performed. In each stage a linear combination is found for each vector such that the resulting canonical variates achieve maximum overall correlation [90].

The techniques mentioned on this section have been applied to a limited number of pathologies concerning brain (e.g., epilepsy, schizophrenia) in order to derive results of some clinical value. However, the researchers pinpoint the fact that the methods could have a significant impact for studies of brain networks involving sensory-motor, language and memory processing in healthy brains, and eventually in patients with altered or dysfunctional brains. A core example is AD. Hippocampus is one of the core regions that are affected by the presence of the disease even in the early stages. The dysfunction of this brain region can be expressed through characteristics extracted from fMRI presented in the literature (e.g., decreased activation, hypometabolism, presentation of atrophy) and also can be depicted to the altered changes of the amplitudes of the P3a and P3b components of the P300 potential. The contribution of the hippocampus to P3a and P3b components, revealed with simultaneous recordings of EEG-fMRI data, leads the researches to further investigate the extent to which these combined P300/hippocampal functional markers can be utilized for the earlier detection of dementia [92].

5 Conclusions

The potential of fMRI and EEG in diagnosing AD, in monitoring the progression of the disease and in differentiating AD from other types of dementia are issues that have received important attention in the literature. The fMRI is a noninvasive technique which allows indirect measurement of neuronal activity and imaging of activated cortical areas. The measurements are based on the fact that brain stimulation is correlated with an increased local brain metabolism. This metabolic activity causes local changes of the magnetic properties of blood, which can be imaged by fMRI. It highlights the neuronal centers that are responsible for different cognitive functions and allows researchers to examine, to measure and understand how these functions are affected by the presence of AD. The utilization of fMRI does allow researchers not only to detect changes in the brain function of patients suffering from the disease (changes concerning the intensity and extend of activation, the properties of the BOLD response, the metabolism, the functional synchrony, and connectivity of brain regions) but also to quantify these changes and provide a preclinical marker that will support the diagnosis of the disease. The most important advantage of fMRI is that it has the potential to identify subtle pathological changes early in the course of the disease. The EEG, as a non invasive direct measure of cortical activity, can depict the anatomical and functional changes presented due to AD and thus to provide important information about its neuropathology. The coherence analysis of EEG allows clinicians to study synaptic plasticity, a critical factor of brain functions affected by the presence of AD, such as memory and learning. The application of EEG does not only allow for the detection of disturbances in synaptic connections that underlie AD but also provides evidences regarding the correlation between reduced functional connectivity and degree of dementia. Toward this direction valuable information provides the analysis of non linear EEG dynamics, where decreased complexity of EEG patterns is revealed.

Regarding the combination of the two modalities a number of studies have already been proposed each one having potential advantages and limitations. Although, the approaches are based on computational methods with long history their joint application to neuro-scientific problems is a rather new trend. One question that arises is how the best method should be selected (asymmetric data integration, neurogenerative modeling, and multimodal data fusion techniques). The answer to this question strongly dependents on the research question that is addressed. Regarding AD, the fusion of different modalities (EEG-MRI, fMRI-genetics, etc.) has already been applied. However, the fusion of EEG-fMRI for the assessment of AD is an open issue. The major reason contributing to the limited application of EEG-fMRI integration techniques is the substantial artifacts that are associated with collecting EEG in the scanner environment. For the moment, the literature provides evidence that the utilization of above-mentioned techniques in the field of neurodegenerative disorders such as AD could open new horizons for clinicians (e.g., new markers for the early diagnosis of the disease and for the differentiation of dementia types, further understanding brain mechanisms that underlie AD).

References

1. Lauriks S, Reinersmann A, van der Roest HG, Meiland F, Davies R, Moelaert F, Mulvenna MD, Nugent CD, Dröes R (2010) Chapter 4: review of ICT-based services for identified unmet needs in people with dementia. In: Mulvenna MD, Nugent CD (eds) Supporting people with dementia using pervasive health technologies. Springer, London

2. World Health Organization and Alzheimer's Disease International, Dementia: a public health priority. http://apps.who.int/iris/bitstream/10665/75263/1/9789241564458_eng.pdf

3. Draper B (2004) Dealing with dementia. A guide to Alzheimer's disease and other dementias. Allen & Unwin, Australia

4. Alzheimer's Association (2010) Alzheimer's disease facts and figures. Alzheimer's Dementia 6:98–103

5. Alzheimer's Association (2013) Alzheimer's disease facts and figures. http://www.alz.org/downloads/facts_figures_2013.pdf

6. Alzheimer's Association, Basics of Alzheimer's disease - what it is and what you can do, http://

www.alz.org/national/documents/brochure_basicsofalz_low.pdf

7. International Classification of Diseases (ICD), http://www.who.int

8. Carr DB, Goate A, Phil D, Morris JC (1997) Current concepts in the pathogenesis of Alzheimer's disease. Am J Med 103 (Suppl. 3A): 3S–10S

9. Small GW, Rabins PV, Barry PP et al (1997) Diagnosis and treatment of Alzheimer disease and related disorders: consensus statement of the American Association for Geriatric Psychiatry, the Alzheimer's Association, and the American Geriatrics Society. JAMA 278:1363–1371

10. Khachaturian ZS (1985) Diagnosis of Alzheimer's disease. Arch Neurol 42:1097–1105

11. Jellinger K (1990) Morphology of Alzheimer disease and related disorders. In: Maurer K, Riederer P, Beckmann H (eds) Alzheimer disease: epidemiology, neuropathology and clinics. Springer, New York, NY, pp 61–77

12. Selkoe DJ (1997) Alzheimer's disease: genotypes, phenotypes, and treatments. Science 275:630–631

13. Petrella JR, Coleman RE, Doraiswamy PM (2003) Neuroimaging and early diagnosis of Alzheimer disease: a look to the future. Radiology 226:315–336

14. National Institute of Neurological Disorders and Stroke. http://www.ninds.nih.gov/

15. McKhann G, Drachman D, Folstein M, Katzman R, Price D, Stadlan EM (1984) Clinical diagnosis of Alzheimer's disease: report of the NINCDS-ADRDA Work Group under the auspices of Department of Health and Human Services Task Force on Alzheimer's disease. Neurology 34:939–944

16. Alzheimer's Disease and Related Disorders Association. http://www.alz.org/

17. Diagnostic and Statistical Manual of Mental Disorders. http://www.psych.org/

18. Scheltens P (1999) Early diagnosis of dementia: neuroimaging. J Neurol 246:16–20

19. O'Brien JT (2007) Role of imaging techniques in the diagnosis of dementia. Br J Radiol 80: S71–S77

20. Perrin RJ, Fagan AM, Holtzman DM (2009) Multimodal techniques for diagnosis and prognosis of Alzheimer's disease. Nature 461:916

21. Petersen RC, Aisen PS, Beckett LA, Donohue MC, Gamst AC, Harvey DJ, Jack CR, Jagust WJ, Shaw LM, Toga AW, Trojanowski JQ, Weiner MW (2010) Alzheimer's disease neuroimaging initiative (ADNI): clinical characterization. Neurology 74:201–209

22. Ally BA (2011) Using EEG and MEG to understand brain physiology in Alzheimer's disease and related dementias. In: Budson AE, Kowall NW (eds) The handbook of Alzheimer's disease and other dementias. Blackwell Publishing Ltd., Oxford

23. Jezzard P, Clare S (2001) Principles of nuclear magnetic resonance and MRI. In: Jezzard P, Matthews P, Smith S (eds) Functional MRI: An Introduction to Methods. Oxford University Press, Oxford

24. Lazar N (2008) The statistical analysis of functional MRI data. Springer Science + Business Media, LLC, ISBN 978-0-387-78190-7, USA

25. Tripoliti EE, Fotiadis DI (2009) Recent developments in computer methods for fMRI data processing, In: Carlos Alexandre Barros de Mello (ed) Biomedical Engineering, ISBN 978-953-307-013-1, InTech, Chapters published

26. Nierhaus T, Margulies D, Long X, Villringer A (2012) fMRI for the assessment of functional connectivity. In: Neuroimaging methods, Peter Bright (ed) ISBN 978-953-51-0097-3, InTech, Chapters published

27. Minoshima S, Giordani B, Berent S, Frey KA, Foster NL, Kuhl DE (1997) Metabolic reduction in the posterior cingulate cortex in very early Alzheimer's disease. Ann Neurol 42 (1):85–94

28. Johnson KA, Jones K, Holman BL, Becker JA, Spiers PA, Satlin A, Albert MS (1998) Preclinical prediction of Alzheimer's disease using SPECT. Neurology 50(6):1563–1571

29. Matsuda H (2001) Cerebral blood flow and metabolic abnormalities in Alzheimer's disease. Ann Nucl Med 15(2):85–92

30. Reiman EM, Caselli RJ, Yun LS, Chen K, Bandy D, Minoshima S, Thibodeau SN, Osborne D (1996) Preclinical evidence of Alzheimer's disease in persons homozygous for the epsilon 4 allele for apolipoprotein E. N Engl J Med 334(12):752–758

31. Small GW, Ercoli LM, Silverman DH, Huang SC, Komo S, Bookheimer SY, Lavretsky H, Miller K, Siddarth P, Rasgon NL et al (2000) Cerebral metabolic and cognitive decline in persons at genetic risk for Alzheimer's disease. Proc Natl Acad Sci U S A 97 (11):6037–6042

32. Reiman EM, Caselli RJ, Chen K, Alexander GE, Bandy D, Frost J (2001) Declining brain activity in cognitively normal apolipoprotein E epsilon 4 heterozygotes: a foundation for using positron emission tomography to efficiently test treatments to prevent Alzheimer's disease. Proc Natl Acad Sci U S A 98(6):3334–3339

33. Raichle ME, MacLeod AM, Snyder AZ, Powers WJ, Gusnard DA, Shulman GL (2001) A default mode of brain function. Proc Natl Acad Sci U S A 98(2):676–682

34. Greicius MD, Krasnow B, Reiss AL, Menon V (2003) Functional connectivity in the resting brain: a network analysis of the default mode hypothesis. Proc Natl Acad Sci U S A 100 (1):253–258

35. Rombouts S, Barkhof F, Goekoop R, Stam CJ, Scheltens P (2005) Altered resting state networks in mild cognitive impairment and mild Alzheimer's disease: An fMRI study. Hum Brain Mapp 26:231–239

36. Sorg C, Riedl V, Muhlau M, Calhoun VD, Eichele T, Laer L, Drzezga A, Forstl H, Kurz A, Zimmer C, Wohlschlager AM (2007) Selective changes of resting-state networks in individuals at risk for Alzheimer's disease. Proc Natl Acad Sci U S A 104(47):18760–18765

37. Oghabian MA, Batouli SAH, Norouzian M, Ziaei M, Sikaroodi H (2010) Using functional magnetic resonance imaging to differentiate between healthy aging subjects, mild cognitive impairment, and Alzheimer's patients. J Res Med Sci 15(2):84–93

38. Grunwald P (2004) A tutorial introduction to the minimum description length principle. MIT, Amsterdam

39. Beckmann CF, Jenkinson M, Smith SM (2003) General multi-level linear modeling for group analysis in FMRI. Neuroimage 20 (2):1052–1063

40. Small SA, Perera GM, DeLaPaz R, Mayeux R, Stern Y (1999) Differential regional dysfunction of the hippocampal formation among elderly with memory decline and Alzheimer's disease. Ann Neurol 45:466–472

41. Small SA, Nava AS, Perera GM, Delapaz R, Stern Y (2000) Evaluating the function of hippocampal subregions with high-resolution MRI in Alzheimer's disease and aging. Microsc Res Tech 51:101–108

42. Smith CD, Andersen AH, Kryscio RJ, Schmitt FA, Kindy MS, Blonder LX, Avison MJ (1999) Altered brain activation in cognitively intact individuals at high risk for Alzheimer's disease. Neurology 53:1391–1396

43. Machulda MM, Ward HA, Borowski B, Gunter JL, Cha RH, O'Brien PC, Petersen RC, Boeve BF, Knopman D, Tang-Wai DF, Ivnik RJ, Smith GE, Tangalos EG, Jack CR (2003) Comparison of memory fMRI response among normal, MCI, and Alzheimer's patients. Neurology 61(4):500–506

44. Petrella JR, Wang L, Krishnan S, Slavin MJ, Prince SE, Tran TT, Doraiswamy PM (2007) Cortical deactivation in mild cognitive impairment: high-field-strength functional MR imaging. Radiology 245(1):224–235

45. Gron G, Bittner D, Schmitz B, Wunderlich AP, Riepe MW (2002) Subjective memory complaints: objective neural markers in patients with Alzheimer's disease and major depressive disorder. Ann Neurol 51:491–498

46. Sperling RA, Bates JF, Chua EF, Cocchiarella AJ, Rentz DM, Rosen BR, Schacter DL, Albert MS (2003) fMRI studies of associative encoding in young and elderly controls and mild Alzheimer's disease. J Neurol Neurosurg Psychiatry 74:44–50

47. Sperling RA (2007) Functional MRI studies of associative encoding in normal aging, mild cognitive impairment, and Alzheimer's disease. Ann N Y Acad Sci 1097:146–155

48. Thulborn KR, Martin C, Voyvodic JT (2000) Functional MR imaging using a visually guided saccade paradigm for comparing activation patterns in patients with probable Alzheimer's disease and in cognitively able elderly volunteers. AJNR Am J Neuroradiol 21:524–531

49. Buckner R, Snyder AZ, Sanders AL, Raichle ME, Morris JC (2000) Functional brain imaging of young, non demented, and demented older adults. J Cogn Neurosci 12(2):24–34

50. Prvulovic D, Hubl D, Sack AT, Melillo L, Maurer K, Frolich L, Lanfermann H, Zanella FE, Goebel R, Linden DEJ, Dierks T (2002) Functional imaging of visuospatial processing in Alzheimer's disease. Neuroimage 17:1403–1414

51. Hao J, Li K, Zhang D, Wang W, Yang Y, Yan B, Shan B, Zhouc X (2005) Visual attention deficits in Alzheimer's disease: an fMRI study. Neurosci Lett 385:18–23

52. Johnson SC, Saykin AJ, Baxter LC, Flashman LA, Santulli RB, McAllister TW, Mamourian AC (2000) The relationship between fMRI activation and cerebral atrophy: comparison of normal aging and Alzheimer disease. Neuroimage 11:179–187

53. Rombouts S, Barkhof F, Veltman DJ, Machielsen WCM, Witter MP, Bierlaagh MA, Lazeron RHC, Valk J, Scheltens P (2000) Functional MR imaging in Alzheimer's disease during memory encoding. AJNR Am J Neuroradiol 21:1869–1875

54. Saykin AJ, Flashman LA, Frutiger SA, Johnson SC, Mamourian AC, Moritz CH, O'Jile JR, Riordan HJ, Santulli RB, Smith CA, Weaver JB (1999) Neuroanatomic substrates of semantic memory impairment in Alzheimer's disease: patterns of functional MRI activation. J Int Neuropsychol Soc 5:377–392

55. Bookheimer SY, Strojwas MH, Cohen MS, Saunders AM, Pericak-Vance MA, Mazziotta JC, Small GW (2000) Pattern of brain activation in people at risk for Alzheimer's disease. N Engl J Med 343(7):450–456

56. Grossman M, Koenig P, DeVita C, Glosser G, Moore P, Gee J, Detre J, Alsop D (2003) Neural basis for verb processing in Alzheimer's disease: an fMRI study. Neuropsychology 17 (4):658–674

57. Lustig C, Snyder AZ, Bhakta M, O'Brien KC, McAvoy M, Raichle ME, Morris JC, Buckner RL (2003) Functional deactivations: change with age and dementia of the Alzheimer type. Proc Natl Acad Sci U S A 100 (24):14504–14509

58. Celone KA, Calhoun VD, Dickerson BC, Atri A, Chua EF, Miller SL, DePeau K, Rentz DM, Selkoe DJ, Blacker D, Albert MS, Sperling RA (2006) Alterations in memory networks in mild cognitive impairment and Alzheimer's disease: an independent component analysis. J Neurosci 26(40):10222–10231

59. Liu Y, Wang K, Yu C, He Y, Zhou Y, Liang M, Wang L, Jiang T (2008) Regional homogeneity, functional connectivity and imaging markers of Alzheimer's disease: a review of resting-state fMRI studies. Neurophsychologia 46:1648–1656

60. Hafkenmeijer A, van der Grond J, Rombouts SARB (2012) Imaging of the default mode network in aging and dementia. Biochim Biophys Acta 1822:431–441

61. Greicius MD, Srivastava G, Reiss AL, Menon V (2004) Default-mode network activity distinguishes Alzheimer's disease from healthy aging: evidence from functional MRI. Proc Natl Acad Sci U S A 101:4637–4642

62. Supekar K, Menon V, Rubin D, Musen M, Greicius MD (2008) Network analysis of intrinsic functional brain connectivity in Alzheimer's disease. PLoS Comput Biol 4(6): e1000100

63. Wang K, Jiang T, Liang M, Wang L, Tian L, Zhang X, Li K, Liu Z (2006) Discriminative analysis of early Alzheimer's disease based on two intrinsically anti-correlated networks with resting-state fMRI. In: Larsen R, Nielsen M, Sporring J (eds) Medical image computing and computer-assisted intervention (MICCAI 2006). Lecture notes in computer science, vol 4191, pp 340–347. Springer, Berlin

64. Zhang HY, Wang SJ, Liu B, Ma ZL, Yang M, Zhang ZJ, Teng GJ (2010) Resting brain connectivity: changes during the progress of Alzheimer disease. Neuroradiology 256 (2):598–606

65. Li SJ, Li Z, Wu G, Zhang MJ, Franczak M, Antuono PG (2002) Alzheimer disease: evaluation of a functional MR imaging index as a marker. Radiology 255:253–259

66. Xu Y, Xu G, Wu G, Antuono P, Rowe DB, Li SJ (2008) The phase shift index for marking functional asynchrony in Alzheimer's disease patients using fMRI. Magn Reson Imaging 26:379–392

67. Chen R, Herskovits EH (2007) Clinical diagnosis based on Bayesian classification of functional magnetic resonance data. Neuroinformatics 5:178–188

68. Burge J, Lane T, Link H, Qiu S, Clark VP (2009) Discrete dynamic Bayesian network analysis of fMRI data. Hum Brain Mapp 30:122–137

69. Tripoliti EE, Fotiadis DI, Argyropoulou M, Manis G (2010) A six stage approach for the diagnosis of Alzheimer's disease based on fMRI data. J Biomed Inform 43:307–320

70. Tripoliti EE, Fotiadis DI, Manis G (2013) Modifications of the random forests algorithm. Data Knowl Eng 87:41–65

71. Niedermeyer E, da Silva FL (ed), (2004) Electroencephalography: basic principles, clinical applications, and related fields. Lippincot Williams & Wilkins. ISBN 0-7817-5126-8

72. Lizio R, Vecchio F, Frisoni GB, Ferri R, Rodriguez G, Babiloni C (2011) Electroencephalographic rythms in Alzheimer's disease. Int J Alzheimers Dis 2011:927573

73. van Deursen JA, Vuurman EFPM, Verhey FRJ, Riedel WJ (2009) EEG and other biological markers in Alzheimer's disease: a review. In: Van Deursen JA (ed) Functional and structural brain markers of Alzheimer's disease: clinical studies using EEG and VBM. NeuroPsych Publishers, Maastricht

74. Sakkalis V (2011) Applied strategies towards EEG/MEG biomarker identification in clinical and cognitive research. Biomark Med 5 (1):93–105

75. Waser M, Deistler M, Garn H, Benke T, Dal-Bianco P, Ransmayr G, Grossegger D, Schmidt R (2013) EEG in the diagnostics of Alzheimer's disease. Stat Paper 54:1095–1107

76. Jeong J (2004) EEG dynamics in patients with Alzheimer's disease. Clin Neurophysiol 115:1490–1505

77. Sakkalis V (2011) Review of advanced techniques for the estimation of brain connectivity measured with EEG/MEG. Comput Biol Med 41(12):1110–1117

78. Abarbanel HD, Rabinovich MI (2001) Neurodynamics: nonlinear dynamics and neurobiology. Curr Opin Neurobiol 11:423–430

79. Faure P, Korn H (2001) Is there chaos in the brain? I. Concepts of nonlineardynamics and methods of investigation. C R Acad Sci III 324:773–793

80. Stam CJ, van der Made Y, Pijnenburg YA, Scheltens P (2003) EEG synchronization in mild cognitive impairment and Alzheimer's disease. Acta Neurol Scand 108:90–96

81. Jeong J, Gore JC, Peterson BS (2001) Mutual information analysis of the EEG in patients with Alzheimer's disease. Clin Neurophysiol 112:827–835

82. Stam CJ (2005) Nonlinear dynamical analysis of EEG and MEG: review of an emerging field. Clin Neurophysiol 116(10):2266–2301

83. Wibral M, Bledowski C, Turi G (2010) Integration of separately recorded EEG/MEG and fMRI data. In: Ullsperger M, Debener S (eds) Simultaneous EEG and fMRI, recording, analysis and application. Oxford University Press, New York, NY

84. Rosa MJ, Daunizeau J, Friston KJ (2011) EEG-fMRI integration: a critical review of biophysical modeling and data analysis approaches. J Integr Neurosci 9:453–476

85. Huster RJ, Debener S, Eichele T, Herrmann CS (2012) Methods for simultaneous EEG-fMRI: an introductory review. J Neurosci 32:6053–6060

86. Mangalathu Arumana J (2012), Integration of EEG-FMRI in an auditory oddball paradigm using joint independent component analysis. Dissertations (2009–). Paper 210. http://epublications.marquette.edu/dissertations_mu/210

87. Calhoun VD, Jingyu L, Aduli T (2009) A review of group ICA for fMRI data and ICA for joint inference of imaging, genetic and ERP data. Neuroimage 45:S163–S172

88. Liu J, Demirci O, Calhoun VD (2008) A parallel independent component analysis approach to investigate genomic influence on brain function. IEEE Signal Process Lett 15:413–416

89. Liu J, Calhoun VD (2007) Parallel independent component analysis for multimodal analysis: application to fMRI and EEG data. Proceedings of 4th IEEE International Symposium on Biomedical Imaging: From Nano to Macro, ISBI 2007, pp. 1028–1031. doi: 10.1109/ISBI.2007.357030

90. Correa NM, Adali T, Li YO, Calhoun VD (2010) Canonical correlation analysis for data fusion and group inferences. IEEE Signal Process Mag 27:39

91. Correa N, Li Y-O, Adali T, Calhoun VD (2008) Canonical correlation analysis for feature-based fusion of biomedical imaging modalities and its application to detection of associative networks in schizophrenia. IEEE J Sel Top Signal Process 2(6):998–1007

92. O'Connell RG, Balsters JH, Kilcullen SM, Campbell W, Bokde AW, Lai R, Upton N, Robertson IH (2012) A simultaneous ERP/fMRI investigation of the P300 aging effect. Neurobiol Aging 33(10):2448–2461

Neuromethods (2015) 91: 385–386
DOI 10.1007/7657_2014
© Springer Science+Business Media New York 2015

INDEX

A

B

C

E

Entropy

F

G

I

L

M